D1172177

TABLES OF PHYSICAL AND CHEMICAL CONSTANTS

Tables of Physical and Chemical Constants

and some Mathematical Functions

Originally compiled by
G. W. C. KAYE, O.B.E., M.A., D.Sc., F.R.S.
and
T. H. LABY, M.A., Sc.D., F.R.S.

Now prepared under the direction of an Editorial Committee

Longman

513 27

LONGMAN GROUP LIMITED
London
Associated companies, branches and representatives
throughout the world

© Longman Group Limited 1973

All rights reserved. No part of this publication may be reproduced, stored in a retrieval system, or transmitted in any form or by any means, electronic, mechanical, photocopying, recording, or otherwise, without the prior permission of the Copyright owner.

First published 1911
Fourteenth Edition 1973
ISBN 0 582 46326 2

Printed in Great Britain by
William Clowes & Sons, Limited
London, Beccles and Colchester

PREFACE TO THE FOURTEENTH EDITION

In this edition all tabulated values have been expressed in SI units. This completes the process started earlier and has involved recalculating most of the tables and rewriting much of the text. The first section of the book, which deals with the system of units, has been thoroughly revised to take account of recent developments and the tables of conversion factors have been expanded. The next section contains some new material and brings together data on fundamental constants which were scattered through several sections in previous editions. There is also a new section on astronomy and geophysics. The scope of the remaining material in General Physics is unchanged.

The arrangement of Part II, Chemistry, is also largely unchanged but the increasing use of physical methods in chemical analysis has led us to include sections on nuclear magnetic resonance, on the Mossbauer effect and on crystal structures. The section on the properties of chemical bonds has been recast and that on polymers considerably enlarged.

Major changes have been made in the sections dealing with Atomic Physics, reflecting the steady accumulation of new experimental data in this field in recent years. Most of the tables have been completely recast and new sections have been added on Auger spectroscopy, nuclear fission, neutron cross-sections, neutron moderation and diffusion and on sub-atomic particles.

We have omitted some older material which now seems of little importance. In a few cases we have substituted graphs for tables where the accuracy of the figures does not suffer and there may be a gain in clarity and compactness. These savings are outweighed, however, by the volume of new material we have felt it desirable to include and this edition is considerably longer than its predecessors. To offset this, the publishers have increased the page size and have improved the typography and layout: we hope that the book remains convenient to use as a source of reference data.

In 1948, after the deaths of the original authors, an editorial board was set up and was responsible for the production of the eleventh, twelfth and thirteenth editions before retiring. The present board is thus the third editorial generation. While we have made some inevitable alterations, it has been our care to see that the essential and distinctive character of 'Kaye and Laby' should remain unchanged. In particular, we have tried to ensure that all the new material is of interest in more than one branch of science and not principally to specialists. We have been at pains to continue the inclusion wherever possible of references to sources of more detailed data.

In his preface to the thirteenth edition, Professor Feather wrote of his sense of continuity in the encouragement he had earlier received from G. F. C. Searle, whose help Kaye and Laby had acknowledged in their preface to the first edition. It is with particular pleasure, therefore, that we record that all four members of the present board have recollections of Dr Searle as a familiar Cambridge figure in earlier days: two of us recall vividly the occasional visits he paid to the undergraduate practical physics class and the wise and pertinent advice he gave. We hope that advice continues to inform the present edition.

December 1972   A.E.B.

EXTRACT FROM PREFACE TO FIRST EDITION

The need for a set of up-to-date English physical and chemical tables of convenient size and moderate price has repeatedly impressed us during our teaching and laboratory experience. We have accordingly attempted in this volume to collect the more reliable and recent determinations of some of the important physical and chemical constants.

To increase the utility of the book, we have inserted, in the case of many of the sections, a brief *résumé* containing references to such books and original papers as may profitably be consulted.

Attention has been paid to the setting and accuracy of the mathematical tables; these are included merely to facilitate calculations arising out of the use of the book, and limitations of space have cut out all but a few of the more essential functions.

We began this book while at the Cavendish Laboratory, Cambridge, and Dr G. A. Carse shared in its inception. To Mr G. F. C. Searle, F.R.S., we feel we owe much for his encouragement and suggestions when the scope of the book was under consideration. . . .

September, 1911

G.W.C.K.
T.H.L.

ACKNOWLEDGEMENTS

We are indebted to the following for permission to reproduce copyright material. Each extract is acknowledged by footnote in the appropriate place in the text.

Academic Press Ltd.; The American Chemical Society; British Standards Institution; Butterworth & Co. (Publishers) Ltd.; Chance Pilkington Optical Works; The Chemical Society; The Clarendon Press; Commissariat de l'Energie Atomique; Department of Trade and Industry— National Physical Laboratory (Division of Molecular Science); Professor H. K. Drickamer and Butterworth Inc.; The late Professor Sir Ronald A. Fisher, Dr. Frank Yates, Rothamsted, and Oliver & Boyd Ltd.; W. H. Freeman & Co., Publishers; Mr. F. A. Gould; Mr. T. Vickers and the Institute of Physics; Interscience Publishers Ltd.; Macmillan and Co. Ltd.; McGraw-Hill Book Co. Inc.; Modern Plastics, Inc.; National Academy of Sciences, Washington D.C.; National Bureau of Standards, Washington D.C.; D. Van Nostrand Co. Inc.; Pergamon Press Ltd.; Prentice-Hall Inc.; Reinhold Publishing Co.; Reynolds Metal Co.; Taylor and Francis Ltd.; John Wiley and Sons Inc.; Wykeham Publications Ltd.

PUBLISHERS' NOTE

While all reasonable care has been taken in compiling these tables, neither the publishers nor the editors or contributors can accept any liability or responsibility for the accuracy of the information contained in the tables.

MEMBERS OF THE EDITORIAL BOARD

A. E. Bailey, M.A., C.Eng., F.I.E.E., F.Inst.P., (Chairman), National Physical Laboratory, Teddington

E. J. Gillham, M.A., F.Inst.P., National Physical Laboratory, Teddington

E. F. G. Herington, D.Sc., Ph.D., D.I.C., B.Sc., A.R.C.S., A.R.I.C., National Physical Laboratory, Teddington

B. Rose, M.A., Sc.D., F.Inst.P., Atomic Energy Research Establishment, Harwell

CONTRIBUTORS AND INITIALS

D.A.	D. Ambrose, B.Sc., Ph.D.	D.J.	D. Jakeman, B.Sc., Ph.D.
P.A.	P. Anderton.	O.N.J.	O. N. Jarvis, B.Sc., Ph.D.
A.E.B.	A. E. Bailey, M.A.	O.C.J.	O. C. Jones, M.A., D.Phil.
E.C.B.	Sir Edward Bullard, M.A., Ph.D., D.Sc., F.R.S.	R.J.K.	R. J. King, B.Sc., M.Sc., Ph.D.
		I.J.L.	I. J. Lawrenson, B.Sc., Ph.D.
J.A.C.	J. A. Catterall, B.Sc., Ph.D.	J.W.L.	J. W. Leake, B.Eng., Ph.D.
P.B.C.	P. B. Clapham, B.Sc., Ph.D.	B.R.L.	B. R. Leaton, B.Sc.
F.J.J.C.	F. J. J. Clarke, B.Sc., Ph.D.	M.L.M.	M. L. McGlashan, Ph.D., D.Sc.
A.H.C.	A. H. Cook, M.A., Ph.D., F.R.S.	M.F.M.	M. F. Markham, B.Sc.
T.E.C.	T. E. Cranshaw, M.A., Ph.D.	H.D.M.	H. D. Megaw, M.A., Ph.D., Sc.D.
J.G.C.	J. G. Cuninghame, B.Sc.	L.W.N.	L. W. Nickols, B.Sc.
R.S.D.	R. S. Dadson, M.A.	B.W.P.	B. W. Petley, Ph.D.
M.E.D.	M. E. Delaney, Ph.D.	D.M.P.	D. M. Poole, B.Sc., M.Sc., D.Phil.
J.A.D.	J. A. Dennis, B.Sc.	R.W.P.	R. W. Powell, Ph.D., D.Sc.
L.E.D.	L. E. Drain, M.A., D.Phil.	J.E.P.	The late J. E. Prue, M.A., D.Phil.
K.W.T.E.	K. W. T. Elliott.	T.J.Q.	T. J. Quinn, B.Sc., D.Phil.
L.E.	L. Essen, D.Sc., F.R.S.	J.B.R.	J. B. Rands
A.F.	A. Felton, B.Sc.	E.D.v.R.	E. D. van Rest, M.A.
K.D.F.	K. D. Froome, Ph.D., D.Sc.	J.C.R.	J. C. Riviere, M.Sc., Ph.D.
C.G.G.	C. G. Garton.	B.R.	B. Rose, M.A., Sc.D.
E.J.G.	E. J. Gillham, M.A.	W.R.C.R.	W. R. C. Rowley, B.Sc., Ph.D.
E.T.G.	E. T. Goodwin, M.A., Ph.D.	J.M.C.S.	J. M. C. Scott, M.A.
J.H.S.G.	J. H. S. Green, B.Sc., Ph.D.	J.R.S.	J. R. Sutton, Ph.D.
K.J.H.	K. J. Habell, M.Sc.	P.V.	P. Vigoureux, D.Sc.
E.F.G.H.	E. F. G. Herington, Ph.D., D.Sc.	C.W.	C. Whitehead, M.A., Ph.D.
M.J.H.	M. J. Hickman.	F.J.W.	F. J. Wilkins, B.A., B.Sc.
B.W.H.	B. W. Hooton, B.Sc., M.Sc., Ph.D.	G.A.W.	G. A. Wilkins, Ph.D.
J.T.H.	J. T. Houghton, M.A., D.Phil.		

CONTENTS

ATOMIC AND NUCLEAR PHYSICS

NOTE ON THE USE OF PAGE HEADINGS

The user of this book will find that the Index is the key to what it contains. Headings in heavy type on each page are intended to assist him in finding his way in the tables, within the sections listed briefly above, and without continual reference to the Index. When parts of more than one table are to be found on a single page, the heading refers in general to the first table which *starts* on that page.

General Physics

1.1 Units

1.1.1 The international system of units (SI)

History

In the second half of the nineteenth century the centimetre, gram and second were in fairly general use as base units for scientific work even in such countries as the UK and the USA where the foot and the pound were employed for commerce and engineering. As a result, the units required by the rapidly emerging science of electricity were based on the centimetre, gram and second, with which they formed a coherent system known as the CGS electromagnetic system. A system of units is said to be coherent when derived units are formed from the base units without the insertion of factors of proportionality other than unity. There was also the CGS electrostatic system, but the only quantities frequently expressed in electrostatic units were electric charge, electric potential and capacitance.

The young but fast-growing electrical industry soon found that many CGS electromagnetic units were of an extremely inconvenient size for its needs. Accordingly, in 1881, international agreement was reached to fix the practical unit of potential, to be called the volt, at 10^8 CGS units (which is approximately equal to the e.m.f. of a primary cell), and the unit of resistance, the ohm, at 10^9 CGS units (which is approximately the resistance of a column of mercury 1 m long and 1 mm^2 in cross-section). The unit of electric current, the ampere, was made a tenth of the CGS unit. A coherent system of practical electric units was thus secured which, however, was not coherent with the mechanical units based on the centimetre and gram. The practical electric units suited the needs of telegraphy, which was then the main electrical industry, and they also happen to be convenient for heavy electrical engineering and electronics.

The magnetic units were left at their CGS values. This was presumably because the CGS unit of magnetic flux density, subsequently called 'gauss', is of the order of the flux density of the Earth's field, and, as it was suitable for geomagnetism, there seemed no point in changing it for a unit 10^4 times larger. Coherence was thereby lost to electromagnetism as it had already been lost to the system embracing the mechanical units and the practical electric units.

Whereas the electric units, by the agreement of 1881, were chosen to be of suitable magnitude for everyday use, and whereas the centimetre and the second have acceptable sizes, the gram is too small for the practical needs of man, which are better served by a unit nearer the size of the pound or the kilogram. Moreover, the CGS unit of force, the dyne, and the unit of energy, the erg, are much too small. On the other hand, the unit of energy provided by the practical electric units, the volt-ampere-second, called the joule—which equals 10^7 ergs—is of a satisfactory size.

These considerations—the advantages of coherence and the fortuitous circumstance that a mechanical system based on the metre and the kilogram has precisely the same unit of energy as is provided by the practical electric units—led G. Giorgi in 1902 to propose a system based on the **metre**, the **kilogram**, the **second** and one of the practical electric units. He pointed out that if magnetic field strength were expressed as amperes per metre instead of 4π times amperes per metre, which is the definition corresponding to that of the CGS unit, the number π would disappear from most electric and magnetic formulae involving rectilinear geometry, but would appear, as expected, in those involving cylinders or spheres.

The International Electrotechnical Commission eventually chose the **ampere** as the fourth base unit of the MKSA or 'Giorgi' system, and in 1948 the 9th General Conference of Weights and Measures† recommended it for science and technology, as well as for commerce and industry. This

† The General Conference of Weights and Measures (CGPM) is the authority set up by the Metre Convention of 1875 to promote use and improvement of the metric system and to secure international uniformity in metric units and standards of measurement. It consists of delegations from the member nations (of which there were 41, including the UK, in 1972), which meet every few years, the 12th, 13th and 14th conferences having been held in 1964, 1967–1968 and 1971. The International Bureau of Weights and Measures, Sèvres (near Paris), is the central office and laboratory of the organization which is managed, under the authority of the General Conference, by the International Committee of Weights and Measures consisting of 18 members, each from a different nation. The International Committee meets yearly and has the responsibility for recommending proposals for approval by the General Conference. Seven specialist advisory committees assist the Inter-

system admirably covers mechanics and electromagnetism, but it does not provide for other branches of science such as heat. In 1960, in the hope of securing world-wide uniformity in the units employed in natural science, the 11th CGPM added to the units metre, kilogram, second and ampere, the **kelvin** for thermodynamic temperature, the **candela** for luminous intensity, and the radian and steradian for plane and solid angle. The first two joined the original four in being called 'base' units, and the last two were called 'supplementary' units. Any unit formed from two or more of those eight is called 'derived'. The radian and steradian are regarded as base units or derived units according to convenience. The MKSA system thus broadened is called the International System of Units, often abbreviated to SI, and is the most satisfactory system of units we have had so far, in that it caters for the commercial and industrial activities of man as well as for the needs of science. In 1971, the 14th CGPM added the **mole**, the unit of amount of substance used in chemistry, to the list of base units, thus making them seven in all.

Definitions of some SI units

The seven base quantities, each with its unit and unit symbol, are listed below.

SI base quantities and units

Quantity	Name of unit	Unit symbol
Length	metre	m
Mass	kilogram	kg
Time	second	s
Electric current	ampere	A
Thermodynamic temperature	kelvin	K
Luminous intensity	candela	cd
Amount of substance	mole	mol

The SI base units are defined as follows:

The metre is the length equal to 1 650 763.73 wavelengths in vacuum of the radiation corresponding to the transition between the levels $2p_{10}$ and $5d_5$ of the krypton-86 atom.

The kilogram is the unit of mass; it is equal to the mass of the international prototype of the kilogram.

The second is the duration of 9 192 631 770 periods of the radiation corresponding to the transition between the two hyperfine levels of the ground state of the caesium-133 atom.

The ampere is that constant current which, if maintained in two straight parallel conductors of infinite length, of negligible circular cross-section, and placed 1 metre apart in vacuum, would produce between these conductors a force equal to 2×10^{-7} newton per metre of length.

The kelvin, unit of thermodynamic temperature, is the fraction $1/273.16$ of the thermodynamic temperature of the triple point of water.

The candela is the luminous intensity, in the perpendicular direction, of a surface of $1/600\,000$ square metre of a black body at the temperature of freezing platinum under a pressure of 101 325 newtons per square metre.

The mole is the amount of substance of a system which contains as many elementary entities as there are atoms in 0.012 kilogram of carbon 12. When the mole is used, the elementary entities must be specified and may be atoms, molecules, ions, electrons, other particles or specified groups of such particles.

The SI supplementary units are defined thus:

The radian is the plane angle between two radii of a circle which cut off on the circumference an arc equal in length to the radius.

national Committee in planning co-operative programmes of research and the preparation of recommendations on units of measurement, length (definition of the metre), time (definition of the second), temperature, electricity, photometry and radiometry, and the ionizing radiations.

The steradian is the solid angle which, having its vertex in the centre of a sphere, cuts off an area of the surface of the sphere equal to that of a square with sides of length equal to the radius of the sphere.

Derived units. The table below lists some of the more common SI derived quantities, each with its unit and unit symbol. The composite symbols in the last column are to some extent indicative of the definition of the quantity.

Quantity	Unit		Symbol
Supplementary			
Plane angle	radian	rad	
Solid angle	steradian	sr	
Derived			
Area	square metre		m^2
Volume	cubic metre		m^3
Frequency	hertz	Hz	s^{-1}
Density	kilogram per cubic metre		$kg\ m^{-3}$
Concentration	mole per cubic metre		$mol\ m^{-3}$
Velocity	metre per second		$m\ s^{-1}$
Angular velocity	radian per second		$rad\ s^{-1}$
Acceleration	metre per second squared		$m\ s^{-2}$
Angular acceleration . . .	radian per second squared		$rad\ s^{-2}$
Force	newton	N	$m\ kg\ s^{-2}$
Pressure, stress	pascal	Pa	$N\ m^{-2}$
Viscosity (dynamic) . . .	pascal-second		$Pa\ s$
Viscosity (kinematic) . . .	metre squared per second		$m^2\ s^{-1}$
Energy, work, quantity of heat .	joule	J	$N\ m$
Power, radiant flux	watt	W	$J\ s^{-1}$
Quantity of electricity . . .	coulomb	C	$A\ s$
Potential difference, electro-			
motive force	volt	V	$W\ A^{-1}$
Electric field strength . . .	volt per metre		$V\ m^{-1}$
Electric resistance	ohm	Ω	$V\ A^{-1}$
Electric conductance . . .	siemens	S	Ω^{-1}
Capacitance	farad	F	$C\ V^{-1}$
Magnetic flux	weber	Wb	$V\ s$
Magnetic flux density . . .	tesla	T	$Wb\ m^{-2}$
Inductance	henry	H	$\Omega\ s$
Magnetic field strength . . .	ampere per metre		$A\ m^{-1}$
Magnetomotive force . . .	ampere	A	
Wave number†	1 per metre		m^{-1}
Activity (radioactive) . . .	1 per second		s^{-1}
Luminous flux	lumen	lm	$cd\ sr$
Luminance	candela per square metre		$cd\ m^{-2}$
Illuminance	lux	lx	$lm\ m^{-2}$
Heat flux density, irradiance .	watt per square metre		$W\ m^{-2}$
Heat capacity, entropy . . .	joule per kelvin		$J\ K^{-1}$
Specific heat capacity, specific			
entropy	joule per kilogram kelvin		$J\ kg^{-1}\ K^{-1}$
Thermal conductivity . . .	watt per metre kelvin		$W\ m^{-1}\ K^{-1}$
Molar energy	joule per mole		$J\ mol^{-1}$
Molar entropy, molar heat			
capacity	joule per mole kelvin		$J\ mol^{-1}\ K^{-1}$

† Wave numbers in the infra-red are still often expressed in cm^{-1}.

Multiples. Prefixes may be used, instead of powers of 10, to express certain decimal multiples of the units. Their names and symbols are listed below.

Factor	Name	Symbol	Factor	Name	Symbol
10^{12}	tera	T	10^{-1}	deci	d
10^{9}	giga	G	10^{-2}	centi	c
10^{6}	mega	M	10^{-3}	milli	m
10^{3}	kilo	k	10^{-6}	micro	μ
10^{2}	hecto	h	10^{-9}	nano	n
10	deca	da	10^{-12}	pico	p
			10^{-15}	femto	f
			10^{-18}	atto	a

An exponent attached to a symbol containing a prefix indicates that the multiple of the unit is raised to the power expressed by the exponent.

Example: $1 \text{ cm}^3 = 10^{-6} \text{ m}^3$; $1 \text{ cm}^{-1} = 10^2 \text{ m}^{-1}$.

Compound prefixes should not be used, e.g. p not $\mu\mu$. Names of multiples of the unit of mass are formed by attaching prefixes to the word 'gram'.

1.1.2 Realization of SI units

The metre

According to the definition of the metre, realization of the unit consists in marking off a length containing the stated number of wavelengths. Although, to keep the definition simple, only the particular line and the particular isotope are mentioned in it, the International Committee of Weights and Measures has specified the conditions necessary for accurate reproducibility of wavelength, e.g. type of lamp, current density, temperature, etc.

The metre is defined by an atomic transition—unperturbed, i.e. in the absence of disturbance due to Doppler broadening, isotopic broadening and other causes. One problem in realizing the metre is to find a lamp reproducing the wavelength without serious disturbance. In practice the lamps do affect the wavelength to the extent of at least 1 or 2 in 10^8, depending upon operating conditions. Studies on these lamps have shown that under certain conditions some of the effects cancel, making it possible to obtain light of a wavelength corresponding very closely to that of the ideal unperturbed transition. The krypton-86 orange line at 606 nm was chosen because it was found to be particularly reproducible in wavelength. Its source is operated at the triple point of nitrogen, 63 K, so that Doppler broadening, being proportional to the square root of the absolute temperature, is kept very small.

In most cases what is required is not realization of the unit itself, but rather measurement of a particular distance, e.g. the length of a rod or the separation of two optical flats. Interferometers, usually of the Fabry–Perot or of the Michelson type, are employed to determine the number of waves covering the distance. The krypton-86 line gives satisfactory fringes at path differences less than half a metre, but stabilized lasers, which have much narrower lines, have recently increased the workable distance to over 100 m. It is, of course, necessary to know the wavelength of the light used in terms of that of the krypton-86 line, and the relation for various useful optical and laser lines has been established by careful measurements.

The kilogram

Realization of the kilogram consists in making 'weighing masses' equal in mass to the mass of the prototype of the kilogram. The prototype is kept by the International Bureau of Weights and

Measures at Sèvres, near Paris, and the various nations adhering to the Metre Convention have copies which from time to time they send to Sèvres for comparison with the prototype or with the Bureau's copies of the prototype.

Methods of adjusting weighing masses are well known, as also are balances for comparing them. Similarly, the production of masses which are multiples of the unit, and their calibration by means of balances, are standard. With reliable weighing masses—for example, of platinum–iridium or of stainless steel—and good balances, and with a great deal of care, masses of the order of 1 kg can be compared to 1 in 10^9.

The second

The second of time was formerly defined as $1/86\,400$ of the mean solar day. Its present definition in terms of the radiation corresponding to the hyperfine transition of the caesium-133 atom was adopted at the 13th General Conference of Weights and Measures in 1966, the duration of $9\,192\,631\,770$ periods of this radiation being chosen in order to secure as close agreement as possible with astronomical definitions.

The transition of the caesium atom is observed in an atomic beam equipment in which all the major causes tending to broaden or shift the line are inherently eliminated. The atoms pass through a system of magnets in which they are deflected, and through an electrical resonator supplied with an alternating field at a frequency derived by synthesis from a quartz oscillator. When the applied frequency equals that of the spectral line, transitions between the states are induced and deflection of the atoms is reversed. A resonant line is thus observed and if the frequency deviates from that of the spectral line, an error signal is produced and applied to the quartz oscillator to pull its frequency to the correct value. The quartz oscillator acts as the working standard as it did when time was based on astronomical measurements.

Although the unit has been defined in terms of the caesium line, similar lines of other atoms can be measured with great accuracy and can then be used as standards. The frequency of the hydrogen line, for example, is $1\,420\,405\,751.77$ Hz.

It is not necessary for other than highly specialized organizations to own caesium beams or similar standards with which to realize the second, for many countries broadcast frequency and time signals by which any laboratory having a suitable radio receiver can calibrate its wavemeters and electronic counters (see p. 127). These transmissions are, in general, accurate to 1 in 10^{11}, and corrections to 1 in 10^{12} are published later.

The ampere

The definition of the ampere is simple and satisfies legal requirements, but the configuration it specifies precludes it from being applied as it stands to the measurement of electric current. However, as it is consistent with electromagnetic theory, other more practical configurations can be employed. The force per unit current between current-carrying coils of circular or helical shape can be calculated from the dimensions of the coils by fairly simple, exact formulae, and the current becomes known if the force is also measured with, for instance, a balance. These current balances, or dynamometers, are only used at rare intervals of, say, a few years. As their accuracy of a few p.p.m. is far superior to the precision of pointer or even reflecting ammeters, they are fed with a current stabilized by equalizing the volt drop RI across a standard resistor to the e.m.f. E of a Weston cell. The current I in the balance is then E/R, where *a priori* neither E nor R is known in absolute value. As explained below, however, R can be measured independently in terms of the metre and the second, and the two measurements thus yield E and R. In the intervals between measurements with current balances, the ampere is maintained, to a precision of about 1 p.p.m., by means of standard resistors and Weston cells.

The monitoring, though not the realization, of the ampere—i.e. of the resistors and cells which serve to maintain it—may also be effected by an application of nuclear magnetic resonance. The current, nominally equal to E/R as for the current balance, is carried by a coil of shape such that the magnetic field near its centre is uniform, and the frequency of precession of protons in that field is observed. If the dimensions of the coil do not change from year to year, or if the change is measured and allowed for, the frequency of precession provides a check on the stability of E/R. Measurement of a frequency is less laborious than accurate 'weighing' of a current.

The volt and the ohm

The volt, or watt/ampere, could, according to its definition, be realized from the ampere by a measurement of dissipation of energy in the form of heat. However, as energy is not easily measured to an accuracy of 1 p.p.m., it has been the practice to realize the volt, or the ohm, by an independent measurement involving electromagnetism. Whichever is realized, the other is obtained from it and the ampere by Ohm's law.

In the past the ohm has in general been chosen for the independent measurement. If we remember that the quantities ωL, ωM, $1/\omega C$—where ω, L, M and C stand for angular frequency, self-inductance, mutual inductance and capacitance—all have the dimensions of R, we see that a method of determining R is to compare it with one of those quantities. Self- and mutual inductors and capacitors have all been used. They are designed so that their values may be calculated from their linear dimensions. Accuracies of 1 in 10^7 and even better have been claimed for the calculable capacitor method of realizing the ohm (excluding the uncertainty in the velocity of light, which enters the formula for capacitance). As the metre and the second are defined in terms of atomic constants, the ohm also depends on those constants.

The calculable capacitor, being fairly easy to use once it has been adjusted, is also suitable for monitoring standard resistors. The Weston cells which serve to maintain the volt can be monitored independently by an application of the Josephson effect in superconductivity. The p.d. between two superconductors separated by a narrow gap assumes values which are multiples of $hf/2e$, where f is the frequency of electromagnetic waves irradiating the superconductors, h is Planck's constant and e the electronic charge. For the mere purpose of monitoring the Weston cell from year to year, it is not even necessary to know the value of $h/2e$ as the same value can be used each time.

Standard resistors for preserving the ohm are in general of nominal value 1 Ω. The e.m.f. of Weston cells depends slightly on temperature and on the acidity of the electrolyte. The figures below are representative.

E.M.F. of saturated (0.05 N) Weston cell

Temperature/°C	Electromotive force/V	Temperature/°C	Electromotive force/V
0	1.018 97	25	1.018 40
5	1.018 98	30	1.018 14
10	1.018 92	35	1.017 84
15	1.018 80	40	1.017 51
20	1.018 62		

International electric units. From the beginning of the century until 1948, the electric and magnetic units were not derived directly from the base units of length, mass and time, but on values internationally agreed upon for the electrochemical equivalent of silver and for the specific resistance of mercury. These units were called 'international units'. As one would not, for accurate work, use tables of constants published before 1948, the need for conversion factors hardly ever arises. The figures given by the International Bureau of Weights and Measures are

$$1 \text{ 'mean international ohm'} = 1.000\ 49\ \Omega$$
$$1 \text{ 'mean international volt'} = 1.000\ 34\ V$$

The kelvin

The unit of thermodynamic temperature, the kelvin, is defined by assigning the value 273.16 K to the temperature of the triple point of water. The determination of other temperatures in terms of this unit requires measurements, by means of a gas thermometer for example, which can be evaluated in accordance with the thermodynamic definition of temperature. Absolute measurements of this kind are difficult, and their accuracy is usually less than the reproducibility attainable by non-absolute methods based on the fixed points of various substances, and interpolation or extrapolation by instruments such as resistance thermometers or optical pyrometers. This situation has led to the adoption, by international agreement, of a practical temperature scale, IPTS-68, based on the use of these more reproducible methods, and adjusted to conform to the best available knowledge of the thermodynamic temperatures of fixed points (see section 1.5, p. 42).

The candela

Realization of the candela entails absolute photometric measurements in which the standard is the luminance of a black body at the temperature of freezing platinum, 2045 K† approximately. With great care, an accuracy somewhat better than 1% can be attained. The secondary standards of luminous intensity are special tungsten-filament vacuum lamps adjusted to operate at a temperature at which their spectral power distribution in the visible region matches that of the basic standard. A group of these standard lamps maintains the unit during periods of years between their recalibration against the basic standard; similar groups serve also for the intercomparison of national standards, made every few years at the International Bureau of Weights and Measures.

For sources of light of spectral distribution other than that of the primary standard 'relative spectral luminous efficiencies', $V(\lambda)$, have been recommended by the International Committee of Weights and Measures. By their means photometric quantities are defined in purely physical terms as quantities proportional to the sum or integral of a spectral power distribution, weighted according to the $V(\lambda)$ function (see section 1.7.3, p. 78).

The mole

Although the mole is defined in terms of number of entities, it is usually realized by weighing rather than by counting. A mole of atoms of an element X, for example, is obtained by weighing an amount, in grams, of X, equal to its relative atomic mass (atomic weight); similarly a mole of molecules of a substance Y is obtained by weighing an amount, in grams, of Y, equal to its relative molecular mass (molecular weight).

In the case of perfect gases, 1 mole of molecules occupies the same volume, independently of the particular gas, at any given temperature and pressure. This relation provides a method for measuring equal amounts of substance of perfect gases. The method of volume comparison can be extended to non-perfect gases, because the corrections to apply are well known.

Ratios of amounts of substance liberated in electrolytic reactions can be determined by measuring the corresponding quantities of electricity. For example, 1 mole of Ag and 1 mole of Cu(1/2) are deposited on a cathode by the same quantity, approximately 96 487 C, of electricity.

1.1.3 Relations between SI and other units

CGS units

Many books and papers on electromagnetism were written before the general adoption of SI. The following table gives the relations between the units, and the conversion factor by which the number expressing the value of the quantity in CGS units must be multiplied to express its value in SI units.

† A recent determination gave a value of 2040.8 K.

Electric and magnetic units

Quantity and symbol	SI unit		Conversion factors	
	Name and symbol	Defining equation	CGSm	CGSe†
1. Electric current, I	ampere A	$F_z = 10^{-7} I^2 (dN/dz)$‡	10	$10/c$
2. Current density, J	A m^{-2}		10^5	$10^5/c$
3. Electromotive force, potential difference, V	volt V	$p\S = IV$	10^{-8}	$10^{-8}c$
4. Electric field strength, E . .	V m^{-1}	$E = V/l$	10^{-6}	$10^{-6}c$
5. Resistance, R	ohm Ω	$R = V/I$	10^{-9}	$10^{-9}c^2$
6. Conductance, G	siemens S	$G = 1/R$	10^9	$10^9/c^2$
7. Volume resistivity, ρ . . .	Ω m		10^{-2}	$10^{-11}c^2$
8. Conductivity, γ . .	S m^{-1}	$\gamma = 1/\rho = J/E$	10^2	$10^{11}/c^2$
9. Electric charge, Q	coulomb C	$Q = It$	10	$10/c$
10. Capacitance, C	farad F	$C = Q/V$	10^9	$10^9/c^2$
11. Electric flux density, D . .	C m^{-2}	$D = Q/l^2$‖	10^5	$10^5/c$
12. Permittivity, ϵ	F m^{-1}	$\epsilon = D/E$		$10^{11}/4\pi c^2$
13. Magnetic field strength, H .	A m^{-1}	$\oint H\,dl = nI$	$10^3/4\pi$	
14. Magnetic flux, ϕ	weber Wb	$E = d\phi/dt$	10^{-8}	
15 Magnetic flux density, B . .	tesla T	$B = \phi/l^2$‖	10^{-4}	
16. Inductance, L, M	henry H	$M = \phi/I$	10^{-9}	
17. Permeability, μ	H m^{-1}	$\mu = B/H$	$4\pi \times 10^{-7}$	

† c = velocity of light in free space in cm s^{-1} = 2.997 925 × 10^{10}.

‡ N denotes Neumann's integral for two linear circuits each carrying the current I; F_z is the force between the two circuits in the direction defined by the co-ordinate z, the circuits being in a vacuum.

$$N = \oint_1 \oint_2 \frac{ds_1 \, ds_2 \cos\theta}{r}$$

§ p denotes 'power'.

‖ l^2 denotes area.

The proportionality constants in the following equations are determined by use of the defining equations (1) to (17) in the table above.

For the mutual inductance M of two linear circuits, for which Neumann's integral is N, in a medium of permeability μ:

$$M = \frac{\mu}{4\pi} N \qquad \ldots (18)$$

For the force F_z due to currents I_1 and I_2 in the above two circuits:

$$F_z = I_1 I_2 \frac{dM}{dz} = \frac{\mu}{4\pi} I_1 I_2 \frac{dN}{dz} \qquad \ldots (19)$$

For the capacitance δC of an element of volume of length δl and cross-sectional area δA in a uniform dielectric medium of permittivity ϵ, the electric field having the direction of l:

$$\delta C = \epsilon \frac{\delta A}{\delta l} \qquad \ldots (20)$$

In the SI, Maxwell's equations for propagation in a medium are

$$\nabla \times H = \dot{D} + J, \qquad -\nabla \times E = \dot{B} \qquad \ldots (21)$$

where the dot means d/dt, J is current density, and the other quantities are as in the table. The velocity of propagation of an electromagnetic wave in the medium is given by

$$v^2 = 1/\mu\epsilon \qquad \ldots (22)$$

In free space, where J is zero, and where D and B are linked with E and H by the electric and magnetic constants ϵ_0 and μ_0, the equations may be written

$$c^2 \nabla \times B = \dot{E}, \qquad -\nabla \times E = \dot{B} \qquad \ldots (23)$$

The velocity of propagation c is given by

$$c^2 = 1/\mu_0 \epsilon_0 \qquad \ldots (24)$$

The values of μ_0, ϵ_0 and of the characteristic impedance Z_0 of free space are

$$\mu_0 = 4\pi \times 10^{-7} \text{ H m}^{-1} = 1.256\ 637\ \mu\text{H m}^{-1}$$
$$\epsilon_0 = 8.854\ 16 \text{ pF m}^{-1}$$
$$Z_0 = \sqrt{(\mu_0/\epsilon_0)} = 376.731\ \Omega$$

Coulomb's law of force f between electric charges Q, Q' a distance r apart, and its analogue for magnetic poles ϕ, ϕ', can be deduced from Maxwell's equations. The expressions are

$$f = \frac{QQ'}{4\pi\epsilon r^2} \qquad \ldots (25)$$

$$f = \frac{\phi\phi'}{4\pi\mu r^2} \qquad \ldots (26)$$

Other mechanical units

In the gravitational systems of units sometimes still used in engineering and aerodynamics, the units of length and time are base units, but the third base unit is a unit of force, namely the force due to a conventional value of gravity, called standard gravity, acting on a specified mass. The unit of mass is thus a derived unit, and since unit mass is given unit acceleration when acted on by unit force, the unit of mass is equal to the standard value of gravity multiplied by the specified mass.

In the metric gravitational system, standard gravity is $9.806\ 65 \text{ m s}^{-2}$, the unit of force is called kilogram-force, symbol kgf, the specified mass is the kilogram, and the unit of mass is therefore $9.806\ 65$ kg. In Germany and some other European countries the same unit of force is called kilopond, symbol kp.

In the foot–pound–second gravitational system the base unit of force is the pound-force, symbol lbf, standard gravity is 32.174 ft s^{-2}, and the derived unit of mass, the slug, is now 32.174 lb although originally it was 32.2 lb, based on 32.2 ft s^{-2} for gravity.

Conversion factors

The following table gives conversion factors for British and other units which are still used in some fields, but which, for the most part, are likely to disappear in course of time. Exact conversion factors are in bold type.

Unit name and symbol		*SI equivalent*		*Reciprocal*
Length				
ångström . . . . . . Å	**0.1**	nm	**10**	
fermi	**1**	fm		
micron . . . . . . μ	**1**	μm		
yard . . . . yd	**0.914 4**	m	1.093 61	
foot . . . . ft	**30.48**	cm	0.032 808 4	
inch . . . . in	**2.54**	cm	0.393 701	
statute mile	**1.609 344**	km	0.621 371	
nautical mile (international) .	**1.852**	km	0.539 957	
fathom (6 ft)	**1.828 8**	m	0.546 807	
X unit	0.100 2	pm		
astronomical unit (AU) . .	0.149 6	Tm		
parsec . . . . . . pc	30 857	Tm		

Conversion factors (*contd*)

Unit name and symbol	SI equivalent		Reciprocal
Area			
hectare . . . . ha	1	hm²	
barn . . . . b	100	fm²	
square yard . . . . . . yd²	0.836 127	m²	1.195 99
square foot . . . . . . ft²	9.290 30	dm²	0.107 639
square inch . . . . . . in²	**6.451 6**	cm²	0.155 000
square statute mile	2.589 99	km²	0.386 102
acre	0.404 686	hm²	2.471 05
Volume			
stere . . . . st	1	m³	
litre l	1	dm³	
cubic yard . . . . . . yd³	0.764 555	m³	1.307 95
cubic foot . . . . . . ft³	0.028 316 8	m³	35.314 7
cubic inch . . . . . . in³	16.387 1	cm³	0.061 023 7
gallon (UK) . . . . . gal	4.546 09	dm³	0.219 969
gallon (US) = 231 in³ . . .	3.785 41	dm³	0.264 172
pint . . . . pt	0.568 261	dm³	1.759 75
barrel (US) = 42 gal (US) . . bbl	0.159 0	m³	6.290
Plane angle			
right angle = π/2 rad . . .	1.570 796	rad	0.636 620
degree = 1/90 right angle . . °	π/180	rad	57.295 8
minute = (1/60)° . . . . ′	π/10 800	rad	3437.75
second = (1/60)′ . . . . ″	π/648	mrad	206.265
grade = 0.01 right angle . .	π/200	rad	63.662 0
Mass			
tonne . . . . t	1	Mg	
metric carat	0.2	g	5
pound (avdp) . . . . . lb	**0.453 592 37**	kg	2.204 62
grain . . . . gr	**0.064 798 91**	g	15.432 4
ounce (avdp) . . . . . oz	28.349 5	g	0.035 274 0
ounce (troy) = 480 gr . . .	31.103 5	g	0.032 150 7
ton = 2240 lb	1.016 05	Mg	0.984 207
slug	14.593 9	kg	0.068 521 8
Specific volume			
cubic foot per pound . . . ft³ lb⁻¹	0.062 428	m³ kg⁻¹	16.018 5
Density			
pound per cubic foot . . . lb ft⁻³	16.018 5	kg m⁻³	0.062 428
pound per cubic inch . . . lb in⁻³	27.679 9	Mg m⁻³	0.036 127 3
pound per gallon (UK) . . lb gal⁻¹	99.776 3	kg m⁻³	0.010 022 4
slug per cubic foot . . . slug ft⁻³	0.515 379	Mg m⁻³	1.940 32
Consumption, yield			
gallon (UK) per mile . . .	2.824 8	dm³ km⁻¹	0.354 01
bushel per acre	0.089 87	m³ ha⁻¹	11.127
Moment of inertia			
lb in²	2.926 40	kg cm²	0.341 717
slug ft²	1.355 82	kg m²	0.737 562
Moment of section			
in⁴	41.623 1	cm⁴	0.024 025 1

Conversion factors (*contd*)

Unit name and symbol	SI equivalent		Reciprocal
Time			
minute . . . . . . min	**60**	s	
hour = 60 min . . . . h	**3600**	s	
day = 24 h . . . . . d	**86 400**	s	
Velocity			
foot per second . . . . ft s^{-1} {	**0.304 8**	m s^{-1}	3.280 84
	1.097 28	km h^{-1}	0.911 344
mile per hour . . . . mile h^{-1}	**1.609 344**	km h^{-1}	0.621 371
knot (international)	**1.852**	km h^{-1}	0.539 957
Acceleration			
gal . . . . . . Gal	**1**	cm s^{-2}	
Force			
dyne . . . . . . dyn	**10**	μN	**0.1**
poundal . . . . . pdl	0.138 255	N	7.233 01
pound-force . . . . lbf	4.448 22	N	0.224 809
ton-force . . . . . tonf	9.964 02	kN	0.100 361
kilogram-force . . . . kgf	**9.806 65**	N	0.101 972
Force per unit length			
lbf in^{-1}	0.175 127	kN m^{-1}	5.710 15
tonf ft^{-1}	32.690 3	kN m^{-1}	0.030 590 1
Torque			
pound-force foot . . . . lbf ft	1.355 82	N m	0.737 562
Pressure, stress			
bar . . . . . . bar	**0.1**	MPa	**10**
lbf in^{-2} . . . . . in	6.894 76	kPa	0.145 038
lbf ft^{-2}	47.880 3	Pa	0.020 885 4
tonf in^{-2}	15.444 3	MPa	0.064 749 0
standard atmosphere . . . atm	**0.101 325**	MPa	9.869 23
torr } mmHg } . . . . . torr	0.133 322	kPa	7.500 62
inHg	3.386 39	kPa	0.295 300
Work, energy, heat			
erg . . . . . . erg	**0.1**	μJ	**10**
electronvolt . . . . eV	0.160 219	aJ	6.241 46
calorie (IT) . . . . cal	**4.186 8**	J	0.238 846
British thermal unit . . . Btu	1.055 06	kJ	0.947 817
therm = 10^5 Btu	0.105 506	GJ	9.478 17
foot poundal . . . . ft pdl	0.042 140 1	J	23.730 4
foot pound-force . . . . ft lbf	1.355 82	J	0.737 562
Power			
horse-power = 550 ft lbf s^{-1} . hp	0.745 700	kW	1.341 02
cheval-vapeur = 75 m kgf s^{-1} . ch	0.735 499	kW	1.359 62
Temperature			
degree Celsius . . . . °C	**1**	K	
t °C	**273.15** + *t*	K	
degree Fahrenheit . . . . °F	5/9	K	1.8
t °F	5(*t* − 32)/9	°C	
degree Rankine	5/9	K	1.8
Heat transmission			
Btu h^{-1}	0.293 071	W	3.412 14

Conversion factors (*contd*)

Unit name and symbol		SI equivalent		Reciprocal
Viscosity (*dynamic*)				
poise	P	0.1	Pa s	10
lb ft^{-1} s^{-1} = pdl s ft^{-2} . . .		1.488 16	Pa s	0.671 969
slug ft^{-1} s^{-1} = lbf s ft^{-2} . .		47.880 3	Pa s	0.020 885 4
Viscosity (*kinematic*)				
stokes	St	1	cm^2 s^{-1}	
ft^2 s^{-1}		9.290 30	dm^2 s^{-1}	0.107 639
Electric stress				
kilovolt per inch ⎱				
volt per mil ⎰	kV in^{-1}	0.393 701	kV cm^{-1}	2.54
Magnetism				
gauss	Gs, G	100	μT	0.01
oersted	Oe	(1/4π)	kA m^{-1}	4π
maxwell	Mx	10	nWb	0.1
gamma	γ	1	nT	
Light				
stilb	sb	10	kcd m^{-2}	0.1
phot	ph	10	klx	0.1
Radioactivity				
curie	Ci	37	ns^{-1}	0.027 027
röntgen	R	0.258	mC kg^{-1}	3.875 97
rad	rad	10	mJ kg^{-1}	0.1

References

(1) *The International System of Units*, 1970 (translation of a BIPM publication) (HMSO).
(2) BS 3763: 1970, *The International System of Units*.
(3) P. Vigoureux, *Units and Standards of Electromagnetism*, 1971 (Wykeham, London).
(4) P. Anderton and P. H. Bigg, *Changing to the Metric System: Conversion Factors, Symbols and Definitions*, 1969 (HMSO).
(5) BS 350: 1972, Part 1, *Conversion Factors and Tables*.

P.V.

1.2 Fundamental constants

1.2.1 Velocity of electromagnetic waves

In 1941 Birge derived the value 299 776 ± 4 km s^{-1} for the velocity of propagation of light *in vacuo* (c_0) from a statistical review of all previous determinations; other reviewers assigned wider uncertainty limits of ± 10 or ± 20 km s^{-1} to this value. Beginning with Essen and Gordon-Smith's measurement in 1946, a number of precise determinations of c_0 have yielded results lying for the most part within the range 299 792 and 299 793 km s^{-1}, using optical systems (principally Bergstrand's geodimeter) as well as microwave devices, i.e. Essen's improved cavity resonator, microwave interferometers, and Wadley's tellurometer (*Empire Survey Review*, 1956, **14**, nos. 105 and 106). Values of c_0 for the period 1946–1956 are summarized in *J. Br. Instn Radio Engrs*, 1956, **16**, 497. In 1957 the International Union of Scientific Radio (URSI) and the International Union of Geodesy and Geophysics (IUGG) adopted the value $c_0 = 299$ 792.5 ± 0.4 km s^{-1}, which agrees with the latest value obtained by microwave interferometry (Froome, *Nature*, 1958), namely 299 792.50 ± 0.10 km s^{-1}, probably the most precise determination to date.

K.D.F.

1.2.2 The constant of gravitation

Newton's law of gravitation states that the force between two masses M_1 and M_2 at a distance d is GM_1M_2/d^2. The constant G is not known to better than 1 in 1000; the most recent value is that of Heyl (*J. Res. Nat. Bur. Stand. Wash.*, 1930):

$$G = 6.67 \times 10^{-11} \text{ m}^3 \text{ kg}^{-1} \text{ s}^{-2}$$

A.H.C.

1.2.3 Atomic constants

The values of e, h, N etc. given on pp. 16, 17 are based on the investigation by Taylor, Parker and Langenberg ('TPL'), *Rev. Mod. Phys.*, 1969, **41**, 375, which is the most recent discussion at the time of writing.

    The best values of constants such as the mass and charge of the electron are not values found by measuring each separately (e.g. as in Millikan's oil-drop experiment), but come from much more accurate measurements of certain combinations such as $m_e e^4 c^3/h^3$. The values recommended in the above-mentioned paper were obtained by a 'least-squares' statistical treatment of selected experimental data; but, since the final values depend mainly on only a few results, it may be useful to give a simplified derivation to indicate the connection between the values of e, etc. and what is actually measured.

    Before doing this, it is convenient to introduce the *'fine structure constant'* α. Since $(\mu_0 c^2/4\pi)e^2/\hbar$ is a velocity, $(\mu_0 c^2/4\pi)e^2/\hbar c = \alpha$ is a dimensionless number, approximately $1/137$. The symbol $\hbar$ denotes Planck's constant h divided by 2π.

    TPL in their paper take the results of some twenty experimental investigations. They reject a few on the ground that accepting them with the full accuracy claimed would require rejection of too many others; they also reject those in which the theory includes disputable quantum-electrodynamic corrections[†] of appreciable magnitude. They then determine α, e and N by a least-squares analysis. (Actually they also adjust a fourth variable, the ratio of the officially maintained N.B.S. standard of current to the true (MKS) ampere, but in what follows it will simply be taken to be 1.000 009). The velocity of light, c, and other constants known to better than 1 in 10^6, will here be treated as exact.

Evidence for e, m_e and h. Temporarily disregarding N, the first problem is to determine e, m_e and $\hbar$. One combination of these constants is known to better than 1 part in 10^6, namely the Rydberg constant specifying the wavelengths in the H spectrum,

$$R_\infty = \mu_0^2 c^3 m_e e^4/8h^3 = 10\ 973\ 731 \text{ m}^{-1}$$

Using this relation, the problem can be reduced to finding the values of e/m_e and α, and from these one can deduce, for example,

$$e = (\alpha^3/\mu_0 R_\infty)(e/m_e)^{-1}, \qquad h = (c\alpha^5/2\mu_0 R_\infty^2)(e/m_e)^{-2}$$

    Taking e/m_e first, the experiments which determine the frequency with which an electron describes a circle in a uniform magnetic field, in relation to the precession frequency of a proton[‡] in the same field, are capable of high precision (ratio of two frequencies). They give

$$\mu_B/\mu_p' = \tfrac{1}{2}e\hbar/m_e\mu_p' = 657.465\ 2$$

Measurements of the precession of a proton in a known field then provide a link with the standard ampere, and so (via measurements of gravity, g) with the standard kilogram. The mean of recent values,

$$\gamma_p' = \mu_p'/\tfrac{1}{2}\hbar = 2.675\ 126 \times 10^8 \text{ Hz T}^{-1}$$

[†] The full theory of some phenomena requires series in powers of α ($\doteqdot 1/137$), and the higher order terms are very difficult to calculate. E.g. the magnetic moment of the electron is not exactly $e\hbar/2m_e$ but

$$[1 + \tfrac{1}{2}\alpha/\pi - 0.3285(\alpha/\pi)^2 + \ldots]e\hbar/2m_e = 1.001\ 159\ 6\ \mu_B$$

[‡] The true value μ_p of the magnetic moment differs from the apparent value μ_p' measured in water by a small diamagnetic correction, which here cancels out.

(s.e. 3 p.p.m.) leads to

$$e/m_e = 1.758\ 802 \times 10^{11}\ \text{C kg}^{-1}$$

In determining α, TPL differ from earlier authors. They first use $2e/h$ from the a.c. Josephson effect—the alternating current which is observed when two superconductors separated by a thin film are maintained at differing d.c. potentials. The measured constant, namely

$$2e/h = 4R_\infty e/mc\alpha^2 = 4.835\ 93 \times 10^{14}\ \text{Hz V}^{-1}$$

leads to $\alpha = 1/137.036\ 1$. They also include a value $137.035\ 9$ from the *hyperfine* structure† in hydrogen, but do not use the *fine* structure spacing because of uncertainty about higher order terms in the theory.

The remaining constant, N, can be found from the *Faraday constant* of electrolysis, $F = Ne$. A check on this is obtained from the proton moment in nuclear magnetons, $\mu_p'/\mu_N = \mu_p'/(e\hbar/2m_p)$; in conjunction with the value of μ_B/μ_p' it gives the ratio of the mass of the proton to that of the electron which has already been determined, and hence the absolute mass of the H atom, and the value of N.

Values accepted formerly. Other conventional values of e, h, etc. may be encountered in books, and some of these earlier values are given below. The first is essentially Millikan's oil-drop value of e, 4.77×10^{-10} esu, and the second the 'X-ray grating' value 4.802. The numbers in brackets are standard errors. It will be seen that the accuracies claimed have sometimes appeared over-optimistic in retrospect, and that besides the supposed random errors of observation some unsuspected systematic error has been present. (In Millikan's value of e, accepted from 1917 to 1935, it was an inaccuracy in the assumed viscosity of air; in the case of the 1963 values the quantum-electrodynamic corrections in the formulae for fine structure are suspected.) Consequently, it may be thought prudent to regard the standard errors given here with caution.

Author of discussion	$e \times 10^{19}/\text{C}$	$h \times 10^{34}/(\text{J s})$	*Kaye and Laby*
Birge, 1929	1.591 1	6.547	7th ed.
	(0.002 4)	(0.012)	
Birge, 1942	1.602 03	6.624 2	10th ed.
	(0.000 50)	(0.003 5)	
DuMond and Cohen, 1963. . .	1.602 10	6.625 6	13th ed.
	(0.000 02)	(0.000 16)	
Taylor *et al.*, 1969	1.602 192	6.626 20	14th ed.
	(0.000 007)	(0.000 05)	

† Since the measured hyperfine structure separation in hydrogen is $1420.405\ 751\ 786$ MHz within about 5×10^{-9} MHz, it is not surprising that the main uncertainty is in the theory.

Table of fundamental constants

			SI unit	*Standard error (parts in 10^6)*
Principal constants				
Velocity of light	c	$2.997\ 925 \times 10^8$	m s^{-1}	1
Planck's constant	h	$6.626\ 2\ \times 10^{-34}$	J s	8
Planck's constant ($h/2\pi$)	$\hbar$	$1.054\ 592 \times 10^{-34}$	J s	8
Charge of electron	e	$1.602\ 192 \times 10^{-19}$	C	4.5
Mass of electron	m_e	$9.109\ 56\ \times 10^{-31}$	kg	6
Mass of electron	Nm_e	$5.485\ 93\ \times 10^{-4}$	amu	6
Avogadro's constant (number of atoms in 12 kg of ^{12}C) . .	N	$6.022\ 17\ \times 10^{26}$	kmol^{-1}	7
Mass of any atom $\div$ atomic weight . . .	N^{-1}	$1.660\ 53\ \times 10^{-27}$	kg amu^{-1}	7
Faraday constant of electrolysis . . .	$F\ (= Ne)$	$9.648\ 67\ \times 10^7$	C kmol^{-1}	6
Constant of gravitation	G	$6.673\ \times 10^{-11}$	N m^2 kg^{-2}	500

Table of fundamental constants (*contd*)

			SI unit	*Standard error* (parts in 10^6)
Spectroscopy and atoms				
Planck's constant	h	4.135 71	$\times 10^{-15}$ eV s	4
Planck's constant ($h/2\pi$)	$\hbar$	6.582 18	$\times 10^{-16}$ eV s	4
Charge/mass ratio of electron . . .	e/m_e	1.758 803	$\times 10^{11}$ C kg^{-1}	3
Fine structure constant	α	7.297 35	$\times 10^{-3}$	1.5
Fine structure constant, reciprocal . . .	α^{-1}	137.036 0		1.5
Scale of the fine structure	$\alpha^2 R_\infty$	5.843 66	$\times 10^2$ m^{-1}	3
Rydberg's constant (fixed nucleus) . .	R_∞	1.097 373 1	$\times 10^7$ m^{-1}	0.1
Rydberg's constant (hydrogen H) . . .	R_H	1.096 775 8	$\times 10^7$ m^{-1}	0.1
Bohr radius $(4\pi/\mu_0 c^2)\hbar^2/m_e e^2$	a_0	5.291 771	$\times 10^{-11}$ m	1.5
Compton wavelength of electron . . .	λ_C	2.426 310	$\times 10^{-12}$ m	3
Compton wavelength of electron $\div 2\pi$ . .	$\hbar/m_e c$	3.861 59	$\times 10^{-13}$ m	3
Classical 'radius' of electron $(\mu_0 c^2/4\pi)e^2/mc^2$	r_e	2.817 94	$\times 10^{-15}$ m	5
Thomson cross-section $8\pi r_e^2/3$	σ_T	6.652 45	$\times 10^{-29}$ m	10
Zeeman effect, μ_B/hc		46.686 0	m^{-1} T^{-1}	3
Bohr magneton $e\hbar/2m_e$	μ_B	9.274 10	$\times 10^{-24}$ J T^{-1}	7
Nuclear magneton $e\hbar/2m_p$	μ_N	5.050 95	$\times 10^{-27}$ J T^{-1}	10
Ratio of masses proton/electron . . .	m_p/m_e	1.836 11	$\times 10^3$	6
Gyromagnetic ratio of proton	$\gamma_p = \mu_p/\frac{1}{2}\hbar$	2.675 196	$\times 10^8$ s^{-1} T^{-1}	3
in H$_2$O	γ_p'	2.675 13	$\times 10^8$ s^{-1} T^{-1}	3
in H$_2$O (cycles)	$\gamma_p'/2\pi$	4.257 60	$\times 10^7$ Hz T^{-1}	3
Band-spectra constant	$Nh/8\pi^2 c$	1.685 80	$\times 10^{-17}$ amu m	7
Conversion factors for Mass, Energy and Wavelength				
Energy and mass				
Electron volt		1.602 192	$\times 10^{-19}$ J	4.5
Atomic mass unit in MeV		931.481		6
1 MeV in a.m.u.		1.073 559	$\times 10^{-3}$	6
Rest-mass of electron		0.511 004	MeV	3
Joules per kmol for 1 eV per molecule .		9.648 67	$\times 10^7$	6
Frequency, wavelength and energy				
Quantum energy $\div$ wave number }		{ 1.986 48	$\times 10^{-25}$ J m	8
Energy $\times$ wavelength } . .		{ 1.239 854	$\times 10^{-6}$ eV m	4
Wave number $\div$ energy		8.065 46	$\times 10^5$ eV^{-1} m^{-1}	4
Quantum energy $\div$ frequency . . .		4.135 71	$\times 10^{-15}$ eV Hz^{-1}	4
Frequency $\div$ energy		2.417 966	$\times 10^{14}$ Hz eV^{-1}	4
Thermal constants				
Gas constant	R_0	8.314 3	$\times 10^3$ J K^{-1} kmol^{-1}	42
Loschmidt's constant (number of molecules in 1 m^3 of ideal gas at stp)		2.686 8	$\times 10^{25}$	43
Boltzmann's constant R_0/N	k	1.380 62	$\times 10^{-23}$ J K^{-1}	43
Boltzmann's constant		8.617 1	$\times 10^{-5}$ eV K^{-1}	42
Energy kT for $T = 273.15$ K		0.023 538	eV	42
Stefan's constant $2\pi^5 k^4/15c^2 h^3$† . . .		5.669 6	$\times 10^{-8}$ W m^{-2} K^{-4}	170
Constants in Planck's formula (power flow)†	$2\pi hc^2$	3.741 84	$\times 10^{-16}$ W m^2	8
Constants in Planck's formula (energy density)†	$8\pi hc$	4.992 58	$\times 10^{-24}$ J m	8
Constants in Planck's formula† . . .	hc/k	1.438 83	$\times 10^{-2}$ m K	43

† See also p. 76.

Note: Magnetic moments are defined so that mechanical energy $= -\boldsymbol{\mu}.\mathbf{B}$; the unit is 1 J T^{-1} = 1 J Wb^{-1}m^2 = 1 A m^2 (1 tesla = 1 weber m^{-2} = 10^4 gauss).

1.3 Measurement of mass, pressure and other mechanical quantities

1.3.1 Mass, volume and density

Correction of weighings for buoyancy

If a substance is weighed in air and found to be balanced by mass standards ('weights') of value M, the true mass of the substance is

$$M + Md_a\left[\left(\frac{1}{d}\right) - \left(\frac{1}{d_s}\right)\right]$$

i.e. the correction for air buoyancy is

$$Md_a\left[\left(\frac{1}{d}\right) - \left(\frac{1}{d_s}\right)\right]$$

where d is the density of the substance, d_s is the density of the mass standards and d_a is the density of the air. In practice, the members of an ordinary set of mass standards, including the fractions, are standardized by being assigned values which are equal to the masses of reference standards of density 8.0 g cm^{-3} which they would balance in air. The value 8.0 g cm^{-3} is a conventional value close to the density of the larger weights in a set. The buoyancy of any group of weights from such a set should be calculated by taking this value for d_s. The following table has been prepared on this basis, d_a being taken to be 1.2 kg m^{-3} (or 0.0012 g cm^{-3}).

Density of substance weighed	Buoyancy correction (mg per gram of substance)	Density of substance weighed	Buoyancy correction (mg per gram of substance)
$d/(\text{g cm}^{-3})$	$d_s = 8.0$ g cm^{-3} $d_a = 1.2$ kg m^{-3}	$d/(\text{g cm}^{-3})$	$d_s = 8.0$ g cm^{-3} $d_a = 1.2$ kg m^{-3}
0.5	+2.250	6.0	+0.050
1.0	+1.050	8.0	0
1.5	+0.650	10.0	−0.030
2.0	+0.450	12.5	−0.054
2.5	+0.330	15.0	−0.070
3.0	+0.250	20.0	−0.090
4.0	+0.150	22.0	−0.096

Density of dry CO_2-free air

The density in kg m^{-3} of dry air free from carbon dioxide may be derived from $(1.2928P)/(101.325)(1.000\ 028)(1 + 0.003\ 67t)$ where the air is at a pressure of P kPa and a temperature of t °C (1 kPa $= 10$ mb).

Density of ambient air (unit $= 1$ kg m^{-3})

The figures in the main table relate to air of 50% relative humidity and containing 0.04% carbon dioxide by volume. Derived from $(1.293\ 07P - 0.2443p)/(101.325)(1.000\ 028)(1 + 0.003\ 67t)$, where $P =$ atmospheric pressure in kPa, $p =$ saturated vapour pressure of water in kPa at t °C. For other states of humidity a correction should be applied from the subsidiary table.

Air pressure, P/kPa (1 kPa = 10 mb)	Air temperature, $t/°C$												
	6	8	10	12	14	16	18	20	22	24	26	28	30
80	0.997	0.989	0.982	0.975	0.967	0.960	0.953	0.946	0.939	0.932	0.925	0.918	0.910
81	1.009	1.002	0.994	0.987	0.979	0.972	0.965	0.958	0.951	0.943	0.936	0.929	0.922
82	1.022	1.014	1.007	0.999	0.992	0.984	0.977	0.970	0.962	0.955	0.948	0.941	0.933
83	1.034	1.027	1.019	1.011	1.004	0.996	0.989	0.982	0.974	0.967	0.960	0.952	0.945
84	1.047	1.039	1.031	1.024	1.016	1.008	1.001	0.993	0.986	0.979	0.971	0.964	0.956
85	1.059	1.051	1.043	1.036	1.028	1.020	1.013	1.005	0.998	0.990	0.983	0.975	0.968
86	1.072	1.064	1.056	1.048	1.040	1.033	1.025	1.017	1.010	1.002	0.995	0.987	0.979
87	1.084	1.076	1.068	1.060	1.052	1.045	1.037	1.029	1.021	1.014	1.006	0.999	0.991
88	1.097	1.089	1.080	1.072	1.064	1.057	1.049	1.041	1.033	1.026	1.018	1.010	1.002
89	1.109	1.101	1.093	1.085	1.077	1.069	1.061	1.053	1.045	1.037	1.029	1.022	1.014
90	1.122	1.113	1.105	1.097	1.089	1.081	1.073	1.065	1.057	1.049	1.041	1.033	1.025
91	1.134	1.126	1.117	1.109	1.101	1.093	1.085	1.077	1.069	1.061	1.053	1.045	1.037
92	1.147	1.138	1.130	1.121	1.113	1.105	1.097	1.089	1.080	1.072	1.064	1.056	1.048
93	1.159	1.150	1.142	1.134	1.125	1.117	1.109	1.100	1.092	1.084	1.076	1.068	1.060
94	1.172	1.163	1.154	1.146	1.137	1.129	1.121	1.112	1.104	1.096	1.088	1.080	1.071
95	1.184	1.175	1.167	1.158	1.149	1.141	1.133	1.124	1.116	1.108	1.099	1.091	1.083
96	1.197	1.188	1.179	1.170	1.162	1.153	1.145	1.136	1.128	1.119	1.111	1.103	1.094
97	1.209	1.200	1.191	1.182	1.174	1.165	1.156	1.148	1.140	1.131	1.123	1.114	1.106
98	1.222	1.212	1.203	1.195	1.186	1.177	1.168	1.160	1.151	1.143	1.134	1.126	1.117
99	1.234	1.225	1.216	1.207	1.198	1.189	1.180	1.172	1.163	1.155	1.146	1.137	1.129
100	1.246	1.237	1.228	1.219	1.210	1.201	1.192	1.184	1.175	1.166	1.158	1.149	1.140
101	1.259	1.250	1.240	1.231	1.222	1.213	1.204	1.196	1.187	1.178	1.169	1.161	1.152
102	1.271	1.262	1.253	1.244	1.234	1.225	1.216	1.207	1.199	1.190	1.181	1.172	1.163
103	1.284	1.274	1.265	1.256	1.247	1.237	1.228	1.219	1.210	1.201	1.193	1.184	1.175
104	1.296	1.287	1.277	1.268	1.259	1.250	1.240	1.231	1.222	1.213	1.204	1.195	1.186
105	1.309	1.299	1.290	1.280	1.271	1.262	1.252	1.243	1.234	1.225	1.216	1.207	1.198
106	1.321	1.312	1.302	1.292	1.283	1.274	1.264	1.255	1.246	1.237	1.228	1.218	1.209

Corrections for humidity (unit = 1 kg m^{-3})

Relative humidity, $R\%$	Air temperature, $t/°C$		
	5	10	20
20	+0.001	+0.003	+0.006
25	+0.001	+0.003	+0.005
30	+0.001	+0.002	+0.004
35	0	+0.002	+0.003
40	0	+0.001	+0.002
45	0	+0.001	+0.001
50	0	0	0
55	0	−0.001	−0.001
60	0	−0.001	−0.002
65	−0.001	−0.002	−0.003
70	−0.001	−0.002	−0.004
75	−0.001	−0.003	−0.005
80	−0.001	−0.003	−0.006

Reduction of gaseous volumes to s.t.p.

The volume at s.t.p. ($0\,°C$ and 1 atm) $= vp/[101.325\,(1+\gamma t)]$, where v and $t\,°C$ are the observed volume and temperature, p is the observed pressure in kPa (1 kPa $= 10$ mb) and γ is the coefficient of cubical expansion at constant pressure of the gas concerned. For 'permanent' gases $\gamma = 0.003\,67 \pm 0.000\,01$ approximately. For coefficients of other gases, see section 2.5, p. 183.

If it is desired to find the volume of dry gas at s.t.p., v and p being measured when the gas contains aqueous vapour whose pressure is e kPa, p in the above expression must be replaced by $(p-e)$.

Volumetric calibration of vessels with water or mercury

The volume content (V_t) of a vessel in cm^3 at the temperature of determination ($t\,°C$) is given by

$$V_t = Wf,$$

where W is the apparent mass in grams of contained water or mercury at $t\,°C$ when weighed in air at $t\,°C$ and 1 atm (101.325 kPa) against mass standards adjusted on the basis of density 8.0 g cm^{-3}, and f is a factor which includes allowances for the densities of the liquids and the appropriate buoyancy corrections. Density tables for water and mercury are given in section 1.4.1, p. 29.

Temperature of liquid, $t/°C$	10	11	12	13	14	15
f { Water	1.001 39	1.001 48	1.001 58	1.001 70	1.001 83	1.001 97
{ Mercury	0.073 685	0.073 698	0.073 712	0.073 725	0.073 738	0.073 752

Temperature of liquid, $t/°C$	16	17	18	19	20
f { Water	1.002 13	1.002 29	1.002 47	1.002 65	1.002 85
{ Mercury	0.073 765	0.073 778	0.073 792	0.073 805	0.073 819

Temperature of liquid, $t/°C$	21	22	23	24	25
f { Water	1.003 06	1.003 28	1.003 51	1.003 75	1.004 00
{ Mercury	0.073 832	0.073 845	0.073 859	0.073 872	0.073 886

An additional buoyancy correction is required when weighings with water are made at air pressures other than 101.325 kPa. The values of f (for water) must be increased ($+$), or decreased ($-$), by the following amounts (unit $0.000\,01$):

Pressure kPa (1 kPa $= 10$ mb)	97	98	99	100	101	102	103	104	105
Correction to f (water)	-5	-4	-3	-2	0	$+1$	$+2$	$+3$	$+4$

No additional buoyancy correction is required for mercury. The above tables give the volume content of the vessel at the temperature of determination (t). At any other temperature (T) the volume V_T is given by

$$V_T = V_t\,(1 + \gamma(T-t))$$

where γ is the coefficient of cubical expansion of the material of the vessel.

Example. Let the apparent mass of water contained in a vessel at $10\,°C$, when weighed in air at $10\,°C$ and pressure 102 kPa, be 100.00 g.

Then volume of vessel at 10 °C is

$$100.00\ (1.001\ 39 + 0.000\ 01) = 100.14\ \text{cm}^3$$

The same vessel, if made of glass ($\gamma = 0.000\ 027\ \text{K}^{-1}$, assumed), would contain at 20 °C

$$100.14(1 + 0.000\ 027(20 - 10)) = 100.17\ \text{cm}^3$$

Comprehensive correction tables are given in BS 1797:1968.

Measurement of density

The density of a solid or liquid specimen is normally found by determining in air the apparent mass of a particular volume of the specimen and the apparent mass of an equal volume of water, the latter being obtained either by hydrostatic displacement by the solid or by using the same container (bottle or pyknometer) for the liquid weighings. The ratio, r, of the two apparent masses thus determined is approximately equal to the density of the specimen in g cm^{-3}; the true density is equal to $r(d_w - d_a) + d_a$, where d_w is the density of the water at its observed temperature t °C (see p. 29) and d_a is the density of the air (see p. 19). The correction to be applied to r at 20 °C and 1 atm is of the order of 1 part in 350 of the density. For temperatures ranging from 12 °C to 25 °C and pressures from 97 kPa (970 mb) to 104 kPa, the correction ranges over about 1 part in 400 of the density.

r is usually termed the relative density in air with reference to water, and is denoted by $d\ t/t'$ *in air*, where t and t' are the temperatures of the material and water respectively (usually $t = t'$). The true relative density, viz. $d\ t/t'$ *in vacuo*, is equal to $r(1 - d_a) + d_a$.

Hydrometers

Density hydrometers usually indicate density, g ml^{-1} or g cm^{-3} at 20 or 15 °C when readings are taken at 20 or 15 °C respectively. In hot countries 27 °C may be used.

Relative density (or specific gravity) hydrometers usually indicate density at 60 °F relative to water at 60 °F ($d\ 60/60$ °F) when readings are taken at 60 °F.

Twaddle hydrometers indicate ($d\ 60/60$ °F $- 1$)200 when readings are taken at 60 °F.

Sikes' hydrometers have an arbitrary scale in which 1° Sikes is roughly equivalent to 0.002 g cm^{-3} on the average and 100° Sikes corresponds to water. See P. H. Bigg, *J. Sci. Instrum.*, 1961.

Baumé scales are related to relative density by various arbitrary formulae.

Temperature correction. To obtain from the reading R (density or relative density) of a soda-glass hydrometer, standard at t °C, the density or relative density $d\ \theta$ °C/t °C, of the liquid at the temperature θ °C of observation, subtract $R \times 0.000\ 025(\theta - t)$.

Surface tension. The surface tensions of aqueous solutions are often reduced considerably (20 mN m^{-1} or more) by contaminating films. Change of reading due to -1 mN m^{-1} change in surface tension $= +0.004/nld$, where $n =$ density or relative density, $l =$ length of scale (mm) corresponding to 0.01 density or relative density and $d =$ stem diameter (mm).

J.B.R.

1.3.2 Barometry

Barometric units

The pressure units at present used in barometry are, in order of preference:

   (i) the **millibar** (mb), equal to 100 Pa;
  (ii) the conventional **millimetre of mercury** (mmHg), defined as the pressure exerted by a 1-mm column of liquid of uniform density 13.5951 g cm^{-3} under standard acceleration 9.806 65 m s^{-2} (1 mmHg = 133.322 Pa);

(iii) the conventional **inch of mercury** (inHg), defined as the pressure exerted by a 1-in column of liquid of uniform density 13.5951 g cm^{-3} under standard acceleration 9.806 65 m s^{-2} (1 inHg = 3386.39 Pa).

The two units, mmHg and inHg, are included because of the large number of barometers in current use which have scales graduated in terms of these units. It is expected that such barometers will become obsolete in due course.

Under the barometer conventions which received international approval in 1953, mercury barometers bearing scales representing the above pressure units should be graduated to measure pressures directly in these units when the whole barometer is at 0 °C and under standard acceleration (see BS 2520: 1967).

Note: The **standard atmosphere** (atm) is defined as 101 325 Pa, which is equal to 760 mmHg within 1 part in 7 million.

Capillary corrections for mercury columns

Capillary forces tend to depress the summit of a mercury column by an amount C, called the **capillary depression**, which depends on the column diameter D, the surface tension γ and the angle of contact θ of the mercury. γ may lie between 500 mN m^{-1} (exceptional cleanliness) and 400 mN m^{-1} (slight contamination), and θ varies widely. γ and θ are not easily measured, but in a given tube they condition the meniscus height h, reckoned from the summit to the plane of the ring of contact. It is therefore convenient to determine C by reference to D and h. The following table of computed values of C is based on Bashforth and Adams (1883), *Capillary Action* (Cambridge University Press), for $D = 1$ to 10 mm and on B. E. Blaisdell, *J. Math. Phys.*, 1940, for $D = 10$ to 22 mm, for the values of γ and h indicated. Density of mercury = 13.546 g cm^{-3} at 20 °C. Standard acceleration = 9.806 65 m s^{-2}.

For $D = 40$ mm, C is approximately 0.0001 mm.

Note: Allowance is made for the effect of the capillary depression in mercury barometers as adjusted by the manufacturers. For mercury U-tubes having limbs of the same diameter, it should not be assumed that C is the same in each limb.

Capillary depression

(mm of mercury at 20 °C, standard acceleration)

Bore of tube	Meniscus height/mm (Surface tension = 400 mN m^{-1})										Bore of tube
mm	0.2	0.4	0.6	0.8	1.0	1.2	1.4	1.6	1.8	2.0	mm
8	0.108	0.214	0.315	0.409	0.494	0.569	0.633	0.684	0.723	0.749	8
10	0.058	0.115	0.170	0.222	0.270	0.314	0.352	0.384	0.410	0.430	10
16	0.011	0.022	0.032	0.042	0.052	0.060	0.068	0.076	0.082	0.086	16
22	0.002	0.004	0.006	0.009	0.010	0.012	0.014	0.015	0.016	0.018	22

Capillary depression (*contd*)

Bore of tube mm	Meniscus height/mm (Surface tension = 450 mN m⁻¹)										Bore of tube mm
	0.2	0.4	0.6	0.8	1.0	1.2	1.4	1.6	1.8	2.0	
1	9.3										1
2	2.56	4.54									2
3	1.13	2.15	2.96	3.52							3
4	0.62	1.21	1.72	2.13	2.44						4
5	0.38	0.75	1.09	1.38	1.62	1.81	1.94				5
6	0.25	0.50	0.73	0.94	1.12	1.27	1.38	1.47			6
7	0.18	0.35	0.51	0.66	0.79	0.91	1.01	1.08	1.14		7
8	0.126	0.250	0.368	0.478	0.579	0.668	0.744	0.807	0.855	0.890	8
9	0.093	0.183	0.271	0.353	0.429	0.498	0.557	0.608	0.649	0.679	9
10	0.069	0.137	0.202	0.264	0.322	0.375	0.422	0.462	0.495	0.522	10
11	0.052	0.104	0.153	0.200	0.245	0.285	0.322	0.354	0.380	0.402	11
12	0.040	0.079	0.117	0.153	0.187	0.218	0.247	0.272	0.293	0.311	12
13	0.030	0.060	0.089	0.117	0.144	0.168	0.190	0.210	0.227	0.241	13
14	0.023	0.046	0.069	0.090	0.111	0.130	0.147	0.163	0.176	0.187	14
15	0.018	0.036	0.053	0.070	0.086	0.101	0.114	0.126	0.137	0.146	15
16	0.014	0.028	0.041	0.054	0.067	0.078	0.089	0.098	0.106	0.113	16
17	0.011	0.022	0.032	0.042	0.052	0.061	0.069	0.076	0.083	0.088	17
18	0.008	0.017	0.025	0.033	0.040	0.047	0.054	0.059	0.064	0.069	18
19	0.006	0.013	0.019	0.026	0.031	0.037	0.042	0.046	0.050	0.054	19
20	0.005	0.010	0.015	0.020	0.024	0.029	0.032	0.036	0.039	0.042	20
21	0.004	0.008	0.012	0.015	0.019	0.022	0.025	0.028	0.030	0.032	21
22	0.003	0.006	0.009	0.012	0.015	0.017	0.020	0.022	0.024	0.025	22

(Surface tension = 500 mN m⁻¹)

	0.2	0.4	0.6	0.8	1.0	1.2	1.4	1.6	1.8	2.0	
8	0.143	0.286	0.421	0.547	0.663	0.766	0.855	0.929	0.989	1.031	8
10	0.080	0.159	0.235	0.308	0.375	0.436	0.491	0.542	0.582	0.615	10
16	0.017	0.034	0.051	0.067	0.082	0.097	0.110	0.122	0.133	0.142	16
22	0.004	0.008	0.012	0.016	0.020	0.023	0.026	0.029	0.032	0.034	22

Table reproduced by permission of Mr F. A. Gould, Mr T. Vickers and the Institute of Physics from *J. Sci. Instrum.*, 1952, **29**, 86.

Correction of Fortin barometer readings to 0 °C

The corrected reading $r = r_t\{1 - (\beta - \alpha)t/(1 + \beta t)\}$, where r_t = reading at temperature t; $\beta = 1.818 \times 10^{-4}$ °C⁻¹, the coefficient of cubical expansion of mercury; $\alpha = 0.184 \times 10^{-4}$ °C⁻¹, the coefficient of linear expansion of brass. The simpler expression $r = r_t(1 - \beta't)$, where $\beta' = 1.63 \times 10^{-4}$ °C⁻¹, gives the correction to within about 0.01 mb over the range 0–35 °C. The correction of 'fixed cistern' or Kew type barometers requires an additional term due to the expansion of the whole of the mercury in the barometer. This term, independent of the barometer reading, may amount at 20 °C to 0.1 or 0.2 mb (BS 2520: 1967, *Barometer Conventions and Tables*).

Correction of barometer readings to standard gravity

The corrected reading $r = r_0 g_0/g_n$ where r_0 = observed reading, g_0 = value of gravity at the point of observation, and $g_n = 9.806\ 65$ m s⁻². If g_0 is not known, a value in terms of geographical latitude may be obtained from the formula in section 1.9.4, p. 134. The correction for diminution of g with increasing height above sea-level may be taken as 0.031 mb per 100 mb per 1000 m, correction to be subtracted from reading (BS 2520: 1967, *Barometer Conventions and Tables*).

Correction of barometer readings for small differences in altitude

A mercury barometer measures the atmospheric pressure at the level of its lower mercury surface. If the point where the pressure is required to be known is at a different height, the hydrostatic effect of the intervening air column can be appreciable. At sea-level the pressure decreases by about 1.2 mb for an increase in height of 10 m. More correctly, the correction to the barometer reading $= -g\rho_a H$, or $0.098\rho_a H$ mb, where ρ_a is the ambient air density in kg m^{-3} and H is the height in metres of the point above the zero of the barometer. This expression gives correction to 0.01 mb accuracy up to $H = 30$ m approximately.

K.W.T.E.

1.3.3 The measurement of high pressures

The practical pressure scale at high pressures

The use of fixed points as an aid to the definition of the pressure scale above the range conveniently accessible to the primary standards is now an established technique, and at an important symposium held in 1968 at the US National Bureau of Standards (ref. 6) a wide measure of agreement was recorded on the pressure values associated with certain phase transitions of pure substances in the range of pressure below 10 GPa (100 kbar). Recommendations were also made regarding the use of equations of state as a basis for the specification of high pressures.

Improved pressure values for many transitions above 10 GPa are also now available, but it is still advisable to regard reference points in this higher range as provisional, and it should be recognized that their degree of reproducibility will depend on the experimental conditions and may be especially sensitive to non-hydrostatic distributions of stress.

To maintain consistency all the pressure values in this section of the Tables have been expressed in terms of the 10^9 multiple of the coherent SI unit, the newton per square meter (pascal), viz. the giganewton per square meter (GN/m^2), i.e. the gigapascal (GPa). The following conversion factors to non-SI metric units are relevant:

$$1 \text{ kilobar} = 0.1 \text{ GPa}$$
$$1 \text{ megabar} = 100 \text{ GPa}$$

Basic group of reference pressures associated with phase transitions of pure substances recommended for general adoption

(From ref. 6: Lloyd, Beckett and Boyd, p. 1; H. T. Hall *et al.*, p. 313)

Substance	Type of transition and means of identification	Pressure/GPa	Temperature/°C
Mercury . . .	Solid–liquid equilibrium; elec. res. or vol. change	0.7569	0
Bismuth	Polymorphic (Bi I–II); elec. res. or vol. change	2.550	25
Thallium . . .	Polymorphic (Tl II–III); elec. res. or vol. change	3.67	25
Barium	Polymorphic (Ba I–II); elec. res. or vol. change	5.5	25
Bismuth	Polymorphic (Bi III–V); elec. res. or vol. change	7.7	25

Mercury solid–liquid equilibrium at other temperatures. It was also recommended that the following form of the Simon equation should be adopted for the expression of the mercury solid–

liquid equilibrium pressures at other temperatures:

$$p = 3.8227 \left\{ \left(\frac{T}{234.29} \right)^{1.1772} - 1 \right\}$$

where p is the pressure in GPa and T the temperature in K on the 1948 IPTS. (*NB*: Small adjustments will be needed to convert the equation to IPTS-68.)

Additional reference pressures associated with phase transitions

Substance	Nature of transition	Pressure/GPa	Temperature/°C
Carbon dioxide . .	Liquid–vapour equilibrium	$\begin{cases} 3.4856 \times 10^{-3}\dagger \\ 3.4847 \times 10^{-3} \end{cases}$	0.01 (H_2O tr. pt.) 0
Water	Ice I–Ice II–liq. triple pt.	0.2093	−22
Ammonium fluoride	NH_4F I–II; vol. change	$\begin{cases} 0.3538\ddagger \\ 0.3605 \\ 0.3664 \end{cases}$	0 25 50
Ammonium fluoride	NH_4F II–III; vol. change	$\begin{cases} 1.0977\ddagger \\ 1.1531 \\ 1.2011 \end{cases}$	0 25 50
Potassium bromide .	Polymorphic; vol. change	1.80	∼25
Potassium chloride .	Polymorphic; vol. change	2.02	∼25
Bismuth	Bi II–III; elec. res. or vol. change	2.69§	∼25
Caesium	Cs II–III (increasing pressure)	4.25§	∼25
Caesium	Cs III–IV (increasing pressure)	4.30§	∼25
Tin	Sn I–II elec. res. change	9.2–9.6‖	∼25
Iron	Fe α–ϵ $\begin{cases} \text{elec. res. change} \\ \text{shock-wave vel. change} \end{cases}$	11.0–11.3‖	∼25
Barium	Ba II–III; elec. res. change	11.8–12.2‖	∼25
Europium . . .	Elec. res. change	12.2–13.0‖	∼25
Lead	Pb I–II; elec. res. change	12.8–13.2‖	∼25
Rubidium . . .	Rb II–Liq.; elec. res. change	14.2–15.3‖	∼25
Caesium	Max. in elec. res. curve	13.3–14.2‖	∼25
Calcium	Max. in elec. res. curve	23.5–25.5‖	∼25
Rubidium . . .	Max. in elec. res. curve	29–32‖	∼25
Cadmium sulphide .	Max. in elec. res. curve	33–34‖	∼25
Zinc sulphide . .	Max. in elec. res. curve	41–42‖	∼25

† Derived from Greig and Dadson, *Br. J. appl. Phys.*, 1966; and Michels *et al.*, *Physica*, 1950.
‡ Derived from Kaneda, Yamamoto and Nishibata, ref. 6, p. 257.
§ Derived from H. T. Hall *et al.*, ref. 6, p. 313.
‖ Derived from H. T. Hall *et al.*, ref. 6, p. 313 and Drikamer, *Rev. Sci. Insts.*, 1970.

For information on possible reference points combining high pressures and high temperatures see the following papers from ref. 6:

Strong and Bundy, p. 283: Fe; Au
Akimoto and Syono, p. 273: Coesite-Stishovite
Committee Report, p. 317: Various phase transitions.

Pressure scale defined by equations of state of pure substances

The Committee Report of ref. 6, p. 315, recommended that the following three materials be selected as potential 'equation-of-state' standards, in this order of preference:

sodium chloride; copper; aluminium.

These and other possible reference materials are extensively treated in ref. 6:

(a) Weaver, Takahashi and Bassett, p. 189: NaCl; MgO
(b) Fritz *et al.*, p. 201:   NaCl
(c) Weaver, p. 325:   NaCl
(d) Carter *et al.*, p. 147:   2024 Al; Cu; Mo; Ag; Pd; Na; Mg; MgO
(e) Pastine and Carroll, p. 91:   Na; Al
(f) Holt and Grover, p. 131:   Al
(g) Barsch and Chang, p. 173:   Caesium halides

The following table, which gives some of the isothermal compression data for the 293 K isotherm, is an abridged version of tables given in papers (b) and (d) above, the values being calculated from shock-wave measurements by means of the Rankine–Hugoniot equations.

Pressure/GPa	Relative density for 293 K isotherm for:								
	NaCl	2024 Al	Cu	Mo	Ag	Pd	Na	Mg	MgO
0	1.000	1.000	1.000	1.000	1.000	1.000	1.000	1.000	1.000
5	1.155	1.059	1.035	1.018	1.043	1.026	1.516	1.123	1.031
10	1.262	1.110	1.065	1.036	1.080	1.049	1.708	1.217	1.059
15	1.348	1.154	1.093	1.052	1.112	1.071	1.896	1.297	1.085
20	1.424	1.195	1.118	1.068	1.141	1.090	2.046	1.366	1.109
25	1.489	1.232	1.142	1.084	1.168	1.109	2.172	1.427	1.132
30	1.547	1.266	1.164	1.098	1.192	1.126	2.279	1.483	1.153
40		1.329	1.204	1.127	1.237	1.159		1.583	1.192
50		1.385	1.241	1.153	1.276	1.188		1.669	1.228
100		1.608	1.384	1.268	1.426	1.304			1.372
150			1.491	1.364	1.534	1.393			
200				1.448		1.464			

Pressure dependence of melting temperatures of pure metals in range 0–8 GPa

(Derived from Kennedy and Newton in ref. 4)

Substance	Melting temperature (°C) at pressure:								
	0 GPa†	1 GPa	2 GPa	3 GPa	4 GPa	5 GPa	6 GPa	7 GPa	8 GPa
Sodium	97.8	166	207	240	269	290			
Indium	156.4	200	240	275	310	345			
Cadmium	320.9	365	415	456	500	535			
Lead	327.3	390	447	505	560	615			
Zinc	419.5	462	508	545	580	615			
Magnesium . . .	650	720	790	860	940	1000			
Silver	960.8	1010	1060	1110	1160	1210			
Nickel	1453	1490	1525	1555	1590	1620	1650	1675	1700
Iron	1539	1555	1590	1615	1640	1665	1690	1710	1730
Platinum	1769	1825	1875	1920	1960	2000	2030	2050	2065
Rhodium	1960	2040	2100	2150	2205	2255	2310	2360	2405

† Strictly one standard atmosphere (= 101 325 Pa).
For information on the melting line of copper to 100 GPa see McQueen *et al.*, ref. 6, p. 219.

References

(1) P. W. Bridgman, *The Physics of High Pressure*, 2nd ed., London, Bell, 1950.
(2) R. H. Wentorf, ed., *Modern Very High Pressure Techniques*, London, Butterworths, 1962.
(3) A. A. Giardini and E. C. Lloyd, eds., *High Pressure Measurement*, Washington, Butterworths, 1963.
(4) W. Paul and D. M. Warschauer, eds., *Solids under Pressure*, New York, McGraw-Hill, 1962.
(5) R. S. Bradley, ed., *High Pressure Physics and Chemistry*, London, Academic Press, 1963.
(6) E. C. Lloyd, ed., *Accurate Characterisation of the High-Pressure Environment*, Proceedings of Symposium at the National Bureau of Standards, Gaithersburg, Md., Oct. 1968, NBS Special Publication 326, 1971.

R.S.D.

1.3.4 Hygrometry

Relative humidity

The relative humidity is the ratio (expressed as a percentage) of the pressure of the water vapour actually present to the saturation pressure at the same air temperature. For most practical purposes, this is equal to the ratio of the mass of water vapour actually present in unit volume of moist air (absolute humidity) to the mass in unit volume of saturated air at the same temperature. (For definitions see BS 1339.) For a table of saturation pressures see section 2.4.1, p. 173.

Chemical hygrometer

The values below are grams of water contained in a cubic metre (m^3) of saturated air at a total pressure of 101 325 Pa (1013 mb).

Temperature/°C	0	1	2	3	4	5	6	7	8	9
0	4.85	5.20	5.55	5.95	6.35	6.80	7.25	7.75	8.25	8.80
10	9.40	10.00	10.65	11.35	12.05	12.80	13.60	14.45	15.35	16.30
20	17.30	18.35	19.40	20.55	21.75	23.05	24.35	25.75	27.20	28.75
30	30.35	32.05	33.80	35.60	37.55	39.55	41.65	43.90	46.20	48.60

Wet- and dry-bulb hygrometer

For this instrument the water-vapour pressure p is given in terms of the actual or dry-bulb temperature t and the wet-bulb temperature t_w by the relation

$$p = p_w - AH(t - t_w)$$

where p_w is the saturation water-vapour pressure at temperature t_w, A is a constant and H is the total atmospheric pressure. The relative humidity expressed as a percentage is $100p/p_s$, where p_s is the saturation water-vapour pressure at temperature t. The values of A depend on the speed of the air passing the wet bulb and are

$A = 0.000\ 666/°C$ for moving air as in the Assmann ventilated psychrometer.

$A = 0.000\ 80/°C$ in a Stevenson screen as used by the Meteorological Office.

(See BS 1339: 1964 for other values of A.)

Wet- and dry-bulb humidity tables

The tables below are for use with forced-ventilated instruments.

For more complete tables see *Hygrometric Tables for Aspirated Psychrometer Readings*, 1961 (M.O. 265c, London, HMSO): for range −35 °C to 60 °C see *Aspirations-Psychrometer-Tafeln*, 1963 (Braunschweig, F. Vieweg & Sohn); for range −30 °C to 60 °C see 'Institution of Heating and

Ventilating Engineers Guide 1970—Book C' (London), properties of humid air from − 10 °C to 60 °C. See also *Measurement of Humidity*, Notes on Applied Science, No. 4, 1970 (HMSO).

Dry-bulb temperature/°C	Wet-bulb depression/°C										
	0.5	1.0	1.5	2.0	2.5	3.0	3.5	4.0	4.5	5.0	5.5
	Relative humidity (%)										
10	94	88	82	76	71	65	60	54	49	44	39
12	94	89	83	78	73	68	63	57	53	48	43
14	95	90	84	79	74	70	65	60	56	51	47
16	95	90	85	81	76	71	67	62	58	54	50
18	95	91	86	82	77	73	69	65	60	56	52
20	96	91	86	83	78	74	70	66	62	59	55
22	96	92	87	83	79	76	72	68	64	61	57
24	96	92	88	84	80	77	73	69	66	62	59
	6.0	6.5	7.0	7.5	8.0	8.5	9.0	9.5	10.0	10.5	11.0
10	34	29	24	19	14	9	5	—	—	—	—
12	38	34	29	24	20	16	11	7	3	—	—
14	42	38	33	29	25	24	17	13	9	5	2
16	46	41	37	34	30	26	22	18	15	11	8
18	49	45	41	37	34	30	27	23	20	16	13
20	51	48	44	41	37	34	30	27	24	21	18
22	54	50	47	44	40	37	34	31	28	25	22
24	56	52	49	46	43	40	37	34	31	28	26

Humidities over saturated salt solutions

Saturated solutions of various salts in water are used to obtain known relative humidities of air circulated in sealed enclosures which are maintained at a uniform and constant temperature.

See BS 3718: 1964, *Laboratory Humidity Ovens (Non-injection Type)*.

Saturated salt solution	Temperature/°C										
	0	5	10	15	20	25	30	35	40	50	60
	Relative humidity (%)										
Potassium sulphate	99	98	98	97	97	97	96	96	96	96	96
Potassium nitrate	97	96	95	94	93	92	91	89	88	85	82
Potassium chloride	89	88	88	87	86	85	84	83	82	81	80
Ammonium sulphate . . .	83	82	82	81	81	80	80	80	79	79	78
Sodium chloride	76	76	76	76	76	75	75	75	75	75	75
Sodium nitrite	—	—	—	—	66	65	63	62	62	59	59
Ammonium nitrate . . .	77	74	72	69	65	62	59	55	53	47	42
Sodium dichromate . . .	60	59	58	56	55	54	52	51	50	47	—
Magnesium nitrate	60	58	57	56	55	53	52	50	49	46	43
Potassium carbonate . . .	—	—	47	44	44	43	43	43	42	—	—
Magnesium chloride . . .	35	34	34	34	33	33	33	32	32	31	30
Potassium acetate	25	24	24	23	23	22	22	21	20	—	—
Lithium chloride	15	14	13	13	12	12	12	12	11	11	11
Potassium hydroxide . . .	—	14	13	10	9	8	7	6	6	6	5

M.J.H.

1.4 Mechanical properties of materials

1.4.1 Densities

Density of water

Pure air-free water under 1 atm. Unit 1 g cm^{-3}

Temp./°C	0	2	4	6	8	10	12	14	16	18
0	0.999 84	0.999 94	0.999 97	0.999 94	0.999 85	0.999 70	0.999 50	0.999 24	0.998 94	0.998 59
20	0.998 20	0.997 77	0.997 30	0.996 78	0.996 23	0.995 65	0.995 03	0.994 37	0.993 69	0.992 97
40	0.992 22	0.991 44	0.990 63	0.989 79	0.988 93	0.988 04	0.987 12	0.986 18	0.985 21	0.984 22
60	0.983 20	0.982 16	0.981 09	0.980 01	0.978 90	0.977 77	0.976 61	0.975 44	0.974 24	0.973 03
80	0.971 79	0.970 53	0.969 26	0.967 96	0.966 65	0.965 31	0.963 96	0.962 59	0.961 20	0.959 79
100	0.958 36									

Values up to 40 °C, Chappuis (1907) and Thiesen, Scheel and Diesselhorst (1900), recalculated by P. H. Bigg, *Br. J. appl. Phys.*, 1967. Above 40 °C, Owen, White and Smith (1956) and Kell and Whalley (1965), recalculated by Kell, *J. Chem. Engng. Data*, 1967.

The temperature ($t_m/$°C) of maximum density at different pressures (p) measured in atm is given by

$$t_m = 3.98 - 0.0225(p - 1)$$

Density of heavy water

Calculated by Kell, *J. Chem. Engng Data*, 1967, from measurements by Chang and Tung (1949), Shrader and Wirtz (1951) and Steckel and Szapiro (1963). Unit 1 g cm^{-3}.

°C	5	10	15	20	25	30	35	40	45	50
g cm^{-3}	1.105 62	1.105 99	1.105 87	1.105 34	1.104 45	1.103 23	1.101 73	1.099 96	1.097 94	1.095 70

°C	55	60	65	70	75	80	85	90	95	100
g cm^{-3}	1.093 25	1.090 60	1.087 77	1.084 75	1.081 58	1.078 24	1.074 75	1.071 12	1.067 36	1.063 46

Density of mercury

In grams per cubic centimetre under pressure of 1 atmosphere. Based on:

Density at 20 °C (IPTS-48) = 13.545 884 g cm^{-3}

(A. H. Cook, *Phil. Trans. A*, 1961) and thermal expansion data of Beattie *et al.*
(J. A. Beattie *et al.*, *Proc. Am. Acad. Arts Sci.*, 1941, and A. H. Cook, *Br. J. appl. Phys.*, 1956).

Temp./°C	0	2	4	6	8	10	12	14	16	18
−20	13.644 59	13.639 62	13.634 66	13.629 70	13.624 75	13.619 79	13.614 85	13.609 90	13.604 96	13.600 02
0	13.595 08	13.590 15	13.585 22	13.580 29	13.575 36	13.570 44	13.565 52	13.560 60	13.555 70	13.550 79
20	13.545 87	13.540 97	13.536 06	13.531 17	13.526 26	13.521 37	13.516 47	13.511 58	13.506 70	13.501 82
40	13.496 93	13.492 07	13.487 18	13.482 29	13.477 42	13.472 56	13.467 68	13.462 82	13.457 96	13.453 09
60	13.448 23	13.443 37	13.438 52	13.433 67	13.428 82	13.423 97	13.419 13	13.414 28	13.409 43	13.404 60
80	13.399 77	13.394 92	13.390 09	13.385 26	13.380 42	13.375 60	13.370 77	13.365 94	13.361 12	13.356 30

°C	0	20	40	60	80	100	120	140	160	180
100	13.351 48	13.303 4	13.255 4	13.207 6	13.159 8	13.112 0	13.064 5	13.016 9	12.969 2	12.921 5
300	12.873 7	12.825 8	12.777 8	—	—	—	—	—	—	—

Approximate densities of commonly used materials

Substance	Density g cm⁻³	Substance	Density g cm⁻³	Substance	Density g cm⁻³
Acetone	0.8	Emery	4.0	Paraffin oil	0.8
Agate	2.6	Gas carbon	1.9	Paraffin wax	0.9
Alcohol	0.8	Gelatine	1.3	Permalloy C	8.6
Alni	6.9	German silver	8.4	Petroleum	0.8
Alnico	7.1	Glass (soda)	2.5	Phosphor-bronze	8.9
Aluminium-bronze		,, (Pyrex)	2.23	Pine (white)	0.5
(8% Al)	7.7	,, (lead)	3–4	Pitch	1.1
Amber	1.1	,, (optical)	see p. 90	Plaster of Paris	1.8
Asbestos	2.4	Glycerine	1.3	Plastics	see p. 251
Ash (timber)	0.75	Gold (22 carat)	17.5	Platinum-iridium	
Asphalt	1.4	,, (9 carat)	11.3	(90/10)	21.5
Balsa wood	0.2	Granite	2.7	Porcelain	2.3
Bamboo	0.4	Graphite	2.3	Quartz (crystal)	2.6
Bearing metal (80% Sn)	7.3	Gunmetal	8.2	Resin	1.1
Beech	0.75	'Heavy alloy'†	16.8–18.0	Sand (dry)	1.6
Beeswax	0.95	Ice	0.92	Sealing wax	1.8
Benzine	0.7	Inconel	8.5	Sea-water	1.03
Beryllium-copper	8.2	Invar	8.0	Silica (fused)	
Bone	1.9	Ivory	1.8	translucent	2.1
Borax	1.7	Keramot	1.6	transparent	2.2
Boxwood	1.0	Lard	0.9	Silicon iron	6.9
Brass (60/40)	8.4	Lignum vitae	1.3	Silver sand	2.6
,, (70/30)	8.5	Linseed oil	0.95	Slate	2.8
Brightray	8.4	Lo Ex	2.7	Soft solder (70% Sn	
Butter	0.9	Magnalium	2.6	30% Pb)	8.3
Carbon steel (<1% C)	7.8	Mahogany	0.8	Stainless iron (12% Cr)	7.7
Cast iron	7.0–7.4	Manganin	8.5	Stainless steel	7.8
Castor oil	0.95	Marble	2.7	Supermalloy	8.9
Cedarwood	0.55	Mazak (No. 2)	6.7	Tar	1.0
Charcoal	0.4	Methylated spirit	0.8	Teak	0.85
China clay (kaolin)	2.6	Mica	2.8	Thiokol	1.4
Coal (anthracite)	1.6	Mild steel	7.9	Tungsten carbide	
,, (bituminous)	1.4	Milk	1.03	(6% Co)	15.0
Constantan	8.9	Monel	8.8	Tungsten carbide	
Cork	0.25	Mumetal	8.8	(12% Co)	14.2
Corundum	4.0	Mycalex	2.4	Turpentine	0.85
Cronite	8.1	Naphtha	0.8	Wax (soft red)	1.0
Diamond	3.5	Nickel-chromium	8.4	White spirit	0.85
Duralumin	2.8	Nickel-silver	8.8	Wrought iron	7.8
Ebonite	1.2	Nimonic	8.2	Xylol	0.85
Ebony	1.2	Oak	0.7	Y-alloy	2.8
Elinvar	8.1	Olive oil	0.9		

† Tungsten with metallic additives.

J.B.R.

1.4.2 Elasticities

The following is a list of constants in common use, any two of which are sufficient to define the elastic properties of a homogeneous isotropic solid.

The two fundamental constants are those which relate change of volume and change of shape to external stress. They are, respectively, the bulk modulus K and the shear modulus G.

For many practical purposes, the following constants are commonly used:

     Young's modulus, or longitudinal elasticity, E.
     Poisson's ratio, $\sigma =$ lateral contraction per unit breadth divided by the longitudinal extension per unit length.
     Compressibility, $\kappa = 1/K$.
     Longitudinal modulus, M, which is the longitudinal modulus for zero lateral strain and determines the velocity of ultrasonic stress pulses in solids.

For a homogeneous isotropic solid, the following relations between these constants exist:

$$G = \frac{E}{2(1+\sigma)} \quad \cdots \cdots \quad \text{(a)} \qquad\qquad K = \frac{E}{3(1-2\sigma)} \quad \cdots \cdots \quad \text{(b)}$$

$$K = \frac{1}{3}\frac{EG}{(3G-E)} \quad \cdots \cdots \quad \text{(c)} \qquad\qquad M = K + \frac{4}{3}G \quad \cdots \cdots \quad \text{(d)}$$

The value of Poisson's ratio is usually positive and lies between 0 and $+\frac{1}{2}$, but in some cases it may be negative.

Elasticities of metals and alloys

Material 20 °C	E $\mathrm{GN\,m^{-2}}$	G $\mathrm{GN\,m^{-2}}$	σ	K $\mathrm{GN\,m^{-2}}$
Aluminium	70.3	26.1	0.345	75.5
Bismuth	31.9	12.0	0.330	31.3
Cadmium	49.9	19.2	0.300	41.6
Chromium	279.1	115.4	0.210	160.1
Copper	129.8	48.3	0.343	137.8
Gold	78.0	27.0	0.44	217.0
Iron (soft)	211.4	81.6	0.293	169.8
Iron (cast)†	152.3	60.0	0.27	109.5
Lead†	16.1	5.59	0.44	45.8
Magnesium	44.7	17.3	0.291	35.6
Nickel (unmag., soft)†	199.5	76.0	0.312	177.3
,, (,, hard)†	219.2	83.9	0.306	187.6
Niobium	104.9	37.5	0.397	170.3
Platinum	168.0	61.0	0.377	228.0
Silver	82.7	30.3	0.367	103.6
Tantalum	185.7	69.2	0.342	196.3
Tin	49.9	18.4	0.357	58.2
Titanium	115.7	43.8	0.321	107.7
Tungsten	411.0	160.6	0.280	311.0
Vanadium	127.6	46.7	0.365	158.0
Zinc	108.4	43.4	0.249	72.0
Brass (70 Zn, 30 Cu)	100.6	37.3	0.350	111.8
Constantan	162.4	61.2	0.327	156.4
Hidurax Special†‡	144.5	54.4	0.333	144.1
Invar (36 Ni, 63.8 Fe, 0.2 C)	144.0	57.2	0.259	99.4
Nickel Silver§	132.5	49.7	0.333	132.0

Elasticities of metal and alloys (contd)

Material 20 °C	$\dfrac{E}{\text{GN m}^{-2}}$	$\dfrac{G}{\text{GN m}^{-2}}$	σ	$\dfrac{K}{\text{GN m}^{-2}}$
Steel (Mild)	211.9	82.2	0.291	169.2
,, ($\frac{3}{4}$% C)	210.0	81.1	0.293	168.7
,, ($\frac{3}{4}$% C hardened) . .	201.4	77.8	0.296	165.0
,, Tool‖	211.6	82.2	0.287	165.3
,, Tool (hardened)‖ . .	203.2	78.5	0.295	165.2
,, Stainless†† . .	215.3	83.9	0.283	166.0
Tungsten Carbide† . . .	534.4	219.0	0.22	319.0

† Approx. or values for materials of variable composition.
‡ Cu–Ni alloy with Al, Fe and Mn additions.
§ Approx. % composition: Cu 55, Ni 18, Zn 27.
‖ Oil hardening non-deforming tool steel of approx. % composition:
      C 0.98, Mn 1.03, Cr 0.65, W 1.01, V 0.1.
†† Approx. % composition: C 0.2, Si 0.5, Mn 0.7, Ni 2, Cr 18.

Elasticities of miscellaneous materials

Material 20 °C	$\dfrac{E}{\text{GN m}^{-2}}$	$\dfrac{G}{\text{GN m}^{-2}}$	σ	$\dfrac{K}{\text{GN m}^{-2}}$
Glass (Heavy Flint) . . .	80.1	31.5	0.27	57.6
Glass (Crown)	71.3	29.2	0.22	41.2
Quartz (fused)	73.1	31.2	0.170	36.9

Note. Many of the values in the above tables are from measurements made at the NPL and are reproduced from Notes on Applied Science No. 30, *Use in Industry of Elasticity Measurements with the Help of Mechanical Vibrations*, G. Bradfield, London, H.M.S.O., 1964.

Elasticities for woods

All woods are elastically anisotropic and in general there are nine independent elastic constants. The values in the table below are for some common woods and give the three principal values of Young's modulus measured along the grain E_L, in a radial direction E_R and tangential direction E_T. The values have been supplied by R. F. S. Hearmon (Forest Products Research Laboratory). Further information is given in a report by R. F. S. Hearmon, *Elasticity of Wood and Plywood*, Forest Products Research Special Report No. 7, London, H.M.S.O., 1948.

Wood	Relative density	$\dfrac{E_L}{\text{GN m}^{-2}}$	$\dfrac{E_R}{\text{GN m}^{-2}}$	$\dfrac{E_T}{\text{GN m}^{-2}}$
Ash	0.7	16	1.6	0.9
Balsa	0.2	6	0.3	0.1
Beech	0.7	14	2.2	1.1
Birch	0.6	16	1.1	0.6
Mahogany	0.5	12	1.1	0.6
Oak	0.7	11	—	—
Walnut	0.6	11	1.2	0.6
Teak	0.6	13	—	—
Douglas Fir	0.5	16	1.1	0.8
Scots Pine	0.5	16	1.1	0.6
Spruce	0.4–0.5	10–16	0.4–0.9	0.4–0.6

Elasticities for plastics

All plastics are visco-elastic and consequently the elasticity varies considerably with temperature and strain rate. The table below gives approximate values at 20 °C for very slow rates of strain.

Plastic	$\dfrac{E}{\mathrm{GN\ m^{-2}}}$	*Plastic*	$\dfrac{E}{\mathrm{GN\ m^{-2}}}$
Nylon 6 (cast) . . .	2.4–3.1	Polyethylene (high density)	0.4–1.3
Nylon 6 (moulded) . .	0.8–3.1	Polyimide	~3.1
Nylon 66	1.2–2.9	Polymethylmethacrylate .	2.4–3.4
Polybenzoxazole . . .	~3.5	Polypropylene . . .	1.1–1.6
Polycarbonate . . .	2.4	Polystyrene	2.7–4.2

Temperature coefficients of elastic constants for some materials

At 15 °C	*Temperature coefficient α in* $E_t = E_{15}\{1 - \alpha(t-15)\}$ $G_t = G_{15}\{1 - \alpha'(t-15)\}$	
	α for E	*α' for G*
Aluminium	4.8×10^{-4}	5.2×10^{-4}
Brass	3.7×10^{-4}	4.6×10^{-4}
Copper	3.0×10^{-4}	3.1×10^{-4}
German silver		6.5×10^{-4}
Gold	4.8×10^{-4}	3.3×10^{-4}
Iron	2.3×10^{-4}	2.8×10^{-4}
Phosphor-bronze		3.0×10^{-4}
Platinum	0.98×10^{-4}	1.0×10^{-4}
Quartz fibre	-1.5×10^{-4}	-1.1×10^{-4}
Silver	7.5×10^{-4}	4.5×10^{-4}
Steel	2.4×10^{-4}	2.6×10^{-4}
Tin		5.9×10^{-4}

Tensile strengths of materials

The elastic limit is always exceeded before the breaking stress is reached. The process of drawing the material into a wire increases the strength, and the finer the wire the greater the breaking stress. (See Poynting and Thomson's *Properties of Matter.*) Cold working generally tends to increase the breaking stress of the material at the expense of its ductility.

For crushing and shearing strengths see Ewing's *Strength of Materials* or one of the Engineering *Pocket Books.*

For engineering properties (i.e. proof stress, breaking stress, fatigue and creep properties, etc.) of materials at normal and elevated temperatures see S. L. Hoyt's *Metals and Alloys Data Book.*

Substance	Tensile strength MN m^{-2}	Substance	Tensile strength MN m^{-2}
Metals		**Miscellaneous**	
Aluminium (cast)	90–100	Catgut	420
,, (rolled)	90–150	Glass	30–90
Brass ⎰ 66% Cu ⎱ (cast) . . .	150–190	Hemp rope	60–100
,, ⎱ 34% Zn ⎰ (rolled) . . .	230–270	Leather belt	30–50
Calcium	42–60	Silk fibre	260
Cobalt	260–750	Spider thread	180
Copper (cast)	120–170	Woods† :	
,, (rolled)	200–400	Ash, beech, oak, teak, mahogany .	60–110
Gun metal (90% Cu, 10% Sn) . .	190–260	Fir, pitch-pine	40–80
Iron (cast)	100–230	Red or white deal	30–70
,, (wrought)	290–450	White or yellow pine	20–50
Lead (cast)	12–17	Quartz fibre (fused)	∼1000
Magnesium (cast)	60–80		
,, (extruded)	170–190	**Wires**	
Phosphor-bronze (cast)	180–280	Aluminium	200–450
Steel (castings)	400–600	Brass	350–550
,, (mild) (0.2% C)	430–490	Copper (hard-drawn)	400–460
High-carbon spring steel:		,, (annealed)	280–310
(annealed)	700–770	Duralumin	400–550
(tempered)	930–1080	German silver	460
(nickel) (5% Ni)	800–1000	Gold	200–250
(nickel-chromium)	1000–1500	Iron (charcoal, hard-drawn) . . .	540–620
Soft solder	55–75	,, (annealed)	460
Tin (cast)	20–35	Molybdenum	1100–3000
Zinc (rolled)	110–150	Nickel	500–900
		Palladium	350–450
Plastics		Phosphor-bronze (hard-drawn) . .	690–1080
Nylon 6	76–97	Platinum	330–370
Nylon 66	62–83	Pt + 10% Rh	630
Polyacetal	∼69	Silver	290
Polybenzoxazole	82–117	Steel (ordinary)	∼1100
Polycarbonate	55–65	,, (tempered)	1550
Polyethylene	21–35	,, (pianoforte, hard-drawn) . .	1860–2330
Polyimide	69–104	Tantalum	800–1100
Polymethylmethacrylate	50–76	Tungsten	1500–3500
Polypropylene	30–40	Zirconium (annealed)	260–390
Polystyrene	34–52	,, (hard-drawn)	1000

† Along the grain.

Bulk moduli of elements

Element	K GN m^{-2}	Element	K GN m^{-2}	Element	K GN m^{-2}	Element	K GN m^{-2}
Aluminium† .	75.5	Chlorine		Molybdenum	231.0	Selenium . .	8.3
Antimony . .	42.0	(liq) . . .	1.1	Nickel†		Silicon . . .	100.0
Arsenic . . .	22.0	Chromium† .	160.1	(soft) . .	177.3	Silver . . .	103.6
Bismuth . . .	31.3	Copper . .	137.8	(hard) . .	187.6	Sodium . .	6.3
Bromine . . .	1.9	Gold . . .	217.0	Palladium . .	182.0	Sulphur . .	7.7
Cadmium . .	41.6	Iodine . . .	7.7	Phosphorus		Thallium . .	43.0
Caesium . . .	1.6	Iron † . . .	169.8	(red) . .	10.9	Tin . . .	58.2
Calcium . . .	17.2	Lead† . . .	45.8	Phosphorus		Zinc† . . .	72.0
Carbon‡		Lithium . .	11.1	(white) . .	4.9		
(diamond) .	542.0	Magnesium .	44.7	Platinum . .	228.0		
Carbon		Manganese .	118.0	Potassium . .	3.1		
(graphite) .	33.0	Mercury . .	25.0	Rubidium . .	2.5		

† G. Bradfield, *Notes on Applied Science*, No. 30, 1964 (HMSO).
‡ Measurements made by M. F. Markham, NPL.

Bulk moduli of liquids

As the pressure increases, K increases. In general a rise in temperature decreases the bulk modulus of a liquid; water, however, shows a maximum value of K at about 50 °C. (See Poynting and Thomson's *Properties of Matter* and Bridgman's *The Physics of High Pressure*).

Liquid	Temp. °C	K GN m^{-2}	Liquid	Temp. °C	K GN m^{-2}
Acetic acid, 1–16 atm . . .	20	1.45	Mercury:		
Amyl alcohol, 8 atm	17.7	1.12	8–37 atm	20	26.2
Benzene, 8 atm	17.9	1.10	100–200 atm	15	30.0
Butyl alcohol, 8 atm	17.4	1.13	Methyl acetate, 8–37 atm . .	14.3	1.04
Butyl alcohol, *iso-*, 8 atm . .	17.9	1.03	Methyl alcohol, 37 atm . .	14.7	0.97
Carbon bisulphide, 8–37 atm .	15.6	1.16	Olive oil	20.5	1.60
Carbon tetrachloride	20	1.12	Paraffin oil	14.8	1.62
Chloroform, 100–200 atm . .	20	1.1	Pentane	20	0.318
Ether:			Petroleum	16.5	1.46
1–50 atm	0	0.689	Propyl alcohol, 8 atm . . .	17.7	1.04
900–1000 atm	0	1.56	Propyl alcohol, *iso-*, 8 atm .	17.8	0.983
900–1000 atm	198	0.703	Turpentine	19.7	1.280
Ethyl acetate, 8–37 atm . .	13.3	0.974	Water:		
Ethyl alcohol:			1–25 atm	15	2.05
1–500 atm	0	1.32	900–1000 atm	15	2.75
150–200 atm	310	0.024	900–1000 atm	198	1.81
Ethyl bromide, 8–37 atm . .	99.3	0.343	2500–3000 atm	14.2	3.88
Ethyl chloride, 8–37 atm . .	15.2	0.662	Water (sea)	—	2.32
Glycerine	20.5	4.03			

M.F.M.

1.4.3 Viscosities

Viscosities of liquids

The **dynamic viscosity**, η, of a (Newtonian) fluid is given by $\eta = \tau \div dv/dr$; τ = shearing stress between two planes parallel with the direction of flow, dv/dr = velocity gradient at right angles to the direction of flow. The dimensions of dynamic viscosity are $ML^{-1}T^{-1}$, and the SI unit is $N\,s\,m^{-2}$.

Kinematic viscosity, v, is the ratio of the dynamic viscosity to the density, ρ. The dimensions of kinematic viscosity are $L^2 T^{-1}$ and the SI unit is $m^2\,s^{-1}$.

Fluidity, ϕ, is the reciprocal of the dynamic viscosity. η of liquids decreases with the temperature approximately according to the equation $\log \eta = A + B/T$, where A and B are characteristic constants and T is the absolute temperature. Values of A and B for a large number of liquids are given by Barrer, *Trans. Farad.*, 1943, **39**, 48.

(i) Viscosity of water

Data from Bingham and Jackson, *U.S. Bur. Stand. Bull.*, 1918, **14**, 75, corrected for national standard in the US and the UK as from July 1, 1953: η at 20 °C = 1.0019 mN s m^{-2}.

Temp. °C	η mN s m^{-2}	Temp. °C	η mN s m^{-2}	Temp. °C	η mN s m^{-2}	Temp. °C	η mN s m^{-2}
0	1.7865	20	1.0019	50	0.5477	90	0.3155
5	1.5138	25	0.8909	60	0.4674	100	0.2829
10	1.3037	30	0.7982	70	0.4048	125	0.220†
15	1.1369	40	0.6540	80	0.3554	150	0.183†

† Mean of several observers.

(ii) Viscosities of various liquids, η in mN s m^{-2}

Liquid	0 °C	10 °C	20 °C	30 °C	40 °C	50 °C	60 °C	70 °C	100 °C
Acetic acid	—	—	1.219	1.037	0.902	0.794	0.703	0.629	0.464
Acetone	0.397	0.358	0.324	0.295	0.272	0.251	—	—	—
Aniline	—	6.53	4.39	3.18	2.40	1.91	1.56	1.29	0.76
Benzene	—	0.757	0.647	0.560	0.491	0.435	0.389	0.350	—
Bromobenzene . . .	1.556	1.325	1.148	1.007	0.889	0.792	0.718	0.654	0.514
Carbon bisulphide . .	0.436	0.404	0.375	0.351	0.329	—	—	—	—
Carbon dioxide (liq.) .	0.099	0.085	0.071	0.053	—	—	—	—	—
Carbon tetrachloride .	1.348	1.135	0.972	0.845	0.744	0.660	0.591	0.533	0.400
Chloroform	0.704	0.631	0.569	0.518	0.473	0.434	0.399	—	—
Di-ethyl ether . . .	0.294	0.267	0.242	0.219	0.199	0.183	0.168	0.154	0.119
Ethanol	1.767	1.447	1.197	1.000	0.830	0.700	0.594	0.502	—
Ethyl acetate . . .	0.581	0.510	0.454	0.406	0.366	0.332	0.304	0.278	—
Ethyl formate . . .	0.508	0.453	0.408	0.368	0.335	0.307	—	—	—
Formic acid	—	2.241	1.779	1.456	1.215	1.033	0.889	0.778	0.547
Mercury	1.681	1.661	1.552	1.499	1.450	1.407	1.367	1.327	1.232
Methanol	0.814	0.688	0.594	0.518	0.456	0.402	0.356	—	—
n-Octane	0.710	0.618	0.545	0.485	0.436	0.394	0.358	0.326	0.255
Oil castor	—	2420	986	451	231	125	74	43	16.9
Oil olive	—	138	84	52	36	24.5	17	12.4	—
n-Pentane	0.278	0.254	0.234	0.215	0.198	0.184	0.172	0.161	0.130
Sulphuric acid . . .	56	49	27	20	14.5	11.0	8.2	6.2	—
Toluene	0.771	0.668	0.585	0.519	0.464	0.418	0.379	0.345	0.268

(iii) Relative viscosities of some aqueous solutions

Strength of solutions 1 normal. Viscosities relative to that of water at same temperature. For a complete list, see *International Critical Tables* (1928).

Substance	Temp./°C	Relative viscosity	Substance	Temp./°C	Relative viscosity
Ammonia	25	1.02	Potassium chloride . .	17.6	0.98
Ammonium chloride .	17.6	0.98	Potassium iodide . .	17.6	0.91
Calcium chloride . .	20	1.31	Sodium hydroxide . .	25	1.24
Hydrochloric acid . .	15	1.07	Sulphuric acid . . .	25	1.09

Viscosities of liquids at high pressures

(i) Relative viscosity of water

Pressure/(MN m^{-2})	2.2 °C	10 °C	20 °C	30 °C	50 °C	75 °C	100 °C
0.1	1.000	1.000	1.000	1.000	1.000	1.000	1.000
49	0.946	0.969	0.990	0.998	1.021	1.029	1.043
98	0.926	0.957	0.990	1.008	1.046	1.063	1.085
196	0.940	0.982	1.023	1.053	1.104	1.137	1.170
294	0.993	1.037	1.081	1.116	1.174	1.217	1.256
392	1.072	1.185	1.163	1.195	1.253	1.302	1.349
588	1.296	1.330	1.367	1.386	1.439	1.492	1.546
784	—	—	1.629	1.642	1.664	1.708	1.757
981	—	—	—	1.950	1.936	1.948	1.986

(ii) Relative viscosities of various liquids

$\eta = 1.000$ at 30 °C and 0.1 mN m^{-2}

Liquid	Temp./°C	Pressures/(MN m^{-2})			
		98	392	784	1177
Acetone	30	1.68	4.03	9.70	—
	75	1.30	2.79	5.78	10.7
Carbon bisulphide .	30	1.45	3.23	6.92	15.5
	75	1.12	2.35	4.69	8.83
Di-ethyl ether . .	30	2.11	6.20	18.2	46.8
	75	1.41	3.99	9.69	20.5
Ethanol	30	1.59	4.14	10.5	24.5
	75	0.747	1.95	4.30	8.28
Methanol	30	1.47	2.96	5.62	9.95
	75	0.857	1.61	2.80	4.52
n-Pentane	30	2.07	7.03	22.9	70.2
	75	1.46	4.74	13.2	31.1

(iii) Viscosity of aqueous glycerol solutions

Data from Sheely, *Industr. Engng. Chem.*, 1932, **24**, 1060, corrected to recent value for the viscosity of water (p. 36).

Relative density 25°/25 °C	% weight glycerol	$\eta/(\mathrm{N\ s\ m^{-2}})$		
		20 °C	25 °C	30 °C
1.262 01	100	1.495	0.942	0.622
1.259 45	99	1.194	0.772	0.509
1.256 85	98	0.971	0.627	0.423
1.254 25	97	0.802	0.521 5	0.353
1.251 65	96	0.659	0.434	0.295 8
1.249 10	95	0.543 5	0.365	0.248
1.209 25	80	0.061 8	0.045 72	0.034 81
1.127 20	50	0.006 032	0.005 024	0.004 233
1.061 15	25	0.002 089	0.001 805	0.001 586
1.023 70	10	0.001 307	0.001 149	0.001 021

(iv) Viscosity of aqueous sucrose solutions

Data from Bingham and Jackson, *National Bureau of Standards, Bulletin*, 1918, **14**, 59, corrected to recent value for the viscosity of water (p. 36).

Relative density 20°/4 °C	% weight sucrose	$\eta/(\mathrm{N\ s\ m^{-2}})$		
		15 °C	20 °C	25 °C
1.3790	75	4.039	2.328	1.405
1.3472	70	0.746 9	0.481 6	0.321 6
1.3163	65	0.211 3	0.147 2	0.105 4
1.2865	60	0.079 49	0.058 49	0.040 03
1.2296	50	0.019 53	0.015 43	0.012 40
1.1764	40	0.007 463	0.006 167	0.005 164
1.1270	30	0.003 757	0.003 187	0.002 735

Viscosities of glasses and minerals

$\log_{10} \eta$ in N s m^{-2}

Material	900 °C	1000 °C	1100 °C	1200 °C	1300 °C	1400 °C	1600 °C	1800 °C	2000 °C
Plate glass	4.00	3.03	2.41	1.87	1.46	1.07	—	—	—
Medium flint glass . .	3.9	2.8	1.9	1.4	0.9	0.7	—	—	—
Silica	—	—	14.6	12.7	11.8	9.7	8.2	4.7	3.4
Olivine	—	—	—	2.5	1.5	1.2	—	—	—
Diorite	—	—	—	3.1	2.3	1.8	—	—	—
Diopside.	—	—	—	—	0.52	0.43	—	—	—

Viscosities of gases and vapours

(i) Viscosities at normal pressure

μN s m^{-2}

Gas	0 °C	20 °C	50 °C	100 °C	200 °C	300 °C	400 °C	500 °C	600 °C
Air	17.3	18.2	19.6	22.0	26.1	29.8	33.2	36.4	39.4
Argon	21.0	22.3	24.2	27.3	32.8	37.7	42.2	46.4	50.4
Benzene	7.0	7.5	8.1	9.4	12.0	—	—	—	—
Carbon dioxide . . .	13.6	14.7	16.2	18.6	23.0	27.0	30.7	34.1	37.3
Carbon monoxide . .	16.6	17.5	18.9	21.0	24.9	28.3	31.4	34.4	37.2
Chlorine	12.3	13.2	14.5	16.9	21.0	25.0	—	—	—
Chloroform	9.4	10.2	11.2	12.9	16.0	—	—	—	—
Ethylene	9.7	10.3	11.2	12.8	15.4	17.9	—	—	—
Helium	18.7	19.6	21.0	23.2	27.3	31.2	34.8	38.4	41.8
Hydrogen	8.4	8.8	9.4	10.4	12.1	13.7	15.3	16.9	18.4
Krypton	23.4	25.0	27.4	31.2	38.0	44.2	49.9	55.2	60.2
Methane	10.3	11.0	11.9	13.5	16.3	18.8	21.1	23.3	25.3
Neon	29.8	31.3	33.6	37.0	43.2	48.9	54.3	59.4	64.4
Nitrogen	16.6	17.6	18.9	21.2	25.1	28.6	31.9	34.9	37.8
Nitrous oxide . . .	13.7	14.6	16.0	18.3	22.5	26.5	—	—	—
Oxygen	19.5	20.4	21.8	24.4	29.3	33.7	37.6	41.3	44.7
Steam	—	9.7	9.9	12.1	16.2	20.2	24.0	28.0	30.0
Sulphur dioxide . .	11.7	12.6	14.0	16.3	20.7	24.6	—	—	—
Xenon	21.2	22.8	25.1	28.8	35.7	42.0	47.9	53.4	58.6

For viscosity of gases μN s m^{-2} is a convenient size of unit. From the kinetic theory the viscosity is expected to be independent of pressure and to vary as the square root of the absolute temperature. The first is true except at very low and at high pressures; the second requires correction. Dividing the kinetic theory expression by a correction factor which is a linear function of the reciprocal of the absolute temperature leads to Sutherland's formula $\eta = K T^{3/2}/(T+C)$, where K and C are constants characteristic of the gas. For higher accuracy a polynomial can be used instead of a linear factor.

(ii) Viscosities of some gases at high pressure

η in μN s m^{-2} (all at 20 °C)

Gas	Pressures/(MN m^{-2})		
	2	5	10
Air	18.5	19.1	20.6
Argon	22.7	23.5	25.6
Helium	19.6	19.6	19.6
Hydrogen	8.9	8.9	9.0
Neon	31.5	31.6	31.7
Nitrogen	17.8	18.4	19.9
Oxygen	20.6	21.4	23.4

(iii) Viscosity of nitrogen at high pressure

η in μN s m^{-2}

Temperature/°C	Pressure/(MN m^{-2})			
	5	10	20	30
50	20.0	21.0	23.2	25.5
100	22.1	22.9	24.7	26.8
200	25.7	26.4	27.8	29.2
300	29.2	29.7	30.8	31.7

J.R.S.

1.4.4 Mean velocity, free path and size of molecules

From the Maxwell–Boltzmann distribution law for molecular velocities, the mean square and mean velocities are

$$\overline{v^2} = 3kT/m; \qquad \overline{v} = (8\overline{v^2}/3\pi)^{1/2}$$

From the kinetic theory of gases, assuming that molecules interact like hard spheres, it follows that:

$$\eta = (5/16\,\sigma^2)(mkT/\pi)^{1/2} \qquad \tau = l/\overline{v} = 4\eta/5p$$
$$l = m/(\pi\rho\sigma^2\sqrt{2})$$

where
- k = Boltzmann's constant
- T = absolute temperature
- p = pressure
- ρ = density
- η = viscosity
- σ = molecular diameter
- m = mass of molecule
- l = mean free path
- τ = mean time between collisions.

A more exact theory uses the Lennard-Jones intermolecular potential. (See Hirschfelder, Curtiss and Bird's *Molecular Theory of Gases and Liquids*, 1954 (Wiley, New York).)

Equations are obtained for viscosity and second virial coefficient, B (i.e. first order departure from the perfect gas law) from which the molecular diameter, σ, can be estimated more accurately than from the simple equation for viscosity given above.

Gas	$v/(\text{m s}^{-1})$	l/nm	τ/ps	σ/pm	
	at 0 °C and atmospheric pressure			η	B
Argon	380	62.6	165	342	340
Benzene	272	148.2	545	527	—
Carbon dioxide . .	362	39.0	108	390	407
Carbon monoxide .	454	58.6	129	371	376
Chlorine	285	27.4	96	440	—
Chloroform . . .	220	161.0	732	543	—
Ethylene	454	34.3	75	423	452
Helium	1202	173.6	144	258	256
Hydrogen	1694	110.6	65	297	293
Methane	600	48.1	80	380	382
Neon	535	124.0	232	279	275
Nitrogen	454	58.8	130	375	370
Nitrous oxide . . .	362	38.7	107	388	459
Oxygen	425	63.3	149	354	358
Sulphur dioxide . .	300	27.4	91	429	—

J.R.S.

1.4.5 Surface tensions

Surface tension is caused by the inward attraction of molecules at a boundary. This attraction produces curvature of free liquid surfaces, and causes a pressure difference to exist at the curved boundary: $\Delta p = \gamma(1/R_1 + 1/R_2)$, where Δp = pressure difference, R_1 and R_2 = principal radii of curvature. γ, the *surface tension* at a liquid-gas boundary, is usually measured in mN/m. At liquid–liquid boundaries there is *interfacial tension*; at liquid–solid boundaries *adhesion tension*.

Temperature variation. γ decreases with increase of temperature and vanishes at the critical temperature. For many liquids $-\mathrm{d}[\gamma(M/\rho)^{2/3}]/\mathrm{d}t = 2.12$, where M = molecular weight; ρ = density (g cm^{-3}); t = temperature (°C) (Eötvös, 1886). (See Adam, *The Physics and Chemistry of Surfaces.*)

Surface tensions of various liquids

A = against air V = against own vapour

Substance	Against	Temp. °C	γ mN m^{-1}	Substance	Against	Temp. °C	γ mN m^{-1}
Inorganic							
Aluminium . . .	A	700	840	Lead chloride . .	A	490	138
Cadmium . . .	N_2	320	630	Potassium chloride .	N_2	909	88
Gold	A	1130	1102	Sodium chloride . .	N_2	908	106
Lead	CO_2	350	453	Sodium fluoride . .	N_2	1010	200
Mercury . . .	V	20	472	Sodium iodide . .	N_2	861	77.6
Mercury . . .	V	100	456	Sodium sulphate .	N_2	900	195
Potassium . . .	CO_2	64	410	Oxygen	A	−183	13.1
Sodium	CO_2	90	290	Nitrogen	V	−183	6.2
Organic							
Acetic acid . . .	V	20	27.8	Ethanol	V	10	23.6
Acetone	V	20	23.7	Ethanol	V	20	22.8
Acetone	V	50	20.0	Ethanol	V	30	21.9
Aniline	V	20	42.9	Ethyl acetate . .	V	20	23.6
Benzene	A	10	30.22	Ethyl bromide . .	V	20	24.2
Benzene	A	20	**28.88**	Di-ethyl ether . .	V	20	**16.96**
Bromoform . . .	V	20	41.5	Di-ethyl ether . .	V	50	13.4
n-Butyl alcohol .	A	20	24.6	Glycerol	A	20	63.4
n-Butyric acid . .	A	20	26.8	n-Hexane	A	20	18.4
Carbon				n-Octane	A	20	21.8
bisulphide . .	V	20	32.3	Methanol	A	0	24.5
Carbon tetra-				Methanol	A	20	22.6
chloride . . .	V	20	27.0	Methanol	V	50	20.2
Carbon tetra-				Oleic acid . . .	A	20	32.5
chloride . . .	V	100	17.3	Phenol	A	20	40.9
Chloroform . . .	A	20	**27.14**	n-Propyl alcohol .	V	20	23.8
Ethanol	A	0	24.1	Toluene	V	20	28.4

Bold figures in the γ columns denote values, accurately known, which can be used for calibration purposes.

Surface tension of water against air

Temp./°C	0	10	15	20	25	30	40	50	60	70	80	100
γ/(mN m^{-1})	75.7	74.2	73.5	72.75	72.0	71.2	69.6	67.9	66.2	64.4	62.6	58.8

Surface tensions of aqueous salt solutions

Usually greater than that of water. Approximately $\gamma = \gamma H_2O + M . \Delta\gamma$, where M is concentration. Values below are for $M = 1$ mole dm^{-3} at 20 °C.

Salt	$\Delta\gamma/(mN\ m^{-1})$
KCl	1.4
NaCl	1.64
Na_2CO_3	2.7
$NaNO_3$	1.2
Na_2SO_4	2.7

Interfacial tensions of liquids at 20 °C

Liquids	$\gamma/(mN\ m^{-1})$	Liquids	$\gamma/(mN\ m^{-1})$
Water against:		Mercury against:	
Benzene	35	Acetone	390
Carbon tetrachloride . . .	45	Benzene	357
Chloroform	28	Chloroform	357
Di-ethyl ether	10	Di-ethyl ether	379
Heptylic acid	7	n-Heptane	379
n-Heptane	51	Oleic acid	322
n-Octane	51		
Olive oil	20		
Paraffin oil	48		

J.R.S.

1.5 Temperature and heat

1.5.1 The International Practical Temperature Scale of 1968 (IPTS-68)

The history of the IPTS-68

The International Temperature Scale was adopted in 1927 to overcome the practical difficulties of the direct realization of thermodynamic temperatures by gas thermometry and to unify existing temperature scales. It was introduced by the Seventh General Conference on Weights and Measures with the intention of producing a practical scale of temperature which was easily and accurately reproducible and which gave as nearly as possible thermodynamic temperatures. The Scale was revised in 1948, amended in 1960 (the numerical values of temperature remaining the same as in 1948) and revised again in 1968. (See *The International Practical Temperature Scale of 1968*, 1969 (HMSO).) The 1968 revision reduced the lower limit of the Scale from 90.18 K to 13.81 K and the values formerly assigned to the defining fixed points were modified where necessary to conform as nearly as possible to thermodynamic temperatures. The approximate difference between the values of temperature given by IPTS-68 and its predecessor, IPTS-48 are given below.

The unit of temperature

The basic temperature is the thermodynamic temperature, symbol T, the unit of which is the kelvin, symbol K. The kelvin is defined as the fraction 1/273.16 of the thermodynamic temperature of the triple point of water.

The Celsius temperature, symbol t, is defined by

$$t = T - T_0$$

where $T_0 = 273.15$ K. The unit employed to express a Celsius temperature is the degree Celsius, symbol °C, which is equal to the kelvin. A difference of temperature is expressed in kelvins, it may also be expressed in degrees Celsius.

The IPTS-68 distinguishes between the International Practical Kelvin Temperature with the symbol T_{68} and the International Practical Celsius Temperature with the symbol t_{68}, the relation between T_{68} and t_{68} being

$$t_{68} = T_{68} - 273.15 \text{ K}$$

The units of T_{68} and t_{68} are the kelvin, symbol K, and the degree Celsius, symbol °C, as in the case of the thermodynamic temperature T and the Celsius temperature t.

Approximate differences $(t_{68} - t_{48})$, in kelvins, between the values of temperature given by the IPTS-68 and the IPTS-48

t_{68}/°C	0	-10	-20	-30	-40	-50	-60	-70	-80	-90
-100	0.022	0.013	0.003	-0.006	-0.013	-0.013	-0.005	0.007	0.012	
-0	0.000	0.006	0.012	0.018	0.024	0.029	0.032	0.034	0.033	0.029

t_{68}/°C	0	10	20	30	40	50	60	70	80	90
0	0.000	-0.004	-0.007	-0.009	-0.010	-0.010	-0.010	-0.008	-0.006	-0.003
100	0.000	0.004	0.007	0.012	0.016	0.020	0.025	0.029	0.034	0.038
200	0.043	0.047	0.051	0.054	0.058	0.061	0.064	0.067	0.069	0.071
300	0.073	0.074	0.075	0.076	0.077	0.077	0.077	0.077	0.077	0.076
400	0.076	0.075	0.075	0.075	0.074	0.074	0.074	0.075	0.076	0.077
500	0.079	0.082	0.085	0.089	0.094	0.100	0.108	0.116	0.126	0.137
600	0.150	0.165	0.182	0.200	0.23	0.25	0.28	0.31	0.34	0.36
700	0.39	0.42	0.45	0.47	0.50	0.53	0.56	0.58	0.61	0.64
800	0.67	0.70	0.72	0.75	0.78	0.81	0.84	0.87	0.89	0.92
900	0.95	0.98	1.01	1.04	1.07	1.10	1.12	1.15	1.18	1.21
1000	1.24	1.27	1.30	1.33	1.36	1.39	1.42	1.44		

t_{68}/°C	0	100	200	300	400	500	600	700	800	900
1000		1.5	1.7	1.8	2.0	2.2	2.4	2.6	2.8	3.0
2000	3.2	3.5	3.7	4.0	4.2	4.5	4.8	5.0	5.3	5.6
3000	5.9	6.2	6.5	6.9	7.2	7.5	7.9	8.2	8.6	9.0
4000	9.3									

Principle of the IPTS-68

The IPTS-68 is based on the assigned values of the temperatures of a number of reproducible equilibrium states (**defining fixed points**) and on standard instruments calibrated at those temperatures. Interpolation between the fixed point temperatures is provided by formulae used to establish the relation between indications of the standard instruments and values of International Practical Temperature.

The defining fixed points are established by realizing specified equilibrium states between phases of pure substances. These equilibrium states and the values of the International Practical Temperature assigned to them are given below.

The standard instrument used from 13.81 K to 630.74 °C is the platinum resistance thermometer. The thermometer resistor must be strain-free, annealed pure platinum. The resistance ratio $W(T_{68})$, defined by

$$W(T_{68}) = R(T_{68})/R(273.15 \text{ K})$$

where R is the resistance, must not be less than 1.392 50 at $T_{68} = 373.15$ K. Below 0 °C the resistance–temperature relation of the thermometer is found from a reference function† and specified deviation equations. From 0 °C to 630.74 °C two polynomial equations provide the resistance–temperature relation.

The standard instrument used from 630.74 °C to 1064.43 °C is the platinum–10% rhodium/platinum thermocouple, the e.m.f.–temperature relation of which is represented by a quadratic equation.

Above 1337.58 K (1064.43 °C) the International Practical Temperature of 1968 is defined by the Planck law of radiation with 1337.58 K as the reference temperature and a value of 0.014 388 metre kelvin for c_2.

Defining fixed points of the IPTS-68

Equilibrium state	Assigned value of International Practical Temperature	
	T_{68}/K	$t_{68}/°\text{C}$
Triple point of equilibrium hydrogen	13.81	−259.34
Equilibrium between the liquid and vapour phases of equilibrium hydrogen at a pressure of 33 330.6 N m^{-2}	17.042	−256.108
Boiling point of equilibrium hydrogen . . .	20.28	−252.87
Boiling point of neon	27.102	−246.048
Triple point of oxygen	54.361	−218.789
Boiling point of oxygen	90.188	−182.962
Triple point of water	273.16	0.01
Boiling point of water	373.15	100
Freezing point of zinc	692.73	419.58
Freezing point of silver	1235.08	961.93
Freezing point of gold	1337.58	1064.43

Note: 1. The boiling points relate to equilibrium states at a pressure of one standard atmosphere (101 325 N m^{-2}).
2. The freezing point of tin has the assigned value of $t_{68} = 231.9681$ °C and may be used as an alternative to the boiling point of water.

T.J.Q.

1.5.2 Thermoelectric thermometry

The tables below are given to provide a basis for interpolation between calibration points of an individual thermocouple. They represent very accurate determinations of the e.m.f.–temperature relationship of a batch of thermocouples of each kind. The calibration of a particular thermocouple may be obtained by observation at a few points over the temperature range and by plotting a curve of difference from these reference tables.

† The reference function and full details of the deviation equations are given in *The International Practical Temperature Scale of 1968*, 1969 (HMSO).

Thermocouple tables

Note that the temperature for a given entry is obtained by adding the corresponding temperature in the top row to that in the left or right-hand column, regardless of whether the latter is positive or negative. In every case the reference junction is at 0 °C. See also B.S. 4937, 1973.

Temp. °C	0	10	20	30	40	50	60	70	80	90	Temp. °C
					Absolute e.m.f. in microvolts						

Platinum 10% rhodium/platinum

Temp. °C	0	10	20	30	40	50	60	70	80	90	Temp. °C
−100						−236	−194	−150	−103	−53	−100
0	0	55	113	173	235	299	365	432	502	573	0
100	645	719	795	872	950	1 029	1 109	1 190	1 273	1 356	100
200	1 440	1 525	1 611	1 698	1 785	1 873	1 962	2 051	2 141	2 232	200
300	2 323	2 414	2 506	2 599	2 692	2 786	2 880	2 974	3 069	3 164	300
400	3 260	3 356	3 452	3 549	3 645	3 743	3 840	3 938	4 036	4 135	400
500	4 234	4 333	4 432	4 532	4 632	4 732	4 832	4 933	5 034	5 136	500
600	5 237	5 339	5 442	5 544	5 648	5 751	5 855	5 960	6 064	6 169	600
700	6 274	6 380	6 486	6 592	6 699	6 805	6 913	7 020	7 128	7 236	700
800	7 345	7 454	7 563	7 672	7 782	7 892	8 003	8 114	8 225	8 336	800
900	8 448	8 560	8 673	8 786	8 899	9 012	9 126	9 240	9 355	9 470	900
1000	9 585	9 700	9 816	9 932	10 048	10 165	10 282	10 400	10 517	10 635	1000
1100	10 754	10 872	10 991	11 110	11 229	11 348	11 467	11 587	11 707	11 827	1100
1200	11 947	12 067	12 188	12 308	12 429	12 550	12 671	12 792	12 913	13 034	1200
1300	13 155	13 276	13 397	13 519	13 640	13 761	13 883	14 004	14 125	14 247	1300
1400	14 368	14 489	14 610	14 731	14 852	14 973	15 094	15 215	15 336	15 456	1400
1500	15 576	15 697	15 817	15 937	16 057	16 176	16 296	16 415	16 534	16 653	1500
1600	16 771	16 890	17 008	17 125	17 243	17 360	17 477	17 594	17 711	17 826	1600
1700	17 942	18 056	18 170	18 282	18 394	18 504	18 612				1700

Platinum 13% rhodium/platinum

Temp. °C	0	10	20	30	40	50	60	70	80	90	Temp. °C
−100						−226	−188	−145	−100	−51	−100
0	0	54	111	171	232	296	363	431	501	573	0
100	647	723	800	879	959	1 041	1 124	1 208	1 294	1 380	100
200	1 468	1 557	1 647	1 738	1 830	1 923	2 017	2 111	2 207	2 303	200
300	2 400	2 498	2 596	2 695	2 795	2 896	2 997	3 099	3 201	3 304	300
400	3 407	3 511	3 616	3 721	3 826	3 933	4 039	4 146	4 254	4 362	400
500	4 471	4 580	4 689	4 799	4 910	5 021	5 132	5 244	5 356	5 469	500
600	5 582	5 696	5 810	5 925	6 040	6 155	6 272	6 388	6 505	6 623	600
700	6 741	6 860	6 979	7 098	7 218	7 339	7 460	7 582	7 703	7 826	700
800	7 949	8 072	8 196	8 320	8 445	8 570	8 696	8 822	8 949	9 076	800
900	9 203	9 331	9 460	9 589	9 718	9 848	9 978	10 109	10 240	10 371	900
1000	10 503	10 636	10 768	10 902	11 035	11 170	11 304	11 439	11 574	11 710	1000
1100	11 846	11 983	12 119	12 257	12 394	12 532	12 669	12 808	12 946	13 085	1100
1200	13 224	13 363	13 502	13 642	13 782	13 922	14 062	14 202	14 343	14 483	1200
1300	14 624	14 765	14 906	15 047	15 188	15 329	15 470	15 611	15 752	15 893	1300
1400	16 035	16 176	16 317	16 458	16 599	16 741	16 882	17 022	17 163	17 304	1400
1500	17 445	17 585	17 726	17 866	18 006	18 146	18 286	18 425	18 564	18 703	1500
1600	18 842	18 981	19 119	19 257	19 395	19 533	19 670	19 807	19 944	20 080	1600
1700	20 215	20 350	20 483	20 616	20 748	20 878	21 006				1700

Thermocouple tables (contd)

Temp. °C	0	10	20	30	40	50	60	70	80	90	Temp. °C
				Absolute e.m.f. in microvolts							

Platinum 30% rhodium/platinum 6% rhodium

Temp. °C	0	10	20	30	40	50	60	70	80	90	Temp. °C
0	0	−2	−3	−2	−0	2	6	11	17	25	0
100	33	43	53	65	78	92	107	123	140	159	100
200	178	199	220	243	266	291	317	344	372	401	200
300	431	462	494	527	561	596	632	669	707	746	300
400	786	827	870	913	957	1 002	1 048	1 095	1 143	1 192	400
500	1 241	1 292	1 344	1 397	1 450	1 505	1 560	1 617	1 674	1 732	500
600	1 791	1 851	1 912	1 974	2 036	2 100	2 164	2 230	2 296	2 363	600
700	2 430	2 499	2 569	2 639	2 710	2 782	2 855	2 928	3 003	3 078	700
800	3 154	3 231	3 308	3 387	3 466	3 546	3 626	3 708	3 790	3 873	800
900	3 957	4 041	4 126	4 212	4 298	4 386	4 474	4 562	4 652	4 742	900
1000	4 833	4 924	5 016	5 109	5 202	5 297	5 391	5 487	5 583	5 680	1000
1100	5 777	5 875	5 973	6 073	6 172	6 273	6 374	6 475	6 577	6 680	1100
1200	6 783	6 887	6 991	7 096	7 202	7 308	7 414	7 521	7 628	7 736	1200
1300	7 845	7 953	8 063	8 172	8 283	8 393	8 504	8 616	8 727	8 839	1300
1400	8 952	9 065	9 178	9 291	9 405	9 519	9 634	9 748	9 863	9 979	1400
1500	10 094	10 210	10 325	10 441	10 558	10 674	10 790	10 907	11 024	11 141	1500
1600	11 257	11 374	11 491	11 608	11 725	11 842	11 959	12 076	12 193	12 310	1600
1700	12 426	12 543	12 659	12 776	12 892	13 008	13 124	13 239	13 354	13 470	1700
1800	13 585	13 699	13 814								1800

Nickel–chromium/copper–nickel

Temp. °C	0	10	20	30	40	50	60	70	80	90	Temp. °C
−300				−9 835	−9 797	−9 719	−9 604	−9 455	−9 274	−9 063	−300
−200	−8 824	−8 561	−8 273	−7 963	−7 631	−7 279	−6 907	−6 516	−6 107	−5 680	−200
−100	−5 237	−4 777	−4 301	−3 811	−3 306	−2 787	−2 254	−1 709	−1 151	−581	−100
0	0	591	1 192	1 801	2 419	3 047	3 683	4 329	4 983	5 646	0
100	6 317	6 996	7 683	8 377	9 078	9 787	10 501	11 222	11 949	12 681	100
200	13 419	14 161	14 909	15 661	16 417	17 178	17 942	18 710	19 481	20 256	200
300	21 033	21 814	22 597	23 383	24 171	24 961	25 754	26 549	27 345	28 143	300
400	28 943	29 744	30 546	31 350	32 155	32 960	33 767	34 574	35 382	36 190	400
500	36 999	37 808	38 617	39 426	40 236	41 045	41 853	42 662	43 470	44 278	500
600	45 085	45 891	46 697	47 502	48 306	49 109	49 911	50 713	51 513	52 312	600
700	53 110	53 907	54 703	55 498	56 291	57 083	57 873	58 663	59 451	60 237	700
800	61 022	61 806	62 588	63 368	64 147	64 924	65 700	66 473	67 245	68 015	800
900	68 783	69 549	70 313	71 075	71 835	72 593	73 350	74 104	74 857	75 608	900
1000	76 358										1000

Iron/copper–nickel

Temp. °C	0	10	20	30	40	50	60	70	80	90	Temp. °C
−300										−8 096	−300
−200	−7 890	−7 659	−7 402	−7 122	−6 821	−6 499	−6 159	−5 801	−5 426	−5 036	−200
−100	−4 632	−4 215	−3 785	−3 344	−2 892	−2 431	−1 960	−1 481	−995	−501	−100

Thermocouple tables (*contd*)

Temp. °C	0	10	20	30	40	50	60	70	80	90	Temp. °C
					Absolute e.m.f. in microvolts						

Iron/copper–nickel (*contd*)

Temp. °C	0	10	20	30	40	50	60	70	80	90	Temp. °C
0	0	507	1 019	1 536	2 058	2 585	3 115	3 649	4 186	4 725	0
100	5 268	5 812	6 359	6 907	7 457	8 008	8 560	9 113	9 667	10 222	100
200	10 777	11 332	11 887	12 442	12 998	13 553	14 108	14 663	15 217	15 771	200
300	16 325	16 879	17 432	17 984	18 537	19 089	19 640	20 192	20 743	21 295	300
400	21 846	22 397	22 949	23 501	24 054	24 607	25 161	25 716	26 272	26 829	400
500	27 388	27 949	28 511	29 075	29 642	30 210	30 782	31 356	31 933	32 513	500
600	33 096	33 683	34 273	34 867	35 464	36 066	36 671	37 280	37 893	38 510	600
700	39 130	39 754	40 382	41 013	41 647	42 283	42 922	43 563	44 207	44 852	700
800	45 498	46 144	46 790	47 434	48 076	48 716	49 354	49 989	50 621	51 249	800
900	51 875	52 496	53 115	53 729	54 341	54 948	55 553	56 155	56 753	57 349	900
1000	57 942	58 533	59 121	59 708	60 293	60 876	61 459	62 039	62 619	63 199	1000
1100	63 777	64 355	64 933	65 510	66 087	66 664	67 240	67 815	68 390	68 964	1100
1200	69 536										1200

Copper/copper–nickel

Temp. °C	0	10	20	30	40	50	60	70	80	90	Temp. °C
−300				−6 258	−6 232	−6 181	−6 105	−6 007	−5 889	−5 753	−300
−200	−5 603	−5 439	−5 261	−5 069	−4 865	−4 648	−4 419	−4 177	−3 923	−3 656	−200
−100	−3 378	−3 089	−2 788	−2 475	−2 152	−1 819	−1 475	−1 121	−757	−383	−100
0	0	391	789	1 196	1 611	2 035	2 467	2 908	3 357	3 813	0
100	4 277	4 749	5 227	5 712	6 204	6 702	7 207	7 718	8 235	8 757	100
200	9 286	9 820	10 360	10 905	11 456	12 011	12 572	13 137	13 707	14 281	200
300	14 860	15 443	16 030	16 621	17 217	17 816	18 420	19 027	19 638	20 252	300
400	20 869										400

Nickel–chromium/nickel–aluminium

Temp. °C	0	10	20	30	40	50	60	70	80	90	Temp. °C
−300				−6 458	−6 441	−6 404	−6 344	−6 262	−6 158	−6 035	−300
−200	−5 891	−5 730	−5 550	−5 354	−5 141	−4 912	−4 669	−4 410	−4 138	−3 852	−200
−100	−3 553	−3 242	−2 920	−2 586	−2 243	−1 889	−1 527	−1 156	−777	−392	−100
0	0	397	798	1 203	1 611	2 022	2 436	2 850	3 266	3 681	0
100	4 095	4 508	4 919	5 327	5 733	6 137	6 539	6 939	7 338	7 737	100
200	8 137	8 537	8 938	9 341	9 745	10 151	10 560	10 969	11 381	11 793	200
300	12 207	12 623	13 039	13 456	13 874	14 292	14 712	15 132	15 552	15 974	300
400	16 395	16 818	17 241	17 664	18 088	18 513	18 938	19 363	19 788	20 214	400
500	20 640	21 066	21 493	21 919	22 346	22 772	23 198	23 624	24 050	24 476	500
600	24 902	25 327	25 751	26 176	26 599	27 022	27 445	27 867	28 288	28 709	600
700	29 128	29 547	29 965	30 383	30 799	31 214	31 629	32 042	32 455	32 866	700
800	33 277	33 686	34 095	34 502	34 909	35 314	35 718	36 121	36 524	36 925	800
900	37 325	37 724	38 122	·38 519	38 915	39 310	39 703	40 096	40 488	40 879	900
1000	41 269	41 657	42 045	42 432	42 817	43 202	43 585	43 968	44 349	44 729	1000
1100	45 108	45 486	45 863	46 238	46 612	46 985	47 356	47 726	48 095	48 462	1100
1200	48 828	49 192	49 555	49 916	50 276	50 633	50 990	51 344	51 697	52 049	1200
1300	52 398	52 747	53 093	53 439	53 782	54 125	54 466	54 807			1300

Thermal e.m.f. of chemical elements relative to platinum

E.M.F. in microvolts; cold junctions at 0 °C
(At the hot end, the current flows from the platinum to the element when the sign is positive)

Element	− 100 °C	+ 100 °C	Element	− 100 °C	+ 100 °C	Element	− 100 °C	+ 100 °C
Lithium	− 1000	+ 1820	Aluminium	− 60	+ 420	Iron	− 1940	+ 1980
Sodium	+ 290	—	Carbon	—	+ 700	Cobalt	—	− 1330
Potassium	+ 780	—	Silicon	+ 37 170	− 41 560	Nickel	+ 1220	− 1480
Rubidium	+ 460	—	Germanium	− 26 620	+ 33 900	Iridium	− 350	+ 650
Calcium	− 130	− 510	Tin	− 120	+ 420	Rhodium	− 340	+ 700
Cerium	—	+ 1140	Lead	− 130	+ 440	Palladium	+ 480	− 570
Magnesium	− 90	+ 440	Antimony	—	+ 4 890	Molybdenum	—	+ 1450
Zinc	− 330	+ 760	Bismuth	+ 7 540	− 7 340	Tungsten	− 150	+ 1120
Cadmium	− 310	+ 900	Copper	− 370	+ 760	Tantalum	− 100	+ 330
Mercury	—	+ 45	Silver	− 390	+ 740	Thorium	—	− 130
Indium	—	+ 690	Gold	− 390	+ 780			
Thallium	—	+ 580						

Thermal e.m.f. of alloys relative to platinum

E.M.F. in microvolts; cold junctions at 0 °C and hot junctions at 100 °C

Alloy	e.m.f.
Chromel	+ 2810
Alumel	− 1290
Constantan	− 3510
Manganin (84 Cu, 4 Ni, 12 Mn)	+ 610
Phosphor-bronze (96 Cu, 3.5 Sn, 0.3 P)	+ 550
Solder (50 Sn–50 Pb)	+ 460
Stainless steel (18–8)	+ 440
Nickel–chromium (80 Ni, 20 Cr)	+ 1140
„ „ (60 Ni, 24 Fe, 16 Cr)	+ 850
Yellow brass (70 Cu, 30 Zn)	+ 600
Beryllium copper (97½ Cu, 2½ Be)	+ 670

Absolute thermoelectric power of platinum

Temperature/K	300	400	500	600	700	800	900	1000	1100	1200
Thermoelectric power/$(\mu\text{V K}^{-1})$	− 5.05	− 7.66	− 9.69	− 11.33	− 12.87	− 14.38	− 15.97	− 17.58	− 19.03	− 20.56

T.J.Q.

1.5.3 Total radiation and spectral pyrometry

Total radiation and spectral pyrometers are normally calibrated in terms of black-body radiation and when sighted on to an unenclosed radiating surface measure an apparent or black-body temperature. The apparent temperature, T_r, measured by a total radiation pyrometer is related to the true temperature, T, by the formula: $T^4 = T_r^4/\epsilon_t - T_0^4(1 - \epsilon_t)/\epsilon_t$ where T_0 is the (kelvin) temperature of the pyrometer and its surroundings, and ϵ_t is the total emissivity of the surface. Similarly, the apparent temperature T_s measured by a spectral pyrometer is related to the true temperature T by the equation: $(1/T) - (1/T_s) = (\lambda \log \epsilon_y)/c_2$,† where ϵ_y is the spectral emissivity and c_2 is the constant in the Planck equation for the distribution of energy in the spectrum.

† This formula derives from the Wien approximation to the Planck distribution function for black-body radiation, i.e. $E_\lambda = c_1 \lambda^{-5} e^{-c_2/\lambda T}$ (see section 1.7.2, p. 76). The error due to the approximation is negligible for ordinary optical pyrometry.

Emissivity corrections (°C) for the total radiation pyrometer †

Observed temp./°C	Total emissivity								
	0.1	0.2	0.3	0.4	0.5	0.6	0.7	0.8	0.9
200	337	211	148	107	78	56	38	23	11
400	513	326	230	168	123	89	61	37	17
600	675	429	304	223	163	118	80	49	23
800	833	530	375	275	202	145	100	61	28
1000	989	630	446	327	240	173	118	73	34
1200	1146	729	517	379	278	200	137	84	39
1400	1302	828	587	430	316	228	156	96	45
1600	1457	928	658	482	354	255	175	107	50
1800	1613	1027	728	534	392	282	193	119	55

† The value of T_0 is barely significant in the temperature range given in the table, but has been taken as 298.15 K (25 °C).

Emissivity corrections (°C) for spectral pyrometer having effective wavelength of 0.65 μm
$c_2 = 0.014\ 388\ \text{m K}$

Observed temp./°C	Spectral emissivity								
	0.1	0.2	0.3	0.4	0.5	0.6	0.7	0.8	0.9
600	87	59	44	33	25	18	12	8	4
800	135	91	67	50	37	27	19	12	6
1000	194	130	95	71	53	39	27	17	8
1200	267	177	128	96	71	52	36	22	10
1400	353	232	168	125	93	67	46	29	13
1600	453	295	212	157	117	85	58	36	17
1800	570	368	263	195	144	104	72	44	21
2000	704	450	321	236	174	126	86	53	25
2500	1124	700	493	360	264	190	130	80	37
3000	1690	1022	709	513	374	267	182	112	52

Normal spectral emissivities

Material	Emissivity			T/K
	Wavelength			
	0.65 μm	1.0 μm	5.0 μm	
Aluminium (polished surface) . .	(0.1 at 2 μm)		0.05	500
Aluminium (rough surface) . . .	(0.3 at 2 μm)		0.25	500
Aluminium oxide	0.1	0.06	0.39	1200
(recrystallized alumina) . .	0.15	0.07	0.43	1400
	0.25	0.1	0.46	1600
	0.4	0.2	0.6	1800

Normal spectral emissivities (*contd*)

Material	Emissivity			T/K
	Wavelength			
	0.65 μm	1.0 μm	5.0 μm	
Beryllia†	0.5	0.35	0.8	1100
Chromium	0.35	—	—	1550
Cobalt	0.35	0.25	—	1300
Gold	0.15	0.05	0.03	1000
Hafnium	0.45	—	—	1200
Iridium	0.3	0.23	0.1	1500
Iron	0.35	0.3	0.15	1400
Stainless steel	0.33	0.3	0.2	1200
,, ,, (oxidized)	0.8	0.8	0.7	1200
Kanthal A (oxidized)	0.85	0.85	0.75	1300
Molybdenum	0.4	0.3	0.15	2000
Magnesia†	0.25	0.2	0.37	1400
Nickel	0.45	0.35	0.15	1100
Nickel (oxidized)	0.88	0.84	0.75	1300
Nickel/20% chromium	0.4	—	—	1200
,, ,, (oxidized) .	0.9	(0.85)	(0.8)	1200
Niobium	0.4	0.32	0.2	2000
Osmium	0.43	—	—	1500
Palladium	0.35	—	—	1500
Platinum	0.35	0.25	0.08	1200
Platinum 13% rhodium	0.28	—	—	1100
Rhenium	0.4	0.35	0.2	1400
Rhodium	0.2	—	—	1400
Ruthenium	0.34	—	—	1500
Silicon	0.4	0.25	0.25	1500
Silver	0.05	0.05	0.05	1000
Tantalum	0.41	0.3	0.18	2200
Thoria†	0.4	0.35	0.37	1400
Titanium	0.6	0.3	0.18	1000
Tungsten	0.43	0.38	0.12	2400
Uranium	0.3	—	—	1100
Vanadium	—	0.65	0.28	800
Zirconium	0.5	0.45	0.3	1600
Zirconia†	0.4	0.2	0.45	1400

† These materials have an emissivity which increases with temperature in a way similar to that of aluminium oxide.

T.J.Q.

1.5.4 Thermal expansion

Coefficients of cubical expansion of gases

The volume coefficient, α, at constant pressure is defined by $v_t = v_0(1 + \alpha t)$; the pressure coefficient, β, at constant volume is defined by $p_t = p_0(1 + \beta t)$, where v_t and p_t are the volume and pressure respectively corresponding to t °C, the initial volume and pressure (v_0, p_0) being measured at 0 °C. Both α and β depend on the initial pressure of the gas, and for a perfect gas $\alpha = \beta$.

(a) **Gases used in thermometry.** $p_0 = 0.1333$ MN/m^2. Temperature interval 0–100 °C.

Coefficient	Gas				
	He	H$_2$	N$_2$	Air	Ne
$\alpha/(10^{-3}$ K$^{-1})$	3.6580	3.6588	3.6735	3.6728	3.6600
$\beta/(10^{-3}$ K$^{-1})$	3.6605	3.6620	3.6744	3.6744	3.6617

(b) **Other gases** and various initial pressures. Temperature interval 0–100 °C except where otherwise stated.

Gas	$p_0/(\text{MN m}^{-2})$	$\alpha/(10^{-3}$ K$^{-1})$	$\beta/(10^{-3}$ K$^{-1})$
Air	0.031 (0–1067 °C)	—	3.6643
	2.67	—	3.887
O$_2$	0.024–0.031 (0–1067 °C)	—	3.6652
	0.088	—	3.6738
	10.1	4.86	—
CO	0.031 (0–1067 °C)	—	3.6648
	0.1013	3.669	3.667
CO$_2$	0.032 (0–1067 °C)	—	3.6756
	0.069	3.7073	3.6981
	0.133	3.7410	3.7262
N$_2$O	0.1013	3.719	3.676
SO$_2$	0.1013	3.903	3.845
NH$_3$	0.1013 (0–50 °C)	3.854	—
A	0.069	—	3.6680
N$_2$	20.3	4.34	—
	101	2.18	—

T.J.Q.

Coefficients of cubical expansion of liquids

The following table gives values of $a = (1/V_t)(dV_t/dt)$, where $t = 20$ °C. The expansion coefficient generally increases with rising temperature.

Liquid	$10^5 a/°C$ (cubical)	Liquid	$10^5 a/°C$ (cubical)
Acetic acid	107	Glycerol	47
Acetone.	143	Mercury†	18.1
Alcohol, methyl	119	Methyl iodide	120
,, ethyl	108	n-Pentane	155
Aniline	85	Phenol	79
Benzene.	121	Toluene.	107
Carbon bisulphide	119	Turpentine	96
Carbon tetrachloride . . .	122	m-Xylene	99
Chloroform	127	Water	21
Ether, ethyl	163	H_2SO_4 100%	56
Ethyl bromide	140		

† See also section 1.4.1, p. 29 (density of Hg).

Coefficients of linear expansion of solids

Over small ranges of temperature, the variation of length (l) with temperature (t) of most solid substances is represented by the equation $l_{t_1} = l_{t_0}\{1 + \alpha(t_1 - t_0)\}$, where α is the mean coefficient of linear expansion over the temperature range ($t_1 - t_0$). The values of α below apply to normal ambient temperatures unless otherwise stated.

Substance	$10^6 a/°C$ (linear)	Substance	$10^6 a/°C$ (linear)
Metals		**Metals** (contd)	
Aluminium	23	Magnesium 0 to 400 °C	29
Aluminium 0 to 600 °C	~29	Nickel	12.8
Antimony	~11	Nickel 0 to 1000 °C	18
Antimony ‖ axis	~16	Palladium	~11
Antimony ⊥ axis	~8	Palladium 0 to 1000 °C	~14
Bismuth	~13	Platinum	8.9
Bismuth ‖ axis	~16	Platinum at 800 °C	~11
Bismuth ⊥ axis	~12	Rhodium	8.4
Cadmium	~30	Silver	19
Cadmium ‖ axis	~52	Silver 0 to 900 °C	~22
Cadmium ⊥ axis	~20	Tantalum	6.5
Chromium	~7	Tantalum 0 to 400 °C	6.6
Chromium 0 to 900 °C	~11	Thallium	~28
Cobalt	~12	Tin	~21
Cobalt 25 to 350 °C	~14	Tin ‖ axis	~30
Copper	16.7	Tin ⊥ axis	~16
Copper 0 to 1000 °C	20	Titanium	~9
Gold	14	Tungsten	4.5
Gold 0 to 500 °C	15	Tungsten 600 to 1400 °C	~6
Iridium	6.5	Tungsten 1400 to 2200 °C	~7
Iridium 0 to 1750 °C	9	Vanadium	~8
Iron (electrolytic)	11.7	Zinc (cast)	~30
,, (electrolytic) 0 to 700 °C	15	,, (crystal); ‖ axis	~60
,, (wrought)	12	,, (crystal); ⊥ axis	~13
Lead	29	,, (rolled sheet; ‖ rolling)	~31
Lead 0 to 300 °C	32	,, (rolled sheet; ⊥ rolling)	~20
Magnesium	25		

Coefficients of linear expansion of solids (*contd*)

Substance	$10^6 a/°C$ (*linear*)	Substance	$10^6 a/°C$ (*linear*)
Alloys		**Alloys** (*contd*)	
Aluminium bronze, ~92 Cu, 7 Al, 0.4 Zn	16–17	Nickel-steels (including Invar) 10% Ni	13
Brass, ~68 Cu, 32 Zn	18–19	36% Ni at about 20 °C	0–1.5
Yellow brass (Muntz metal), ~60 Cu, 40 Zn, 20 to 100 °C	19–23	36% Ni at about 500 °C	9
Bronze (gunmetal, bell metal),		43% Ni	7.9
~80 Cu, 20 Sn	17–18	58% Ni	11.4
Constantan, ~60 Cu, 40 Ni	15–17	Phosphor bronze	17
Cupro-nickel (Monel metal),	~14	Platinum-iridium, 90 Pt, 10 Ir	8.7
~63 Ni, 30 Cu, with about		Speculum metal, ~68 Cu, 32 Sn	~19
2% each of Fe, Mn and Pb		Stainless steel, typical, 18 Cr, 8 Ni	~16
between 500 and 600 °C	~19	12% Cr, 20 to 100 °C	~10
Delta metal, 60 Cu, 36 Zn, 2 Sn,		Steel, carbon, typical	~11
2 Fe	~20	Stellite, ~80 Co, 20 Cr	
Duralumin	23	20 to 100 °C	14
~95 Al, 4 Cu, 20 to 500 °C	27	20 to 600 °C	16
German silver (nickel silver)	~18	Stellite No. 2, ~55 Co, 35 Cr, 10 W	
Magnalium, ~90 Al, 10 Mg	~23	20 to 100 °C	11
Monel metal (see Cupro-nickel)	~14	20 to 600 °C	14
Nichrome, ~90 Ni, 10 Cr	13	Y-alloy	22
87 Ni, 9 Cr, 1.5 Fe	12.5		
85 Ni, 10 Cr, 3 Fe	13		
Miscellaneous		**Miscellaneous** (*contd*)	
Alundum Al_2O_3 at 100 °C	~8	flint, lead, 46% PbO	~8
Building materials, etc.		borate	3–7
Brick	3–10	Pyrex	3
Cement and concrete	10–14	Hysil (Chance)	3.3
Granite	5–10	Porcelain, insulators	3–6
Limestone	4–10	Porcelain, refractory	2–5
Marble	3–15	Porcelain, laboratory	3–4
Portland stone	~3	Quartz, ‖ axis	8
Sandstone	5–12	Quartz, ⊥ axis	14
Slate	6–12	Silica, fused	0.4
Fluorspar, CaF_2	19	Silica, fused 0 to 1000 °C	0.5
Plastics	see p. 251	Woods, typical, along grain	~3–5
Glass†		Woods, typical, across grain	~35–60
soft soda 14% Na_2O	~8.5		
hard potash 20% K_2O	~10		

† Most glasses have coefficients ranging from 8 to $10 \times 10^{-6}/°C$ at ordinary ambient temperatures, the exceptions being those containing boron. See also p. 90.

Except in a few cases, the coefficient of linear expansion increases with the temperature. Over wider ranges of temperature, the variation of length with temperature of most metals and some alloys follows a parabolic or cubic law; however, many alloys expand irregularly, and in general the coefficient of expansion of an alloy cannot accurately be calculated from the known coefficients of its constituents.

P.A.

1.5.5 Specific heat capacities

Ratio of the principal specific heat capacities for gases and vapours

For specific heat capacities of gases and vapours see p. 232.

γ = the ratio of the specific heat capacity at constant pressure to that at constant volume. γ is usually determined directly by some method involving an adiabatic expansion, such as the determination of the velocity of sound in the gas. From a knowledge of either (1) the pressure or (2) the temperature immediately following an adiabatic expansion (Clément and Desormes', Lummer and Pringsheim's methods respectively), γ can be deduced from pv^{γ} = const., or $\theta v^{\gamma-1}$ = const. (See J. K. Roberts, *Heat and Thermodynamics*, 5th edn., London, Blackie, 1960.)

Gas	Temp./°C	γ	Gas	Temp./°C	γ
Monatomic gases			Nitrogen $\{N_2O_4$ . . .	20	1.172
Helium	0	1.63	peroxide $\{NO_2$	150	1.31
Argon.	0	1.667	H_2S	—	1.340
Neon	19	1.642	CS_2	—	1.239
Krypton	19	1.689	Sulphur dioxide . . . $\{$	16–34	1.26
Xenon	19	1.666		500	1.2
Mercury vapour	310	1.666			
			Polyatomic gases		
Diatomic gases			Methane, CH_4	—	1.313
Air (dry)	−79.3	1.405	Ethane, C_2H_6	—	1.22
Air (dry)	0–17	1.401/2	Propane, C_3H_8	—	1.130
Air (dry)	500	1.357	Acetylene, C_2H_2. . . .	—	1.26
Air (dry)	900	1.32	Ethylene, C_2H_4	—	1.264
Air (dry) (200 atm) . . $\{$	0	1.828	Benzene C_6H_6	20	1.40
	−79.3	2.333	Benzene C_6H_6	99.7	1.105
Hydrogen	4–17	1.407/8	Chloroform $CHCl_3$ . . $\{$	24–42	1.110
Nitrogen	20	1.401		99.8	1.150
Oxygen	5–14	1.400	CCl_4	—	1.130
Carbon monoxide . . .	1800	1.297	Me. alcohol	99.7	1.256
Nitric oxide	—	1.394	Me. bromide	—	1.274
			Me. chloride	19–30	1.279
Triatomic gases			Me. iodide	—	1.286
Ozone	—	1.29†	Et. alcohol	53	1.133
Water vapour	100	1.334	Et. alcohol	99.8	1.134
Carbon dioxide	4–11	1.300	Et. bromide	—	1.188
Carbon dioxide	300	1.22	Et. chloride	22.7	1.187
Carbon dioxide	500	1.20	Et. ether	12–20	1.024
Ammonia, NH_3	—	1.336	Et. ether	99.7	1.112
Nitrous oxide, N_2O . . .	—	1.324	Acetic acid	136.5	1.147

† Extrapolated.

Specific heat capacity of water

The International Committee of Weights and Measures, Paris, 1950, accepted W. J. de Haas's recommended values: $c_p(15\ ^\circ C) = 4.1855\ J\ g^{-1}\ ^\circ C^{-1}$ (agreeing with Birge's 1941 value), $c_p(t\ ^\circ C)$ then deduced from:

$$\frac{c_p(t\ ^\circ C)}{c_p(15\ ^\circ C)} = 0.996\ 185 + 0.000\ 287\ 4\left(\frac{t+100}{100}\right)^{5.26} + 0.011\ 160 \times 10^{-0.036t}$$

due to Osborne, Stimson and Ginnings.

In the following table the unit is $J\ g^{-1}\ °C^{-1}$.

°C	0	1	2	3	4	5	6	7	8	9
0	4.2174	4.2138	4.2104	4.2074	4.2045	4.2019	4.1996	4.1974	4.1954	4.1936
10	4.1919	4.1904	4.1890	4.1877	4.1866	4.1855	4.1846	4.1837	4.1829	4.1822
20	4.1816	4.1810	4.1805	4.1801	4.1797	4.1793	4.1790	4.1787	4.1785	4.1783
30	4.1782	4.1781	4.1780	4.1780	4.1779	4.1779	4.1780	4.1780	4.1781	4.1782
40	4.1783	4.1784	4.1786	4.1788	4.1789	4.1792	4.1794	4.1796	4.1799	4.1801
50	4.1804	4.1807	4.1811	4.1814	4.1817	4.1821	4.1825	4.1829	4.1833	4.1837
60	4.1841	4.1846	4.1850	4.1855	4.1860	4.1865	4.1871	4.1876	4.1882	4.1887
70	4.1893	4.1899	4.1905	4.1912	4.1918	4.1925	4.1932	4.1939	4.1946	4.1954
80	4.1961	4.1969	4.1977	4.1985	4.1994	4.2002	4.2011	4.2020	4.2029	4.2039
90	4.2048	4.2058	4.2068	4.2078	4.2089	4.2100	4.2111	4.2122	4.2133	4.2145

Specific heat capacities of various substances

In most cases, the values given must only be regarded as average values. Unit $= J\ g^{-1}\ °C^{-1}$.

Substance	Temp./°C	Sp. ht.
Alloys		
Brass, red	0	0.377
,, yellow . . .	0	0.368
Eureka (constantan) .	18	0.410
German silver . . .	0–100	0.398
Solder†		0.176
Steel, carbon . . .	50–100	0.48
,, stainless . . .	50–100	0.51
Liquids		
Alcohol, amyl . . .	18	2.30
,, ethyl . . .	0	2.29
,, ,, . . .	40	2.71
,, methyl . .	12	2.52
Aniline	15	2.15
Benzene	10	1.42
,, ...	40	1.77
Brine	−20	2.89
Brine‡	0	2.97
,, . . .	15	3.01
Ether, ethyl . . .	18	2.34
Glycerine . . .	18–50	2.43
Oil, castor . . .	20	2.13
Oil, linseed . . .	20	1.84
,, olive . . .	7	1.97
,, paraffin . .	20–60	2.13–2.26
,, rape . . .	20	2.04
,, sperm . . .	20	2.06
Sea-water . . .	17	3.93
Toluene	18	1.67
Turpentine . . .	18	1.76

Substance	Temp./°C	Sp. ht.
Miscellaneous		
Air (dry), C_p . . .	20	1.006
,, ...	100	1.011
,, ...	500	1.092
,, ...	1000	1.192
,, ...	−100	1.008
,, (100 atm) . . .	−80	1.902
Asbestos	20–100	0.84
Basalt	20–100	0.84–1.00
Ebonite	20–100	1.38
Fluorspar, CaF_2 . .	30	0.88
Glass, crown . . .	10–50	0.67
,, flint . . .	10–50	0.50
,, Pyrex . . .	26	0.78
Granite	20–100	0.80–0.84
Ice	−250	0.15
,, ...	−160	1.0
,, ...	−21 to −1	2.0–2.1
Indiarubber . . .	15–100	1.13–2.01
Marble, white . . .	18	0.88–0.92
NaCl	−248	0.0414
,, ...	−38	0.825
,, ...	+10	0.88
KCl	−250	0.0653
,, ...	−187	0.490
,, ...	277	0.741
Paraffin wax . . .	0–20	2.9
Porcelain	15–1000	1.07
,, . . .	15–200	0.75
Quartz, SiO_2 . . .	0	0.73
Quartz	350	1.17
Sand	20–100	0.80
Silica (fused) . . .	15–200	0.84
,, ,, . . .	15–800	1.04

† Sp. ht. $= 0.1766 + 0.000\ 159t$; Sn 54%, Pb 46%.

‡ Density 1.2.

Specific heat capacity C_p of aluminium oxide

(Synthetic sapphire used as a standard)

Temp./K	100	150	200	250	273.16	300	350	400
$C_p/(\text{J kg}^{-1}\text{ K}^{-1})$	125.9	313.7	501.6	657.2	717.5	779	872	943
Temp./K	450	500	600	700	800	900	1000	1100
$C_p/(\text{J kg}^{-1}\text{ K}^{-1})$	997	1040	1103	1148	1180	1204	1223	1238

See Furukawa *et al.*, 'Thermal properties of Aluminium oxide from 0 °K to 1200 °K', *J. Res. Nat. Bur. Stand.*, 1956, **57**, 67.

M.J.H.

1.5.6 Thermal conductivities

Definition and units

The thermal conductivity, k, of a substance may be defined as the quantity of heat transmitted, due to unit temperature gradient, in unit time under steady conditions in a direction normal to a surface of unit area, when the heat transfer is dependent only on the temperature gradient. A related quantity, thermal diffusivity $= k/(\text{density} \times \text{specific heat capacity})$.

The values of k given in this section are expressed in the SI unit $\text{W m}^{-1}\text{ K}^{-1}$. Factors for converting to other units are as follows:

$$
\begin{aligned}
1 \text{ W m}^{-1}\text{ K}^{-1} &= 0.01 \text{ J cm cm}^{-2}\text{ s}^{-1}\text{ K}^{-1} \\
&= 0.002\,388 \text{ cal cm cm}^{-2}\text{ s}^{-1}\text{ K}^{-1} \longleftarrow \\
&= 0.859\,8 \text{ kcal m m}^{-2}\text{ h}^{-1}\text{ K}^{-1} \\
&= 0.001\,926 \text{ Btu in ft}^{-2}\text{ s}^{-1}\text{ °F}^{-1} \\
&= 6.933 \text{ Btu in ft}^{-2}\text{ h}^{-1}\text{ °F}^{-1} \\
&= 0.577\,8 \text{ Btu ft ft}^{-2}\text{ h}^{-1}\text{ °F}^{-1} \\
&= 0.000\,160\,5 \text{ Chu ft ft}^{-2}\text{ s}^{-1}\text{ K}^{-1}
\end{aligned}
$$

The above factors are calculated using 1 cal $= 4.1868$ J, 1 Btu $= 1055.06$ J and 1 in $= 2.54$ cm. Thermal conductivity determinations have seldom been of sufficient accuracy to justify differentiation between the various Internationally adopted temperature scales, calories and other thermal units, when converting to the units now used.

More extensive collections of thermal conductivity data will be found in three volumes of the series *Thermophysical Properties of Matter*, IFI/Plenum Data Corp., New York, Washington, 1970: Vol. 1, *Thermal Conductivity: Metallic Elements and Alloys*, by Y. S. Touloukian, R. W. Powell, C. Y. Ho and P. G. Klemens; Vol. 2, *Thermal Conductivity: Nonmetallic Solids*, by Y. S. Touloukian, R. W. Powell, C. Y. Ho and P. G. Klemens; and Vol. 3, *Thermal Conductivity: Nonmetallic Liquids and Gases*, by Y. S. Touloukian, P. E. Liley and S. C. Saxena.

Thermal conductivities of metals and alloys

Many metals satisfy the relation $k = 2.443 \times 10^{-8}\,\sigma T$, where T is the temperature (K) and σ the electrical conductivity (S m^{-1}). This relationship, which holds best for good electrical conductors towards absolute zero and above the Debye temperature, can be used for the estimation of thermal conductivity.

For many alloys, near room temperature and above, the relation $k = L\sigma T + C$ can be used for the estimation of values of k to within about 6%. L and C have the following values:

Main constituent metal	L $(10^{-8}\,\mathrm{W\,S^{-1}\,K^{-2}})$	C $(\mathrm{W\,m^{-1}\,K^{-1}})$
Aluminium	2.22	10.5
Copper	2.39	7.5
Alpha-iron	2.43	9.2
Gamma-iron.	2.39	4.2
Magnesium	2.21	9.6
Nickel	2.13	8.4
Nickel–chromium (nimonic type)	2.20	6.0
Titanium	2.30	2.9
Zirconium	2.50	2.2

Thermal conductivities of metallic elements

$$k/(\mathrm{W\,m^{-1}\,K^{-1}})$$

Metal	Temperature/K					Metal	Temperature/K				
	173.2	273.2	373.2	573.2	973.2		173.2	273.2	373.2	573.2	973.2
Aluminium . .	241	236	240	233	92	Niobium . .	53	53	55	58	64
Antimony . .	33	25.5	22	19	27	Osmium . .	93	88	87	87	—
Beryllium. . .	367	218	168	129	93	Palladium. .	72	72	73	79	93
Bismuth . . .	11	8.2	7.2	13	17	Platinum . .	73	72	72	73	78
Cadmium . . .	100	97	95	89	45	Plutonium. .	4	6	8	—	—
Caesium . . .	37	36	20	20.6	17.7	Potassium . .	105	104	53	45	32
Cerium . . .	8	11	13	16	—	Praesodymium	9.9	12	13.4	—	—
Chromium . .	120	96.5	92	82	66	Rhenium . .	52	49	47	44	45
Cobalt . . .	130	105	89	69	53	Rhodium . .	156	151	147	137	—
Copper . . .	420	403	395	381	354	Rubidium. . .	59	58	32	29	22
Dysprosium . .	9	10.5	—	—	—	Ruthenium . .	123	117	115	108	98
Erbium . . .	14	15	—	—	—	Samarium. . .	10	13	13	14	—
Gadolinium . .	12	10	—	—	—	Scandium . .	15	16	—	—	—
Gallium† . . .	43	41	33	45	—	Silver . . .	432	428	422	407	377
Gold	324	319	313	299	272	Sodium . . .	141	142	88	78	60
Hafnium . . .	25	23	22	21	21	Tantalum . . .	58	57	58	58.5	60
Holmium. . . .	14	16	17	—	—	Technetium . .	—	51	50	50	—
Indium . . .	92	84	76	42	—	Terbium . . .	11	10.5	—	—	—
Iridium . . .	156	147	145	139	—	Thallium . . .	51	47	44	—	—
Iron	99	83.5	72	56	34	Thorium . . .	55	54	54	56	58
Lanthanum . .	12	13	14.5	—	—	Thulium . . .	16	17	—	—	—
Lead	37	36	34	32	21	Tin	76	68	63	32	40
Lithium . . .	94	86	82	47	59	Titanium . . .	26	22	21	19	21
Lutetium. . . .	18	17	—	—	—	Tungsten . . .	188	177	163	139	119
Magnesium . .	160	157	154	150	—	Uranium . . .	24	27	29	33	43
Manganese . .	7	8	—	—	—	Vanadium . . .	32	31	31	33	38
Mercury . . .	29.5	7.8	9.4	11.7	—	Yttrium . . .	16.5	17	—	—	—
Molybdenum . .	145	139	135	127	113	Zinc	117	117	112	104	66
Nickel . . .	113	94	83	67	71	Zirconium. . .	26	23	22	21	23

† Values for the a-axis which approximate to the polycrystal; those for the b-axis are 91 and 88 and for the c-axis are 16.5 and 16 at 173.2 and 273.2 K.

Thermal conductivities of metallic elements

Values of the thermal conductivity given in the preceding table are for the purest polycrystalline condition for which reliable measurements have been reported.

Entries in italics relate to the liquid phase.

Thermal conductivity invariably decreases with decrease in purity. Whereas such dependence is relatively weak for the listed temperature range it becomes very strong at lower temperatures. Below 173.2 K, the thermal conductivity of each sample will tend to differ inversely as its residual electrical resistivity, ρ_0. The Lorenz function, $k/\sigma T$, is also strongly temperature dependent, but often tends towards the theoretical value in the liquid helium range, so here the thermal conductivity can often be derived from the relation $k = 2.443 \times 10^{-8} \, T \rho_0^{-1}$.

Thermal conductivities of elements which are semi-conductors or insulators

Substance	$k/(\mathrm{W\ m^{-1}\ K^{-1}})$ Temperature/K				
	173.2	273.2	373.2	573.2	973.2
Boron	72	32	19	11	10
Carbon:					
Amorphous	1.1	1.5	1.8	2.2	2.5
Diamond	17–49	10–26	7–17	—	—
Graphite	70–220	80–230	75–195	50–130	35–70
Pyrolytic graphite:					
Parallel to planes	3870	2130	1510	936	549
Normal to planes	10.8	6.4	4.4	2.8	1.6
Germanium	113	67	46.5	29	17.5
Iodine	—	0.5	0.4	*0.09*	—
Phosphorus:					
Black	20	13	—	—	—
White (or yellow)	—	0.25	*0.18*	*0.16*	—
Selenium:					
Amorphous	0.23	0.43	—	—	—
Crystalline:					
Parallel to c-axis	6.8	4.8	4.8	—	—
Normal to c-axis	2.0	1.4	1.4	—	—
Silicon	330	168	108	65	32
Sulphur:					
Amorphous	0.18	0.20	—	*0.17*	—
Polycrystalline	0.39	0.29	0.15	*0.17*	—
Tellurium:					
Parallel to c-axis	5.1	3.6	2.9	2.4	*6.3*
Normal to c-axis	2.9	2.1	1.7	1.5	*6.3*

Thermal conductivities of single crystals of some non-cubic metals at normal temperature

	$k/(\text{W m}^{-1}\,\text{K}^{-1})$		
Metal	*Thermal conductivity in direction of*		
	c-axis	*a-axis*	*b-axis*
Bismuth	5.4	9.3	
Cadmium	83.05	104	
Dysprosium	11.65	10.25	
Erbium	18.4	12.6	
Gadolinium	10.7	10.3	
Gallium	16.0	40.8	88.3
Holmium	22.1	13.6	
Lutetium	23.3	13.8	
Mercury (at 227.7 K)	33.0	25.9	
Terbium	14.5	9.45	
Thulium	24.2	14.1	
Tin	51.8	74.0	

Thermal conductivities of alloys

	$k/(\text{W m}^{-1}\,\text{K}^{-1})$															
Alloy	*Composition*											*Temperature/K*				
	Al	C	Cr	Cu	Fe	Mg	Mn	Ni	Si	W	Zn	173.2	273.2	373.2	573.2	973.2
Alpax gamma† . . .	87	—	—	—	0.3	0.3	0.3	—	12.0	—	—	—	188	188	184	—
Alumel	2	—	—	—	—	—	2.0	95.0	1.0	—	—	—	30	32	35	—
Brass	—	—	—	70	—	—	—	—	—	—	30	89	106	128	146	—
Bronze	—	—	—	90‡	—	—	—	—	—	—	—	—	53	60	80	—
Chromel P	—	—	10	—	—	—	—	90	—	—	—	—	—	19	23	—
Constantan	—	—	—	60	—	—	—	40	—	—	—	19	22	24	27	—
German silver	—	—	—	62	—	—	—	15	—	—	22	20	23	29	45	—
Lo-Ex†	85	—	—	1	0.5	0.9	—	1	11.8	—	—	—	172	175	173	—
Manganin	—	—	—	84	—	—	12	4	—	—	—	16	21	26	—	—
Monel	—	0.2	—	29.2	1.7	0.1	1.0	67.1	—	—	—	19	21	24	30	43
Nichrome	—	0.1	21	—	0.6	—	0.65	77.3	0.4	—	—	11	13	14	17	21
RR 59†	93	—	—	2.3	1.2	1.5	—	1.2	0.9	—	—	—	168	176	186	—
RR 77†	89	—	—	2.2	0.3	2.5	0.5	—	0.3	—	5	—	161	171	178	—
Steel, carbon	—	0.8	0.1	—	98.4	—	0.3	0.1	0.1	—	—	48	50	48.5	41.5	30.5
,, 18/8	—	0.15	17.9	—	73.5	—	0.3	8.0	0.2	—	—	12	15	16.5	19	24.5
,, Era ATV	—	0.5	15.2	—	52.2	—	1.2	26.9	1.3	2.8	—	9	11	12.5	15.5	21.5
,, Ni–Cr	—	0.4	0.8	—	94.4	—	0.6	3.6	0.2	—	—	31	33	36	37.5	28
,, silicon	—	0.5	—	0.6	95.7	—	0.9	0.15	2.0	—	—	—	25	28.5	31	28
,, stainless	—	0.3	13.7	—	85.1	—	0.3	0.4	0.3	—	—	—	24.5	25	25.5	24.5
Y-alloy†	92	—	—	3.8	0.4	1.3	—	1.8	0.4	—	—	—	180	188	194	—
Platinum 90%, iridium 10%												31				
Platinum 90%, rhodium 10%												38				
Platinum 60%, rhodium 40%												46		51	58	69
Titanium 92.5%, aluminium 5%, tin 2.5%												7		8.3	10.5	
Titanium 96%, aluminium 2%, manganese 2% . . .												9.3		10.5	10.7	
Zirconium 93.1%, tin 6.7%, carbon 0.1%														8.7	12	
Zirconium 97.5%, tin 2.3%, carbon 0.1%														11.3	13	

† After heat treatment at 573.2 K followed by air cooling.
‡ Remaining 10% tin.

Thermal conductivities of refractory materials

Since composition and density may vary, values are mainly typical.

Material	$k/(\text{W m}^{-1}\text{ K}^{-1})$			
	273.2	373.2	873.2	1473.2
Alumina (Al_2O_3)	40	28	9.2	5.7
Beryllia (BeO)	300	213	61	22
Brick, fireclay	—	—	1.1	1.3
Magnesia (MgO)	53	38.6	11.8	6.6
Mica:				
Indian Muscovite, normal to layers	—	0.6	0.6	—
Phlogopite normal to layers	—	0.5	0.3	—
Quartz (SiO_2 single crystal):				
along c-axis	11.6	8.25	4.1	—
normal to c-axis	6.85	5.0	3.0	—
Silica, fused (SiO_2)	1.33	1.48	2.4	—
Thoria (ThO_2)	14	10.8	4.2	2.8
Titania (TiO_2)	8.9	7.4	3.7	3.2
Titanium carbide (TiC)	—	32	37	40
Titanium diboride (TiB_2)	—	70	63	—
Titanium nitride (TiN)	—	25	27	—
Urania (UO_2)	—	7.5	4.0	2.8
Zirconia (ZrO_2)	—	1.8	2.0	2.2
Zirconium carbide (ZrC)	—	—	33	37
Zirconium diboride (ZrB_2)	—	73	66	—

Thermal conductivities of miscellaneous solids

The values below are for normal temperature, except where stated (K) and should be regarded as average values for the type of material specified. Values for the commoner polymers will be found on page 251.

Substance	k $\overline{\text{W m}^{-1}\text{ K}^{-1}}$	Substance	k $\overline{\text{W m}^{-1}\text{ K}^{-1}}$
Asbestos cloth	0.125	Cotton wool	0.03
,, insulating board	0.11	Cork, baked slab	0.038–0.046
,, paper	0.104	,, granular	0.04
,, wool	0.055	Diatomaceous powder	0.07
Beeswax	0.25	Ebonite, solid	0.17
Bitumen	0.17	,, cellular	0.03
Brickwall	0.8–1.2	Felt	0.04
Cardboard	0.21	Fibreboard, insulating	0.055
Charcoal	0.2	,, hardboard	0.125
Coal (1400)	0.2	Glass, borosilicate crown	1.1
Concrete, ballast 1:2:4	1.5	,, cloth	0.07
,, cellular	0.1–0.2	,, double extra dense flint . . .	0.55

Thermal conductivities of miscellaneous solids (*contd*)

Substance	k / W m^{-1} K^{-1}	Substance	k / W m^{-1} K^{-1}
Glass, light flint	0.85	Plywood	0.125
,, Pyrex	1.1	Porcelain	1.5
,, wool	0.037	Pyrophyllite, normal to plane . .	2.0
Ice (268)	2.3	Rubber, cellular	0.045
,, (173)	3.9	,, natural	0.15
Kapok	0.035	,, silicone	0.25–0.4
Mineral wool	0.04	Sand, silver	0.3–0.4
Paper	0.06	Silica aerogel powder	0.024
Paraffin wax	0.25	Timber, ordinary	0.14–0.17
Plaster	0.13	,, balsa	0.055
Plastics, solid (see p. 251) . . .		Vermiculite granules	0.065
Polystyrene, cellular	0.035	Wool	0.05

Thermal conductivities of liquids

Values below are for the thermodynamic temperature (K) shown in brackets. Linear relationships mostly hold for range covered. Liquid metal values are given on p. 57.

Liquid	$k/(\text{W m}^{-1}\text{ K}^{-1})$	Liquid	$k/(\text{W m}^{-1}\text{ K}^{-1})$
Acetone	0.198 (193), 0.146 (333)	Glycerine . . .	0.286 (273), 0.292 (333)
Aniline	0.172 (293)	Medicinal paraffin	0.127 (273), 0.125 (423)
Benzene	0.147 (293), 0.137 (323)	Methyl alcohol .	0.223 (233), 0.186 (333)
n-Butyl alcohol . .	0.167 (213), 0.106 (353)	Oil, cylinder . .	0.152 (293), 0.142 (473)
Carbon tetrachloride	0.115 (253), 0.102 (333)	, transformer .	0.136 (273), 0.127 (373)
Dichlorodifluoro-		n-Propyl alcohol .	0.168 (233), 0.148 (353)
methane† . . .	0.09 (253), 0.073 (293)	Toluene . . .	0.159 (193), 0.119 (353)
Ethyl alcohol . . .	0.189 (233), 0.150 (353)	Water	0.561 (273), 0.673 (353)
Ethyl benzene . .	0.152 (193), 0.117 (353)		0.686 (378–433), 0.598 (542)
Ethyl glycol . . .	0.252 (273), 0.264 (373)	Xenon	0.07 (173), 0.05 (223)

† R12 (CF$_2$Cl$_2$).

Thermal conductivities of gases

The thermal conductivity of a gas is independent of pressure at normal pressures. It increases at high pressures and decreases at low pressures, e.g. for air below about 1 mmHg. Values are given for a pressure of 1 atm.

	$k/(10^{-1}\ \mathrm{W\ m^{-1}\ K^{-1}})$									
Gas	Temperature/K					Gas	Temperature/K			
	73.2	173.2	273.2	373.2	1273.2		173.2	273.2	373.2	1273.2
Argon . .	—	1.09	1.63	2.12	5.0	Air	1.58	2.41	3.17	7.6
Bromine . .	—	—	0.4	0.6	—	Ammonia . .	—	2.18	3.38	—
Chlorine .	—	—	0.79	1.15	—	Carbon dioxide	—	1.45	2.23	7.9
Fluorine . .	—	1.56	2.54	3.47	—	Carbon monoxide	1.51	2.32	3.04	—
Helium . .	5.95	10.45	14.22	17.77	41.9	Ethane . . .	—	1.80	—	—
Hydrogen .	5.09	11.24	16.82	21.18	—	Ethylene . .	—	1.64	—	—
Krypton . .	—	0.57	0.87	1.15	2.9	R12 (CF_2Cl_2) .	—	0.85	1.35	—
Neon . .	1.74	3.37	4.65	5.66	12.8	Hydrogen sulphide	—	1.2	—	—
Nitrogen .	—	1.59	2.40	3.09	7.4	Methane . .	1.88	3.02	—	—
Oxygen . .	—	1.59	2.45	3.23	8.6	Nitric oxide .	1.54	2.38	—	—
Radon . .	—	—	0.33	0.45	—	Nitrous oxide .	—	1.51	—	—
Xenon . .	—	0.34	0.52	0.70	1.9	Sulphur dioxide	—	0.77	—	—
						Water vapour .	—	1.58	2.35	—

R.W.P.

1.6 Acoustics

1.6.1 The velocity and attenuation of sound

Gases and vapours

The attenuation of plane sound waves in neper/unit length is $\alpha = (1/2d)\log_e(I_0/I_d)$, where the initial intensity I_0 has decreased to I_d after traversing distance d. When expressed in decibels per unit length, convenient for practical application, the value is 8.686α. In normal monatomic gases α varies classically as f^2, f being the frequency, but in polyatomic gases relaxation phenomena may dominate over classical absorption (see Cottrell and McCoubrey, *Molecular Energy Transfer in Gases*, 1961 (Butterworths). For gases at moderate pressure the velocity of sound $c = \sqrt{(\gamma p/\rho)}$ or alternatively $c = \sqrt{(\gamma RT/M)}$, p being the ambient gas pressure, γ the ratio of the specific heat at constant pressure to that at constant volume, R the gas constant, T the absolute temperature and M the molecular weight. Thus, to first order, c is proportional to $\sqrt{T}$ and is independent of gas pressure; at higher pressures, however, the value is changed due to contributions from the second and higher virial coefficients. The velocity of sound waves bounded by walls or tubes is less than the free-space value. Velocity dispersion is observed for sound propagation in many polyatomic gases due to molecular relaxation: at higher frequencies the rotational and vibrational degrees of freedom are not excited, the specific heats are modified and the velocity of sound is increased. The table below gives experimentally determined values of sound velocity in gases and vapours, selected from the literature. Where known, both the low-frequency and high-frequency limiting values of velocity are given, the latter being in brackets.

Velocity of sound in gases and vapours

Gas	$t/°C$	$c/(\text{m s}^{-1})$	Gas	$t/°C$	$c/(\text{m s}^{-1})$
Air	\multicolumn{2}{c}{(see text)}		Hydrogen	0	1286
Acetaldehyde	0	244.5	Hydrogen bromide . .	0	200
Acetylene.	0	329	Hydrogen chloride . . .	0	296
Ammonia	0	415	Hydrogen iodide	0	157
Argon.	0	307.8	Hydrogen sulphide . . .	0	310
Benzene	0	177	Iodine	180	138 (144)
Bromine	58	149 (155)	Krypton	0	213
Carbon dioxide	0	259 (268)	Methane	0	430
Carbon disulphide . . .	0	192	Methanol	97	335
Carbon monoxide . . .	0	337	Neon	0	434
Carbon tetrachloride . .	97	145	Nitric oxide	16	334
Chlorine	20	219	Nitrogen	0	337
Cyclohexane	30	181 (200)	Nitrous oxide	25	268 (281)
Deuterium	0	890	Oxygen	30	332
Ethane	31	316 (335)	Propane	0	238
Ethyl alcohol.	53	258	Sulphur dioxide	0	211
Ethylene	0	318	Sulphur hexafluoride. . .	11	133 (147)
Ethyl ether	97	206	Thiophene	0	195 (215)
Fluorine	100	332 (339)	Water	100	405
Freon 113	53	124 (139)	Water (heavy)	100	451
Helium	0	971.9	Xenon.	0	170

Attenuation of sound in air

The attenuation of sound in air at 20 °C due to viscous and thermal loss mechanisms is $1.61 \times 10^{-10} f^2$ dB m^{-1} (see Sivian, *J. Acoust. Soc. Am.*, 1947, **19**, 914). However, losses due to vibrational relaxation of oxygen molecules are generally much greater than those due to the classical processes and the attenuation of sound varies significantly with temperature, water-vapour content and frequency. The table gives values of attenuation in dB km^{-1} for a temperature of 20 °C and a pressure of 1 atmosphere; these are arithmetic mean values derived from the normalized data of Evans and Bazley (1956) and Harris (1966)—see Delany and Bazley, *NPL Report*, Ac. 45, 1970. For the variation of attenuation with ambient pressure see Harris, *J. Acoust. Soc. Am.*, 1967, **43**, 530.

Attenuation of sound in air (dB km^{-1})

f/kHz	\multicolumn{10}{c}{Relative humidity (%)}
	0	10	20	30	40	50	60	70	80	90
1	0	12	6	4	4	4	4	4	3	3
1.25	0	17	8	6	5	5	5	5	5	4
1.6	0	27	13	9	7	7	6	6	6	6
2	1	40	19	13	10	9	9	8	8	8
2.5	1	58	29	19	15	13	12	11	10	10
3.15	2	82	44	28	21	18	16	15	14	14
4	3	110	68	44	32	27	23	21	20	19
5	4	140	99	66	49	39	34	30	28	26
6.3	6	170	150	100	75	59	50	44	40	38
8	10	200	210	150	120	92	77	66	60	55
10	16	220	280	220	170	140	120	99	88	80
12.5	25	250	360	320	250	210	170	150	130	120
16	41	280	460	450	380	320	270	230	210	190
20	64	320	550	600	540	460	400	350	310	280
25	100	370	650	760	740	660	580	520	470	420

Velocity of sound in air

At low frequencies the velocity of sound c_0 in dry air containing 0.03% CO_2 by volume is 331.46 m s^{-1}; due to dispersion this velocity is 0.09 m s^{-1} higher at frequencies above that corresponding to maximum molecular attenuation for given humidity; see Hardy, Telfair and Pielmeyer (1941) and Smith, *J. Acoust. Soc. Am.*, 1953, **25**, 81. The effect of water-vapour content on sound velocity is complex. As moisture is added to dry air the velocity at first decreases and is minimum at 14% relative humidity at $0.999\,50c_0$; above 30% relative humidity the velocity increases linearly with increasing moisture content and at 100% relative humidity is $1.003\,35c_0$. See Harris, *J. Acoust. Soc. Am.*, 1971, **49**, 890.

Velocity and attenuation of sound in liquids

The velocity of dilatational waves in unbounded fluids is $c = (\beta_a \rho)^{-1/2}$, β_a being the adiabatic compressibility and ρ the density. In normal fluids the classical attenuation is proportional to f^2, f being the frequency, but relaxation phenomena may dominate; the attenuation generally decreases with increasing temperature. Values for various liquids are given below, selected from numerous sources; f is in Hz and α is in neper m^{-1}. The frequency at which the attenuation was measured is given in brackets, in MHz.

Velocity and attenuation of sound in liquids

Liquid	t °C	c m s^{-1}	dc/dt m s^{-1} K^{-1}	(α/f^2) 10^{-15} neper m^{-1} Hz^{-2}
Acetic acid	20	1173	—	—
Acetic anhydride	24	1384	—	—
Acetone	20	1190	−4.3	54 (9)
Acetylene tetrabromide	28	1007	—	—
Acetylene tetrachloride	28	1155	—	—
Amyl acetate	29	1173	—	74 (15)
n-Amyl alcohol	29	1224	−3.7	106 (15)
iso-Amyl alcohol	20	1252	—	—
iso-Amyl ether	26	1153	—	—
Aniline	20	1656	−4.0	—
Argon	−186	836	−6.9	10 (44)
Benzene	20	1320	−5.0	840 (200)
Benzophenone	100	1316	—	—
Bismuth	280	1650	—	8 (15)
Blood (horse)	37	1571	—	—
Bromobenzene	50	1074	—	184 (9)
Bromoform	25	908	−2.2	230 (10)
Butyl acetate	30	1172	−3.2	58 (20)
n-Butyl alcohol	20	1258	—	77 (22)
iso-Butyl bromide	−104	1450	—	130 (15)
Cadmium	360	2150	—	14 (20)
Caesium	130	967	—	—
Carbon disulphide	25	1142	—	—
Carbon tetrachloride	20	940	−3.0	540 (100)
Chlorine	20	850	−3.8	—
m-Chlornitrobenzene	40	1386	—	—
Chlorobenzene	20	1290	−4.3	130 (20)
Chloroform	20	990	−3.3	400 (100)
Copper	1350	3350	−0.46	—
Cyclohexane	20	1275	−5.4	330 (9)
Cyclohexanol	30	1622	—	440 (4)

Velocity and attenuation of sound in liquids (*contd*)

Liquid	t °C	c m s^{-1}	dc/dt m s^{-1} K^{-1}	(α/f^2) 10^{-15} neper m^{-1} Hz^{-2}
1-Decene	20	1250	—	—
Diethyl ether	25	985	−4.7	140 (10)
Diethylene glycol	20	1601	−2.5	306 (30)
Diethylene glycol monoethyl ether	30	1296	−3.1	—
Dimethyl polysiloxane	20	912	−3.9	—
Diphenyl ether	30	1462	—	—
Ethanol amide	25	1724	−3.4	—
Ethyl acetate	30	1133	−3.9	52 (20)
Ethyl alcohol	20	1162	−3.6	52 (10)
Ethyl bromide	10	932	−3.4	62 (15)
Ethyl iodide	20	876	−2.7	40 (15)
Ethylene dibromide	24	1014	—	—
Ethylene dichloride	23	1240	—	—
Ethylene glycol monomethyl ether	30	1339	−3.4	—
Eugenol	30	1482	—	113 (14)
Formaldehyde	25	1587	−1.0	—
Formamide	20	1680	−2.5	57 (10)
Formic acid	20	1360	−3.5	1340 (10)
Gallium	50	2740	—	—
Glycerol	20	1860	−1.8	2000 (4)
Helium 1	−269	211	—	230 (15)
Heptane	20	1160	−4.5	80 (15)
l-Heptane	20	1128	—	—
Heptyl alcohol	20	1343	—	—
Heptyne	30	1159	—	—
n-Hexane	30	1060	—	54 (32)
n-Hexyl alcohol	20	1331	—	—
Hydrogen	−258	1242	—	6 (44)
Indium	260	2215	—	—
Kerosene	25	1315	−3.6	110 (10)
Lead	340	1800	—	9.4 (20)
Menthol	50	1271	—	—
Mercury	20	1454	−0.61	5.5 (100)
Methyl acetate	30	1131	−3.7	60 (20)
Methyl alcohol	20	1121	−3.5	43 (5)
Methyl bromide	2	905	—	304 (15)
Methyl iodide	30	815	−2.6	334 (15)
Methylene bromide	24	971	—	570 (30)
Methylene chloride	25	1070	—	1100 (30)
Milk (evaporated)	20	1558	—	—
Nitrobenzene	24	1462	−3.7	79 (15)
Nitrogen	−189	745	−10.6	79 (15)
Nonane	20	1248	—	—
1-Nonene	20	1218	—	—
Oil (camphor)	25	1390	−3.8	—
Oil (castor)	19	1500	−4.1	10 900 (3)
Oil (condenser)	20	1432	−3.5	—
Oil (olive)	22	1440	−2.8	1350 (3)
Oil (sperm)	32	1411	—	—
Octane	20	1197	—	—
1-Octane	20	1184	—	—
Oxygen	−186	950	−6.9	8.6 (44)

Velocity and attenuation of sound in liquids (*contd*)

Liquid	t	c	dc/dt	(α/f^2)
	°C	m s^{-1}	m s^{-1} K^{-1}	10^{-15} neper m^{-1} Hz^{-2}
Paraldehyde	28	1197	—	—
l-Pentadecene	20	1351	—	—
n-Pentane	20	1044	−4.2	—
iso-Pentanol	21	1255	—	—
Phenol	100	1274	—	—
Polymethyl siloxane	20	1019	−3.0	—
Potassium	150	2500	−0.8	14 (20)
n-Propyl acetate	26	1182	—	60 (20)
n-Propyl alcohol	20	1223	−3.7	63 (23)
Pyridine	20	1440	−4.1	432 (24)
Rubidium	160	1260	—	—
Selenium	350	990	−0.75	—
Silicon tetrachloride	30	766	—	—
Silver	1150	2630	−0.41	—
Sodium	150	2500	−0.8	13.5 (20)
Sodium chloride (fused)	850	1991	−0.54	—
Sulphur	130	1332	−1.1	120 (10)
Tetralin	20	1484	—	—
Thiophene	20	1296	—	—
Tin	240	2470	−0.3	5.6 (20)
Toluene	20	1320	−4.3	80 (200)
o-Toluidine	22	1669	—	—
1-Tridecene	20	1313	—	—
Triethylamine	20	1140	—	586 (8)
Turpentine	25	1225	—	150 (10)
Water (distilled)	see text			
Water (heavy)	19.8	1383.4	+3.0	—
Water (sea)	see text			
o-Xylene	22	1352	—	52 (10)
Zinc	450	2700	—	3.7 (20)

Velocity and attenuation in distilled water

Distilled water is non-dispersive but the temperature dependence of sound velocity is anomalous. Values are given in the table for the temperature range 0–100 °C and a pressure of 1 atmosphere; the maximum velocity occurs at a temperature of 74.2 °C. The pressure coefficient of velocity is +0.156 m s^{-1} atm^{-1} at 20 °C. See Greenspan and Tschiegg (1957), Brooks (1960), Wilson (1959) and Del Grosso, *J. Acoust. Soc. Am.*, 1970, **47**, 946. The attenuation of sound in distilled water is proportional to f^2 for the range 3–70 MHz but relatively greater at lower frequencies. The value of α/f^2 decreases with temperature as shown in the table below, in which the frequency is taken to be in Hz and α in neper m^{-1}. See Pinkerton, *Nature*, 1947, **160**, 128; also Tait (1957) and Schulkin and Marsh (1963).

Velocity and attenuation of sound in distilled water

$t/$°C	0	10	20	30	40	50	60	70	80	90	100
$c/$(m s^{-1})	1402.4	1447.2	1482.3	1509.1	1528.8	1542.5	1551.0	1554.8	1554.7	1550.2	1542.2
$\dfrac{(\alpha/f^2)}{(10^{-15}\text{ neper m}^{-1}\text{ Hz}^{-2})}$	57	36	25	19	15	12	10.5	9.7	8.0	7.2	—

Velocity and attenuation in sea water

The velocity of sound in sea water can be expressed in the form of a power series involving temperature t, pressure p and salinity S (their relative importance being in that order), valid for the range $-3 < t < 30$ °C, $10^4 < p < 10^7$ kgf m^{-2}, $3.1 < S < 3.7\%$. See Frye and Pugh, *J. Acoust. Soc. Am.*, 1971, **50**, 384. Values in the table relate to surface depth ($p = 10^4$ kgf m^{-2}) and a salinity of 3.5%. Data for the attenuation of sound in sea water as a function of frequency are based on Sheehy (1950) and Schulkin and Marsh, *J. Acoust. Soc. Am.*, 1962, **34**, 864 (with correction in *J. Acoust. Soc. Am.*, 1963, **35**, 739) and relate to surface depth and a salinity of 3.5%. It should be noted that the attenuation decreases with increasing pressure.

Velocity of sound in sea water

$t/$°C	-4	0	5	10	15	20	25	30
$c/(\text{m s}^{-1})$	1430.2	1449.5	1471.1	1490.2	1507.1	1521.9	1534.7	1545.9

Attenuation of sound in sea water

$t/$°C	Frequency/kHz					
	1	2	5	10	20	50
0	0.144	0.57	3.6	14.0	52	216
10	0.091	0.37	2.3	9.0	35	180
20	0.060	0.24	1.5	6.0	24	134
30	0.040	0.16	1.0	4.0	16	96

Velocity and attenuation in solids

In isotropic solids, both shear (transverse) and longitudinal waves can be propagated. The velocity of shear waves in an extensive medium is $c_S = \sqrt{(G\rho)} = \sqrt{\{E/2\rho(1+\sigma)\}}$, E being Young's modulus, G the rigidity modulus and σ Poisson's ratio, and this is also the velocity of torsional waves in thin cylindrical bars. The velocity of longitudinal or irrotational waves in an extensive medium is $c_L = \sqrt{\{(K + \frac{4}{3}G)/\rho\}} = \sqrt{\{E(1-\sigma)/\rho(1-2\sigma)(1+\sigma)\}}$, K being the bulk modulus. In straight uniform bars and in tubes thin compared with a wavelength, the velocity of longitudinal waves is $c_E = \sqrt{(E/\rho)}$. Surface waves propagating along the surface of an extensive solid are generally known as Rayleigh waves and propagate with velocity $c_{SR} = ac_S$, where a is the least positive root of the equation

$$\frac{a^6}{8(1-a^2)} + a^2 = \frac{1}{1-\sigma}$$

See Bradfield, *NPL Notes on Applied Science No. 30*, 1964 (HMSO).

Variation of a with Poisson's ratio

σ	0.20	0.25	0.30	0.35	0.40
a	0.9110	0.9194	0.9274	0.9350	0.9422

In anisotropic solids, which may have as many as 21 independent elastic constants, there may exist, for a given direction of the wave normal, three distinct displacement vectors each associated with a distinct plane wave velocity. Of the three waves, one is analogous to the longitudinal and the others to transverse waves in the isotropic case. The directions of the respective displacement vectors are mutually orthogonal and are in general oblique to the wave normal. A generalized Rayleigh-type wave may be propagated, in a limited number of directions, along the surface of an extensive anisotropic medium. See Kolsky, *Stress Waves in Solids*; Musgrave, *Proc. Roy. Soc., A*, 1954, **226**, 339; Synge, *Proc. Roy. Irish Acad., A*, 1956, **58**, 13; Stoneley, *Proc. Roy. Soc., A*, 1955, **232**, 447.

Velocities and attenuation constants for various solids in the region of 20 °C are given below. In the case of metals, factors such as texture, cold work, stress, hardening, tempering and aging can cause significant departures from the values given in the table. Properties of plastics vary considerably with molecular weight, with additives and with temperature. In view of this, many values are approximate and relate to materials of variable composition; the velocities given should therefore only be regarded as typical. The bracketed figures following the attenuation are frequencies in MHz. Composition of materials is given as percentage by weight.

Velocity and attenuation of waves in solids

Material	Velocity/(m s^{-1})				α longit. waves neper m^{-1}
	c_L longitudinal bulk waves	c_E irrot. rod waves	c_S shear waves	c_{SR} Rayleigh waves	
Aluminium	6374	5102	3111	2906	0.40 (10)
ADP crystal, X-cut	6250	—	—	—	10.8 (10)
,, Y-cut	6250	—	—	—	—
,, Z-cut	4300	3500	—	—	9.69 (10)
Barium titanate ceramic . . .	4000	—	—	—	—
Beryllium.	12890	—	8880	—	—
Bone, human tibia	4000	—	1970	—	460 (2.9)
Brass	4372	3451	2100	1964	—
Brick	—	3650	—	—	—
Butyl rubber/carbon (100/40) . .	1600	—	—	—	133 (0.35)
Cadmium	2780	2400	—	—	—
Cellulose acetate butyrate . . .	2080	—	—	—	103 (2.5)
Chromium	6608	6229	4005	3655	—
Concrete	4250–5250	—	—	—	—
Constantan	5177	4276	2625	2445	—
Copper	4759	3813	2325	2171	—
Cork	—	500	—	—	—
Duralumin	6398	5120	3122	2917	1.23 (10)
Ebonite	2500	—	—	—	—
Glass (crown)	5660	5342	3420	3127	2 (10)
,, (heavy flint)	5260	4717	2960	2731	—
,, (pyrex).	5640	5170	3280	—	—
Gold (hard-drawn)	3240	2030	1200	—	—
Granite	5400	—	—	—	60 (0.6)
Ice (maximum density, polar ice sheets, firn temperature − 20 °C)	3840	—	—	—	—
Invar (36 Ni, 63.8 Fe, 0.2 C) . .	4657	4216	2658	2447	—
Iron (soft)	5957	5189	3224	2986	—
,, (cast)	4994	4477	2809	2590	—

Velocity and attenuation of waves in solids (*contd*)

Material	Velocity/(m s^{-1})				α longit. waves
	c_L longitudinal bulk waves	c_E irrot. rod waves	c_S shear waves	c_{SR} Rayleigh waves	neper m^{-1}
Lead	2160	1188	700	—	—
Magnesium	5823	5082	3163	2930	—
Manganese	4600	3830	—	—	—
Marble	—	3810	—	—	—
Molybdenum	6475	5636	3505	3248	—
Monel metal	5350	4400	2720	—	—
Neoprene.	1510	—	—	—	230 (2.5)
Neoprene/carbon (100/60) . .	1690	—	—	—	—
Nickel (unmag. soft)	5608	4787	2929	2722	—
,, (unmag. hard)	5814	4974	3078	2857	—
Niobium	5068	3497	2092	1970	—
Ni-Span-C	—	4831	2799	—	—
Nylon	2680	—	—	—	11.5 (1)
Perspex	2700	2177	1330	1242	57 (2.5)
Platinum	3260	2800	1730	—	—
Polythene (polyethylene) . . .	2000	—	—	—	54 (1)
Polystyrene	2350	1840	1120	1047	23 (2.5)
Polyvinyl chloride	2300	—	—	—	3.5 (0.35)
Polyvinyl chloride acetate . . .	2250	—	—	—	1270 (10)
Polyvinyl formal	2680	—	—	—	115 (2.5)
Polyvinylidene chloride	2400	—	—	—	207 (2.5)
Quartz (crystal) X-cut . . .	5720	5440	—	—	0.0127 (10)
,, (fused)	5970	5759	3765	3410	—
Rubber (natural)	1600	—	—	—	15 (0.35)
Rubber/carbon (100/40) . . .	1680	—	—	—	36.6 (0.35)
Sandstone	2920	2820	1840	—	—
Silica (fused)	5968	5760	3764	—	—
Silver	3704	2806	1698	1592	—
Slate	—	4510	—	—	—
Steel (mild)	5960	5196	3235	2996	—
,, (tool) hardened	5874	5116	3179	2945	4.94 (10)
,, (stainless)	5980	5282	3297	3049	—
Tantalum	4159	3337	2036	1902	—
Tin	3380	2626	1594	1491	—
Titanium	6130	5164	3182	2958	—
Tourmaline (crystal) Z-cut . . .	7250	7170	—	—	—
Tungsten (annealed)	5221	4619	2887	2668	—
,, carbide	6655	6223	3984	3643	—
,, (drawn)	5410	4320	2640	—	—
Uranium	3370	—	1940	—	—
Vanadium	6023	4584	2774	2600	—
Wood (Ash) with grain	—	4670	—	—	—
,, across grain	—	1390	—	—	—
,, (Oak) with grain	—	4100	—	—	—
,, (Pine) with grain	—	3600	—	—	—
Zinc (rolled)	4187	3826	2421	2225	—
Zirconium	4650	—	2250	—	—

M.E.D.

1.6.2 Physiological and subjective acoustics

The decibel (dB). Two sounds having intensities I_1 and I_2 differ in intensity level by n decibels where $n = 10 \log_{10} (I_1/I_2)$. Measurement in terms of intensity is seldom convenient and it is usual to express differences in terms of sound pressure, in which case $n = 20 \log_{10} (p_1/p_2)$, p_1 and p_2 being the r.m.s. sound pressures of the two sounds. The **sound pressure level** (s.p.l.) of a sound or noise is $20 \log_{10} (p_1/p_0)$ where p_1 is the r.m.s. sound pressure and p_0 is the reference sound pressure 2×10^{-5} N m^{-2}.

Threshold of hearing, for a sound of given character, is the minimum value of the r.m.s. sound pressure which excites the sensation of hearing. Normal threshold of hearing is the modal value of the thresholds of hearing of a group of otologically normal subjects. Normal threshold for binaural listening in a free progressive wave incident from directly ahead of the listener is called **minimum audible field** (MAF) and defined in terms of the sound pressure in the undisturbed wave. Values of MAF in the following table are taken from ISO Recommendation R 226–1961. See Robinson and Dadson, *Br. J. appl. Phys.*, 1956, **7**, 166. Normal threshold for monaural listening to a sinusoidal sound pressure produced by an earphone is called **minimum audible pressure** (MAP), the sound pressure being measured at the entrance to the ear canal. See Dadson and King, *J. Laryng. Otol.*, 1952, **66**, 366.

Hearing level is measured in decibels by the difference between a listener's threshold of hearing and the audiometric zero. This zero, specified in BS 2497 : 1954, is based on MAP determinations for normal subjects in the age-group 18–25, and is given in the following table. Values are sound pressure level (dB relative to 2×10^{-5} N m^{-2}). See also BS 2497 : 1972.

Normal thresholds of hearing

Frequency/Hz	Normal binaural min. audible field MAF	Normal threshold of hearing MAP	Frequency/Hz	Normal binaural min. audible field MAF	Normal threshold of hearing MAP ·
20	74.3	—	500	6.0	12
30	58.1	—	700	4.7	—
40	48.4	—	1 000	4.2	9
60	36.8	—	1 500	—	10.5
80	29.8	44	2 000	1.0	11.0
100	25.1	—	3 000	−2.9	7.5
125	—	30	4 000	−3.9	9.5
140	18.9	—	6 000	4.6	14
200	13.8	—	8 000	15.3	18.5
250	11.2	19	10 000	16.4	19.5
300	9.4	—	12 000	12.0	22.5
400	7.2	—	15 000	24.1	36.5

The **threshold of hearing** for frequencies above approximately 1 kHz increases with age at a rate which increases with frequency. See Hinchcliffe, *Gerontologia*, 1959, **2**, 211.

The subjective quality known as the **loudness** of a noise or sound is measured by reference to the free field s.p.l. of a pure tone of frequency 1000 Hz which is judged by an observer to be equally loud. The **loudness level** is the value of the s.p.l. so found by a group of normal listeners, and is expressed in **phon**, the value being numerically equal to the free field s.p.l. of the reference tone, a 1000 Hz tone presented as a free progressive plane wave reaching the listener from directly in front. A loudness level of 4 phon corresponds to the threshold of hearing, and above 130 phon pain is felt. For typical loudness levels of common noises see Parkin, *Acustica*, 1957, **7**, 57.

Equal-loudness contours. Loudness levels are frequently shown in the form of parametric curves of constant loudness known as equal-loudness contours. Values in the table below refer to otologically normal listeners between the ages of 18 and 25; see Robinson and Dadson, *Br. J. appl. Phys.*, 1956, **7**, 166, and ISO Recommendation R 226–1961.

Sone scale. The sone scale is arithmetic and provides a numerical designation of loudness of sounds or noises that is proportional to the subjective magnitude, as estimated by normal observers. For all practical purposes it has been found experimentally that over the range 20 to 120 phons the loudness is given to sufficient accuracy by $S = 2^{(P-40)/10}$, S being the loudness in sones and P the loudness level in phons. The loudness of a sound of 1 sone corresponds to a loudness level of 40 phons, and a twofold change in loudness corresponds to an interval of 10 phons.

Loudness of complex sounds. For experimental data on the loudness of frontally incident and diffuse sound fields for bands of noise, see Robinson and Whittle, *Acustica*, 1964, **14**, 24, and ISO Recommendation R 454–1965. For complex sounds the phon is not a convenient unit since a team of normal observers is required to assess each noise. When direct measurement of loudness level is impracticable, the value may be calculated from the sound pressure levels in a number of contiguous bands of the frequency spectrum. Several methods are available, of varying degrees of complexity; see Zwicker, *Frequenz*, 1959, **13**, 234, Stevens, *J. Acoust. Soc. Am.*, 1961, **33**, 1577 and ISO Recommendation R 532–1966.

Loudness level of pure tones in phons

Sound pressure level	Frequency/Hz												
	25	50	100	200	500	1000	2000	3000	4000	6000	8000	10 000	15 000
0	—	—	—	—	—		3	7	8	—	—	—	—
10	—	—	—	—	9	10	13	16	18	10	—	—	—
20	—	—	—	12	20	20	22	26	27	20	10	9	—
30	—	—	11	25	32	30	32	36	37	30	21	21	14
40	—	1	25	38	43	40	42	46	48	40	32	33	30
50	—	18	38	49	53	50	52	57	58	50	42	44	45
60	—	34	51	61	64	60	62	67	69	61	53	55	58
70	15	49	63	72	74	70	73	78	79	71	63	65	69
80	34	63	75	82	84	80	84	90	90	81	74	75	79
90	53	76	86	92	93	90	95	101	102	92	84	85	88
100	69	88	97	102	102	100	106	113	113	102	94	94	94
110	84	100	107	110	111	110	117	124	125	112	103	103	100
120	98	111	116	119	120	120	129	137	136	123	113	112	103
130	109	121	125	127	128	130	—	—	—	—	—	—	—

M.E.D.

1.6.3 Preferred frequencies for acoustical measurements

For many acoustical measurements it is convenient to adopt constant-percentage increments of frequency. To simplify intercomparison of results it is recommended that when data are to be presented at discrete frequencies, appropriate frequencies from the following series be adopted; the order of preference is indicated by the type of print, in the order, bold, italic and roman. When concerned with frequency bands, the frequencies listed in the table should be the geometric centre frequencies of those bands. Extension of the table in either direction is effected by successive multiplication or division by 1000. See ISO Recommendation 266–1962.

Preferred frequencies/Hz	1/1 oct.	1/2 oct.	1/3 oct.	Preferred frequencies/Hz	1/1 oct.	1/2 oct.	1/3 oct.	Preferred frequencies/Hz	1/1 oct.	1/2 oct.	1/3 oct.
16	×	×	×	180		×		**2 000**	×	×	×
20			×	200			×	2 500			×
22.4		×		**250**	×	×	×	2 800		×	
25			×	315			×	3 150			×
31.5	×	×	×	355		×		**4 000**	×	×	×
40			×	400			×	5 000			×
45		×		**500**	×	×	×	5 600		×	
50			×	630			×	6 300			×
63	×	×	×	710		×		**8 000**	×	×	×
80			×	800			×	10 000			×
90		×		**1000**	×	×	×	11 200		×	
100			×	1250			×	12 500			×
125	×	×	×	1400		×		**16 000**	×	×	×
160			×	1600			×				

M.E.D.

1.6.4 Architectural acoustics

The sound absorption coefficient of a material is $\alpha = (1 - r)$, where r, the **sound energy reflection coefficient**, is the ratio of sound energy reflected from the surface of the material to that incident upon it. Values for a specific material depend upon frequency and upon the angle of incidence of the sound. When the sound field is approximately diffuse the corresponding quantity is called the **reverberation absorption coefficient**; this may be determined in accordance with BS 3638:1963 when the cofficient is designated α(ISO). Since values obtained in accordance with the new procedure are not yet available, those in the following table have been taken from *Sound Absorbing Materials*, by Evans and Bazley, 1960 (HMSO). Absorption depends on mounting and other details of construction and the following values should be regarded only as typical.

Reverberation absorption coefficients

Material	Thickness (mm)	Frequency/Hz					
		125	250	500	1000	2000	4000
Acoustic plaster	13	0.15	0.20	0.35	0.60	0.60	0.50
Acoustic tiles (perforated fibreboard) . .	18	0.10	0.35	0.70	0.75	0.65	0.50
Asbestos (sprayed)	25	0.10	0.30	0.65	0.85	0.85	0.80
Brickwork	—	0.02	0.02	0.03	0.04	0.05	0.07
Carpet (Axminster)	8	—	0.05	0.15	0.30	0.45	0.55
Carpet on underlay	14	—	0.05	0.20	0.40	0.60	0.65
Curtain (velour, draped)	—	0.14	0.35	0.55	0.72	0.70	0.65
Glass fibre (resin-bonded)	25	0.10	0.25	0.55	0.70	0.80	0.85
Glass wool (uncompressed)	25	0.10	0.25	0.45	0.60	0.70	0.70
Mineral wool	25	0.10	0.25	0.50	0.70	0.85	0.85
Polystyrene, expanded (rigid backing) . .	13	0.05	0.05	0.10	0.15	0.15	0.20
Polystyrene, expanded (on 50 mm battens) .	13	0.05	0.15	0.40	0.35	0.20	0.20
Polyurethane foam (flexible)	50	0.25	0.50	0.85	0.95	0.90	0.90
Snow	25	0.15	0.40	0.65	0.75	0.80	0.85
Wood panelling (oak, on 25 mm battens) .	13	0.20	0.10	0.05	0.05	0.05	0.05

Insulation against airborne noise. The **sound reduction index** (SRI) is the ratio of sound energy incident on a partition to that which is transmitted through the partition, expressed in decibels. The values vary with frequency and angle of incidence; for comparative purposes the mean value over a frequency range 100–3150 Hz is often used. See BS 2750:1956. Simple dependence of SRI on mass and frequency above the fundamental resonance is not generally observed due to 'coincidence' effects associated with flexural waves in the partition. See Cremer, *Akustische Zeit*, 1942, **7**, 81. For single homogeneous partitions the mean sound reduction index (100–3150 Hz) is given roughly by $10 + 15 \log_{10} m$, m being the superficial density in kg m^{-2}.

    Typical values given in the table are taken from Bazley, *The Airborne Sound Insulation of Partitions*, 1966 (HMSO).

Sound reduction indices

Type of partition (figures in brackets are thicknesses in mm)	m kg m^{-2}	Frequency/Hz						
		100	200	400	800	1600	3150	*Mean*
Brick, unplastered (115) . .	220	34	37	40	41	51	57	43
Brick (225) plastered both faces	480	48	40	45	50	61	67	52
Cavity wall, brick (50 × 115) airspace (50) plastered both faces, butterfly ties . . .	480	30	36	45	58	65	82	53
Clinker concrete wall (75) unplastered	100	16	15	20	23	29	38	23
Clinker concrete wall (75) plastered both faces . . .	150	22	33	36	47	55	56	42
Concrete floor (100) reinforced	230	36	40	38	47	55	65	47
Wood joist floor, boarding (22) nailed to joists (225 × 50) plasterboard ceiling (9.5) skim coat plaster	80	16	26	34	35	42	41	32
Fibreboard (12)	4	11	15	18	22	27	30	21
Plasterboard (9.5)	10	13	19	22	28	34	35	25
Plasterboard (9.5) each side of (100 × 50) studs (airspace 100)	25	12	23	32	36	44	46	32
Plywood (6)	4	8	10	15	20	25	28	18
Steel, 16G	12	14	20	25	31	37	43	28
Plate glass (6)	17	18	23	24	30	30	35	27
Window glass (3)	7	18	12	20	26	30	28	22
Window glass, double, airspace (50)	14	19	19	27	36	47	52	33
Window glass, double, airspace (180)	14	28	25	36	43	50	53	39

M.E.D.

1.6.5 Musical acoustics

Standard musical pitch is based on a frequency of 440 Hz for A in the treble stave (a′). See ISO Recommendation R 16–1955 and BS 880:1950. Wind instruments should be constructed to conform to this frequency at 20 °C room temperature after being blown long enough to reach thermal equilibrium. Tuning C for keyboard instruments (c″) is a minor third of equal temperament above 440 Hz, namely 523.25 Hz.

Diatonic major scale. The frequency ratios of musical intervals in the diatonic major scale are shown below; the upper row shows the ratio between adjacent steps, the lower row the ratio to the keynote.

Diatonic major scale

	Major tone 9/8		Minor tone 10/9		Diatonic semitone 16/15		Major tone 9/8		Minor tone 10/9		Major tone 9/8		Diatonic semitone 16/15	
1 Keynote		1.1250 Major 2nd		1.2500 Major 3rd		1.3333 Perfect 4th		1.5000 Perfect 5th		1.6667 Major 6th		1.8750 Major 7th		2 Octave

The interval between a major and a minor tone, 81/80, is called a *comma*.

A *chromatic semitone* = 135/128 or 25/24 according to whether it divides a major or minor tone.

A *minor interval* is a major interval diminished by a chromatic semitone. An *augmented interval* is a major interval increased by a chromatic semitone. A *diminished interval* is a perfect or minor interval diminished by a chromatic semitone.

Equally tempered major scale. A semitone of equal temperament $= 2^{1/12} = 1.059\,46$. A tone of equal temperament $= 2$ semitones of ET $= 2^{1/6} = 1.122\,46$. The table below gives the frequencies (Hz) of musical notes in equal temperament for the octave starting at c′.

Frequencies of notes in equal temperament (Hz)

c′	c′♯	d′	d′♯	e′	f′	f′♯	g′	g′♯	a′	a′♯	b′
261.63	277.18	293.66	311.13	329.63	349.23	369.99	392.00	415.30	440.00	466.16	493.88

Any pitch interval may be expressed in *cents*, a *cent* being defined as the ratio $2^{1/1200}:1$. Thus there are 100 cents to the ET semitone. See Wood, *The Physics of Music*, 1962.

M.E.D.

1.7 Radiation and optics

1.7.1 The electromagnetic spectrum

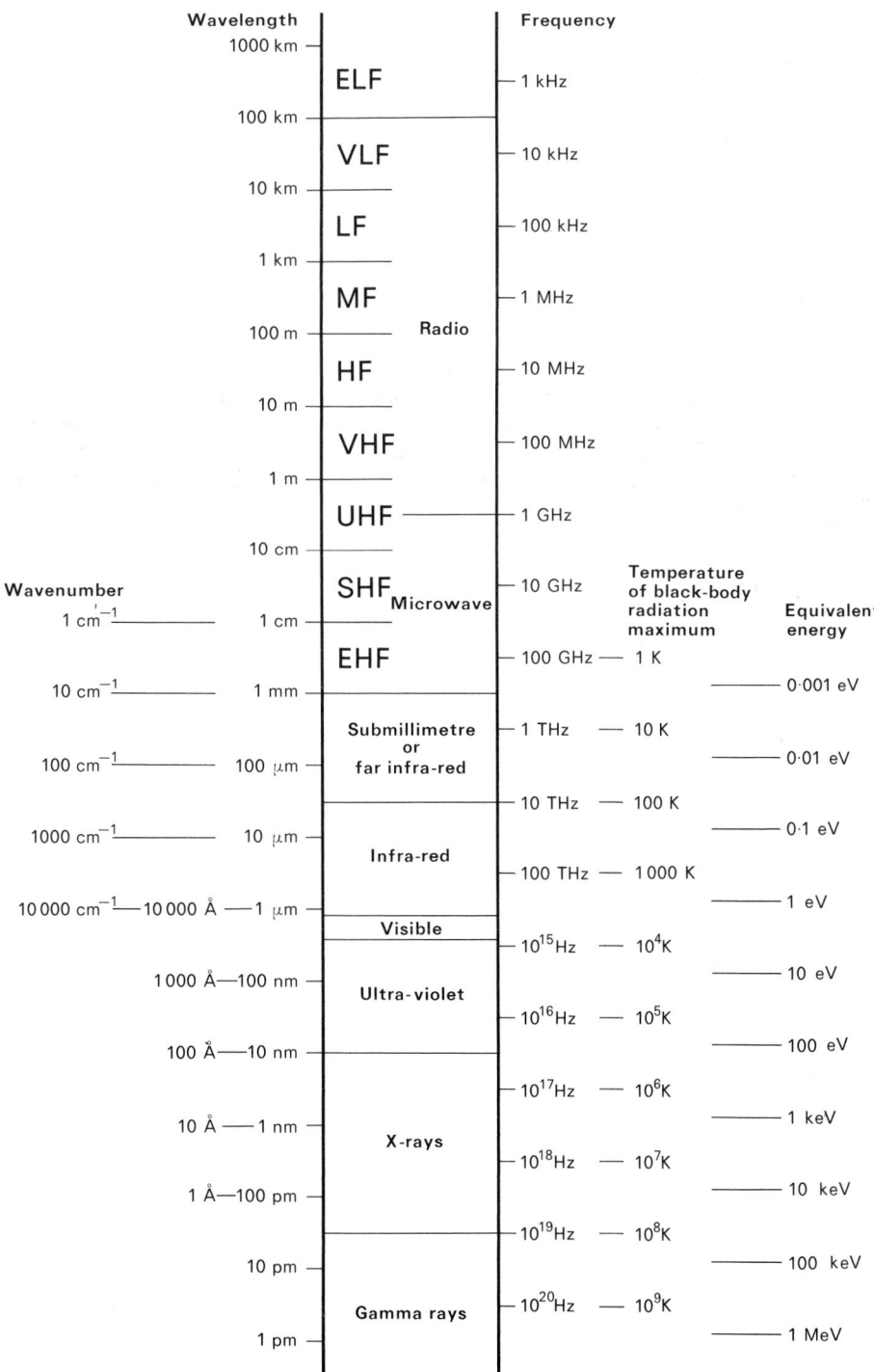

A.E.B.

1.7.2 Thermal radiation

In a field of radiation the radiant power $\delta^2\phi$ falling on an infinitesimal area δA, in the range of directions contained by an infinitesimal cone of solid angle $\delta\Omega$, is given by $\delta^2\phi = L\,\delta A\,\delta\Omega\cos\theta$, where θ is the angle between the cone and the normal to δA. The quantity L is known as the **radiance** of the field at δA in the direction of the cone. Radiance has the property that, in a transparent medium of constant refractive index, the radiance at any point along a ray path in the direction of the ray path is constant.†

The total radiant power $\delta\phi$ falling on δA from one side is given by

$$\delta\phi = \delta A \int L\cos\theta\,d\Omega = E\,\delta A \qquad \ldots (1)$$

where the integration is carried out over the appropriate hemisphere on δA. The quantity E which, unlike L, usually depends on the orientation of δA, is known as **irradiance**. If δA is an element of a radiating surface, the equivalent quantity in terms of the radiation passing out through δA is known as **exitance**. The radiation emitted by the whole of a radiation source in a given direction is described by the quantity I known as **radiant intensity**, defined by the equation $\delta\phi = I\,\delta\Omega$, where $\delta\phi$ is the power radiated in a cone of directions of solid angle $\delta\Omega$ about the specified direction.

If the radiation has a continuous spectrum, the power in the frequency interval $\delta\nu$ about ν or the wavelength interval $\delta\lambda$ about λ is proportional to the interval, and is therefore denoted by a spectral power density $\phi(\nu) = \partial\phi/\partial\nu$ or $\phi(\lambda) = \partial\phi/\partial\lambda$. These and derived quantities such as $L(\nu)$ are commonly distinguished from ordinary functions of ν or λ by writing the ν or λ as a subscript. If vacuum wavelengths are used, then $\nu\lambda = c$ and $\phi_\lambda = (c/\lambda^2)\phi_\nu$, etc.

Black-body radiation

The thermal radiation inside a closed cavity, with opaque walls at a uniform temperature T, is known as full or black-body radiation. It has a continuous spectrum, and is uniform and isotropic, i.e. the spectral radiance L_ν within the cavity is independent of position and direction. Moreover, L_ν is determined solely by T.‡

This radiation may be observed by making a hole in the cavity wall small enough not to reduce appreciably the radiance of the emerging radiation. A means of testing whether this condition is satisfied is provided by the extended form of Kirchhoff's law, which says that the factor (known as emissivity) by which the radiance in a particular direction is reduced, equals the absorption factor for radiation incident on the hole in the opposite direction. A radiation source of this kind with emissivity sensibly equal to unity is known as a black-body radiator.

Spectral distribution of black-body radiation

According to Planck's formula

$$L_\nu = \frac{2h}{c^2}\,\nu^3\left(e^{h\nu/kT} - 1\right)^{-1} \qquad \ldots (2)$$

The spectral radiance is usually expressed in terms of L_λ:

$$L_\lambda = c_1'\lambda^{-5}\left(e^{c_2/\lambda T} - 1\right)^{-1} \qquad \ldots (3)$$

where $\qquad\qquad c_1' = 2hc^2 = 1.1911 \times 10^{-16}\ \text{W m}^2\ \text{sr}^{-1}$

and $\qquad\qquad c_2 = hc/k = 1.4388 \times 10^{-2}\ \text{m K}$

---

† More generally, if the radiation passes without loss through a medium of varying refractive index n, then L/n^2 is constant.

‡ Strictly speaking, L_ν is proportional to n^2, where n is the refractive index inside the cavity, and the subsequent formulae refer to an evacuated cavity.

For E_λ we have, since $E_\lambda = \pi L_\lambda$ for isotropic radiation,

$$E_\lambda = c_1 \lambda^{-5} (e^{c_2/\lambda T} - 1)^{-1}$$

where

$$c_1 = \pi c_1' = 3.7418 \times 10^{-16} \text{ W m}^2 \qquad \ldots (4)$$

The constants c_1 and c_2 are known as the 1st and 2nd radiation constants.

The total radiance over the spectrum, $L = \int_0^\infty L_\lambda \, d\lambda$, is given by $L = \sigma' T^4$, where

$$\sigma' = \frac{2\pi^4 k^4}{15 c^2 h^3} = 1.8047 \times 10^{-8} \text{ W m}^{-2} \text{ sr}^{-1} \text{ K}^{-4} \qquad \ldots (5)$$

Likewise the total exitance is given by $E = \sigma T^4$, where

$$\sigma = \pi \sigma' = 5.670 \times 10^{-8} \text{ W m}^{-2} \text{ K}^{-4} \qquad \ldots (6)$$

σ is known as the Stefan–Boltzmann radiation constant.

The functions L_ν for different T are of the form $f(T) g(\nu/T)$, so that if normalized to a given maximum value and plotted against ν/T, they reduce to a single curve. The maximum of this curve occurs at

$$\frac{\nu_m}{T} = 5.8787 \times 10^{10} \text{ Hz K}^{-1}$$

This result is more conveniently expressed in terms of the wave number σ used by spectroscopists,† $\sigma = 1/\lambda$:

$$\sigma_m / T = 1.9609 \text{ K}^{-1} \text{ cm}^{-1}$$

Likewise the distribution functions in terms of λ reduce to a single function of λT with a maximum at

$$\lambda_m T = 2.8979 \times 10^{-3} \text{ m K}$$

The following table gives the fraction f of the total radiation for temperature T which lies between zero wavelength and that corresponding to various values of λT. It may be used in conjunction with formulae (5) and (6) above to obtain the radiance or exitance in extended spectral bands.

Example: for a black-body at 1000 K the exitance in the band 0.6–0.7 µm ($\lambda T = 600$–700 µm K) is

$$(1.84 \times 10^{-6} - 9.29 \times 10^{-8}) \times 5.67 \times 10^{-8} \times 1000^4 \text{ W m}^{-2}$$

Fraction of total radiation from o to λT

$\lambda T / (\mu\text{m K})$	f	$\lambda T / (\mu\text{m K})$	f	$\lambda T / (\mu\text{m K})$	f
600	9.29×10^{-8}	2 500	0.1613	12 000	0.945 05
700	1.84×10^{-6}	3 000	0.2732	14 000	0.962 85
800	1.64×10^{-5}	3 500	0.3829	16 000	0.973 77
900	8.70×10^{-5}	4 000	0.4808	18 000	0.980 81
1 000	3.21×10^{-4}	4 500	0.5643	20 000	0.985 56
1 200	2.13×10^{-3}	5 000	0.6337	30 000	0.995 29
1 400	7.79×10^{-3}	6 000	0.7378	40 000	0.997 92
1 600	1.97×10^{-2}	7 000	0.8081	50 000	0.998 91
1 800	3.93×10^{-2}	8 000	0.8562	75 000	0.999 67
2 000	6.67×10^{-2}	10 000	0.9142	100 000	0.999 86

E.J.G.

Colour-temperature

The colour of an illuminant often resembles that of a black-body radiator at a particular temperature, and can conveniently be specified in terms of this temperature. If an exact colour match can be obtained the temperature is known as the **colour-temperature**. If only an approximate match can be obtained, then the term **correlated colour-temperature** is used. This can be given a precise

† Not to be confused with σ denoting the Stefan–Boltzmann constant.

meaning by representing the colours of the illuminant and of the black-body radiation as points on the CIE (U–V) Uniform Chromaticity Space diagram. The correlated colour-temperature is then defined as the black-body temperature for which the two points are closest together. For correlated colour-temperatures of various illuminants, see p. 79.

With some illuminants, such as the tungsten-filament lamp, both the colour and the relative spectral power distribution resemble that of a black-body radiator at a suitable temperature. In such cases, the term '**distribution temperature**' is proposed for the temperature which gives the best spectral fit. As yet, however, there is no agreement as to what constitutes a best fit, and where this term is used, the criterion of best fit should be stated.

Colour and distribution temperatures are usually expressed in kelvins (K). The reciprocal of the temperature in megakelvins may also be used, the unit being known as the mired:
$1 \text{ mired} = 10^{-6} \text{ K}^{-1}$.

O.C.J.

1.7.3 Photometry

The results of photometric matching experiments show that radiation may be evaluated in terms of visual effect by means of a quantity known as luminous flux, ϕ_v, which is related to $\phi_{e\lambda}$, the spectral power distribution of the radiation, by an equation of the form

$$\phi_v = K_m \int_0^\infty \phi_{e\lambda} V(\lambda) \, d\lambda \qquad \ldots (1)$$

where K_m is a constant, and $V(\lambda)$ a function representing the wavelength-dependent sensitivity of the eye. The exact shape of $V(\lambda)$ depends on the observer, but in order to provide a precise quantitative basis for photometry, an agreed set of values was recommended for adoption by the Commission International de l'Éclairage (CIE) in 1971, and accepted by the Comité International des Poids et Mesures (CIPM) (see p. 9).

The following table gives the values of this function at 10 nm intervals. For the complete tabulation at 1 nm interval refer to CIE 18 (E-1.2) 1970.

Photopic relative luminous efficiency function $V(\lambda)$†

λ/nm	$V(\lambda)$	λ/nm	$V(\lambda)$	λ/nm	$V(\lambda)$
360	0.000 004	530	0.862 000	680	0.017 000
370	0.000 012	540	0.954 000	690	0.008 210
380	0.000 039				
390	0.000 120	550	0.994 950	700	0.004 102
		555	**1.000 000**	710	0.002 091
400	0.000 396	560	0.995 000	720	0.001 047
410	0.001 210	570	0.952 000	730	0.000 520
420	0.004 000	580	0.870 000	740	0.000 249
430	0.011 600	590	0.757 000		
440	0.023 000			750	0.000 120
		600	0.631 000	760	0.000 060
450	0.038 000	610	0.503 000	770	0.000 030
460	0.060 000	620	0.381 000	780	0.000 015
470	0.090 980	630	0.265 000	790	0.000 007
480	0.139 020	640	0.175 000		
490	0.208 020			800	0.000 004
		650	0.107 000	810	0.000 002
500	0.323 000	660	0.061 000	820	0.000 001
510	0.503 000	670	0.032 000	830	0.000 000
520	0.710 000				

† Yet to be formally recommended.

The spatial properties of light are described in photometry by quantities analogous to those used in radiometry (see p. 76), with luminous flux replacing radiant power. The same symbols are used for corresponding quantities, distinguished where necessary by the suffixes e (radiation) and v (photometric). These quantities are listed in the following table, together with their radiometric equivalents to indicate the relation between the two systems of measurement.

Photometric quantities

Quantity	Radiometric equivalent	Symbol	Definition	Unit name and symbol	
Luminous flux	Radiant power	ϕ		lumen	lm
Luminous intensity	Radiant intensity	I	$\partial\phi/\partial\Omega$	candela	$cd \equiv lm\ sr^{-1}$
Luminance	Radiance	L	$\partial^2\phi/\partial A\ \partial\Omega$	(nit)	$lm\ sr^{-1}\ m^{-2}$
Illuminance	Irradiance	E	$\partial\phi/\partial A$	lux	$lm\ m^{-2}$
Luminous exitance	Radiant exitance	M			$lm\ m^{-2}$

Luminance and colour temperature of selected light sources

Light source	Luminance $10^4\ lm\ sr^{-1}\ m^{-2}$	Correlated colour temp. K
Candle	0.5	1930
Paraffin flame (flat wick)	1.25	2055
Acetylene (Kodak burner)	10.8	2360
Tungsten strip lamp (vacuum)	124	2400
,, ,, ,, (gas-filled)	540	2800
,, ,, ,, (,,)	990	3000
,, coil ,, (quartz-halogen)	~ 2800	3300
Mercury vapour (low pressure, in glass)	2.3	—
,, ,, (high pressure, compact, 250 W)	2×10^4	—
,, ,, (,, ,, ,, 1000 W)	4×10^4	—
,, ,, (tubular fluorescent, colour matching 40 W)	1.2	6050
,, ,, (tubular fluorescent, warm white 40 W)	1.9	3000
Arc crater (solid carbon)	1.72×10^4	3780
,, ,, (high intensity, 150 A)	8.0×10^4	5000–5500
Xenon arc (compact) 250 W	2.5×10^4	6000
,, ,, ,, 2 kW	1.2×10^5	6000
Clear blue sky	~ 0.4	12 000–24 000
Starlit sky	$\sim 5 \times 10^{-8}$	—
Zenith Sun (through atmosphere)	$\sim 1.6 \times 10^5$	5400
Moon (through atmosphere)	~ 0.4	—
Perfect diffuser in sunlight	~ 4	
,, ,, ,, moonlight	$\sim 10^{-5}$	

The magnitude of the unit of luminous flux, the lumen, is determined, not by assigning a value to K_m of equation (1), but by the definition of the candela† in terms of the radiation from a black-body radiator at the temperature of freezing platinum (see p. 4). If a value for T_{Pt} is assumed, then the value of K_m corresponding to this definition may be obtained by substituting for $\phi_{e\lambda}$ in equation (1) the spectral power distribution of a black-body radiator at temperature T_{Pt}, calculated in accordance with Planck's formula and the radiation constants $c_1 = 3.7418 \times 10^{-16}$ W m² and $c_2 = 1.4388 \times 10^{-2}$ m K (see p. 76).

† The choice of the candela rather than the lumen for the definition is an historical accident.

For $T_{Pt} = 2045$ K:

$$K_m = 673 \text{ lm W}^{-1}$$

For $T_{Pt} = 2040.8$ K (see footnote, p. 9):

$$K_m = 690 \text{ lm W}^{-1}$$

Practical measurements of K_m using units and scales maintained in several national laboratories have yielded values ranging from 647 to 686 lm W^{-1}. The CIE (1970) suggestion is that the value $K_m = 680$ lm W^{-1} should be used until the situation is clarified. Since the $V(\lambda)$ function is normalized to a maximum value of unity at $\lambda = 555$ nm, K_m gives the relation between the lumen and the watt for light of this wavelength. The reciprocal of K_m has been called the mechanical equivalent of light.

Scotopic system of photometry

At low levels of luminance the relative luminous effects of radiations of different wavelength are in general no longer represented even approximately by the values of $V(\lambda)$ given in the above table. This is because a different visual mechanism known as the scotopic or dark-adapted mechanism, as opposed to the photopic or light-adapted mechanism, is dominant. To evaluate electromagnetic radiation with respect to the effect it produces on the scotopic mechanism, an alternative set of weighting factors—the **scotopic relative luminous efficiency function** $V'(\lambda)$—must be used, the values of which were agreed by the CIE in 1951.

Scotopic relative luminous efficiency function $V'(\lambda)$

λ/nm	0	10	20	30	40
300	—	—	—	—	—
400	0.009 29	0.034 84	0.096 6	0.1998	0.3281
500	0.982	0.997	0.935	0.811	0.650
600	0.033 15	0.015 93	0.007 37	3.335×10^{-3}	1.497×10^{-3}
700	1.780×10^{-5}	9.14×10^{-6}	4.78×10^{-6}	2.546×10^{-6}	1.379×10^{-6}

λ/nm	50	60	70	80	90
300	—	—	—	5.89×10^{-4}	2.209×10^{-3}
400	0.455	0.567	0.676	0.793	0.904
500	0.481	0.3288	0.2076	0.1212	0.0655
600	6.77×10^{-4}	3.129×10^{-4}	1.480×10^{-4}	7.15×10^{-5}	3.533×10^{-5}
700	7.60×10^{-7}	4.25×10^{-7}	2.413×10^{-7}	1.390×10^{-7}	—

In the alternative system of photometry obtained by taking $V'(\lambda)$ as basis instead of $V(\lambda)$, all the formal definitions of the previous section are retained, with terms modified by the adjective 'scotopic' and symbols by the addition of a prime. In particular, the primary standard radiator has a scotopic luminance L' of 6×10^5 scotopic cd m^{-2}. Calculated values for the scotopic maximum luminous efficacy K'_m are:

For $T_{Pt} = 2045$ K: $K'_m = 1725$ scotopic lm W^{-1}
For $T_{Pt} = 2040.8$ K: $K'_m = 1774$ scotopic lm W^{-1}.

In the so-called mesopic region between the photopic and the scotopic, the concept of luminous flux as an additive quantity does not apply. The CIE (1971) has suggested the following procedure.

For any given luminous measurement in the mesopic region, both the photopic and the scotopic values are determined. The appropriate mesopic quantity, e.g. E_{mesopic}, is then obtained by means of the formula

$$E_{\text{mesopic}} = \frac{E' + 17E^2}{1 + 17E}$$

O.C.J.

1.7.4 Colorimetry

Evaluation of colour

The recommended method of describing the colour of radiation or of irradiated objects is the tri-chromatic system adopted by the CIE in 1931. The colour of any radiation is specified by three **tristimulus values** X, Y, Z which represent the amounts of three standard reference stimuli (X), (Y), (Z) which when mixed together produce the same sensation of colour. The tristimulus values for radiation of spectral power distribution $P(\lambda)$ are given by the formulae:

$$X = K \int P(\lambda) . \bar{x}(\lambda) \, d\lambda; \qquad Y = K \int P(\lambda) . \bar{y}(\lambda) \, d\lambda; \qquad Z = K \int P(\lambda) . \bar{z}(\lambda) \, d\lambda$$

where K is a scaling constant and $\bar{x}(\lambda)$, $\bar{y}(\lambda)$ and $\bar{z}(\lambda)$ are known as **spectral tristimulus values**, i.e. the tristimulus values of unit powers of monochromatic radiation as given in the table below. The 'colour quality' of the radiation, as distinct from its amount, is expressed in terms of the **chromaticity co-ordinates** $x, y,, z$, given by:

$$x = X/(X + Y + Z); \qquad y = Y/(X + Y + Z); \qquad z = Z/(X + Y + Z).$$

Of the three tristimulus values, Y represents the luminosity of a colour, and if K_m is used for the constant K above, then Y becomes the photometric flux, since the function $\bar{y}(\lambda)$ is identical to $V(\lambda)$. (See p. 78.)

The colour of a reflecting surface is specified in terms of the tristimulus values X, Y, Z of the radiation reflected when the surface is illuminated with radiation of specified spectral power distribution $P(\lambda)$. Here the absolute values of X, Y, Z are usually just as important as the chromaticity co-ordinates x, y, z, and they are defined as

$$X = \frac{\int P(\lambda) . \bar{x}(\lambda) . \rho(\lambda) \, d\lambda}{\int P(\lambda) . \bar{y}(\lambda) \, d\lambda}$$

$$Y = \frac{\int P(\lambda) . \bar{y}(\lambda) . \rho(\lambda) \, d\lambda}{\int P(\lambda) . \bar{y}(\lambda) \, d\lambda}$$

$$Z = \frac{\int P(\lambda) . \bar{z}(\lambda) . \rho(\lambda) \, d\lambda}{\int P(\lambda) . \bar{y}(\lambda) \, d\lambda}$$

where $\rho(\lambda)$ is the reflectance expressed as a percentage. These definitions give Y the value 100 for a perfect reflector, or more generally Y equals the luminous reflectance (the photometric value); in addition the chromaticity co-ordinates x, y, z for a spectrally non-selective reflector will be those of the illuminant. Similar expressions hold for a transmitting object, with the spectral transmittance $\tau(\lambda)$ replacing $\rho(\lambda)$ above.

The table of spectral tristimulus values allows the chromaticity co-ordinates $x(\lambda), y(\lambda), z(\lambda)$ of the spectral radiations to be calculated immediately since

$$x(\lambda) = \bar{x}(\lambda)/(\bar{x}(\lambda) + \bar{y}(\lambda) + \bar{z}(\lambda)), \text{ etc.}$$

λ/nm	Spectral tristimulus values			λ/nm	Spectral tristimulus values		
	$\bar{x}(\lambda)$	$\bar{y}(\lambda)$	$\bar{z}(\lambda)$		$\bar{x}(\lambda)$	$\bar{y}(\lambda)$	$\bar{z}(\lambda)$
380	0.0014	0.0000	0.0065	580	0.9163	0.8700	0.0017
385	0.0022	0.0001	0.0105	585	0.9786	0.8163	0.0014
390	0.0042	0.0001	0.0201	590	1.0263	0.7570	0.0011
395	0.0076	0.0002	0.0362	595	1.0567	0.6949	0.0010
400	0.0143	0.0004	0.0679	600	1.0622	0.6310	0.0008
405	0.0232	0.0006	0.1102	605	1.0456	0.5668	0.0006
410	0.0435	0.0012	0.2074	610	1.0026	0.5030	0.0003
415	0.0776	0.0022	0.3713	615	0.9384	0.4412	0.0002
420	0.1344	0.0040	0.6456	620	0.8544	0.3810	0.0002
425	0.2148	0.0073	1.0391	625	0.7514	0.3210	0.0001
430	0.2839	0.0116	1.3856	630	0.6424	0.2650	0.0000
435	0.3285	0.0168	1.6230	635	0.5419	0.2170	0.0000
440	0.3483	0.0230	1.7471	640	0.4479	0.1750	0.0000
445	0.3481	0.0298	1.7826	645	0.3608	0.1382	0.0000
450	0.3362	0.0380	1.7721	650	0.2835	0.1070	0.0000
455	0.3187	0.0480	1.7441	655	0.2187	0.0816	0.0000
460	0.2908	0.0600	1.6692	660	0.1649	0.0610	0.0000
465	0.2511	0.0739	1.5281	665	0.1212	0.0446	0.0000
470	0.1954	0.0910	1.2876	670	0.0874	0.0320	0.0000
475	0.1421	0.1126	1.0419	675	0.0636	0.0232	0.0000
480	0.0956	0.1390	0.8130	680	0.0468	0.0170	0.0000
485	0.0580	0.1693	0.6162	685	0.0329	0.0119	0.0000
490	0.0320	0.2080	0.4652	690	0.0227	0.0082	0.0000
495	0.0147	0.2586	0.3533	695	0.0158	0.0057	0.0000
500	0.0049	0.3230	0.2720	700	0.0114	0.0041	0.0000
505	0.0024	0.4073	0.2123	705	0.0081	0.0029	0.0000
510	0.0093	0.5030	0.1582	710	0.0058	0.0021	0.0000
515	0.0291	0.6082	0.1117	715	0.0041	0.0015	0.0000
520	0.0633	0.7100	0.0782	720	0.0029	0.0010	0.0000
525	0.1096	0.7932	0.0573	725	0.0020	0.0007	0.0000
530	0.1655	0.8620	0.0422	730	0.0014	0.0005	0.0000
535	0.2257	0.9149	0.0298	735	0.0010	0.0004	0.0000
540	0.2904	0.9540	0.0203	740	0.0007	0.0002	0.0000
545	0.3597	0.9803	0.0134	745	0.0005	0.0002	0.0000
550	0.4334	0.9950	0.0087	750	0.0003	0.0001	0.0000
555	0.5121	1.0000	0.0057	755	0.0002	0.0001	0.0000
560	0.5945	0.9950	0.0039	760	0.0002	0.0001	0.0000
565	0.6784	0.9786	0.0027	765	0.0001	0.0000	0.0000
570	0.7621	0.9520	0.0021	770	0.0001	0.0000	0.0000
575	0.8425	0.9154	0.0018	775	0.0001	0.0000	0.0000
580	0.9163	0.8700	0.0017	780	0.0000	0.0000	0.0000

The table is normalized so that the chromaticity co-ordinates of the equi-energy white radiation are each equal to 0.3333. In many cases good accuracy can be obtained by computation at 10 nm intervals from 380 nm provided that the tristimulus values X, Y, Z so obtained are multiplied by 1.0002, 1.0000 and 1.0008 respectively; this has the effect of renormalizing the shortened table for 10 nm intervals to give the same chromaticity co-ordinates for equi-energy white, and hence to minimize spectral sampling errors on average.

The (x, y) chromaticity diagram is subjectively non-uniform as far as colour differences are concerned, and in 1960 the CIE recommended a chromaticity diagram (u, v) that is more uniform for colour spacing. This is derived by transformation as follows:

$$u = 4x/(-2x + 12y + 3); \qquad v = 6y/(-2x + 12y + 3)$$

The CIE (X, Y, Z) system of 1931 is valid for observations with uniform areas subtending up to $4°$; for uniform areas subtending larger angles there is a supplementary (X_{10}, Y_{10}, Z_{10}) system recommended by CIE in 1964, based on observations with a $10°$ field.

Standard illuminants for colorimetry

The following standard illuminants have been specified by the CIE for measuring the colour of materials.

Illuminant A. Black-body radiation corresponding to a value of c_2/T (see p. 76) of $(1.4350 \times 10^{-2}/2848)$ m, or on IPTS-68, to a temperature of approximately 2856 K. The chromaticity co-ordinates of Illuminant A are $(x = 0.4476, y = 0.4074)$.

Illuminant A is realized in the laboratory by Source A, the radiation from a gas-filled tungsten-filament lamp of the same correlated colour temperature.

Illuminant B. The radiation from Illuminant A after selective attenuation in accordance with the published CIE data on the transmittance of the filter described below. The chromaticity co-ordinates of Illuminant B are $(x = 0.3484, y = 0.3516)$ and its correlated colour temperature 4874 K (IPTS-68).

Illuminant B is realized in the laboratory by Source A combined with a colour filter consisting of a layer 10 mm thick of each of two solutions B_1 and B_2, contained in a double cell made of colourless optical glass.

Solution B_1

Copper sulphate $(CuSO_4.5H_2O)$	2.452 g
Mannite $(C_6H_8(OH)_6)$	2.452 g
Pyridine (C_5H_5N)	30.0 ml
Distilled water, to make 1 litre.	

Solution B_2

Cobalt ammonium sulphate $(CoSO_4.(NH_4)_2SO_4.6H_2O)$	21.71 g
Copper sulphate $(CuSO_4.5H_2O)$	16.11 g
Sulphuric acid (density 1.835 g ml^{-1})	10.0 ml
Distilled water, to make 1 litre.	

Illuminant C. The radiation from Illuminant A after selective attenuation in accordance with the published CIE data on the transmittance of the filter described below. The chromaticity co-ordinates of Illuminant C are $(x = 0.3101, y = 0.3162)$ and its correlated colour temperature is 6774 K (IPTS-68).

Illuminant C is realized in the laboratory by Source A combined with a colour filter consisting of a layer 10 mm thick of each of two solutions C_1 and C_2, contained in a double cell made of colour-less optical glass.

Relative spectral power distributions of CIE standard illuminants

λ/nm	(A) $P(\lambda)$	(B) $P(\lambda)$	(C) $P(\lambda)$	(D$_{65}$) $P(\lambda)$	λ/nm	(A) $P(\lambda)$	(B) $P(\lambda)$	(C) $P(\lambda)$	(D$_{65}$) $P(\lambda)$
375	8.77	18.80	27.50	51.0	575	110.80	101.90	100.15	96.1
380	9.80	22.40	33.00	50.0	580	114.44	101.00	97.80	95.8
385	10.90	26.85	39.92	52.3	585	118.08	100.07	95.43	92.2
390	12.09	31.30	47.40	54.6	590	121.73	99.20	93.20	88.7
395	13.35	36.18	55.17	68.7	595	125.39	98.44	91.22	89.3
400	14.71	41.30	63.30	82.8	600	129.04	98.00	89.70	90.0
405	16.15	46.62	71.81	87.1	605	132.70	98.08	88.83	89.8
410	17.68	52.10	80.60	91.5	610	136.35	98.50	88.40	89.6
415	19.29	57.70	89.53	92.5	615	139.99	99.06	88.19	88.6
420	20.99	63.20	98.10	93.4	620	143.62	99.70	88.10	87.7
425	22.79	68.37	105.80	90.1	625	147.24	100.36	88.06	85.5
430	24.67	73.10	112.40	86.7	630	150.84	101.00	88.00	83.3
435	26.64	77.31	117.75	95.8	635	154.42	101.56	87.86	83.5
440	28.70	80.80	121.50	104.9	640	157.98	102.20	87.80	83.7
445	30.85	83.44	123.45	110.9	645	161.52	103.05	87.99	81.9
450	33.09	85.40	124.00	117.0	650	165.03	103.90	88.20	80.0
455	35.41	86.88	123.60	117.4	655	168.51	104.59	88.20	80.1
460	37.81	88.30	123.10	117.8	660	171.96	105.00	87.90	80.2
465	40.30	90.08	123.30	116.3	665	175.38	105.08	87.22	81.2
470	42.87	92.00	123.80	114.9	670	178.77	104.90	86.30	82.3
475	45.52	93.75	124.09	115.4	675	182.12	104.55	85.30	80.3
480	48.24	95.20	123.90	115.9	680	185.43	103.90	84.00	78.3
485	51.04	96.23	122.92	112.4	685	188.70	102.84	82.21	74.0
490	53.91	96.50	120.70	108.8	690	191.93	101.60	80.20	69.7
495	56.85	95.71	116.90	109.1	695	195.12	100.38	78.24	70.7
500	59.86	94.20	112.10	109.4	700	198.26	99.10	76.30	71.6
505	62.93	92.37	106.98	108.6	705	201.36	97.70	74.36	73.0
510	66.06	90.70	102.30	107.8	710	204.41	96.20	72.40	74.3
515	69.25	89.65	98.81	106.3	715	207.40	94.60	70.40	68.0
520	72.50	89.50	96.90	104.8	720	210.36	92.90	68.30	61.6
525	75.79	90.43	96.78	106.2	725	213.27	91.10	66.30	65.7
530	79.13	92.20	98.00	107.7	730	216.12	89.40	64.40	69.9
535	82.52	94.46	99.94	106.0	735	218.92	88.00	62.80	72.5
540	85.95	96.90	102.10	104.4	740	221.67	86.90	61.50	75.1
545	89.41	99.16	103.95	104.2	745	224.36	85.90	60.20	69.3
550	92.91	101.00	105.20	104.0	750	227.00	85.20	59.20	63.6
555	96.44	102.20	105.67	102.0	755	229.59	84.80	58.50	55.0
560	100.00	102.80	105.30	100.0	760	232.12	84.70	58.10	46.4
565	103.58	102.92	104.11	98.2	765	234.59	84.90	58.00	56.6
570	107.18	102.60	102.30	96.3	770	237.01	85.40	58.20	66.8

Solution C_1

Copper sulphate ($CuSO_4.5H_2O$)	3.412 g
Mannite ($C_6H_8(OH)_6$)	3.412 g
Pyridine (C_5H_5N)	30.0 ml
Distilled water, to make 1 litre.	

Solution C_2

Cobalt ammonium sulphate ($CoSO_4.(NH_4)_2SO_4.6H_2O$)	30.58 g
Copper sulphate ($CuSO_4.5H_2O$)	22.52 g
Sulphuric acid (density 1.835 g ml^{-1})	10.0 ml
Distilled water, to make 1 litre.	

Illuminant D_{65}. A relative spectral power distribution defined and recommended by CIE as representing a phase of daylight with a correlated colour temperature of approximately 6504 K (IPTS-68). This is a much better representation of an average overcast sky than Illuminant C. The chromaticity co-ordinates of Illuminant D_{65} are ($x = 0.3127$, $y = 0.3290$). At present Illuminant D_{65} cannot be realized with enough accuracy for many applications, and there is no recommended source specification as yet.

Methods of specifying standard illuminants D_T of different correlated colour-temperatures have been published by CIE: these are spectral power distributions which represent different phases of daylight.

General reference: *Colorimetry* (CIE Publication No. 15, 1971).

F.J.J.C.

1.7.5 Wavelength standards

Primary standard. Wavelength values of optical radiations may be expressed either as the values in standard air (see below) or in vacuum. The fundamental standard is the same as the primary standard of length, the wavelength in vacuum of the orange radiation corresponding to the transition between the unperturbed levels $5p[0\frac{1}{2}]_1 - 6d[0\frac{1}{2}]_1^0$ of the krypton-86 atom. There are, by definition (see p. 4), 1 650 763.73 wavelengths of this radiation in 1 metre, so that the derived value of the wavelength in vacuum is 605.780 211 nm.

Secondary standards. Twelve secondary standards, recommended by the International Committee of Weights and Measures in 1963 for use in interferometric length measurement, have also been recognized by the International Astronomical Union for use in spectroscopy (*Trans. IAU*, vol. XII A, 1965). The table shows their vacuum wavelengths in nanometres.

Krypton-86	*Mercury-198*	*Cadmium-114*
645.807 20	579.226 83	644.024 80
642.280 06	577.119 83	508.723 79
565.112 86	546.227 05	480.125 21
450.361 62	435.956 24	467.945 81

Recommended sources

Krypton-86. The wavelength of the fundamental standard is emitted with an uncertainty of not more than 1 part in 10^8 by lamps conforming to the specification below. The secondary standard radiations emitted under similar conditions have uncertainties estimated to be not more than 2 parts in 10^8.

The source is a hot-cathode discharge lamp containing krypton-86 (purity $\not< 99\%$) in sufficient quantity to assure the presence of solid krypton at 64 K, the lamp having a capillary portion with

the dimensions: internal diameter 2–4 mm, wall thickness 1 mm. The conditions of operation are:
  (i) the capillary is observed end-on from the anode side of the lamp;
  (ii) the lower part of the lamp, including the capillary, is immersed in a refrigerant bath maintained within 1 K of the triple point of nitrogen;
  (iii) the current density in the capillary is 3 ± 1 mA mm^{-2}.

Mercury-198. The uncertainties of the wavelengths are 5 parts in 10^8 when emitted by a high-frequency electrodeless discharge lamp, operated at moderate power with the radiation observed through the side of the capillary. The lamp should be maintained at a temperature below 10 °C and contain mercury-198 (purity $\not< 98\%$) with argon as carrier gas at a pressure between 65 and 133 N m^{-2}. The internal diameter of the capillary should be about 5 mm, with the volume of the lamp preferably > 20 cm^3.

Cadmium-114. The standard wavelengths have an estimated uncertainty of 7 parts in 10^8 when emitted by an electrodeless discharge lamp source, maintained at a temperature such that the green line is not reversed and containing cadmium-114 (purity $\not< 95\%$) with argon as carrier gas (pressure about 150 N m^{-2} at ambient temperatures). The radiations should be observed through the side of the capillary part of the tube, having an internal diameter of about 5 mm.

Laser sources. Gas lasers can emit highly monochromatic and intense radiations which are useful for interferometric measurement and spectroscopic investigations. In order to achieve reproducible wavelength values, the laser can be frequency stabilized by electronic feedback control. The visible helium–neon laser containing neon-20, stabilized to the Lamb dip, emits radiation with the vacuum wavelength 632.991 4 nm with an uncertainty of 1 part in 10^7 due to variations in gas pressure and design of the laser tube (Mielenz *et al.*, *Appl. Opt.*, 1968, **7**, 289).

Lasers of very much better wavelength reproducibility can be made by stabilization with absorption techniques. Measurements of the vacuum wavelength of the infra-red helium–neon laser stabilized by saturated absorption in methane, give a value of 3.392 231 38 μm with a measurement uncertainty of 1 part in 10^8.

W.R.C.R.

1.7.6 Refractive index of gases

Refractive index of air

The wavelength λ_{air} of a radiation in air is related to its vacuum value λ_{vac} by $\lambda_{vac} = n\lambda_{air}$, where n is the refractive index. For standard air (dry air at 15 °C and 101 325 N m^{-2}, containing 0.03% by volume of carbon dioxide) the refractive index n_s is given by the dispersion equation (Edlén, *Metrologia*, 1966, **2**, 71)

$$(n_s - 1) \times 10^8 = 8\,342.13 + 2\,406\,030(130 - \sigma^2)^{-1} + 15\,997(38.9 - \sigma^2)^{-1}$$

where $\sigma = 1/\lambda_{vac}$ and λ_{vac} is expressed in μm. This equation is based upon observations within the range 200 nm to 2 μm, and is in better agreement with recent measurements than the equation (Edlén, *J. Opt. Soc. Am.*, 1953, **43**, 339) which was recommended by the Joint Commission for Spectroscopy in 1952, although the difference is at most 1.4×10^{-8}.

In the visible region (405–705 nm) the following approximate expression is more convenient and gives a maximum discrepancy of only 1.2×10^{-8},

$$n_s - 1 = 0.047\,2300(173.3 - \sigma^2)^{-1}$$

For air at a temperature t °C and a pressure p N m^{-2}, the refractivity is given by the equation

$$n_{tp} - 1 = (n_s - 1) \times \frac{p[1 + p(61.3 - t) \times 10^{-10}]}{96\,095.4(1 + 0.003\,661\,t)}$$

The refractivity of water vapour is less than that of air, so that if the air is moist its refractive index will be smaller than the value calculated for dry air. This water vapour term is dependent upon wavelength. In the visible region (405–644 nm) the relationship is

$$n_{tpf} - n_{tp} = -f(4.2918 - 0.0342\,\sigma^2) \times 10^{-10},$$

where n_{tpf} is the refractive index of air containing water vapour at a partial pressure of f N m^{-2}, the total pressure still being p. This equation is valid only for conditions not deviating very much from normal laboratory conditions ($t = 20\ ^\circ$C, $p = 100\ 000$ N m^{-2}, $f = 1500$ N m^{-2}).

W.R.C.R.

Refractive indices of gases

Refractive index for mean of sodium D lines (589.3 nm) reduced to s.t.p. (101 325 N m^{-2} and 20 °C)

Gas	Refractive index	Gas	Refractive index
Air	1.000 292	Hydrobromic acid	1.000 570
Acetone	1.001 090	Hydrochloric acid	1.000 447
Acetylene	1.000 606	Hydrogen	1.000 132
Ammonia	1.000 376	Hydrogen sulphide	1.000 634
Argon	1.000 281	Krypton	1.000 427
Benzene	1.001 762	Mercury	1.000 933
Bromine	1.000 132	Methane	1.000 444
Carbon dioxide	1.000 451	Methyl alcohol	1.000 586
Carbon disulphide	1.001 481	Methyl chloride	1.000 865
Carbon monoxide	1.000 338	Methyl fluoride	1.000 449
Carbon tetrachloride	1.001 768	Neon	1.000 067
Chlorine	1.000 773	Nitric oxide	1.000 297
Chloroform	1.001 450	Nitrogen	1.000 297
Ether	1.001 533	Nitrous oxide	1.000 516
Ethyl alcohol	1.000 878	Oxygen	1.000 272
Ethylene	1.000 696	n-Pentane	1.001 711
Fluorine	1.000 195	Sulphur dioxide	1.000 686
Helium	1.000 036	Sulphur trioxide	1.000 737
Hydriodic acid	1.000 906	Water vapour	1.000 254
		Xenon	1.000 702

K.J.H.

Refractive indices of gases at radio frequencies

Values below are for dry gases at 0 °C, 101.325 kN m^{-2}. The quoted uncertainty limits are about 3σ (σ = standard error of the mean).

Gas	$(n-1)/10^{-6}$
Air (CO$_2$ free)	288.15 ± 0.1
Deuterium	134.8 ± 0.3
Helium	35.0 ± 0.2
Carbon dioxide	494 ± 1.0
Hydrogen	136.0 ± 0.2
Nitrogen	294.1 ± 0.1
Oxygen	266.4 ± 0.2
Water vapour†	60.7 ± 0.1

† At 20 °C, 1.333 kN m^2 (10 mmHg).

Refractive index of moist air

The following formula has been derived from measured values and the gas laws, and holds over a wide range of conditions:

$$(n-1) \times 10^6 = \frac{103.49}{T} p_1 + \frac{177.4}{T} p_2 + \frac{86.26}{T} \left(1 + \frac{5748}{T}\right) p_3$$

where p_1 = partial pressure of dry air in mmHg
p_2 = partial pressure of carbon dioxide in mmHg
p_3 = partial pressure of water vapour in mmHg
T = thermodynamic temperature in K.

L.E.

1.7.7 Refractive index of optical materials

Spectral lines for refractometry

Refracting optical materials are specified by their refractive indices for the series of wavelengths listed opposite which were recommended by the International Commission of Optics in 1962 (*Optica Acta*, 1963, 10, 217). This series differs from the ICO recommendations of 1950 (also shown), but includes those proposed to the Commission by Franke in 1953. The table includes in the central column various additional lines used by other workers in recent years. Wavelengths in nm and for standard air.

Spectral line sources

Gas discharge lamps are available as convenient sources throughout a wide range of the spectrum; in some instances one lamp may contain two or three elements. The following table gives a list of (mainly) visible wavelengths obtained from lamps which are suitable for refractometry and other purposes, e.g. calibration of spectrometers. Interference filters give the most complete suppression of unwanted lines, but for many purposes complete suppression is not necessary and simpler filters, of which some examples are quoted, may be used.

Lamp	Wave-length nm	Filter	Lamp	Wave-length nm	Filter
Mercury . . .	365.0⎫ 365.5⎬ 366.3⎭	Wratten 18A or Chance OX1	Helium . . .	587.6 667.8	Ilford 626 —
	404.7	Chance OV1 + OY10		706.5	⎧Chance OR1 + OX7 or ⎨Ilford 302 + 608
	435.8	Wratten 50	Potassium . .	766.5⎱	Chance OR1 or
	546.1	Wratten 77A or Ilford 807		769.9⎰	Ilford 207
	577.0⎱ 579.0⎰	Chance OY1 or Ilford 808	Hydrogen . .	434.0	Wratten 47B or Ilford 601
Cadmium . .	467.8	Chance OB10		486.1	Wratton 45
	480.0	—		656.3	Chance OR1
	508.6	Wratten 58	Rubidium . .	780.0⎱ 794.7⎰	Chance OX7 +OR1
	643.8	Wratten 25	Sodium . . .	589.0† 589.6	

† Where figures are quoted for the mean of the sodium doublet, the wavelength is taken as 589.3 nm and the refractive index is denoted by D.

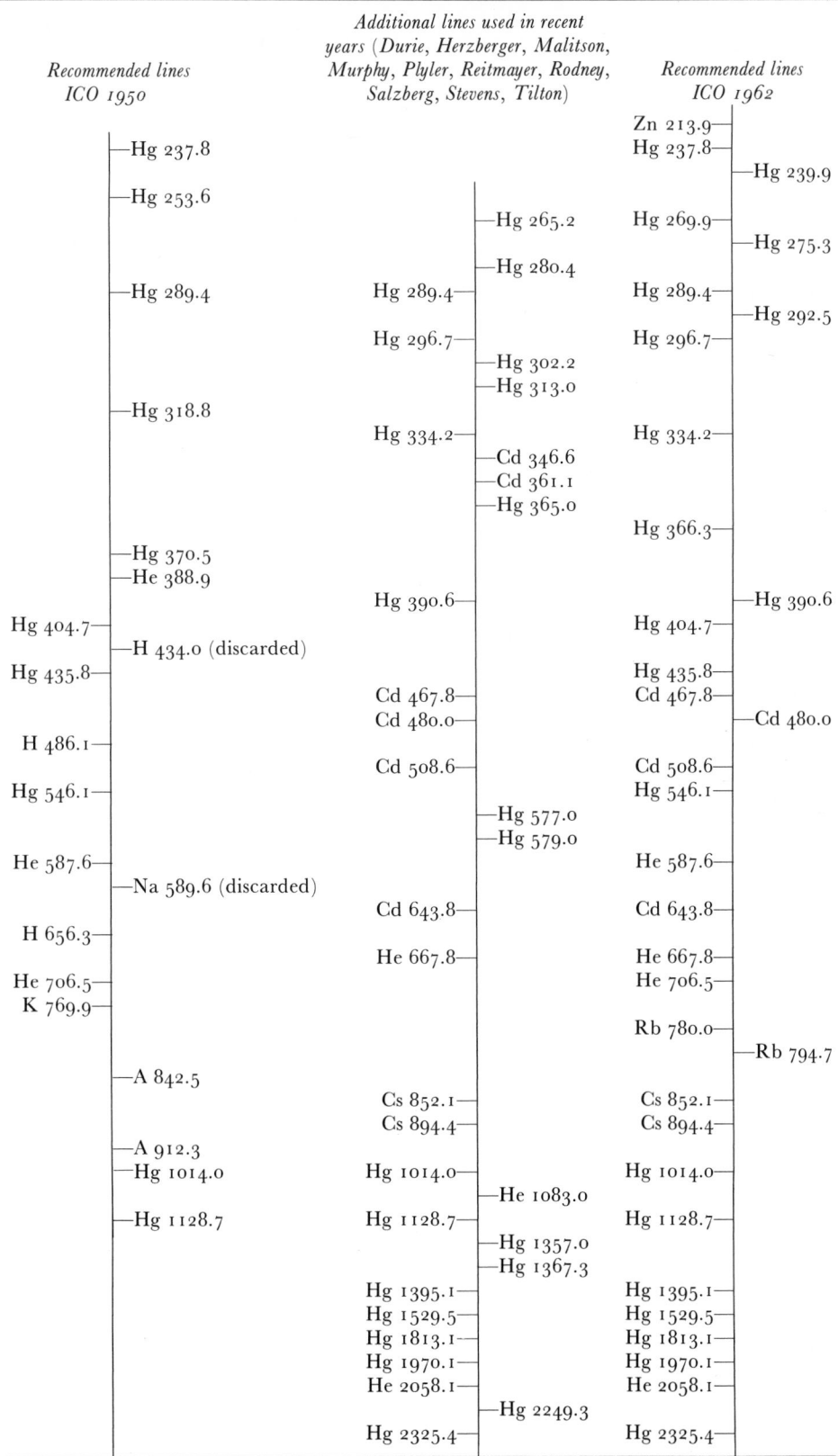

*Recommended lines
ICO 1950*

*Additional lines used in recent
years (Durie, Herzberger, Malitson,
Murphy, Plyler, Reitmayer, Rodney,
Salzberg, Stevens, Tilton)*

*Recommended lines
ICO 1962*

Zn 213.9
Hg 237.8
Hg 239.9
Hg 253.6
Hg 265.2
Hg 269.9
Hg 275.3
Hg 280.4
Hg 289.4
Hg 289.4
Hg 289.4
Hg 292.5
Hg 296.7
Hg 296.7
Hg 302.2
Hg 313.0
Hg 318.8
Hg 334.2
Hg 334.2
Cd 346.6
Cd 361.1
Hg 365.0
Hg 366.3
Hg 370.5
He 388.9
Hg 390.6
Hg 390.6
Hg 404.7
H 434.0 (discarded)
Hg 404.7
Hg 435.8
Hg 435.8
Cd 467.8
Cd 467.8
Cd 480.0
Cd 480.0
H 486.1
Cd 508.6
Cd 508.6
Hg 546.1
Hg 546.1
Hg 577.0
Hg 579.0
He 587.6
He 587.6
Na 589.6 (discarded)
Cd 643.8
Cd 643.8
H 656.3
He 667.8
He 667.8
He 706.5
He 706.5
K 769.9
Rb 780.0
Rb 794.7
A 842.5
Cs 852.1
Cs 852.1
Cs 894.4
Cs 894.4
A 912.3
Hg 1014.0
Hg 1014.0
Hg 1014.0
He 1083.0
Hg 1128.7
Hg 1128.7
Hg 1128.7
Hg 1357.0
Hg 1367.3
Hg 1395.1
Hg 1395.1
Hg 1529.5
Hg 1529.5
Hg 1813.1
Hg 1813.1
Hg 1970.1
Hg 1970.1
He 2058.1
He 2058.1
Hg 2249.3
Hg 2325.4
Hg 2325.4

Optical glasses

A selection of Chance Optical Glasses (for complete range of optical glasses, see Chance Catalogue)

Glass type	n_d	V	F − C	C − A' / (C − A')/(F − C)	d − C / (d − C)/(F − C)	e − d / (e − d)/(F − C)	F − e / (F − e)/(F − C)	g − F / (g − F)/(F − C)	h − g / (h − g)/(F − C)	Density g cm⁻³	Atmospheric resistance
Borosilicate Crown BSC 510644	1.509 70	64.44	0.007 91	0.002 79 / 0.353	0.002 43 / 0.307	0.001 88 / 0.238	0.003 60 / 0.455	0.004 23 / 0.535	0.003 48 / 0.440	2.49	S1
Hard Crown HC 519604	1.518 99	60.42	0.008 59	0.003 00 / 0.353	0.002 62 / 0.307	0.002 05 / 0.238	0.003 92 / 0.455	0.004 65 / 0.535	0.003 86 / 0.449	2.53	S1
Light Barium Crown LBC 541595	1.540 65	59.54	0.009 08	0.003 10 / 0.341	0.002 77 / 0.305	0.002 17 / 0.239	0.004 14 / 0.456	0.004 93 / 0.543	0.004 06 / 0.447	2.87	S2
Zinc Crown ZC 508612	1.507 59	61.16	0.008 30	0.002 93 / 0.353	0.002 55 / 0.307	0.001 98 / 0.239	0.003 77 / 0.456	0.004 46 / 0.537	0.003 68 / 0.443	2.49	S2
Medium Barium Crown MBC 569558	1.569 38	55.77	0.010 21	0.003 49 / 0.342	0.003 09 / 0.303	0.002 43 / 0.239	0.004 69 / 0.459	0.005 60 / 0.548	0.004 68 / 0.458	3.12	S3
Dense Barium Crown DBC 612585	1.612 30	58.54	0.010 46	0.003 57 / 0.341	0.003 18 / 0.304	0.002 49 / 0.238	0.004 79 / 0.459	0.005 69 / 0.544	0.004 72 / 0.451	3.56	S5
Soft Crown SC 515564	1.515 07	56.35	0.009 14	0.003 15 / 0.345	0.002 78 / 0.304	0.002 18 / 0.239	0.004 18 / 0.457	0.005 00 / 0.547	0.004 18 / 0.457	2.58	S3
Telescope Flint TF 530512	1.530 33	51.19	0.010 36	0.003 50 / 0.338	0.003 13 / 0.302	0.002 47 / 0.238	0.004 76 / 0.459	0.005 73 / 0.553	0.004 82 / 0.465	2.70	S1
Barium Light Flint LBF 574520	1.574 27	52.02	0.011 04	0.003 70 / 0.336	0.003 33 / 0.302	0.002 62 / 0.238	0.005 09 / 0.461	0.006 14 / 0.556	0.005 15 / 0.466	3.21	S3
Extra Light Flint ELF 548456	1.547 69	45.60	0.012 01	0.003 97 / 0.330	0.003 58 / 0.298	0.002 85 / 0.237	0.005 58 / 0.465	0.006 78 / 0.565	0.005 78 / 0.481	2.95	S1
Barium Flint BF 605438	1.604 83	43.83	0.013 80	0.004 50 / 0.326	0.004 10 / 0.297	0.003 27 / 0.237	0.006 43 / 0.466	0.007 87 / 0.570	0.006 71 / 0.486	3.48	S2
Light Flint LF 579411	1.578 60	41.12	0.014 07	0.004 57 / 0.325	0.004 17 / 0.296	0.003 33 / 0.237	0.006 57 / 0.467	0.008 07 / 0.574	0.006 93 / 0.493	3.23	S2
Dense Flint DF 620362	1.620 49	36.24	0.017 12	0.005 42 / 0.317	0.005 03 / 0.294	0.004 05 / 0.237	0.008 04 / 0.470	0.009 97 / 0.582	0.008 65 / 0.505	3.63	S4
Extra Dense Flint EDF 700303	1.700 35	30.28	0.023 13	0.007 20 / 0.311	0.006 71 / 0.290	0.005 47 / 0.236	0.010 95 / 0.473	0.013 81 / 0.597	0.012 18 / 0.527	4.33	S4
Double Extra Dense Flint DEDF 748278	1.748 42	27.85	0.026 87	0.008 26 / 0.307	0.007 75 / 0.288	0.006 35 / 0.236	0.012 77 / 0.475	0.016 20 / 0.603	0.014 38 / 0.535	4.76	S6
SPECIAL GLASSES											
Double Extra Dense Flint DEDF 927210	1.927 07	21.01	0.044 12	0.013 06 / 0.296	0.012 50 / 0.283	0.010 36 / 0.235	0.021 26 / 0.482	0.027 79 / 0.630	—	6.11	S6
Borate Flint BoF 612449	1.612 00	44.90	0.013 63	0.004 63 / 0.340	0.004 10 / 0.301	0.003 24 / 0.238	0.006 29 / 0.461	0.007 62 / 0.559	0.006 46 / 0.474	3.18	S6
Special Barium Crown SBC 651586	1.651 00	58.60	0.011 11	0.003 83 / 0.345	0.003 38 / 0.304	0.002 65 / 0.239	0.005 08 / 0.457	0.006 04 / 0.544	0.004 98 / 0.448	3.97	S6
Special Barium Flint SBF 717479	1.717 00	47.90	0.014 97	0.004 97 / 0.332	0.004 49 / 0.300	0.003 56 / 0.238	0.006 92 / 0.462	0.008 34 / 0.557	0.007 05 / 0.471	4.45	S3

The above extracts from the Chance–Pilkington optical glass catalogue and the data on light transmission characteristics of a selected range of optical glasses (see p. 96) are reproduced by kind permission of Chance–Pilkington Optical Works.

Refractive indices

Data on refraction and dispersion are usually given in the following way: refractive index for the helium d line (n_d) is stated (replacing n_{D_1}), the mean dispersion between the hydrogen F and C lines ($n_F - n_C$) and the reciprocal dispersive power (or constringence) V calculated from these figures, thus

$$V = (n_d - 1)/(n_F - n_C)$$

Based on the series of spectral lines for refractometry recommended by the ICO, a new basis has been introduced employing the mercury line e and the cadmium lines C′ and F′, from which the modified constringence is calculated,

$$V_e = (n_e - 1)/(n_{F'} - n_{C'})$$

Partial dispersions $(n_C - n_{A'})$, $(n_d - n_C)$, etc., are stated, and also the relative partial dispersions $(n_C - n_{A'})/(n_F - n_C)$, $(n_d - n_C)/(n_F - n_C)$, etc.

For brevity, A′ is written for $n_{A'}$, C for n_C, etc.

Optical materials vary widely in their resistance to atmospheric attack. Chance–Pilkington classify their optical glasses in the following manner, indicating the probable behaviour of a glass surface under adverse conditions:

S1   Excellent resistance; likely to remain free from attack.
S2   Good resistance, but dull patches may appear on the surface after approximately two years.
S3   Fairly good resistance, but patches of stain may appear within a two-year period.
S4   Fair resistance; tarnish spots may appear within two years.
S5   Poor resistance; general tarnish appears within one year.
S6   Very poor resistance; tarnishes heavily within a few months.

Optical cements

Refractive indices (n_D^{20}) of polymerized cements (*Optica Acta*, 1967, **14**, 401).

Solid Canada balsam	1.54	H.T.	1.45
Soft Canada balsam	1.53	Ross No. 24 . . .	1.46
Cellulose caprate	1.47	Beetle 4128	1.55
Cellulose caprate (plasticized) .	1.49	B. & S. No. 8 . . .	1.58
Gelva	1.48	Epikote 817 . . .	1.57

Plastic materials

Material	V	A′	C	D	e	F	g	Transmission range/μm
Polystyrene 15 °C .	31.0	1.581	1.587	1.592		1.606	1.617	
35 °C .		1.578	1.584	1.589		1.603	1.614	0.34 to 2
55 °C .		1.575	1.581	1.586		1.600	1.612	
Polymethylmethacrylate (Perspex) 20 °C . .	57.8		1.489	1.491	1.493	1.497	1.501	0.34 to 2

Temperature coefficient of refractive index for above two polymers is -14×10^{-5} °C^{-1}.
See *Proceedings of the London Conference on Optical Instruments* (1950).
See also p. 251 for refractive indices of other plastic materials.

Refractive indices of calcite (Iceland Spar)

(From 0.198 to 0.795 μm taken at 18 °C, from 0.8007 to 3.324 μm at 20 °C)

Wavelength μm	Refractive index		Wavelength μm	Refractive index	
	O-ray	E-ray		O-ray	E-ray
0.200	1.902 84	1.576 49	1.422	1.635 90	
0.303	1.719 59	1.513 65	1.497	1.634 57	1.477 44
0.410	1.680 14	1.496 40	1.609	1.632 61	
0.508	1.665 27	1.489 56	1.682	1.631 27	
0.643	1.655 04	1.484 90	1.749		1.476 38
0.706	1.652 07	1.483 53	1.761	1.629 74	
0.801	1.648 69	1.482 16	1.849	1.628 00	
0.905	1.645 78	1.480 98	1.909		1.475 73
1.042	1.642 76	1.479 85	1.946	1.626 02	
1.159	1.640 51	1.479 10	2.053	1.623 72	
1.229	1.639 26	1.478 70	2.100		1.474 92
1.307	1.637 89	1.478 31	2.172	1.620 99	
1.396	1.636 37	1.477 89	3.324		1.473 92

Refractive indices of crystal quartz at 20 °C

Wavelength μm	Refractive index		Wavelength μm	Refractive index	
	O-ray	E-ray		O-ray	E-ray
0.185	1.675 78	1.689 88	1.5414	1.527 81	1.536 30
0.198	1.650 87	1.663 94	1.6815	1.525 83	1.534 22
0.231	1.613 95	1.625 55	1.7614	1.524 68	1.533 01
0.340	1.567 47	1.577 37	1.9457	1.521 84	1.530 04
0.394	1.558 46	1.568 05	2.0531	1.520 05	1.528 23
0.434	1.553 96	1.563 39	2.30	1.515 61	
0.508	1.548 22	1.557 46	2.60	1.509 86	
0.5893	1.544 24	1.553 35	3.00	1.499 53	
0.768	1.539 03	1.547 94	3.50	1.484 51	
0.8325	1.537 73	1.546 61	4.00	1.466 17	
0.9914	1.535 14	1.543 92	4.20	1.456 9	
1.1592	1.532 83	1.541 52	5.00	1.417	
1.3070	1.530 90	1.539 51	6.45	1.274	
1.3958	1.529 77	1.538 32	7.00	1.167	
1.4792	1.528 65	1.537 16			

Refractive indices of synthetic materials at 20 °C

Wave-length nm	Fused silica (Suprasil) (Spectrosil)	Calcium fluoride	Lithium fluoride	Wave-length nm	Fused silica (Suprasil) (Spectrosil)	Calcium fluoride	Lithium fluoride
213.86	1.534 30	1.484 96	1.432 29	780.02	1.453 67	1.430 79	1.389 24
237.83	1.514 72	1.472 15	1.422 56	794.76	1.453 41	1.430 63	1.389 07
239.94	1.513 36	1.471 25	1.421 86	852.11	1.452 47	1.430 07	1.388 49
269.89	1.498 04	1.460 98	1.413 93	894.35	1.451 84	1.429 71	1.388 09
275.28	1.495 91	1.459 54	1.412 80	1014.0	1.450 24	1.428 84	1.387 07
289.36	1.490 99	1.456 20	1.410 18	1128.7	1.448 87	1.428 15	1.386 17
292.54	1.489 99	1.455 52	1.409 65	1395.1	1.445 83	1.426 80	1.384 10
296.73	1.488 73	1.454 67	1.408 97	1529.5	1.444 27	1.426 17	1.383 02
334.15	1.479 77	1.448 52	1.404 10	1813.1	1.440 71	1.424 83	1.380 56
366.33	1.474 35	1.444 79	1.401 11	1970.1	1.438 53	1.424 06	1.379 09
404.66	1.469 62	1.441 52	1.398 47	2058.1	1.437 23	1.423 61	1.378 22
435.84	1.466 69	1.439 50	1.396 83	2325.4	1.432 93	1.422 16	1.375 30
467.82	1.464 29	1.437 85	1.395 47	2800.0		1.419 23	1.369 37
479.99	1.463 50	1.437 31	1.395 02	3400.0		1.414 87	1.360 32
486.13	1.463 12	1.437 05	1.394 80	4000.0		1.409 63	1.349 37
508.58	1.461 86	1.436 18	1.394 08	4600.0		1.403 57	1.336 40
546.07	1.460 08	1.434 97	1.393 05	5000.0		1.399 08	1.326 56
587.56	1.458 46	1.433 89	1.329 11	5600.0			1.309 88
643.85	1.456 70	1.432 72	1.391 07	5893.2		1.387 12	
656.28	1.456 37	1.432 50	1.390 87	6000.0			1.297 40
667.82	1.456 07	1.432 31	1.390 69	7071.8		1.368 05	
706.52	1.455 15	1.431 71	1.390 14	8250.5		1.344 40	
769.90	1.453 86	1.430 91	1.389 35	9429.1		1.316 05	

Refractive indices of ammonium dihydrogen phosphate (ADP) and potassium dihydrogen phosphate (KDP) at 24.7 °C

Wavelength µm	ADP refractive index		KDP refractive index	
	O-ray	E-ray	O-ray	E-ray
0.2139	1.625 91	1.567 30	1.601 71	1.546 12
0.2288	1.607 94	1.551 48	1.585 44	
0.2537	1.586 91	1.532 95	1.566 36	1.515 94
0.3021	1.562 69	1.511 63	1.544 37	1.497 08
0.3650	1.546 13	1.497 15	1.529 32	1.484 25
0.4047	1.539 65	1.491 57	1.523 44	1.479 32
0.4358	1.535 75	1.488 26	1.519 90	1.476 40
0.5461	1.526 59	1.480 76	1.511 60	1.469 85
0.5791	1.524 65	1.479 25	1.509 86	1.468 57
0.6328	1.521 93	1.477 23	1.507 41	1.466 86
1.0140	1.508 25	1.468 89	1.495 23	1.460 43
1.1287	1.504 46	1.467 04	1.491 89	1.459 18
1.3571			1.484 80	
1.5295				1.455 12

Refractive indices, dispersions and useful transmission range

Material	Refractive index D	V	Temp./°C	Transmission range/μm
Ice	1.31			
Sodium fluoride	1.3255	85.2	20	to 10
Strontium fluoride	1.43			to 11
Sodium aluminium sulphate (soda alum)	1.4388	57.8		
Potassium aluminium sulphate (potash alum)	1.4564	58.3		
Ammonium aluminium sulphate (ammonia alum)	1.4594	58.3		
Barium fluoride	1.474	105.0	25	<0.2 to 13
Potassium chloride	1.4904	44.0	18	0.2 to 21
Sodium chloride: natural	1.5443	42.8	20	} 0.2 to 15
synthetic	1.5498	35.2		
Potassium bromide	1.5599	33.4	20	0.2 to 30
Potassium iodide	1.6666	23.2	20	0.25 to 31
Caesium bromide	1.6971	34.1	24	to 40
Magnesium aluminium oxide, natural spinel, $MgO.Al_2O_3$	1.715 to 1.723			
Magnesium aluminium oxide, synthetic spinels: $MgO._3Al_2O_3$	1.7266	66.0		0.2 to 6
$MgO._5Al_2O_3$	1.729			
Magnesium oxide (Periclase) . . .	1.7373	53.6	20	0.22 to 7
Lead fluoride	1.76			to 11
Silver chloride, synthetic	2.09		24	to 25
Zinc sulphide	2.370	29.2		
Strontium titanate	2.4069			
Diamond	2.4195			
Caesium iodide	1.8064†		24	
Thallium bromo-iodide (KRS 5) . .	2.395‡		25	0.5 to 40
Arsenic trisulphide	2.400– 2.412‡			
Silicon	3.448§		26	
Germanium	4.092§		27	
Birefringent				
Magnesium fluoride: O-ray	1.3777			
E-ray	1.3895			
Sodium nitrate: O-ray	1.5848			
E-ray	1.336			
Sapphire (Aluminium oxide): O-ray .	1.7686	71.8		0.2 to 6
E-ray .	1.7604	73.1		

† Indicates at 0.5 μm;   ‡ indicates at 2.0 μm;   § indicates at 2.153 μm.

Refractive indices of liquids

Suitable media for refractometry; values at or near 20 °C (see also p. 167).

Liquid	D	V
Water	1.333	56
Paraldehyde	1.405	
Carbon tetrachloride	1.46	49
Glycerol	1.47	
Liquid paraffin	1.48	
Toluene	1.497	
Benzene	1.501	30
Ethyl salicylate	1.523	
Chlorobenzene	1.525	30.6
Methyl salicylate	1.538	
Ethyl cinnamate	1.559	20
Benzyl benzoate	1.568	
Aniline	1.586	
Quinoline	1.627	20
α-Monobromonaphthalene	1.660	
Mercury potassium iodide	1.717	
Methylene iodide	1.737	
Methylene iodide and sulphur (saturated)	1.78	
Barium mercuric iodide aq.	1.793	
Potassium iodide and mercuric iodide aq.	1.82†	
Solution 35% by weight CH_2I_2		
,, 31% ,, ,, SnI_4		
,, 16% ,, ,, AsI_3	1.868	
,, 8% ,, ,, SbI_3		
,, 10% ,, ,, S		
Hydrogen disulphide	1.885	
Phosphorus in carbon disulphide	1.95†	
Yellow phosphorus, 8 parts by weight		
,, sulphur 1 ,, ,, ,,	2.06	
Methylene iodide 1 ,, ,, ,,		
Mercuric iodide in aniline or quinoline	2.2†	
Oil, paraffin	1.44	
,, olive	1.46	
,, turpentine	1.47	
,, cedar	1.516	
,, cloves	1.532	
,, cinnamon	1.601	

† Maximum value obtainable.

Refractive index of water at 20 °C for various wavelengths

Wavelength nm	1256.0	670.8	656.3	643.8	589.3	546.1	508.6	486.1	480.0	404.7	303.4	214.4
Refractive index	1.3210	1.3308	1.3311	1.3314	1.3330	1.3345	1.3360	1.3371	1.3374	1.3428	1.3581	1.4032

Temperature coefficient -8.0×10^{-5} °C^{-1}.

Light transmission, Chance optical glass

Extinction coefficient $\alpha = \{\log_e(I_0/I)\}/d$, where $I_0 =$ initial light intensity, $I =$ transmitted light intensity for thickness d. Reflection losses ignored.

Glass type	ZC	BSC	HC	LBC	MBC	LF	DBC	DF	EDF	DEDF	DEDF
n_d	1.5076	1.5097	1.5190	1.5407	1.5694	1.5786	1.6123	1.6205	1.7004	1.7484	1.9271
V	61.2	64.4	60.4	59.5	55.8	41.1	58.5	36.2	30.3	27.8	21.0

Wavelength/nm	α/m^{-1}										
300	254.0	240.0	288.0	260.4	—	—	—	—	—	—	—
325	72.0	55.2	44.4	41.9	117.0	104.1	253.5	276.6	—	—	—
350	16.8	8.4	7.4	6.4	25.3	14.9	56.3	42.9	178.0	403.0	—
375	4.8	3.1	3.8	3.1	7.8	6.7	13.4	13.6	45.6	71.3	—
400	1.4	0.7	1.0	1.3	3.4	1.5	4.3	4.2	13.2	20.7	84.0
425	1.0	0.5	1.0	1.1	3.4	1.2	2.5	2.6	5.0	7.4	27.4
450	0.7	0.5	0.7	0.8	3.0	0.9	2.0	1.9	2.4	4.4	13.4
500	0.4	0.4	0.4	0.7	1.4	0.5	1.3	0.5	0.7	3.4	7.6
550	0.5	0.5	0.4	0.6	0.9	0.4	1.3	0.5	0.7	3.4	5.0
600	0.6	0.4	0.5	1.4	1.8	0.5	1.0	0.4	1.1	3.4	3.4
650	0.2	0.2	0.8	2.0	2.2	0.7	1.0	0.8	1.7	3.9	2.5
700	0.4	0.2	1.0	2.1	2.1	0.9	0.9	0.5	2.2	4.1	2.8
750	0.5	0.2	1.3	2.1	1.8	1.0	0.8	0.5	2.2	4.1	2.8
800	0.7	0.5	1.3	2.2	1.4	0.7	0.8	0.5	2.2	4.1	2.8
850	0.7	0.7	1.5	2.0	1.1	0.9	0.9	0.4	2.2	3.9	2.5
900	0.7	1.0	1.9	2.0	0.9	1.0	0.8	0.3	2.3	3.7	2.2
950	0.7	1.2	1.9	1.7	1.1	1.1	0.8	0.3	2.3	3.7	1.4
1000	0.1	1.8	1.9	1.5	1.1	1.5	1.0	0.3	2.3	3.4	2.5

K.J.H.

1.7.8 Light reflection

The reflectance of a single polished glass surface in air at normal incidence is given by $[(n-1)/(n+1)]^2$, where n is the refractive index of the glass.

Refractive index	1.46	1.52	1.60	1.70
Reflectance/%	3.5	4.3	5.3	6.7

At oblique incidence the reflectance depends on the state of polarization of the light. It is convenient to consider the reflectances for light polarized parallel (R_p) and perpendicular (R_s) to the plane of incidence. The mean of these (R_m) is the reflectance appropriate to unpolarized light. The table is calculated for a glass refractive index of 1.52.

Angle of incidence	R_p/%	R_s/%	R_m/%	Angle of incidence	R_p/%	R_s/%	R_m/%
0°	4.3	4.3	4.3	50°	0.4	11.7	6.1
10°	4.1	4.4	4.3	56.7°†	0	15.7	7.9
20°	3.6	5.0	4.3	60°	0.2	18.3	9.3
30°	2.7	6.1	4.4	70°	4.1	30.8	17.5
40°	1.6	8.1	4.9	80°	48.5	73.9	61.4

† The angle at which $R_p = 0$ is known as Brewster's angle.

The reflection from a glass surface may readily be modified by the application of either dielectric (non-absorbing) or metallic thin film coatings. Dielectric films are used to provide anti-reflection coatings, beam-splitters, and high efficiency 'multilayer' mirrors and rely on optical interference for their performance. In many cases a single layer of optical thickness equal to one quarter of the wavelength of the light concerned is employed. The normal incidence reflectance of such a 'quarter wave' film is given by

$$R = \left(\frac{n_f^2 - n_a n_g}{n_f^2 + n_a n_g} \right)^2$$

where n_f is the refractive index of the film, n_a that of the surrounding medium (often air) and n_g that of the substrate. Thus, for magnesium fluoride ($n_f = 1.39$), a commonly used anti-reflection or 'blooming' material, applied to glass ($n_g = 1.52$) the reflectance in air is reduced to 1.5% at the design wavelength. For visual applications the coating is usually designed for green light, and the slightly higher reflectances for blue and red light account for the familiar magenta colour of bloomed lenses. More sophisticated multilayer coatings can reduce the reflectance at a glass surface to less than 0.5% throughout the visible spectrum. Conversely, if the film applied has a refractive index greater than that of the substrate the reflectance is enhanced to form a beam-splitter. For example, the same formula shows that a film of zinc sulphide ($n_f = 2.37$) applied to glass raises the reflectance to 33%; since there is no absorption, the transmittance is 67%.

Typical properties of common thin film materials (wavelength 550 nm) and the reflectance of 'quarter-wave' films on glass†

Material	Refractive index	Reflectance/%
Cryolite (Na_3AlF_6)	1.35	0.8
Magnesium fluoride	1.39	1.5
Silicon dioxide.	1.58	6
Aluminium oxide	1.65	8
Lead fluoride	1.75	11
Zirconium oxide	2.05	22
Cerium oxide	2.25	29
Zinc sulphide	2.37	33
Titanium oxide	2.70	43

† For glass of refractive index 1.52.

For metals and other absorbing layers the formulae are far more complex. However, in the special case of opaque layers, the normal incidence reflectance in air is given by

$$R = \left(\frac{(n-1)^2 + k^2}{(n+1)^2 + k^2} \right)^2$$

Here, n and k are the real and imaginary parts of the complex refractive index $\mathbf{n} = n - ik$, usually known as the refractive index and extinction index respectively.

Optical constants and reflectance of metals (wavelength 550 nm)

Metal	n	k	$R/\%$
Aluminium	0.8	6.0	92
Silver	0.05	3.3	98
Copper	0.7	2.4	67
Gold	0.3	2.3	81
Steel	2.4	3.4	58

Semi-transparent metal films are used as beam-splitters. Aluminium films are easy to prepare and typically provide 33% reflectance (less when measured through the glass) and 33% transmittance. Silver is used where a high reflectance is required with low absorption, as with Fabry Perot coatings. Typical properties are 92% reflectance and 5% transmittance. Nichrome is chosen as a spectrally neutral coating in cemented cube beam-splitters, providing an equal reflectance and transmittance of 22%.

P.B.C.

1.7.9 Magnetic rotation of polarized light

Polarized light passing through a medium in which there is a magnetic field H parallel to the direction of propagation suffers a rotation of the plane of polarization $\alpha = rHl$, where l is the length of path and r is a constant for the material known as Verdet's constant. r may be conveniently expressed in terms of rotation per ampere turn. It is considered positive if, as is usually the case, the rotation produced by passage through a solenoid with a core of the material is in the same sense as the circulating current. As a rough guide r may be taken to vary with wavelength λ as $1/\lambda^2$.

Verdet constants

Gases †		Solids		
Gas	$r/(10^{-6}\,\text{min A}^{-1})$	*Solid*	λ/nm	$r/(10^{-2}\,\text{min A}^{-1})$
Hydrogen	8.867	Fused silica at 25 °C . .	546.1	2.175
Helium	0.667		435.8	3.565
Nitrogen	8.861		334.2	6.48
Oxygen	7.598		302.2	8.29
Carbon dioxide	13.22		284.8	9.59
Methane	24.15			
Propane	50.05	Crystalline quartz at 20 °C	589.3	2.091
iso-Butane	68.12	(along axis)	546.1	2.453
			435.8	3.997
Chance–Pilkington Glasses at 20 °C and 589.3 nm			257.3	13.56
		Rock salt	670.8	3.08
Glass	$r/(10^{-2}\,\text{min A}^{-1})$		404.7	9.74
			259.9	34.03
Hard Crown ($n_D = 1.519$)	2.4	Fluorite	589.3	1.127
Medium Barium Crown ($n_D = 1.572$)	2.5	(calcium fluoride)	435.8	2.158
Dense Barium Crown ($n_D = 1.612$)	2.4		253.7	7.526
Light Flint ($n_D = 1.579$)	3.9			
Dense Flint ($n_D = 1.623$)	4.85			
Extra Dense Flint ($n_D = 1.700$)	6.55			
Double Extra Dense Flint ($n_D = 1.920$)	13.3			
Plate Glass ($n_D = 1.52$)	2.3			

† At 0 °C, atmospheric pressure and $\lambda = 546.1$ nm.

Verdet constants (*contd*)

Water at 20 °C		Liquids†		
λ/nm	$r/(10^{-2}\,\text{min A}^{-1})$	*Liquid*	*Temp./°C*	$r/(10^{-2}\,\text{min A}^{-1})$
		Acetone	20	1.42
589.3	1.645	Methyl alcohol	20	1.17
546.1	1.935	Ethyl alcohol	20	1.41
435.8	3.168	Carbon disulphide . . .	18	5.28
334.2	5.931	Carbon tetrachloride . .	15	4.03
302.2	7.677	Chloroform	20	2.06
280.4	9.418	Pentane	15	1.48
248.2	13.567	Hexane	15	1.57
		Toluene	28	3.38

† For $\lambda = 589.3$ nm.

R.J.K.

1.7.10 Electro-optic materials

The electro-optic effect

In general, an electric field E applied to a transparent material modifies the refractive index n_0 for a particular mode of propagation in accordance with the equation

$$\frac{1}{n^2} = \frac{1}{n_0^2} + rE + RE^2 + \cdots \text{higher terms}$$

In amorphous materials or centro-symmetric crystals the coefficients of odd powers of E are zero, and the first non-zero term RE^2 represents the quadratic **Kerr** effect, usually characterized for a particular material by the Kerr constant:

$$B = \frac{\Delta n}{\lambda_0 E^2} = \frac{n_0^3}{2\lambda_0} R$$

where λ_0 is the free-space wavelength.

For the case of a birefringent crystal, the field E modifies the index ellipsoid $\sum_{i=1}^{3} (x_i^2/n_i^2) = 1$, in the principal axis co-ordinate system, to:

$$\sum_{i=1}^{3} \sum_{j=1}^{3} \sum_{k=1}^{3} \sum_{l=1}^{3} \left(\frac{1}{n_{ij}^2} + r_{ijk}E_k + R_{ijkl}E_kE_l + \cdots \right) x_i x_j = 1$$

where $n_{ij} = n_i$ for $i = j$, $n_{ij} = 0$ for $i \neq j$. From a knowledge of the change in orientation and dimensions of the index ellipsoid, as given by this equation, the field-induced birefringence for a ray of given direction and polarization may be calculated.

In practice the electro-optic effect is either predominantly linear or quadratic in E, and is therefore characterized by either r_{ijk} or R_{ijkl}, depending on the material. Since r_{ijk} is symmetrical in i, j, and R_{ijkl}, symmetrical in i, j and in k, l, pairs of values i, j and k, l are denoted by indices m and n respectively, which run from 1 to 6 according to the scheme:

$$1 \leftrightarrow 1, 1 \qquad 2 \leftrightarrow 2, 2 \qquad 3 \leftrightarrow 3, 3 \qquad 4 \leftrightarrow 2, 3 \qquad 5 \leftrightarrow 1, 3 \qquad 6 \leftrightarrow 1, 2$$

When measured at low frequencies ($< 10^4$ Hz), the coefficients may include contributions from the elasto-optic effects of piezo-electric and/or electrostrictive strains. The linear electro-optic effect observed at frequencies sufficiently high for these contributions to be negligible is called the **Pockels** effect. The strain effects may contribute up to 50% of the low-frequency effect.

Properties of selected electro-optic materials

Material	Point-group symmetry	$\dfrac{V_\pi}{\text{kV}}$	Coefficient	Value		Refractive index	Relative† permittivity	Transparency range nm
Nitrobenzene	Isotropic ∞	34	B	4400	fm V^{-2}	1.55	35.7	460–1500
CS_2	∞	370	B	36	,,	1.63	2.62	400–3000
H_2O	∞	310	B	52	,,	1.331	80	300–1300
$BaTiO_3$	Cubic $m3m$	0.66	$g_{11}-g_{12}$	+0.13	m^4C^{-2}	2.4	4500 (hf)	430–7000
$KTaO_3$	$m3m$	0.95 at 4.2 K	$g_{11}-g_{12}$	0.16	,,	2.24	~800 (hf)	
			g_{44}	0.12	,,			
$KTa_{0.65}Nb_{0.35}O_3$	$m3m$	~0.6	g_{11}	0.136	,,	2.29	~10^4	400–5500
			g_{12}	−0.038	,,			
			g_{44}	0.147	,,			
$SrTiO_3$	Cubic $m3m$	~10	$g_{11}-g_{12}$	0.14	,,	2.38	300 (hf)	400–5500
CuCl	$\bar{4}3m$	6.2	r_{41}	6.1	pm V^{-1}	1.93	8.3 (hf)	400–20 000
Hexamine	$\bar{4}3m$	91	r_{41}	0.8	,,	1.59	3.2	300–2000
ZnS	$\bar{4}3m$	10.2	r_{41}	2.1	,,	2.36	16	400–1300
$Ba_{0.25}Sr_{0.75}Nb_2O_6$	Tetragonal $4mm$	~0.1	r_{13}	64.5	,,	$n_o=2.31$	~2×10^4	400–6000
			r_{33}	1340	,,	$n_e=2.30$		
$(NH_4)H_2PO_4$ (ADP)	Tetragonal $\bar{4}2m$	10.5	r_{41}	+24.5	,,	$n_o=1.522$	$\epsilon_1=57$	250–1300
			r_{63}	−8.5	,,	$n_e=1.477$	$\epsilon_3=16$	
KH_2PO_4 (KDP)	$\bar{4}2m$	8.7	r_{41}	8.6	,,	$n_o=1.507$	$\epsilon_1=42$	250–1600
			r_{63}	−10.5	,,	$n_e=1.467$	$\epsilon_3=21$	
KD_2PO_4 (KDDP)	$\bar{4}2m$	3.5	r_{41}	8.8	,,	$n_o=1.51$	$\epsilon_3=50$	250–1800
			r_{63}	−26.4	,,	$n_e=1.47$		
$LiNbO_3$	Trigonal $3m$	2.8	r_{13}, r_{22}	8.6, 3.4 (hf)		$n_o=2.30$	$\epsilon_1=78$	400–4500
			r_{33}, r_{51}	30.8, 28 (hf)		$n_e=2.21$	$\epsilon_3=32$	
$LiTaO_3$	$3m$	2.4	r_{13}, r_{22}	7.9, 1 (hf)		$n_o=2.176$	$\epsilon_3=43$ (hf)	
			r_{33}, r_{51}	35.8, 20 (hf)		$n_e=2.180$		

† (hf) denotes value for high-frequency operation.

The quadratic effect is exhibited most markedly by materials in which the permittivity is high and varies rapidly with temperature. Here, since the R_{mn} vary accordingly, it is more convenient to replace $R_{mn}E_kE_l$ in the equation of the index ellipsoid by $g_{mn}P_kP_l$ (where $\boldsymbol{P}=\boldsymbol{D}-\epsilon_0\boldsymbol{E}$), and to express the properties of the material in terms of g_{mn}.

Typical parameters for selected electro-optic materials at low frequencies are given in the table opposite. Co-ordinate convention: the indicatrix and electro-optic coefficients are referred to the usual crystallographic co-ordinate system as follows. $Ox_3 \equiv Oz$ and is the fourfold axis for cubic and tetragonal symmetries, or the threefold axis for trigonal symmetry; $Ox_1 \equiv Ox$; $Ox_2 \equiv Oy$, except for the trigonal case in which Ox_1 is perpendicular to the mirror plane. For uniaxial crystals $n_1 = n_2 = n_o$, $n_3 = n_e$.

It should be noted that the half-wave voltage V_π is used conventionally to characterize the sensitivity of an electro-optic material. It is the voltage required to obtain one half-wavelength of optical path difference between the two vibration components of a wavefront, using a cube of the material of side 1 cm with specified directions of light and applied field. In the table below, half-wave voltages are quoted for $\lambda = 633$ nm and correspond to the following conditions:

Material symmetry	Light direction	Field direction	Induced birefringence Δn
∞	any	transverse	$B\lambda_0E^2$
$\bar{4}2$	Oz	Oz	$n_0^3 r_{63} E_3$
$\bar{4}3m$	Ox	Oz	$n_0^3 r_{41} E_3$
$m3m$	Ox	Oz	$n_0^3(g_{11}-g_{12})\mathrm{P}_3^2/2$
$3m$	Ox	Oz	$n_3^3\left\{r_{33}-\left(\dfrac{n_1}{n_3}\right)^3 r_{13}\right\}E_3/2$
$4mm$	$45°$ to Oz	$45°$ to Oz longitudinal	$(n_3^3/4\sqrt{2})(r_{13}-r_{33})E$

Further reading

For a clear description of tensor representation see J. F. Nye, *Physical Properties of Crystals*, 1964 (Oxford University Press). For discussion of applications of electro-optic materials see A. F. Harvey, *Coherent Light*, 1970 (John Wiley & Sons).

O.C.J.

1.8 Electricity and magnetism

1.8.1 Electrical resistivities

The electrical resistivity, or specific resistance, ρ, is the resistance of a metre cube of a material, and the values given below are in ohm metre units (Ω m). The reciprocal of ρ is the electrical conductivity.

The electrical resistivity is readily influenced and usually increased by factors such as impurity content, porosity, cold work, irradiation, etc. The values given in the main table relate to samples for which such effects are usually minimal. For similar reasons values are not given below 78.2 K, since the effects of impurities, etc., become increasingly important the lower the temperature and finally determine the so-called residual resistance eventually attained for many metals. Certain metals, however, become superconducting when the temperature is decreased sufficiently (see p. 108), and, for such metals the transition temperature is entered after the name of the metal as (K).

In the main table, values given in italics are for the liquid phase. All other values are for a poly-crystalline solid sample. For some non-cubic metals, a separate table indicates the degree of electrical resistivity anisotropy reported for single crystals at normal temperature. A similar degree of anisotropy can be expected throughout the temperature range of the particular crystal form.

Semi-conducting elements are also listed separately, and again only for normal temperature. Since the electrical resistivities of these poor conductors are so strongly dependent on the small amounts of impurity that may be added intentionally, or be present inadvertently, the published values differ considerably, as may the value for any specific sample. The table indicates an approximate value or a likely range of values.

Other elements which are solid at normal temperature and do not conduct electricity are dealt with under electrical insulators (see p. 106). These include boron, carbon in the form of diamond, iodine, phosphorus and sulphur. Elements which are gases at normal temperature and pressure are also omitted.

Values are also given for a few typical alloys. These are mainly the alloys for which thermal conductivity data were given and further compositional details will be found on page 59.

Resistivities of metallic elements

Metal	$\rho/(10^{-8}\ \Omega\ \text{m})$					
	Temperature/K					
	78.2	273.2	373.2	573.2	973.2	1473.2 (or other)
Aluminium (1.17)	0.21	2.50	3.55	5.9	*24.7*	*32.1*
Antimony	8	39	59	—	*114*	*123.5* (1273)
Arsenic	5.5	26	—	—	—	—
Barium	7	36	—	—	—	—
Beryllium	—	2.8	5.3	11.1	26	—
Bismuth	35	107	156	*129*	*155*	*172* (1273)
Cadmium (0.54)	1.6	6.8	9.8	—	—	*36.3* (873.2)
Calcium	0.7	3.2	4.75	7.8	20	—
Cerium	—	73	80	92	110	123 (1053.2)
Caesium	4.5	18.8	*43.5*	*67*	*134*	*295*
Chromium	0.5	12.7	16.1	25.2	47.2	80
Cobalt	0.9	5.6	9.5	19.7	48	88.5
Copper	0.2	1.55	2.23	3.6	6.7	*21.3* (1356)
Dysprosium	26	89	103	124	156.5	184
Erbium	41	81	103	135	183	216
Europium	60	89	—	—	—	—
Gadolinium	20	126	—	—	—	—
Gallium (1.1)	2.75	13.6	*27.2*	*31*	—	—
Gold	0.5	2.05	2.88	4.63	8.6	*31* (1336)
Hafnium (0.35)	—	29.6	42.6	—	—	—
Holmium	34	90	105	134	175	203
Indium (3.35)	1.8	8.0	12.1	*36.7*	47	*55* (1273)
Iridium (0.14)	0.9	4.7	6.8	10.8	22	33.5
Iron	0.7	8.9	14.7	31.5	85.5	122; *139* (1823)
Lanthanum (4.71)	—	54	66	83	105	126 (1173)
Lead (7.2)	4.7	19.2	27	50	*108*	*126* (1273)
Lithium	1.04	8.55	12.4	*30*	40.5	53
Lutetium	16	54	—	—	—	—
Magnesium	0.62	3.94	5.6	10.0	27.7	*28.7* (1173)
Manganese	—	138	—	—	—	—
Mercury (4.12)	5.8	*94.1*	*103.5*	*128*	*214*	*630*
Molybdenum (0.92)	0.7	5.0	7.6	12.7	23.3	37.2
Neodymium	—	61	74	93	120	138 (1183)
Neptunium	—	—	119.3	—	—	121.3 (811.2)
Nickel	0.55	6.2	10.3	22.5	40	*109* (1773)
Niobium (9.1)	3.0	15.2	19.2	27.1	43	59

Resistivities of metallic elements (*contd*)

Metal	$\rho/(10^{-8}\ \Omega\ m)$					
	Temperature/K					
	78.2	273.2	373.2	573.2	973.2	1473.2 (or other)
Osmium (0.65)	—	8.1	11.4	17.8	30.4	46
Palladium	1.73	10.0	13.8	21	33	42
Platinum	1.96	9.81	13.6	21.0	34.3	48.3
Plutonium	~150	146	142	109	—	—
Polonium	—	~40	—	—	—	—
Potassium	1.38	6.1	*17.5*	*28.2*	*66.4*	*160*
Praseodymium	—	65	78	96	118	134 (1123)
Promethium†	—	50	64	89	126	—
Protactinium (1.4)	6.1	17.7	—	—	—	—
Rhenium (1.7)	2.62	17.2	24.9	39.7	63.5	84.4
Rhodium	0.46	4.3	6.2	10.2	20	33
Rubidium	2.2	11.0	27.5	*48*	99	*260*
Ruthenium (0.49)	1.34	7.1	10.0	15.6	27.8	44.4
Samarium	66	91.4	—	—	—	—
Scandium	—	50.5	75	115	167	198
Silver	0.3	1.47	2.08	3.34	6.1	*19.4*
Sodium	0.8	4.2	*9.7*	*16.8*	*39.2*	*89*
Strontium	5	20	30	52.5	94.5	—
Tantalum (4.48)	2.5	12.3	16.7	25.5	43.0	61.5
Technetium (11.2)	—	—	22 6	33.3	51	65
Terbium	27	113	—	—	—	—
Thallium (2.37)	3.7	15	22.8	38	*85*	*88* (1073)
Thorium (1.37)	3.9	14.7	20.8	32.5	53.6	68
Thulium	31	67	—	—	—	—
Tin (3.69)	2.1	11.5	15.8	*50*	*60*	*72*
Titanium (0.39)	4.6	39	58	90	142	—
Tungsten (0.01)	0.6	4.9	7.3	12.4	24	39
Uranium (0.68)	11	28	35	47	—	—
Vanadium (5.03)	2.6	18.2	25.3	38	—	54 (873.2)
Ytterbium	13	27.7	—	—	—	—
Yttrium	15.5	55	—	—	—	—
Zinc (0.85)	1.1	5.5	7.8	13.0	—	37 (773.2)
Zirconium (0.55)	7.3	40	58	88	125	110 (1273)

† Estimated values based on R. K. Williams and D. L. McElroy, USAEC, ORNL-TM-1424, 1966.

Resistivities of semi-conducting elements at normal temperature

Carbon	$\rho/(\Omega\ m)$	Other elements	$\rho/(\Omega\ m)$
Amorphous	~6×10^{-5}	Germanium	$(1-500)\ 10^{-3}$
Graphite	$(3-60) \times 10^{-6}$	Selenium	~0.1
Pyrolytic graphite, along planes	~5×10^{-6}	Silicon	$(1-600) \times 10^{-1}$
,, ,, normal to planes	~5×10^{-3}	Tellurium	~3×10^{-3}

Resistivities of single crystals of some non-cubic metals at normal temperature

Metal	$\rho/(10^{-8}\ \Omega\ \text{m})$ *in direction of*		
	c-axis	*a-axis*	*b-axis*
Antimony	35.6	42.6	
Beryllium	3.58	3.13	
Bismuth	138	109	
Cadmium	7.79	6.54	
Dysprosium	77.4	98.2	
Erbium	47.0	87.6	
Gadolinium	122	135	
Gallium	55.5	17.3	7.85
Holmium	59.9	101.2	
Lutetium	34.0	75.6	
Magnesium	3.78	4.53	
Mercury (at 227.7 K) . .	17.7	23.5	
Tellurium	56 000	15 400	
Terbium	101	122	
Thulium	46	87	
Tin	13.1	10.0	
Zinc	6.05	5.83	

Resistivities of some typical alloys

Alloy†	$\rho/(10^{-8}\ \Omega\ \text{m})$				
	Temperature/K				
	273.2	373.2	573.2	973.2	1473.2
Alpax gamma‡	3.5	5.0	7.85	—	—
Alumel	28.1	34.8	43.8	53.2	65.1
Brass	6.3	—	—	—	—
Bronze	13.6	—	—	—	—
Chromel P	70.0	72.8	79.3	89.3	100.1
Constantan	49	—	—	—	—
German silver	40	—	—	—	—
Lo-Ex‡	3.95	5.45	8.55	—	—
Manganin	41.5	—	—	—	—
Monel	42.9	50.1	52.5	63.3	—
Nichrome	107.3	108.3	110.0	110.3	—
RR 59‡	3.5	4.9	7.6	—	—
RR 77‡	3.95	5.2	7.8	—	—
Steel, Carbon	17.0	23.2	39.8	93.5	123.1
„ 18/8	66.3	74.3	89.1	109.4	124
„ Era ATV	98.0	102.7	111.0	122.0	—
„ Ni–Cr	27.7	33.7	48.7	99.4	122.2
„ Silicon	41.9	47.0	60.1	105.7	127.1
„ Stainless	55.0	63.4	80.2	114.1	—
Pt 90%, Ir 10%	24.8	28.0	—	—	—
Pt 90%, Rh 10%	18.7	21.8	—	—	—
Ti 92.5%, Al 5%, Sn 2.5% . .	155.6	165.6	179.4	—	—
Ti 96.0%, Al 2.0%, Mn 2.0% . .	110.0	123.1	149.5	—	—
Zr 93.5%, Sn 6.7%, C 0.1% . .	132.8	139.5	148.3	—	—
Zr 97.5%, Sn 2.3%, C 0.1% . .	91.5	105.2	123.9	—	—

† See thermal conductivity table, p. 59, for further compositional details.
‡ After heat treatment at 573.2 K, followed by air cooling.

R.W.P.

1.8.2 Resistance alloys and wire resistances

Copper–manganese alloys ($\sim 84\%$ Cu, 12% Mn with nickel, aluminium or germanium as the remaining constituent). These alloys are sold under various proprietary names, and manganin, the pioneer alloy of this group, was for many years the traditional material for high-grade standard resistors. The resistivity is about $40 \times 10^{-8}\ \Omega$ m with a temperature coefficient which can be as low as $0.000\ 003\ °C^{-1}$ over a temperature range 15–$25\ °C$. The thermo-e.m.f. of the alloys against copper is close to zero and may be positive or negative according to composition and heat treatment. Joints between the copper manganese alloys and copper are made most effectively by welding in an atmosphere of argon, and by hard soldering if welding is impracticable.

Copper–nickel alloys ($\sim 55\%$ Cu, 45% Ni). These alloys are manufactured commercially under a wide range of proprietary names, and are used in the construction of standard resistors. The resistivity is about $50 \times 10^{-8}\ \Omega$ m with a temperature coefficient which may lie between $\pm 0.000\ 04\ °C^{-1}$.

The alloys can be soft-soldered with ease, but their high thermo-e.m.f. against copper ($\sim 40\ \mu V\ °C^{-1}$) is a disadvantage in d.c. resistors, although the effect is usually negligible in a.c.

Properties of resistance wires

S.W.G.	Diameter		Copper	Copper–manganese alloys	Copper–nickel alloys	Quaternary alloys	Nickel–chromium		
	mm	in	Ω m^{-1} at 20 °C	Ω m^{-1}	Ω m^{-1}	Ω m^{-1}	Ω m^{-1}	Current (A) to maintain temp. rise	
								500 °C	1000 °C
12	2.642	0.104	0.003 12	0.076	0.090	0.243	0.197	38	78
14	2.032	0.080	0.005 32	0.128	0.151	0.410	0.333	26	53
16	1.626	0.064	0.008 31	0.200	0.235	0.640	0.520	19	40
18	1.219	0.048	0.014 8	0.355	0.420	1.14	0.92	13	27
20	0.914	0.036	0.026 3	0.630	0.745	2.03	1.65	8.5	18
22	0.711	0.028	0.043 4	1.05	1.23	3.35	2.72	6.3	13
24	0.559	0.022	0.070 3	1.69	2.00	5.40	4.40	4.5	9.5
26	0.457	0.018	0.105	2.53	3.00	8.1	6.60	3.5	7.0
28	0.376	0.0148	0.155	3.75	4.40	12.0	9.7	2.7	5.5
30	0.315	0.0124	0.221	5.30	6.30	17.1	13.9	2.2	4.5
32	0.274	0.0108	0.292	7.00	8.30	22.5	18.3	1.9	3.5
34	0.234	0.0092	0.402	9.7	11.4	31.0	25.2	1.6	3.0
36	0.193	0.0076	0.589	14.2	16.7	45.5	37.0	1.3	2.3
38	0.152	0.0060	0.945	22.7	27.0	73.0	59.0	1.0	1.7
40	0.122	0.0048	1.48	35.5	42.0	114	92	0.8	1.4
42	0.102	0.0040	2.13	51.0	60.5	164	133	0.65	1.1
44	0.0813	0.0032	3.32	80	94	255	208	—	—
46	0.0610	0.0024	5.91	142	168	455	370	—	—
48	0.0406	0.0016	13.3	320	380	1030	835	—	—
50	0.0254	0.0010	34.0	820	970	2620	2130	—	—

resistors dropping 1 volt or more. These alloys are also used for current controlling resistors when constancy is more important than low cost.

Nickel–chromium alloys ($\sim 80\%$ Ni, 20% Cr). These alloys are also available under a variety of trade names and are used for heater elements as well as for resistors where high accuracy is not required. Their resistivity is about $110 \times 10^{-8}\,\Omega$ m with a temperature coefficient (20–500 °C) of 0.000 06 °C^{-1}. They will operate satisfactorily at temperatures of up to 1100 °C. A ternary alloy (65% Ni, 15% Cr, 20% Fe) is less expensive than nickel–chromium and is satisfactory at temperatures up to 900 °C.

Quaternary alloys ($\sim 73\%$ Ni, 21% Cr, 2% Al) with copper, iron, cobalt, manganese or molybdenum as the fourth main constituent). These alloys are again processed and marketed under many proprietary names, each of which is characterized by the fourth constituent; they are increasingly being used for standard resistors, especially those of high value. Their resistivity is about $130 \times 10^{-8}\,\Omega$ m, with a temperature coefficient controllable by heat treatment and which can be made as small as $\pm 0.000\,002$ °C^{-1}. The thermo-e.m.f. of the alloys, like that of the copper–manganese alloys, is close to zero and may be slightly positive or negative. The tensile strength of the alloys is high, making possible the drawing of very fine wire. Welding in argon is by far the best method of joining with copper, with hard soldering as an alternative only if welding is not possible.

The table gives resistance of wires in ohms per metre and approximate currents in amperes required to maintain stated temperature rises in straight horizontal wires of nickel–chromium free to radiate in air.

F.J.W.

1.8.3 Electrical insulating materials

The resistivity of an insulator varies greatly with the purity and surface condition of the material, time of application of the voltage, and in some cases with the magnitude of the applied stress. It usually decreases rapidly as the temperature is increased (in many cases changing by orders of magnitude) and may be reduced temporarily or permanently by irradiation.

Conventionally, resistivity measurements are made one minute after application of the measurement voltage. For almost all insulating materials the current at this time is not a true conduction current, but an absorption current involving storage of charge in the insulator. The equilibrium conduction current may be several orders of magnitude smaller.

Some typical one-minute values for both resistivity and surface resistivity are given below for normal temperatures and 50–60% relative humidity. Surface resistivity is defined as the resistance between opposite edges of a square (the size is immaterial) and the unit is the ohm per square.

Resistivity of insulators

Material	Volume resistivity ohm metre	Surface resistivity ohm per square
Beeswax (fresh surface)	10^{12}–10^{13}	$\sim 10^{14}$
Boron	10^{10}	
Ceramics		
Alumina	10^{9}–10^{12}	
Porcelain	10^{10}–10^{12}	10^{11} glazed 10^{9} unglazed
Pyrophyllite	10^{10}–10^{13}	
Special H.F.	10^{14}	
Steatite	10^{11}–10^{13}	
Titanates	10^{6}–10^{13}	
Diamond	10^{10}–10^{11}	

Resistivity of insulators (*contd*)

Material	Volume resistivity ohm metre	Surface resistivity ohm per square
Glass		
Soda-lime	10^9–10^{11}	10^{10}–10^{12}
Borosilicate (Pyrex)	10^{12}	
Plate	2×10^{11}	5×10^{10}
Guttapercha	10^7	
Hard rubber (Ebonite, etc.)	10^{13}–10^{15}	10^{10}–10^{18}
Iodine	10^{13}	
Ivory	10^6	10^9
Marble	10^7–10^9	10^9
Mica, sheet	10^{11}–10^{15}	10^{10}–10^{13}
,, moulded	10^{13}	5×10^{13}
Mineral oil	$> 10^{10}$	
Mineral insulating oil	10^{11}–10^{15}	
Paper (dry)	$\sim 10^{10}$	10^9–10^{10}
Paraffin (kerosine, coloured)	10^{11}–10^{12}	
Paraffin wax	10^{13}–10^{17}	10^{15}
Plastics		
Acrylic (Perspex, etc.)	$> 10^{13}$	$> 10^{14}$
Alkydes or polyester (no filler)	10^{12}–10^{13}	10^{13}–10^{14}
Aminos, melamines (cellulose)	$> 10^9$	10^{12}–10^{14}
,, ,, (mineral)	10^9	10^{12}–10^{14}
Casein	10^7–10^8	10^{10}–10^{11}
Cellulose acetate	10^8–10^{11}	10^{11}–10^{12}
Epoxy cast resin (no filler)	10^{12}–10^{13}	3×10^7–$> 10^{14}$
Phenolics	10^6–10^{12}	10^8–10^{14}
Polyamides (nylon)	10^8–10^{13}	10^{11}–10^{15}
P.C.T.F.E.†	10^{16}	10^{12}–10^{13}
P.E.T.‡	10^{15}–10^{17}	
Polyformaldehyde	$\sim 6 \times 10^{12}$	$> 2 \times 10^{13}$
Polypropylene	10^{13}–10^{15}	$> 10^{15}$
Polystyrene (general purpose)	10^{15}–10^{19}	$> 10^{14}$
,, (toughened)	10^{10}–10^{15}	$> 10^{14}$
P.T.F.E.§	10^{15}–10^{19}	4×10^{12}–10^{17}
Polythene (high density)	10^{14}–10^{15}	10^{12}–10^{17}
,, (low density)	10^{14}–10^{18}	$> 10^{14}$
Polyurethanes	10^9–10^{12}	10^{13}–10^{15}
P.V.C.‖ (rigid)	5×10^{12}–10^{13}	10^{12}–10^{15}
,, (flexible)	5×10^6–5×10^{12}	$> 10^{14}$
Silicone (glass)	10^8–10^{12}	$> 10^{11}$
Quartz,‖ optic axis	10^{12}	
,, ⊥ optic axis	10^{14}	
,, fused	$> 10^{16}$	3×10^{12}
Shellac	10^7	5×10^{13}
Silicone, oils	10^{12}	
,, rubber	10^9	
Slate	10^5–10^6	10^7
Soil	10^2–10^4	
Sulphur	10^{14}–10^{15}	7×10^{15}
Wood (paraffined)	10^8–10^{11}	10^{12}
Water (distilled)	10^2–10^5	

† Polychlorotrifluoroethylene.   § Polytetrafluoroethylene.
‡ Polyethyleneterephthalate.   ‖ Polyvinylchloride.   B.W.P.

1.8.4 Superconducting properties of materials

The superconducting state is characterized by three parameters, a critical temperature (T_c), a critical magnetic field strength (H_c) and a critical current (J_c). If any of these are exceeded, the material is driven into the 'normal' state. In the superconducting state there are two principal effects: magnetic flux is excluded from the interior of the material (the 'Meissner effect'), and there is no electrical resistance. Many superconducting materials are known with critical temperatures ranging from 0 to 21 K, and most of these are metals or alloys. Critical fields extend to 41 tesla† at 4.2 K, and current densities to 10^6 A cm^{-2}. The table below lists the critical temperatures and critical fields at 0 K or 4.2 K of the pure elements in bulk form at atmospheric pressure. Critical temperatures and fields are affected by the purity of the material, by pressure and by the form in which the material is prepared (for example, thin films may have different properties from the bulk material). Critical currents and fields depend upon the temperature, and both J_c and H_c in general decrease smoothly to zero as T increases to T_c. Three elements in the table are of Type 2 and have an upper and lower critical field (see below).

† It is common practice, where no confusion can arise, to express magnetic field strengths in terms of the corresponding flux density for free space, and to use the tesla rather than A m^{-1} as the unit.

Properties of superconducting elements

Periodic Table Group No.	Element	T_c/K	Type	H_c/mT 0 K	H_{c_1}/mT 4.2 K	H_{c_2}/mT 4.2 K
2A	Be	0.03				
3A	La(α)	4.80	1			
	La(β)	5.91		160		
4A	Ti	0.39		10.0		
	Zr	0.55		4.7		
	Th	1.37		15.0		
5A	V	5.30	2		43	82
	Nb	9.2	2		139	268
	Ta	4.48	1	80.5		
	Pa	1.4				
6A	Mo	0.92		9.8		
	W	0.01		0.1		
	U	0.68				
7A	Tc	7.75	2		118	145
	Re	1.70		19.3		
8A	Ru	0.49	1	6.6		
	Os	0.66	1	6.5		
	Ir	0.14		1.9		
2B	Zn	0.85	1	5.3		
	Cd	0.55	1	2.9		
	Hg(α)	4.15	1	41.5		
	Hg(β)	3.95		33.9		
3B	Al	1.19	1	10.2		
	Ga	1.09	1	5.5		
	In	3.4	1	28.5		
	Tl	2.39	1	17.0		
4B	Sn	3.72	1	30.5		
	Pb	7.19	1	80.3		

Superconductors fall into two classes: Type 1 and Type 2. The principal difference between them is that Type 1 has one critical field H_c, while Type 2 has both a lower (H_{c_1}) and an upper (H_{c_2}) critical field. In applied fields between H_{c_1} and H_{c_2} Type 2 superconductors consist of a mixture of normal and superconducting regions, and above H_{c_2} they are fully normal. The table below shows the upper critical field H_{c_2} of several 'beta-tungsten' compounds and of NbN.

Upper critical field H_{c_2} of various compounds at o K

Values obtained by extrapolation from values at 4.2 K

Compound	H_{c_2}/T
Nb_3Sn	24.5
Nb_3Al	29.5
V_3Si	23.5
V_3Ga	21.0
NbN	15.3
$Nb_{0.79}(Al_{0.73}Ge_{0.27})_{0.21}$.	41.0

In the following table critical temperatures of certain well-defined compounds are shown. In the case of compounds, the values are dependent on the stoichiometry of the material. Critical currents have not been given since the values are highly dependent on the preparation and structure of the material. A typical figure for the compound Nb_3Sn, however, would be 5×10^5 A cm^{-2} in an applied field of 5 tesla.

The critical temperatures shown by various crystal structures

Structure type	Compound	T_c/K	Structure type	Compound	T_c/K
Beta-tungsten . .	Nb_3Sn	18.45	Sodium chloride	NbN	15.5
	V_3Ga	16.8		MoC	11.3
	Nb_3Ga	14.5		NbC	10.5
	Nb_3Al	17.1		ZrN	10.0
	V_3Si	17.0			
	$Nb_{0.79}(Al_{0.73}Ge_{0.27})_{0.21}$[†]	20.9	Alpha-manganese	Re_3Mo	9.8
				Re_3W	9.0
Sigma	Nb_5Pt_3	4.2		Re_3Ta	6.8
	Nb_3Ir_2	9.8			
	Cr_2Ru	2.0	B.C. tetragonal	Mo_3P	5.3

† $Nb_{1-y}(Al_{1-x}Ge_x)_y$.

J.A.C.

1.8.5 Dielectric properties of materials

The relative permittivity or dielectric constant ϵ (formerly known as specific inductive capacity) of a material is defined as the ratio of the capacitance of a capacitor with the material as dielectric to that of one with the same linear dimensions, but with vacuum as dielectric. The term 'relative permittivity' is preferred as being the SI nomenclature, and also because the word 'constant' is inappropriate for a quantity which can vary greatly with frequency and temperature. In what follows 'relative permittivity' is usually abbreviated to 'permittivity', although strictly speaking this denotes the quantity $\epsilon_0\epsilon$ where ϵ_0 is the permittivity, in its proper sense, of free space.

If ϵ is measured with an alternating field, it varies with the frequency used and is associated with loss of energy in the dielectric. The simple formula for the electric displacement in terms of the field $D=\epsilon_0\epsilon E$ is therefore not in general adequate; D is not strictly in phase with E but lags by a phase difference δ, which corresponds to the power dissipation. These facts are described by the more complete equation

$$D = \epsilon_0(\epsilon - j\epsilon')E = \epsilon_0\epsilon(1 - j\tan\delta)E$$

where $\epsilon - j\epsilon'$ is called the complex permittivity, δ the loss angle and $\tan\delta$ the loss tangent. The power factor of the material is strictly $\sin\delta$, but when δ is small, as is often the case, $\sin\delta \simeq \tan\delta$. The symbols ϵ and ϵ' are often written ϵ' and ϵ'' respectively. This is unnecessary, since ϵ in the simple definition for the displacement is identical with the real part of the complex permittivity. The table gives values of ϵ, but ϵ' is given only through $\tan\delta$, which is usually preferred in experimental work. The two are related by $\epsilon' = \epsilon\tan\delta$.

When electromagnetic waves traverse a dielectric, the loss of energy appears as a reduction of their amplitude with distance travelled, and is represented by an absorption coefficient κ, where $e^{-2\pi\kappa}$ is the factor by which the amplitude is reduced in travelling a distance equal to the wavelength in a vacuum. The absorption coefficient and the refractive index, n, together constitute a complex refractive index $n - j\kappa$ analogous to the complex permittivity, and for non-magnetic materials ($\mu = 1$)

$$(\epsilon - j\epsilon') = (n - j\kappa)^2 \quad \text{or} \quad \epsilon = n^2 - \kappa^2$$
$$\epsilon' = 2\kappa n$$

Permittivities at very high frequencies may be calculated from observed values of n and κ by means of these relations. All the quantities vary with frequency (or wavelength) but ϵ and n are sometimes nearly constant over a range of frequency, and κ is negligibly small over any range in which the material is transparent. When these conditions are satisfied, the well-known approximate relation $\epsilon = n^2$ holds good.

Non-polar organic materials (roughly, those with $\epsilon < 2.5$) show nearly constant values of ϵ and ϵ' over the electrical frequency spectrum. In other organic materials, however, and all the ferro-electrics, these properties are very dependent on temperature, or frequency, or both, and cannot be adequately presented in a table. In simple organic materials having polar molecules with a single time (τ) of response to a field, ϵ consists of a constant part ϵ_∞, and a polar part which varies with frequency and temperature according to the Debye equations

$$\epsilon = \epsilon_\infty + \epsilon_p/(1 + \omega^2\tau^2)$$
$$\epsilon' = \omega\tau\epsilon_p/(1 + \omega^2\tau^2)$$

The symbol ϵ_p can also be written $(\epsilon_0 - \epsilon_\infty)$, where ϵ_0 is the permittivity at a low frequency, not so low that there is any contribution to its value from space charges. The temperature variation is represented, in the simplest case, by substituting

$$\tau = \text{constant } e^{v/kT}$$

where v is an activation energy for the material. It follows that within some range of temperature, different for each frequency and material, ϵ will increase sharply from ϵ_∞ to ϵ_0 with increasing temperature, and $\tan\delta$ will pass through a maximum value given by

$$\tan\delta\,(\text{max}) = \epsilon_p/2(\epsilon_\infty\epsilon_0)^{1/2}$$

Many industrial organic insulating materials are polar, but few have a molecular structure so simple as to follow the foregoing behaviour closely. The transition is spread over wider ranges of frequency and temperature, and the value of $\tan\delta$ (max) is correspondingly reduced. Common values range from one half of the theoretical maximum to transitions so widely spread that ϵ and $\tan\delta$ appear nearly constant. The tables should be used with all the foregoing facts in mind.

The properties of ferroelectric substances are so dependent on many factors that tabular data are misleading, and these materials are omitted from the table. In general, practically useful permittivities exceeding 1000 can be obtained, together with a rather high loss tangent, usually >0.01, and these properties are necessarily dependent upon temperature and the field intensity. More detailed information is given in the reference cited later.

References

(1) H. Fröhlich, *Theory of Dielectrics*, 2nd edn, 1958 (Clarendon Press).
(2) A. R. von Hippel, *Dielectric and Waves*, 1954 (Chapman and Hall).
(3) V. V. Daniel, *Dielectric Relaxation*, 1967 (Academic Press).

Relative permittivity(ϵ) and loss tangent (tan δ)

Temperature (t) is in °C, and frequency (f) in Hz, k($\times 10^3$), M($\times 10^6$), G($\times 10^9$). Temperature coefficient of ϵ is denoted by $a = 10^5 \, d\epsilon/\epsilon \, dt$ and density in g cm^{-3} by d. For non-cubic crystals, the symbols $\perp$, $\parallel$, indicate measurements with field respectively perpendicular to and parallel to the c-axis. Ranges of quantities are indicated by the numerical limits of the range, separated by a solidus. For commercial materials, the values should be regarded as examples only, since some vary greatly with composition and purity. This applies also to the loss angle of some pure materials, which may depend on traces of impurity. The ranges of ϵ and tan δ, however, indicate not these variations, but only the variation within the stated ranges of temperature and/or frequency.

Solids

Material	Remarks	t/°C	f/Hz	ϵ	$10^4 \times \tan \delta$
Cellulose (see also paper)					
Cellophane	unplasticized	20	50/1 M	7.6/6.7	100/650
		−30/70	50	7.2/8.0	100/150
Paper fibres	calculated	20	50	6.5	50
Ceramics					
Alumina	pure	20/100	50/1 M	8.5	20/5
	pure, porosity 1%	20	1 M	10.8	
Calcium titanate	$a = -200$	20	1 M	150	3
Lead zirconate	$a = +140$	20	1 M	110	30
Magnesium titanate		20/150	50/1 M	14	1/4
Porcelain	h.v. electrical	20/100	50/1 M	5.5	300/80
Rutile	$a = -80$	20	1 M/1 G	80	3/8
	$a = -40$	20	1 M/1 G	40	15/30
	$a = -2$	20	1 M/100 M	12	30
	$a = +6$	20	1 M/100 M	15	1
Steatite	$a = +13$	20	1 M/1 G	6	20
(low loss)	$a = +13$	20	1 M/1 G	6	2
Strontium titanate	$a = -300$	20	1 M	200	5
Strontium zirconate	$a = +12$	20	1 M	38	3
Crystals (single, inorganic)					
Alkali halides†					
LiF		25	1 k/10 G	9.1	2
LiCl		20	1 M	11.0	
LiBr		20	1 M	12.1	
LiI		20	1 M	11.0	
NaF		20	1 M	6.0	
NaCl		25	1 k/10 G	5.9	5/1

† The concentration around nearly integral values for the permittivities of these 12 alkali halides appears from the available data to be fortuitous and does not imply an uncertainty of ±0.5 units.

Solids (*contd*)

Material	Remarks	$t/°C$	f/Hz	ϵ	$10^4 \times \tan \delta$
Crystals (*contd*)					
NaBr		20	1 M	6.0	
NaI		20	1 M	6.6	
KF		20	1 M	6.0	
KCl		20	1 M/10 G	4.8	
KBr		25	1 k/10 G	4.9	2/7
KI		20	1 M	5.0	
Calcite	$CaCO_3 \perp$	20	1 k	8.5	
	∥	20	1 k	8.0	
Diamond	C	20	500/3 k	5.68	
Fluorite	CaF_2	20	2 M	6.8	
Mica, muscovite (best)		20/100	50/100 M	7.0	10/2
Periclase	MgO	25	100/100 M	9.7	3
Quartz	$SiO_2 \perp$	20	1 k	4.5	2
	∥	20	1 k	4.5	2
Rutile	$TiO_2 \perp$	20	50/100 M	86	100/2
	∥	varies critically with stoichiometric ratio			
Sapphire	$Al_2O_3 \perp$	20	50 /1 G	9.4	2
	∥	20	50/1 G	11.6	2
Sulphur	rhombic (100)	25	1 k	3.8	5
	(010)	25	1 k	4.0	5
	(001)	25	1 k	4.4	5
Glasses					
Borosilicate	normal	20	1 k/1 M	5.3	50/40
	low alkali	20	1 M	5	30
	very low alkali	20	50/100 M	4	15/5
Fused quartz		20/150	50/100 M	3.8	10/1
Lead		20	1 k/1 M	6.9	17/13
Soda	average	20	1 M/100 M	7.5	100/80
Minerals					
Amber		20	1 M/3 G	2.8/2.6	2/90
Asbestos (chrysotile)	purified, 50% R.H.	25	50/1 M	5.8/3.1	1800/250
	board	20	1 M	3	2200
Bitumen	Gilsonite	25	50/100 M	2.7/2.55	60/10
		20	1 k	3.5	300
Granite		20	1 M	8	
Gypsum		20	10 k	5.7	
Marble	pure, dry	20	1 M	8	400
Sand	dry	20	1 M	2.5	
	15% water	20	1 M	9	
Sandstone		20	1 M	10	
Soil	dry	20	1 M	3	
	moist	20	1 M	10	
Sulphur	cast	20	3 G/10 G	3.4	7/14
Paper and Pressboard					
(see also cellulose)					
Unimpregnated, dry					
Kraft (cable)	$d = 0.9$	20/90	50	2.1	15/20
Kraft (tissue)	$d = 0.8$	20/90	1 k	1.8	10/15
	$d = 1.2$	20/90	1 k	3.0	25/35
Kraft		20	1 k	2.9	45
Rag (cotton)	$d = 0.6$	20/90	50/50 k	1.7	8/65

Solids (*contd*)

Material	Remarks	$t/°C$	f/Hz	ϵ	$10^4 \times \tan \delta$
Paper and Pressboard (*contd*)					
Impregnated, mineral oil ($\epsilon = 2.2$)					
Kraft (cable)	$d = 0.8$	20/90	50	3.2	18/20
	$d = 1.2$	20/90	50	4.7	28/33
Kraft (tissue)	$d = 0.9$	20	50	3.6	22
	$d = 1.1$	20	50	4.3	27
Rag (cotton)	$d = 0.9$	20	50	3.5	13
	$d = 1.1$	20	50	4.2	18
Impregnated (Pentachlordiphenyl)					
Kraft (tissue)	$d = 0.9$	20	50	5.7	33
	$d = 1.1$	20	50	6.0	39
Fibre		20	1 M	4.5	500
Pressboard	dry $d = 0.8$	20	50	3.2	80
Plastics (non-polar, synthetic)					
Poly-					
ethylene		20	50/1 G	2.3	2/3
isobutylene		20	50/3 G	2.2	2/5
4-methylpentene (TPX) . .		20	100/10 k	2.1	2/1
(dimethyl)phenyloxide (PPO)		25	100/1 M	2.6	4/7
propylene		20	50/1 M	2.2	5
styrene		20	50/1 G	2.6	2/5
tetrafluorethylene . . .		20	50/3 G	2.1	2
Plastics (polar, synthetic)					
Poly-					
amides	typical Nylon	20	50/100 M	4/3	200
carbonates	typical	20	50/1 M	3.2/3.0	10/100
ethyleneterephthalate . . .		20	50/100 M	3.2/2.9	20/150
imides	typical	20	1 M	3.4	
methylmethacrylate . . .		20	50/100 M	3.4/2.6	600/60
vinylcarbazole		20	50/100 M	2.8	5/10
vinylchloride	unplasticized	20	50/100 M	3.2/2.8	200/100
Plastics (miscellaneous)					
Aniline resin	unfilled	20	3 G	3.5	500
	paper filled	20	1 M/1 G	5/4	600/300
		100	1 M	6	800
Cellulose acetate		20	1 M/1 G	3.5	300/400
Cellulose triacetate		20	50/100 M	3.8/3.2	100/300
Ebonite	unfilled	20	1 k/1 G	3/2.7	90/30
	filled ($MgCO_3$)	20	50/1 G	4.1/3.8	100/180
Melamine resin		20	3 G	4.7	400
Phenolic resin	fabric filled	20	1 M	5.5	500
	paper filled	20	1 M/1 G	5	300/800
		140	1 M/10 M	6	800/400
	wood filled	20	1 M	5	400
Urea resin	paper filled	20	1 M	6	300
Vinyl acetate (poly-) . . .	plasticized	20	1 M/10 M	4	500
Vinyl chloride (poly-) . . .	plasticized	20	1 M/10 M	4	600

Solids (*contd*)

Material	Remarks	$t/°C$	f/Hz	ϵ	$10^4 \times \tan \delta$
Rubbers					
Natural	crepe	20/80	1 M/10 M	2.4	15/100
	vulcan, soft	20	1 M/10 M	3.2	280/200
Butadiene/styrene.	unfilled	20/80	50/100 M	2.5	5/70
(GR-S)	compounded	20/80	50/100 M	2.5	10/200
Butyl	unfilled	20	50/100 M	2.35	35/10
Chloroprene	Neoprene	20	1 k/1 M	6.5/5.7	300/900
Silicone	filled 67% TiO_2	20	50/100 M	8.6/8.5	50/10
Waxes, etc.					
Chlornaphthalene					
(tri and tetrachlor-) . . .		20	50/100 M	5.4/4.2	7/2700
Ozokerite		20	50/100 M	2.25	5/10
Paraffin wax		20	1 M/1 G	2.2	2
Petroleum jelly		20/60	50	2.1/1.9	1/5
Rosin	colophony	20	3 G	2.4	6
Wood (% water)					
Balsa 0%		20	50/3 G	1.4/1.2	40/140
Beech 16%	$d=0.62$	20	1 M/100 M	9.4/8.5	580/830
Birch 10%.	$d=0.63$	20	1 M/100 M	3.1	400/800
Douglas fir 11%	$d=0.45$	15	1 M/10 M	3.2	520/810
compressed	$d=0.64$	15	1 M/10 M	4.3	570/950
Mahogany.		20	3 G	2	340
Scots pine 15%	$d=0.61$	20	1 M/100 M	8.2/7.3	590/940
Walnut 0%		20	10 M	2	350
Walnut 17%		20	10 M	5	1400
Whitewood 10%	American	20	1 M/100 M	3	400/750

References

Solids: A. R. von Hippel, *Dielectric Materials and Applications*, 1954 (Chapman and Hall).
             Handbook of the American Institute of Physics, 2nd edn, 1963 (McGraw-Hill Book Co.).
             The International Critical Tables.

Ferroelectrics: F. Jona and G. Shirane, *Ferroelectric Crystals*, 1962 (Pergamon Press).

Relative permittivity and loss tangent of liquids

The permittivities in this table, except when a frequency is stated, are 'static' values, relating to frequencies high enough to exclude ionic conductivity, but below any region of dispersion. For non-polar liquids ($\epsilon \sim 2$) the lower limit depends only upon ionic impurities, while the higher is usually above 10 GHz. Within these limits, the permittivity is constant, and the loss tangent unlikely to exceed a few times 10^{-4}. For polar liquids, denoted by 'P', the lower frequency limit depends both upon purity and the intrinsic dissociation of the liquid, while the upper limit varies sharply with temperature. The two limits may overlap, and the permittivity is then nowhere constant, nor the loss tangent small. For these reasons, frequency and loss tangent are quoted only for a few liquids of controlled purity which are used for electrical purposes. More extensive data, and on many more liquids, are given in the references below. The permittivity of liquids is easier to determine accurately than that of solids, and probable accuracy is indicated by the number of places quoted in the Table. Temperature coefficients ($a = 10^5 \, d\epsilon/\epsilon \, dt$) are given, but their accuracy is sometimes low.

Material	Remarks	$t/°C$	f/Hz	ϵ	$10^4 \times \tan \delta$
Castor oil		20	1 k	4.5	
Chlordiphenyl (tri-)		−10/100	50/20 k	7/5	2000/2
(penta-)		0/100	50	5.2/4.3	700/3
Paraffin oil	medicinal	20	1 k	2.2	1
Silicone fluid	0.65 cS	20	50/3 G	2.2	2/19
	1000 cS	20	50/3 G	2.78/2.74	1/100
Transformer oil	BS 138	20	50/100 M	2.2	1/42
		20	100 M/10 G	2.2	42/8

Material	$t/°C$	ϵ	a	Material	$t/°C$	ϵ	a
Alcohols (primary)				$CHCl_2F$	28	5.34 P	
Methanol	25	32.6 P	−600	$CHClF_2$	24	6.11 P	
Ethanol	25	24.3 P		$(-CCl_2F)_2$	25	2.52	
Propanol	25	20.1 P		$(-CClF_2)_2$	25	2.26	
Butanol	20	17.8 P	−700	$(-CH_2Cl)_2$†	20	10.66 P	−550
Pentanol	25	13.9 P		$(=CCl_2)_2$	25	2.30	−85
Hexanol	25	13.3 P		$CCl_2=CHCl$	20	3.4 P	
				F-pentane	20	4.24 P	
Hydrocarbons				F-benzene	25	5.42 P	
n-Pentane	20	1.84_4	−87	Cl-benzene†	20	5.70_8 P	−305
n-Hexane	20	1.88_5	−81				
n-Heptane	20	1.92_0	−75	**Miscellaneous**			
n-Octane	20	1.94_8	−70	Aniline	20	6.90 P	−340
n-Nonane	20	1.97_2	−67	Acetone	25	20.7 P	−470
n-Decane	20	1.99_1	−64	Diethylketone	20	17.0 P	−520
n-Undecane	20	2.00_5	−62	Diethylether	20	4.33 P	−46
n-Dodecane	20	2.01_4	−60	Cyclohexanone	20	18.3 P	
Benzene†	20	2.283_6	−85	Nitrobenzene	25	34.8 P	−520
Cyclopentane	20	1.96		CS_2	20	2.64	−100
Cyclohexane†	20	2.025_0	−79				
Toluene	20	2.39	−100	**Liquid gases**	T/K		
				Argon	82	1.53	−220
(Chloro/Fluoro)-				Helium	4.19	1.048	
hydrocarbons					2.06	1.055	
CCl_4	20	2.24	−90	Hydrogen†	20.4	1.22_8	−280
CCl_3F	29	2.28		Nitrogen	70	1.45	−200
CCl_2F_2	29	2.13		Oxygen†	80	1.50_7	−160
$CClF_3$	−30	2.3					
$CHCl_3$	20	4.80 P	−370				

† These liquids have been recommended as standards for calibrations. The values given apply to especially pure and anhydrous samples. Cyclohexane is preferred.

References

(1) *Dielectric Constants of Pure Liquids*, National Bureau of Standards Circular No. 514, 1951.
(2) *Dielectric Dispersion Data for Pure Liquids and Dilute Solutions*, National Bureau of Standards Circular No. 589, 1958.

Relative permittivity and loss tangent of water

Water is strongly polar, with a region of dispersion, at 20 °C, centred around 17 GHz. It is also intrinsically dissociated, so that even de-ionized water cannot be treated as a dielectric at frequencies much below 1 MHz. Measurements at the high frequencies of the dispersion range contain, before about 1953, many errors shown by values of ϵ and ϵ' mutually inconsistent with the simple Debye equations. It is established that water obeys these with some accuracy, the value of the dispersion coefficient α in the Cole–Cole equation not exceeding 0.05.

At frequencies higher than about 1 MHz (where the loss tangent passes through a minimum of about 5×10^{-3}) values of ϵ and ϵ' can be calculated from the Debye equations given in the introduction to this section, with accuracy better than is obtainable by interpolation in a table. The necessary values of ϵ_0, ϵ_∞ and τ follow, chosen to give the best fit with recent internally consistent data.

The static relative permittivity as a function of temperature, within the range 0–60 °C, is given with an accuracy of ± 0.1 unit by

$$\epsilon_0 = 88.15 - 41.4\theta + 13.1\theta^2 - 4.6\theta^3$$

where $\theta =$ Celsius temperature/100 °C.

The relaxation time as a function of temperature is as follows, with an accuracy of about $\pm 2\%$:

$t/°C$	0	10	20	30	40	50	60
τ/ps	17.7	12.6	9.2	7.1	5.7	4.8	3.9

The 'infinite frequency' dielectric constant, ϵ_∞, occurs in the infra-red, and cannot be directly measured electrically. A value of 5.0 is appropriate for use with the foregoing data. It decreases continuously from this value throughout the infra-red to a value of 1.8 in the optical region. The Debye equations therefore become increasingly inaccurate for $\omega\tau \gg 1$.

References

(1) *Dielectric Dispersion Data for Pure Liquids*, National Bureau of Standards Circular No. 589, 1958, Table 3.
(2) E. H. Grant and R. Shack, *Br. J. appl. Phys.*, 1967, **18**, 1807.

Relative permittivity of gases and vapours

The values relate, excepting the final entry, to a pressure of one standard atmosphere, and hold for all frequencies below the start of the infra-red spectrum. Other values may be calculated over a limited range of temperature and pressure, for non-polar permanent gases, by assuming that $(\epsilon - 1)$ is proportional to density. This does not hold for polar gases, but if the polarity is strong (e.g. water vapour) a close approximation is

$$(\epsilon - 1) \propto \text{pressure}/(\text{absolute temperature})^2$$

provided that the vapour is not near its condensation point, under the conditions either of the data used, or of the desired result. This relation can safely be used, for example, to obtain values for damp air, the densities and pressures involved being then the partial values, and the contributions from the two components additive. The relation should not be applied to mixtures of two polar vapours.

Values of relative permittivity may also be obtained from the data on refractive indices at radio frequencies (p. 87) by using the relation $\mu\epsilon = n^2$ which applies to non-absorbing gases. $\mu = 1$ for all gases except O_2 where $\mu = 1 + 1.9 \times 10^{-6}$.

Relative permittivity of gases and vapours (*contd*)

Material	$t/°C$	$10^4(\epsilon-1)$	Material	$t/°C$	$10^4(\epsilon-1)$
Air (dry)	20	5.36_1	Nitrous oxide	25	10.3
Nitrogen	20	5.47_4	Ethylene	25	13.2
Oxygen	20	4.94_3	Carbon disulphide . . .	29	29.0
Argon.	20	5.16_7	Benzene	100	32.7
Hydrogen	0	2.72	Methanol	100	57
Deuterium	0	2.69_6	Ethanol	100	78
Helium	0	0.7	Ammonia	1	71
Neon	0	1.3	Sulphur dioxide	22	82
Carbon dioxide	20	9.21_6	Water	100	60
Carbon monoxide . . .	25	6.4	Water (10 mmHg) . . .	20	1.21_4

C.G.G.

1.8.6 Magnetic properties of materials

The magnetic properties of materials are expressed in terms of the magnetic field strength H, flux density B, and magnetic polarization (intensity of magnetization) J. The SI units of H and B are, respectively, ampere per metre $(A\ m^{-1})$ and tesla (T) and these are the units in which the magnetic properties of the materials listed in this section are defined.

The relation between the quantities expressed in SI units is:

$$B = \mu_0 H + J$$

in which μ_0 is $4\pi \times 10^{-7}\ H\ m^{-1}$, the permeability of free space. The absolute permeability, $\mu\ (=B/H)$ and the volume susceptibility $\kappa\ (=J/\mu_0 H)$, are thus related by the equation:

$$\mu = \mu_0\ (1 + \kappa)$$

The mass susceptibility χ is equal to κ/ρ, where ρ is the density. The relative permeability μ_r, which is equal to μ/μ_0, is the permeability of the material relative to that of a vacuum and is the value given in the tables.

In ferromagnetic materials as H is increased steadily from zero the permeability changes and is at first relatively small, its value being defined as the initial permeability, then reaches a maximum value, and finally decreases towards μ_0 as the polarization tends towards a limiting value $(B-\mu_0 H)_s$. The flux density remaining when H is reduced to zero is the remanent flux density and the negative H needed to reduce B to zero is the coercive force. The remanent flux density and coercive force for a cycle which proceeds to saturation are called the remanence, B_r, and the coercivity H_c.

When a ferromagnetic material is taken through a cycle of magnetization there is a loss of energy as heat due to the combined effects of hysteresis and induced eddy currents. The hysteresis loss per unit volume, $Q_h = \oint H\,dB$, has been shown empirically to vary as $B_{max}^{1.6}$ over a limited range of peak flux density of up to about 1 T for high saturation materials, and 0.5 T for low saturation materials. This relationship, known as the Steinmetz law, is nevertheless only approximate. Some indication of the second loss, namely the eddy current power loss, may be calculated from standard formulae once certain relevant physical parameters are known. In their present forms, however, these formulae are only approximate. The total power losses that will be dissipated in laminar material when an alternating flux is developed in it has a direct bearing on the efficiency that can be realized in equipment such as transformers and electric motors and should therefore be known accurately. Accordingly the power losses of representative forms of typical materials are measured and some of these are given in Table (2) in terms of power loss per unit mass.

Many magnetic properties of ferromagnetic materials depend greatly on previous history, state of strain, temperature, and size, perfection and orientation of crystals, and the effect of small traces of impurity may be enormous.

When heated, ferromagnetic materials become paramagnetic at a temperature known as the (ferromagnetic) Curie point.

Properties of ferromagnetic materials

(Since the properties may vary considerably from specimen to specimen, the values given are only to be regarded as typical of the materials mentioned.)

(1) Soft (low coercivity) materials

Material (Approx. % composition-balance iron)	Flux density B/T for $H/(A\ m^{-1})$ =						
	0.8	8	40	160	800	4000	40 000
Ferromagnetic elements							
Iron-high purity	1.40	1.98	2.01				
(Single crystals in preferred direction)							
Armco iron 99% Fe		0.004	0.05	1.10	1.50	1.70	2.10
Electrolytic iron (annealed)		0.010	0.75	1.40	1.65	1.72	2.10
Cast iron (annealed)				0.06	0.50	0.85	1.40
Carbonyl iron 99.98% Fe							
Carbonyl iron powder cores							
C type high pressure 99.95% Fe . .							
E type low pressure 99.0% Fe . .							
Swedish iron (annealed)		0.005	0.10	1.00	1.48	1.70	2.10
Nickel				0.16	0.43	0.55	0.65
Nickel (wrought) 99% Ni							
Cobalt				0.030	0.17	0.60	1.15
Gadolinium	Properties dependent on temperature at normal temperatures						
Steels							
Carbon steel 1% C (annealed)				0.045	0.65	1.38	1.90
Cast steel		0.003	0.025	0.40	1.30	1.65	2.05
Constructional steels							
0.3% C 1% Ni				0.30	1.24	1.63	2.06
0.4% C 3% Ni, 1.5% Cr				0.05	0.65	1.62	2.02
Mild steel 0.1% C		0.003	0.030	0.60	1.40	1.70	2.10
Silicon–iron							
Dynamo grade sheet 2% Si		0.010	0.12	1.00	1.40	1.60	2.14
Transformer grade sheet 4% Si		0.020	0.45	1.08	1.35	1.54	1.97
Grain oriented sheet 3% Si							
(a) Along preferred direction		0.300	1.40	1.65	1.76	1.90	2.04
(b) Across preferred direction		0.01	0.05	0.65	1.30	1.50	2.00
Nickel–iron alloys							
Supermumetal[1] / Super C[2] / Nilomag 771[3]		0.30	0.70	0.72	0.80		
Mumetal Plus[1] / Mumetal[1] / Permalloy C[2] / Sanbold NA76[4] / EPC 20[3] / Satmumetal (70–80% Ni with small amounts other elements)		0.05	0.40	0.70	0.75		

Symbols used in tables

B = flux density H = magnetic field strength

$J_s = (B - \mu_0 H)_s$ = saturation polarization H_c = coercivity

B_r = remanence Q_h = hysteresis loss per unit volume per cycle

Range of values indicated by a dash.

Relative permeability μ_r				J_s	H_c	B_r	$Q_h(\text{J m}^{-3})$ for $B_{max}=$		t_c	ρ
Initial		Maximum		$\dfrac{}{\text{T}}$	$\dfrac{}{\text{A m}^{-1}}$	$\dfrac{}{\text{T}}$	1.0 T	0.5 T	$\dfrac{}{°\text{C}}$	$\dfrac{}{10^{-8}\ \Omega\text{ m}}$
$\dfrac{\mu_r}{1000}$	$\dfrac{H}{\text{A m}^{-1}}$	$\dfrac{\mu_r}{1000}$	$\dfrac{H}{\text{A m}^{-1}}$							
		1500		2.16	12			30	770	10
0.25		7		2.16	80	1.3	500		770	11
				2.16	30		60		770	10
				1.70	400		1000			
3		30		2.05	8				770	12
0.05	(approx. constant value)							I		
0.01	(approx. constant value)							0.2		
				2.16	70		220		770	
				0.615	400				358	9
0.25		2		0.6	120	0.3			358	7
				1.76	950				1115	9
									16	
				2.00	600					
				2.10	250		700			
				2.10	250					
				2.05	500					
				2.15	150		500			10
1.0		6		2.10	60		250		730	35
2.0		35		1.93	30		140		690	60
5.0		40		1.98	8		30		720	48
									720	48
50–100	0.08	250–1000	1.2	0.80	0.40–0.80	0.35–0.55		0.5–1.0	350	55
20–50	0.80	50–200	2.4	0.75	1.6–4.0	0.25–0.60		5.0–10.0	350	55
65		240		1.5	2.0	0.7			550	60

(1) Soft (low coercivity) materials (contd)

Material (Approx. % composition-balance iron)	Flux density B/T for $H/(\text{A m}^{-1})$ =						
	0.8	8	40	160	800	4000	40 000
Powder form of this group							
Small amount ceramic binder							
Larger amount ceramic binder							
Nilomag 641[3] / Permalloy F[2] } Approx. 65% Ni + small amount other elements + oriented structure							
Nilomag 471[3] / Super Radio-metal[1] / Permalloy B[2] / Radiometal 50 / Sanbold NA 47[4] } Approx. 50% Ni + small amounts other elements	0.02	1.0	1.2	1.5	1.6		
	0.005	0.4	1.0	1.2	1.6		
HCR alloy[1] / Permalloy G } + oriented structure		0.05	1.45	1.53	1.54		
Permalloy D[2] / Rhometal[1] } Approx. 35% Ni		0.10	0.64	0.95	1.2		
Hyperm 36 Approx. 36% Ni	(Constant permeability alloy)						
Nickel Iron 30% Ni					0.25	0.31	
Cobalt–iron alloys							
Permendur 24[1,2] 24% Co.		0.002	0.013	0.12	1.40	1.80	2.22
V Permendur[1,2] Approx. 49% Co . . .		0.008	0.10	0.60	1.80	2.28	2.39
Supermendur[1] 2% V			2.05	2.20	2.30	2.36	2.39
Cobalt iron 35% Co.			0.40	1.50	2.10	2.42	2.45
Hiperco 35% Co, 5% Cr							
Other alloys							
Heusler Alloy 61% Cu, 26% Mn, 13% Al }				0.020	0.23	0.38	0.45
Isoperm 30% Ni, 11% Cu . . . . }	(Constant permeability alloy)						
Perminvar 45% Ni, 25% Co	(Constant permeability alloy)						
Nickel copper 70% Ni, 30% Cu					0.060	0.100	
Sendust 5% Al, 10% Si	0.040	0.60	0.90	1.06	1.10		
Ferrites							
Type A Mn–Zn ferrite	Properties dependent on grade						
Type B Ni–Zn ferrite	,,	,,	,,	,,			

Manufacturers: (1) Telcon Metals, Crawley; (2) ITT Components, Harlow: (3) Henry Wiggin and Co., Hereford; (4) Sanderson Bros and Newbould, Sheffield.

| Relative permeability μ_r | | | | $\dfrac{J_s}{T}$ | $\dfrac{H_c}{A\,m^{-1}}$ | $\dfrac{B_r}{T}$ | $Q_h(J\,m^{-3})$ for $B_{max}=$ | | $\dfrac{t_c}{°C}$ | $\dfrac{\rho}{10^{-8}\,\Omega\,m}$ |
| Initial | | Maximum | | | | | 1.0 T | 0.5 T | | |
$\dfrac{\mu_r}{1000}$	$\dfrac{H}{A\,m^{-1}}$	$\dfrac{\mu_r}{1000}$	$\dfrac{H}{A\,m^{-1}}$							
0.125		(approx. constant value)							0.1	
0.014		(approx. constant value)							0.6	
0.4–1.0		200–400	2.4–8.0	1.4	4	1.35	22		590	48
2–7	0.08	50–120	8	1.6	4–12	0.4–1.2		5–10	530	43
3–5	0.8	20–50	8	1.6	12–24	0.4–0.8		10–30	530	50
0.5–1.0	0.8	50–100	9.6	1.6	8–12	1.55	30–70		475	40
2		15	16	1.3	12	0.35		40–70	180–270	80
1.75	0.8	6	60							
				0.45					70	
0.26		2.5		2.35	400	1.5	160		980	20
0.7		3–6		2.40	180	1.6			980	47
		100		2.40	25	2.3			980	40
					300		350			
0.65		10		2.40	80			330	970	20
				0.48	550				330	
0.06		0.065								
0.30		1.5		1.55	100				715	19
									10–100	
30	0.08	120		1.10	4			3	500	60
0.70–2.50		1.5–3.5		0.4	30				170	
0.02–0.64				0.3	120–40				120–550	

(2) Total power losses of transformer core lamination materials (hysteresis and eddy current losses)

Material	Thickness		Power losses (mW kg⁻¹) at 50 Hz and values of B_{max} (T) of:						Power losses (mW kg⁻¹) at 400 Hz and values of B_{max} (T) of:				Power losses (mW kg⁻¹) at 2.4 kHz and values of B_{max} (T) of:			
	mm	×0.001 in	0.01	0.10	0.50	1.0	1.3	1.5	0.01	0.10	0.50	1.0	0.01	0.10	0.50	1.0
Silicon–iron																
Dynamo sheet, 2% Si, Grade 19	0.51	20				2530	4190	5840		770	13 220					
Transformer sheet, 4% Si, Grade 90	0.33	13				990	1760	2420		250	4850	17 630				
Grain oriented	0.33	13				480	790	1120		200	3530	12 120		3310	88 160	
3% Si, Grade 51	0.10	4				400		2110		90	1050	5510		990	27 550	
Nickel–iron alloys																
Supermumetal (tape cores)	0.05	2	0.007	0.7	16				0.1	7	155		1.3	110	2200	
Super C (tape cores)	0.10	4	0.009	1.0	26				0.2	13	330		3.3	220	6200	
Nilomag 771	0.25	10	0.002	0.2	4				0.6	44	1100					
Mumetal	0.05	2	0.004	0.4	10				0.1	7	155		1.5	175	4410	
EPC 20	0.10	4	0.013	1.3	33				0.2	14	400		3.3	330	8820	
Permalloy C	0.25	10	0.033	3.3	66				1	90	1760					
Sanbold Na77	0.38	15							1.5	130	3300					
Nilomag 471	0.10	4							1.3	44	550					
Super Radiometal																
Permalloy B	0.05	2	0.07	7	155	440			1.0	66	1980	12 120	11	770	16 530	66 120
Radiometal 50	0.38	15	1	15	264	660			3.3	176	9920	44 080				
Sanbold Na46																
Permalloy D	0.33	13	1	198	397	1100										
Cobalt–iron alloys																
Permendur	0.25	10	1.8	66	330	1984			13	485	7050	22 040	132	5510	66 120	198 360

(3) Feebly magnetic steels and cast irons

Material	Approx. % composition (Balance iron)	Condition	Relative permeability
Cast irons			
Ni-Resist	Ni 14, Cu 7	} As cast	1.03
Nomag	Ni 11, Mn 6		1.03
S.G. Ni-Resist	Ni 20/25, Cr 0/5		1.02–1.1
Steels			
Clyde alloy } Non-magnetic steel	Ni 4/8, Mn 8, Cr 8		1.003
N.M.C. steel	Ni 10, Mn 5, Cr 4		1.003
Stainless steels (1) . . .	Ni 8, Cr 18	Austenized	1.005–1.03
		10% Cold reduction	1.08–1.13
		20% Cold reduction	1.4
		50% Cold reduction	10
(2) . . .	Ni 12, Cr 18	Austenized	1.003
		50% Cold reduction	1.11
		90% Cold reduction	1.7
(3) . . .	Ni 12, Cr 25	Austenized	1.003
		90% Cold reduction	1.005

This table was taken from *Nickel-containing Magnetic Materials*, published by The International Nickel Company (Mond) Limited.

(4) Permanent magnet (high coercivity) materials

Material	Approx. % composition (Balance iron)	Remanence B_r/T	Coercivity H_c/(kA m^{-1})	$(BH)_{max}$ kJ m^{-3}	H(kA m^{-1})
Steels					
Carbon steel	C 1.0	0.90	4.4	1.6	2.7
Chrome steel	Cr 2.5, C 1.0	0.95	4.8	2.1	3.2
Cobalt steel					
(1) low and medium .	Co 2, Cr 4, C 0.9	1.00	6.0	2.9	4.2
cobalt	Co 3–15, Cr 9, C 1.0	0.70–0.85	10.0–14.0	3.0–5.0	
(2) high cobalt . . .	Co 35, Cr 6, W 4, C 0.9	0.95	20	8.0	1.3
Tungsten steel . . .	C 0.7, W 6	1.05	5.5	2.5	3.6
Typical alloys in the Al–Ni–Co series					
Alni	Ni 25, Al 13, Cu 4	0.60	45	9.95	28
Alnico I low Co . . .	Ni 21, Al 12, Co 5	0.65	43	11.0	25
Alnico } Reco 3 } medium Co .	Ni 19, Al 10, Co 12, Cu 6	0.73	46	13.5	29
Hynico II } Reco 2 } high Co. .	Ni 20, Al 8, Co 20, Cu 5, Ti 6	0.60	72	14.5	41.5
Alcomax III } high Co } Ticonal C } anisotropic} Hycomax II }	Ni 13.5, Al 8, Co 24, Cu 3 / Ni 14, Al 7, Co 29, Cu 4, Ti 4.5	1.26 / 0.85	51.5 / 95	43 / 32	42 / 59.5
Alcomax III high Co semi-columnar . .	Ni 13.5, Al 8, Co 24, Cu 3	1.32	55.5	48.5	45
Columax } high Co } Ticonal GX } columnar }	Ni 14, Al 8, Co 24, Cu 3	1.35	58	59.5	51

(4) Permanent magnet (high coercivity) materials (contd)

Material	Approx. % composition (Balance iron)	Remanence B_r/T	Coercivity $H_c/(kA\ m^{-1})$	$(BH)_{max}$	
				kJ m^{-3}	$H/(kA\ m^{-1})$
Other alloys and materials					
Cobalt-platinum . . .	Pt 77, Co 23	0.65	360	72	206
Cunife	Cu 60, Ni 20	0.6	40	12.5	31
Silmanal	Ag 87, Mn 9, Al 4	0.055	43	0.65	23
Vicalloy I . . .	Co 52, V 9.5	0.9	24	8.0	
Vicalloy II . . .	Co 52, V 13	1.0	36	24	26.5
Caslox Feroba Magnadur	Mixture of barium and iron oxides	0.20	16	7.2	72
Caslox Feroba anisotropic Magnadur	Barium Ferrite	0.36	110	20	86
Micropowder					
HR	30% Co	0.90	28	12	20
HC	30% Co	0.55	48	9.5	32
Rare-earth cobalt magnets					
(Still under active development in many centres)	(Fe not a main constituent)				
Typical performance figures realized:					
Benz and Martin (G.E.C., Schenectedy, New York)	Co–mischmetal–Sa	0.9	1260	160	
Buschow et al. . . . (Phillips, Eindhoven)	Co–Sa	0.87	1280	148	
Johnson & Fellows . (PMA Sheffield)	Co–Pr	0.96	680	175	
Nesbitt et al. . . . (Bell Lab., New Jersey)	Co–Ce–Cu–Fe	0.63	416	80	

For more detailed list see *Permanent Magnets and Magnetism*, edited by D. Hadfield, Iliffe Books Ltd.

Magnetic susceptibilities of paramagnetic and diamagnetic materials

Values are mass susceptibilities per kilogram χ† at 20 °C

	$\times 10^{-8}$		$\times 10^{-8}$		$\times 10^{-8}$
Common elements ‡					
Hydrogen	−2.49	Aluminium . . .	+0.82	Uranium	+2.19
Oxygen	+133.6	Copper.	−0.107	Germanium . . .	−0.15
Helium	−0.59	Silver	−0.25	Silicon	−0.16
Neon	−0.41	Gold	−0.19	Arsenic. . . .	−0.39
Argon	−0.60	Platinum . . .	+1.22	Indium	−0.14
Krypton	−0.41	Mercury . . .	−0.21	Antimony . . .	−1.09
Xenon	−0.40	Bismuth . . .	−1.70	Tellurium . . .	−0.39
Nitrogen	−0.54	Sulphur . . .	−0.62	Gallium . . .	−0.30
Sodium	+0.75	Lead	−0.15	Phosphorus . . .	−1.13
Potassium . . .	+0.65				

† The values in CGS units per gramme are obtained by multiplying the SI values given by $10^3/4\pi$.

‡ For more complete list of elements, see L. F. Bates, *Modern Magnetism*, CUP.

Magnetic susceptibilities of paramagnetic and diamagnetic materials (*contd*)

Values are mass susceptibilities per kilogram χ† at 20 °C

	$\times 10^{-8}$		$\times 10^{-8}$		$\times 10^{-8}$
Common compounds				*Common materials*	
H_2O	-0.90	$NiSO_47H_2O$ . . .	$+20.1$	Araldite	-0.63
NO	$+59.3$	$NiSO_4K_2SO_47H_2O$.	$+13.9$	P.V.C.	-0.75
CO_2	-0.59	$CuSO_45H_2O$ . . .	$+7.7$	Perspex	-0.5
NH_3	-1.38	$MnSO_44H_2O$ . . .	$+81.2$	Polyethylene . . .	$+0.2$
HCl	-0.75	$FeSO_4(NH_4)_2$			
H_2SO_4	-0.50	SO_46H_2O . . .	$+40.6$		
NaCl	-0.64				
$NiCl_2$					
(anhydrous) . .	$+78.5$				
(in sol)	$+43.0$				

† The values in CGS units per gramme are obtained by multiplying the SI values given by $10^3/4\pi$.
‡ For more complete list of elements, see L. F. Bates, *Modern Magnetism*, CUP.

F.J.W.

1.9 Astronomy and geophysics

1.9.1. Astronomical and atomic time systems

Time systems

An ideal time system would provide both an invariable unit of time and a unique, precise method of identifying any instant of time. Astronomical time systems provide methods of identifying the instants at which observations are made, but it is necessary to use clocks to provide continuity between observations and to make more readily available smaller time units than the natural astronomical units of the day and the year. In recent years it has been possible to construct clocks of sufficient precision and reliability to match the long-term stability of the astronomical systems. In fact, a system of **atomic time** (AT) would be sufficient for most purposes, but there are still valid, practical reasons for continuing to relate the time system in general use to the rotation of the Earth. The system that forms the basis of most standard time systems is usually known as **universal time** (UT), rather than by the more descriptive name of Greenwich mean solar time. UT is based on observations of **sidereal time** (ST) and provides a measure of the state of rotation of the Earth on its axis. A knowledge of this state is an essential requirement for the determination of position on the Earth's surface from astronomical observations and for some other purposes. The rate of rotation of the Earth is, however, not quite constant and so, in the development of the theories of the motions of the Moon and planets, astronomers use a particular system of gravitational time that is known as **ephemeris time** (ET). This provides a uniform time scale with reference to which the past irregularities of universal time may be determined from recorded astronomical observations. The ephemeris times of even current observations can, however, only be determined to low precision and retrospectively.

Sidereal time

Optical observations of stars currently provide the principal method of determining the rotational position of the Earth and hence the corresponding values of sidereal time and of universal time. Local sidereal time is usually defined as the local hour angle of the equinox. An equivalent definition is that it is the right ascension of the local meridian. The direction of the equinox is the line of intersection of the mean plane of the Earth's orbit around the Sun with the plane of the Earth's equator; it is used as the zero of right ascension).

In its simplest form, therefore, the method depends on observing stars, whose right ascensions are assumed to be known, as they cross the observer's meridian. The local sidereal time at a place differs from Greenwich sidereal time by the longitude of the place measured from the Greenwich meridian. Sidereal time is essentially an angle expressed in time measure at the rate of $1^h = 15°$.

The precession of the Earth's axis of rotation causes a slow drift of the equinox with respect to the stars, and so the mean length of the sidereal day is actually slightly shorter than the period of rotation of the Earth with respect to inertial axes (or 'fixed stars'). The forced nutation of the Earth's axis of rotation causes an oscillatory motion that is superimposed on the precessional drift. It is therefore necessary to distinguish between apparent (or true) sidereal time and mean sidereal time, where the latter is free from the quasi-periodic irregularities due to nutation. The corrections for nutation can be computed from theoretical expressions, and so the values of local apparent sidereal time that are given by stellar observations can be reduced to values of **Greenwich mean sidereal time** (GMST), using an adopted value for the longitude of the observing site. Corrections are also required to allow for the motion of the axis of figure of the Earth with respect to the axis of rotation.

Universal time

In order that UT shall correspond closely to the alternation of day and night but be free from the irregularities that would be present if it were based on observations of the actual Sun, it is defined in terms of the hour angle of a fictitious mean sun and calculated from mean sidereal time. In fact,

$$UT = GMST - (6^h \ 38^m \ 45\overset{s}{.}836 + 8 \ 640 \ 184\overset{s}{.}542 \, T_U + 0\overset{s}{.}0929 \, T_U^2),$$

where T_U is the number of Julian centuries of 36 525 days of universal time that have elapsed since 1900 January 0 at 12^h UT. Even so, UT does not provide a uniform time scale since the rate of rotation of the Earth is subject to seasonal, irregular and long-term changes. The seasonal changes of about 1 part in 10^8 in the length of the day are roughly predictable on a short-term basis. The irregular and long-term changes appear to correspond on average to a gradual increase in the length of the day by about 1 ms (i.e. 1 part in 10^8) per century.

Equation of time

The time indicated by a sundial or similar device is known as local apparent (solar) time, and it differs from local mean (solar) time by the so-called equation of time. Local mean time differs from UT by the time-equivalent of the longitude of the place concerned. The average amounts by which apparent solar time is in advance of mean solar time are indicated in the following table; the actual values for a given day may vary by up to $0\overset{m}{.}2$ from these average values.

		m			m			m			m
Jan.	1	− 3.3	Apr.	1	− 3.8	July	1	− 3.8	Oct.	1	+ 10.4
	15	− 9.2		15	0.0		15	− 5.9		15	+ 14.3
Feb.	1	− 13.6	May	1	+ 3.0	Aug.	1	− 6.2	Nov.	1	+ 16.4
	14	− 14.3		15	+ 3.7		15	− 4.4		15	+ 15.3
Mar.	1	− 12.4	June	1	+ 2.0	Sep.	1	+ 0.1	Dec.	1	+ 10.9
	15	− 8.9		15	− 0.4		15	+ 4.9		15	+ 4.7

Ephemeris time

In general terms, ephemeris time is the independent variable of the differential equations for the motions of the Sun, Moon and planets under the influence of their mutual gravitational attractions. Formally, the fundamental epoch and the unit of time interval are defined by the first two coefficients in the adopted expression for the geometric mean longitude of the Sun, referred to the mean equinox of date:

$$L = 279° \ 41' \ 48\overset{''}{.}04 + 129 \ 602 \ 768\overset{''}{.}13 \, T_E + 1\overset{''}{.}089 \, T_E^2,$$

where T_E is the number of Julian centuries of 36 525 days of ephemeris time that have elapsed since 1900 January 0 at 12^h ET. In particular, the ephemeris day is the fraction 1/365.242 199 of the tropical year at 1900 January 0 at 12^h ET; this is close to the length of the mean solar day towards the end of the nineteenth century. The ephemeris second is equal to the SI second to within the present limits of uncertainty of its determination.

Observations of the position of the Moon are used to determine the difference between ET and UT. Current estimates of the mean values of ET – UT for the years indicated are:

1820	+ 43 s	1860	+ 25 s	1900	+ 5 s	1940	+ 27 s
1830	+ 37	1870	+ 16	1910	+ 17	1950	+ 30
1840	+ 32	1880	+ 6	1920	+ 25	1960	+ 34
1850	+ 28	1890	+ 5	1930	+ 27	1970	+ 41

The determinations of ET – UT are liable to revision as improved ephemerides and additional observations become available. ET has now been replaced by **international atomic time** (TAI) for current scientific purposes requiring a uniform, continuous time scale of high precision. The difference ET – TAI is about 32 s.

G.A.W.

International atomic time

Ephemeris time provides a uniform time scale which is not subject to irregularities due to changes in the Earth's motion. In practice it suffers from the disadvantage that it is accurately known only retrospectively (with a delay of years) because of the time needed to take and process astronomical observations. The development of caesium atomic clocks has made possible the immediate realization of a uniform time scale with high accuracy. The Bureau International de l'Heure (BIH), in Paris, and a number of national laboratories have been maintaining atomic time scales for about fifteen years. In 1967 the General Conference of Weights and Measures (CGPM) adopted the atomic definition of the second as the unit of time duration. In 1971 the CGPM gave formal recognition to the scale maintained by the BIH as the International Atomic Time Scale (TAI).

TAI and Universal Time were coincident on 1 January 1958. UT now lags by about 12 seconds. For practical use Coordinated Universal Time (UTC) has been redefined from 1 January 1972 as an approximation to UT in which the second markers coincide with those of TAI but the numerical values assigned to them are those of the nearest seconds in UT. At 0^h 0^m 0^s on 1 January 1972 the difference TAI – UTC was made exactly 10 seconds. As Universal Time drifts with respect to Atomic Time, 'leap seconds' are added to or subtracted from UTC so that the difference between UT and UTC does not normally exceed 0.7 second. The one-second steps, if necessary, are made on the last second of a month, preferably June or December. UTC is now used as the basis of broadcast time signals. The seconds markers provide a uniform time scale. The numerical 'labels' attached to them, with known corrections if needed, provide Universal Time for those who require it.

Time and frequency signals

Time and its reciprocal, frequency, are unique among physical quantities in that the magnitude of the units can be disseminated by radio broadcasts, without the need for direct intercomparison of material standards. A number of countries operate standard time and frequency broadcast services: these are available for use by anyone within range of the transmitters who has a suitable radio receiver. The majority of the transmissions are in the MF and HF bands (p. 75) and have an effective range of some thousands of kilometres. A number of stations, notably GBR, Rugby, and WWVL, Boulder, Colorado, operate in the VLF band and can be received virtually anywhere in the world. However, because variable delays occur when signals are propagated via the ionosphere, the full accuracy of the signals can be used only by a receiver within ground-wave range of a transmitter.

In Britain, the service is provided by the National Physical Laboratory, using Post Office transmitters at Rugby and a BBC transmitter at Droitwich. The carriers of these transmitters are all derived from atomic frequency standards and are maintained within ± 2 in 10^{11} of the nominal frequency. The signals are received at Teddington and are compared with the group of atomic clocks which forms the national frequency standard. Corrections to the radiated frequencies are subsequently published in *The Radio and Electronic Engineer* to ± 1 in 10^{12}.

The Droitwich transmission is on 200 kHz; it operates for about 20 hours per day and carries BBC Radio 2 programmes. The Rugby transmissions are effectively continuous and carry time signals. The frequencies used are: GBR, 16 kHz; MSF, 60 kHz, 2.5, 5 and 10 MHz. In practice the transmissions are interrupted for occasional maintenance (four hours each week for GBR and four hours each month for MSF), and the MSF HF service operates for alternate five-minute periods only, to allow time-sharing of the frequencies with other European transmitters.

The time signals on the Rugby transmissions are derived from the same atomic clocks that provide the carriers and have the same accuracy. They indicate Coordinated Universal Time (UTC—see previous section). They take the form of 5 ms pulses spaced by one second and lengthened to 100 ms at the minutes on HF (100 ms pulses on the seconds lengthened to 500 ms at the minutes on LF and VLF). Coding is used to indicate the difference (UT − UTC) in multiples of 0.1 second, double pulses being transmitted in the period 1–7 seconds after the minute for positive, and in the period 9–15 seconds after the minute for negative values of the difference. The time signals are received by the Royal Greenwich Observatory (at Herstmonceux) and are compared with a group of atomic clocks. Corrections to the signals are subsequently published by the observatory in *Time Service Circulars—Series A* and in *Greenwich Time Reports*.

A.E.B.

1.9.2 Astronomical units and constants

The system of astronomical units

For the computation of motions in the solar system it is customary to use a special system of astronomical units of mass, length and time. The astronomical unit of mass is the mass of the Sun, the astronomical unit of time is the ephemeris day (see above), and the astronomical unit of length (AU) is such that the Gaussian gravitational constant k, has the exact value 0.017 202 098 950. In these units the acceleration of a particle under the gravitational attraction of a mass, m, is $k^2 m/r^2$, where r is the distance of the mass; thus k^2 corresponds to the Newtonian gravitational constant G. The values of M, L, T of these astronomical units in terms of any other system of units must be such that the value of the gravitational constant G in that system is equal to $k^2 L^3 / MT^2$. This relationship is used to calculate the mass of the Sun in terrestrial units from the observationally determined values of the other quantities. Thus in terms of the SI system:

$$G = (6.670 \pm 0.006) \times 10^{-11} \text{ m}^3 \text{ kg}^{-1} \text{ s}^2$$
$$L = (1.495\ 979 \pm 1 \times 10^{-6}) \times 10^{11} \text{ m}$$
$$T = 86\ 400\ (1.0 \pm 2 \times 10^{-9}) \text{ s}$$
$$M = (1.990 \pm 0.002) \times 10^{30} \text{ kg}$$

The value of L is based on the observed value of the ratio of the astronomical unit of length to the light-second, which is the unit normally used to express interplanetary distances that are measured by radar or laser techniques, together with the adopted value for the velocity of light in SI units.

The relative orbit of two isolated particles of mass M and m moving around each other under their mutual gravitational attraction is an ellipse. The mean angular velocity n in radians per day and the semi-major axis a of the orbit (usually called the mean distance of the particles) are related by Kepler's law:

$$n^2 a^3 = k^2(M+m)$$

The sidereal period of a particle of negligible mass moving around the Sun in an orbit with a mean distance of 1 AU is 365.256 898 ($= 2\pi/k$) days; this period is referred to as a Gaussian year.

A convenient unit for the measurement of the distances of nearby stars is the parsec (pc); this is the distance at which 1 AU subtends an angle of 1 second of arc. Hence,

$$1 \text{ parsec} = 1/(\sin 1'') \text{ AU} = 2.06 \times 10^5 \text{ AU} = 3.08 \times 10^{16} \text{ m}$$

The multiple units kiloparsec (kpc) and megaparsec (Mpc) are more appropriate for most galactic and extragalactic objects respectively. The 'light-year' is normally only used in popular astronomical texts:

$$1 \text{ light-year} = 0.307 \text{ pc} = 6.33 \times 10^4 \text{ AU} = 9.46 \times 10^{15} \text{ m}$$

IAU system of astronomical constants

A self-consistent system of values of the principal constants of the solar system was adopted by the International Astronomical Union in 1964. The system has been used in the ephemerides of the Sun, Moon, planets and stars that have been published in the principal national and international almanacs for 1968 onwards, and it will continue in use in the almanacs up to 1979. In general, the values represent the most likely estimates of the constants at the time of adoption of the system, but in a few cases (e.g. the constants of precession and nutation) the advantages of continuity are so great that no changes were made even though more likely values were available.

In the following list the constants for the Moon and the planetary masses have been omitted since values are given on p. 131. Additional notes and derived constants for the Earth are given on p. 132. The values for the solar parallax and the constant of aberration have been converted from radians to seconds of arc; the rounded values given in parentheses should be used except when extra figures are required to ensure numerical consistency. The ephemeris second is equal to the SI second to within the present limits of uncertainty of its determination.

For reviews of recent determinations of the constants and discussions of related topics see the proceedings of IAU Colloquium No. 9 published in *Celestial Mechanics*, 1971, **4** (No. 2), 127–280.

Defining constants

Number of ephemeris seconds in 1 tropical year (1900)	$s = 31\ 556\ 925.974\ 7$
Gaussian gravitational constant	$k = 0.017\ 202\ 098\ 95$

Primary constants

Measure of 1 AU	$A = 1.496\ 00 \times 10^{11}$ m
Velocity of light	$c = 2.997\ 925 \times 10^8$ m s^{-1}
Equatorial radius for the Earth	$a_e = 6.378\ 160 \times 10^6$ m
Dynamical form factor for the Earth†	$J_2 = 1.0827 \times 10^{-3}$
Geocentric gravitational constant (M = mass of Earth)	$GM = 3.986\ 03 \times 10^{14}$ m^2 s^{-2}
General precession in longitude per tropical century (1900)	$p = 5025''.64$
Obliquity of the ecliptic (1900)	$\epsilon = 23°\ 27'\ 08''.26$
Constant of nutation (1900)	$N = 9''.210$

Auxiliary constants

$k/86\ 400$, for use when the unit of time is 1 second	$k' = 1.990\ 983\ 675 \times 10^{-7}$
Number of seconds of arc in 1 radian	$206\ 264.806$
Factor for constant of aberration, depending on elements of Earth's orbit	$F_1 = 1.000\ 142$

† See p. 132 for definition.

Derived constants

Solar parallax	$\arcsin(a_e/A) = \pi_\odot = 8''.794\ 05\ (8''.794)$
Light-time for unit distance	$A/c = \tau_A = 499^s.012$
	$= 1^s/0.002\ 003\ 96$
Constant of aberration	$F_1\ k'\ \tau_A = \kappa = 20''.4958\ (20''.496)$
Flattening factor for Earth	$f = 0.003\ 352\ 9$
	$= 1/298.25$
Heliocentric gravitational constant ($S =$ mass of Sun)	$A^3\ k'^2 = GS = 1.327\ 18 \times 10^{20}\ \text{m}^3\ \text{s}^{-2}$

Constants relating to time

The following relationships hold in all time systems:

1 day = 24 hours = 1440 minutes = 86 400 seconds
1 Julian year = 365.25 days = 8766 hours = 525 960 minutes = 31 557 600 seconds

The following values of the lengths of the year, month and day are formally expressed in intervals of ephemeris time, but may be treated as if in SI units; T is measured in centuries from 1900.

1 tropical year (equinox to equinox)	$= 365^d.242\ 199 - 0^d.000\ 006\ 1\ T$
	$= 365^d\ 05^h\ 48^m\ 46^s.0 - 0^s.53\ T$
1 sidereal year (fixed star to fixed star)	$= 365^d.256\ 360 + 0^d.000\ 000\ 1\ T$
	$= 365^d\ 06^h\ 09^m\ 09^s.5 + 0^s.01\ T$
1 mean synodic month (new moon)	$= 29^d.530\ 859 = 29^d\ 12^h\ 44^m\ 02^s.9$
1 mean tropical month (equinox)	$= 27^d.321\ 582 = 27^d\ 07^h\ 43^m\ 04^s.7$
1 mean sidereal month (fixed star)	$= 27^d.321\ 661 = 27^d\ 07^h\ 43^m\ 11^s.5$
1 mean solar day ($\sim$ 1972)	$= 1^d.000\ 000\ 04 = 86\ 400^s.003$
1 mean sidereal day ($\sim$ 1972)	$= 0^d.997\ 269\ 60 = 86\ 164^s.094$
1 mean period of rotation of Earth ($\sim$ 1972)	$= 0^d.997\ 269\ 70 = 86\ 164^s.102$

Although the Earth's period of rotation with respect to the fixed stars is variable when measured in SI units the following relationships are almost independent of the variability.

1 mean solar day
 = interval between successive transits of the fictitious mean sun
 = $1^d.002\ 737\ 909\ 3 = 24^h\ 03^m\ 56^s.555\ 36 = 86\ 636^s.555\ 36$ of mean sidereal time
1 mean sidereal day
 = interval between successive transits of the mean equinox
 = $0^d.997\ 269\ 566\ 4 = 23^h\ 56^m\ 04^s.090\ 54 = 86\ 164^s.090\ 54$ of mean solar time
1 period of rotation of the Earth with respect to the fixed stars
 = $1^d.000\ 000\ 097\ 1 = 24^h\ 00^m\ 00^s.008\ 39 = 86\ 400^s.008\ 39$ of mean sidereal time
 = $0^d.997\ 269\ 663\ 2 = 23^h\ 56^m\ 04^s.098\ 90 = 86\ 164^s.098\ 90$ of mean solar time

Constants for precession and nutation

The precessional motion of the celestial equator (plane normal to the axis of rotation of the Earth) around the ecliptic (mean plane of the Earth's orbit around the Sun) has a period of about 25 725 years, and gives rise to the luni-solar precession in celestial longitude of 50''.37 per year. There is also a slow precessional motion of the ecliptic due to planetary perturbations; this gives both a change in the obliquity (inclination of the ecliptic to the equator) and a motion of the equinox (direction of the line of intersection of equator and ecliptic) along the equator of $0''.12$ per year. The combined effect is known as the general precession.

Obliquity of ecliptic $= 23° 27' 08''.26 - 46''.84 T = 23°.452\,294 - 0°.013\,01\,T$

Longitude of axis of rotation of ecliptic $= 173° 57'.06 + 54'.77\,T$

Annual rate of rotation of ecliptic $= 0''.4711 - 0''.0007\,T$

Annual general precession in longitude $= 50''.2564 + 0''.0222\,T$

For a star with right ascension α and declination δ the annual precessions in right ascension and declination are $m + n \sin\alpha \tan\delta$ and $n \cos\alpha$, respectively, where

$$m = 3^s.072\,34 + 0^s.001\,86 T, \qquad n = 1^s.336\,46 - 0^s.000\,57 T = 20''.0468 - 0''.0085 T$$

The nutation of the axis of rotation of the Earth is normally specified by its effects in celestial longitude and obliquity. The principal terms in the trigonometric series for the nutation are:

In longitude	In obliquity	Period (days)
$-17''.233 - 0''.017 T$	$+9''.210 + 0''.001 T$	6798
$+0.209$	-0.090	3399
-1.273	$+0.552$	183
-0.204	$+0.088$	13.7

The solar system: The Sun, Moon and planets

Dimensions, etc., of Sun, Moon and planets

Body	Equatorial radius (on scale: Earth = 1)†	Mass	Surface gravity	Mean density $\overline{\text{g cm}^{-3}}$	Flattening	Period of rotation	Inclination of equator to orbit	Number of known satellites
Sun . . .	109.12	332 960	27.96	1.41	0.0	$25^d\ 09^h$	$7°.2$	—
Mercury . .	0.38	0.055	0.38	5.4	0.0	$58^d\ 16^h$	$0°.0$	0
Venus. . .	0.95	0.815	0.90	5.2	0.0	$242^d\ 23^h$	$177°.8$	0
Earth . . .	1.00	1.000	1.00	5.52	0.003	$23^h\ 56^m$	$23°.4$	1
Moon . . .	0.27	0.0123	0.17	3.34	See note 4	$27^d\ 07^h$	$1°.5$	—
Mars . . .	0.53	0.108	0.38	3.9	0.007	$24^h\ 37^m$	$24°.0$	2
Jupiter . .	11.11	317.9	2.58	1.4	0.07	$9^h\ 50^m$	$3°.1$	12
Saturn . .	9.41	95.1	1.07	0.7	0.10	$10^h\ 14^m$	$26°.7$	10
Uranus . .	4.0	14.5	0.9	1.4	0.06	$10^h\ 48^m$	$97°.9$	5
Neptune . .	3.8	17.2	1.2	1.7	0.02	$15^h\ 40^m$	$28°.8$	2
Pluto . . .	0.5?	0.1?	0.2?	5?	?	6^d?	$90°$?	0

† For Earth values see pp. 132, 134.

Notes:
(1) The Sun is a star of spectral type G2 V and absolute magnitude $+4.79$. The effective temperature of its photosphere is 5800 K. The radiation emitted per unit area is $0.641\,\text{J m}^{-2}\,\text{s}^{-1}$, giving a total emission rate of $3.90 \times 10^{26}\,\text{J s}^{-1}$. The radiation received at the Earth under standard conditions (i.e. the solar constant) is $8.33 \times 10^{-4}\,\text{J m}^{-2}\,\text{s}^{-1}$.

(2) The periods of rotation of the Sun, Jupiter and Saturn refer to their equatorial visual regions; the periods increase with latitude.

(3) The inclinations of the equators of the Sun and Moon are referred to the ecliptic; the inclination of the equator of the Moon to the plane of the Moon's orbit around the Earth is about $6°.7$ and the axis of rotation precesses at the same rate as the line of nodes of the orbit (period 18.6 years).

(4) The mean radius of the Moon may be taken as 1738.0 km, but the irregularities in the surface are of the order of 1/500 of the radius; the principal moments of inertia are all different.

Mean elements of the orbits of the planets

Planet	Mean distance from Sun		Sidereal period	Synodic period	Inclination to ecliptic	Eccentricity	Longitude of node	Longitude of perihelion
	AU	10^6 km	year	day	degree		degree	degree
Mercury . .	0.387	57.9	0.241	116	7.004	0.2056	48.0	77.0
Venus. . .	0.723	108.2	0.615	584	3.394	0.0068	76.4	131.1
Earth + Moon	1.000	149.6	1.000	—	0.0	0.0167	—	102.4
Mars . . .	1.524	227.9	1.881	780	1.850	0.0934	49.3	335.5
Jupiter . .	5.203	778.3	11.862	399	1.305	0.0484	100.1	13.8
Saturn . .	9.539	1427.0	29.46	378	2.490	0.0557	113.4	92.5
Uranus . .	19.182	2870	84.01	370	0.773	0.0472	73.8	170.2
Neptune . .	30.058	4497	164.79	367	1.774	0.0086	131.4	44.4
Pluto . . .	39.439	5900	248.43	367	17.14	0.253	109.9	223.0

Notes:

(1) The planetary orbits are subject to both secular and periodic perturbations, so that the osculating elements at any instant may differ in the end figures from the mean elements (for 1970) that are given above for all planets except Pluto, for which osculating elements for 1970 are given.

(2) The values given for Earth + Moon refer to motion of their centre of mass, which lies about 4700 km from the centre of the Earth. The mean distance between the centres of the Earth and Moon is 384 400 km; the mean eccentricity of the relative orbit is 0.055 and its mean inclination to the ecliptic is 5°2. The orbit is, however, subject to considerable perturbations. The line of nodes moves round westwards once in 18.6 years, and so the inclination of the orbit to the Earth's equator (and hence the extremes of declination in any month) varies between 28°6 and 18°3.

References

Annual tabulations of astronomical data and ephemerides will be found in: *The Astronomical Ephemeris* (published by HMSO, London, up to one year in advance), *The Handbook of the British Astronomical Association*, and *Whitaker's Almanac*. Extended lists of astronomical data and tables are given in: *The Explanatory Supplement to the Astronomical Ephemeris*, 1973 (HMSO); C. W. Allen, *Astrophysical Quantities*, London, 1963 (The Athlone Press); Landolt-Börnstein, *Numerical Data and Functional Relationships in Science and Technology*, New Series, Group VI, Vol. 1, *Astronomy and Astrophysics*, 1964 (Springer-Verlag, Berlin).

G.A.W.

The Earth: mechanical properties

Property	Whole Earth	Core
p Equatorial radius, a	6378.160 km	3488 km
Polar radius, c	6356.775 km	3479 km
Polar flattening, $f = (a-c)/a$ . . .	1/298.25	1/390
Mean radius $(a^2c)^{1/3}$	6371.02 km	3485 km
Mass M	5.976×10^{24} kg	1.88×10^{24} kg
Mean density	5518 kg m^{-3}	10 720 kg m^{-3}
Moments of inertia in terms of Ma^2		
Polar C/Ma^2	0.3306	0.380
Equatorial A/Ma^2	0.3295	
p GM	398 603 $\times 10^9$ m^3 s^{-2}	
p $J_2 = (C-A)/Ma^2$	$1.082\ 65 \times 10^{-3}$	
Dynamical ellipticity $(C-A)/C$ . .	3.275×10^{-3}	
Angular velocity	$7.292\ 115\ 2 \times 10^{-5}$ rad/s	
Surface area	5.101×10^{14} m^2	1.52×10^{14} m^2
Volume	1.083×10^{21} m^3	0.176×10^{21} m^3
Quadrant of meridian	10 002.002 km	5640 km

1.9.3 Physical properties of the Earth

A set of astronomical constants, based mainly on data from artificial satellites and radar observations of the Moon and the planets, was adopted by the International Astronomical Union in 1964 (IAU Reports for 1964). While adopted as conventional values, they also represent the best observational data very closely. In the preceding table, values given under 'Whole Earth' are IAU primary constants (p) or are derived from them; 'Core' values are obtained from Bullen, *Introduction to the Theory of Seismology*, 1963 (CUP) and Jeffreys, *The Earth*, 5th edn, 1970 (CUP).

The Earth: variation of properties with depth

It is not possible to construct a unique model of the variation of mechanical properties within the Earth. The following table is based on models constructed by Haddon and Bullen, *Phys. Earth, Planet. Interiors*, 1969, **2**, and Derr, *J. Geophys. Res.*, 1969, **74**.

Zone	Depth km	Density kg m^{-3}	Elastic wave velocity P km s^{-1}	Elastic wave velocity S km s^{-1}	Bulk modulus 10^{11} N m^{-2}	Rigidity modulus 10^{11} N m^{-2}	Pressure 10^{11} N m^{-2}	g m s^{-2}
Mantle	0	2 600	6.00	3.50	0.56	0.32	0	9.82
	60	3 390	7.96	4.62	1.19	0.72	0.018	9.85
	150	3 400	8.15	4.43	1.36	0.66	0.048	9.88
	350	3 580	8.80	4.58	1.76	0.76	0.116	9.95
	500	3 830	9.65	5.24	2.16	1.05	0.174	9.98
	1000	4 560	11.44	6.36	3.50	1.86	0.388	9.96
	1500	4 825	12.22	6.70	4.26	2.16	0.622	9.92
	2000	5 080	12.78	6.89	5.09	2.42	0.868	10.01
	2886 {	5 520	13.66	7.23	6.44	2.88	} 1.353	10.74
		9 900	8.11	0	6.36	0		
Core	3500	10 680	8.88	0	8.30	0	1.936	9.32
	4000	11 410	9.50	0	10.31	0	2.467	7.96
	4500	11 850	9.97	0	11.65	0	2.882	6.64
	5000	12 200	10.30	0	12.96	0	3.231	4.80
	5150 {	12 280	9.86	0	13.05	0	} 3.311	4.62
Inner core		13 000	11.18	1	17.03	0.68		
	6371	13 500	11.80	1	17.81	0.70	3.691	0

P—Compressional wave;   *S*—Shear wave

The Earth: other physical constants

Land:	area	1.49×10^{14} m^2 (29.2 per cent of Earth's surface)
	mean height	840 m
	greatest height	8840 m
Oceans:	area	3.61×10^{14} m^2 (70.8 per cent of Earth's surface)
	volume	1.37×10^{18} m^3
	mass	1.42×10^{21} kg
	mean depth	3800 m
	greatest depth	10 550 m
Atmosphere:	mass	5.27×10^{18} kg (10^{-6} of Earth's mass)

Heat flow from the interior of the Earth

Heat flows out from the interior of the Earth at an average rate of 0.06 W m^{-2}. Most measurements have been made in the deep oceans. There are large variations in heat flow, from 0.01 to 0.2 W m^{-2} but no significant differences between major continents and oceans.

Area	Equal area average heat flow/(W m^{-2})	
	Average	*Standard deviation*
World	0.061	0.031
Continents	0.061	0.020
Oceans	0.061	0.033
Atlantic Ocean	0.056	0.025
Indian Ocean	0.055	0.022
Pacific Ocean	0.063	0.036

See W. H. K. Lee, *Phys. Earth Planet. Interiors* 1970, **2**.

A.H.C.

1.9.4 Gravity

The gravity field of the Earth

The potential of the external gravity field of the Earth is usually expressed as a series of spherical harmonic terms:

$$V = -\frac{GM}{r}\left[1 - \sum_{n=2}^{\infty} J_n \left(\frac{a}{r}\right)^n P_n\left(\cos\theta\right) + \text{terms depending on longitude}\right]$$

where r = distance from the centre of the Earth
    a = Earth's equatorial radius,† θ = geocentric co-latitude,
    M = mass of the Earth† (see below)
$P_n(\cos\theta)$ = Legendre function of degree n
    $10^6 J_2 = 1082.65$†
    $10^6 J_3 = -2.5$
    $10^6 J_4 = 1.6$

The values of J_n are determined from the behaviour of artificial satellites about the Earth (A.H. Cook, *Phil. Trans. Roy. Soc.*, 1967, **262**).

The corresponding expression for the variation of the acceleration due to gravity over the surface of the (spinning) Earth is

$$g = g_e(1 + \beta_1 \sin^2\phi - \beta_2 \sin^2 2\phi) - 3.086 \times 10^{-6} H \text{ m s}^{-2}$$

where ϕ is the geographical latitude, H is the height above sea level (in metres) and g_e is the value of gravity at the equator.

The values recommended by the International Association of Geodesy are

$$g_e = 9.780\ 318\ 4 \text{ m s}^{-2}$$
$$\beta_1 = 0.005\ 302\ 4$$
$$\beta_2 = 0.000\ 005\ 9$$

(*Sp. Pub. Bull. géod.*, 1970).

The above formula gives the best simple method of calculating g at a place where it has not been measured. It will almost always give results within 10^{-3} and usually within 5×10^{-4} m s^{-2}. The agreement with observation is usually made worse by the application of a correction for the attraction of the land above sea level.

For the definition of standard gravity (standard acceleration) see p. 11.

† See section 1.9.3 above.

Absolute value of the acceleration due to gravity, g

The difference between the accelerations due to gravity at two nearby sites may be compared more accurately (by spring-balance gravity meters) than either can be measured separately. The network of world-wide gravity measurements is therefore based on differential measurements which tie sites to a single site at which the absolute value (m s^{-2}) is believed to be well established. The site hitherto adopted is the Potsdam Geodetic Institute where g was measured by Kühnen and Furtwängler (*Veröff. Preuss. Geodät. Inst.*, 1966) using reversible pendulums of Kater's type. Modern determinations use the direct observation of a body freely moving under the action of gravity. The value of Kühnen and Furtwängler (g at Potsdam $= 9.812\ 74$ m s^{-2}) is in error by 14 p.p.m., whereas the following recent values agree to within 0.1 p.p.m.:

Cook:	Teddington, 1965	$9.811\ 818\ 1$ m s^{-2}
Faller:	Teddington, 1969	$9.811\ 818\ 6$
	Washington, 1969	$9.801\ 024\ 0$
	Sèvres, 1969	$9.809\ 259\ 6$
Sakuma:	Sèvres, 1967–70	$9.809\ 259\ 6$

A.H.C.

1.9.5 Geomagnetism

The Earth's magnetic field corresponds approximately to that of a dipole situated near the centre of the Earth, but there are appreciable temporal and spatial departures from uniformity. The smooth geomagnetic or 'normal' field and its slow secular change are ascribed largely to fluid motions in the Earth's core. The influence of crustal deposits superposes on the normal field anomalies whose magnitude may approach four times that of the normal field.

In addition, changes on the Sun and of its position relative to the Earth cause erratic and often rapid fluctuations in the magnetic field (magnetic storms) occasionally exceeding one-tenth the normal field value, as well as small diurnal and seasonal variations.

The field at any point is usually defined by three of seven elements: five intensity components, H (parallel to Earth's surface), Z (vertically downwards), F or T (total, scalar), X (geographic North) and Y (geographic East); and two angles, D ($D = \arctan Y/X$) and I ($I = \arctan Z/H$).

For further details, see the article 'Magnetism, Terrestrial and Solar', in *Chambers's Encyclopaedia*, 1950.

The geomagnetic field, due to internal causes, for points on or above the Earth's surface may be represented by the following series:

$$X = \sum_{n=1}^{k} \sum_{m=0}^{n} (g_n^m \cos m\lambda + h_n^m \sin m\lambda) \left(\frac{a}{r}\right)^{n+2} n X_n^m$$

$$Y = \sum_{n=1}^{k} \sum_{m=0}^{n} (g_n^m \sin m\lambda - h_n^m \cos m\lambda) \left(\frac{a}{r}\right)^{n+2} n Y_n^m$$

$$Z = -\sum_{n=1}^{k} \sum_{m=0}^{n} (g_n^m \cos m\lambda + h_n^m \sin m\lambda) \left(\frac{a}{r}\right)^{n+2} (n+1) P_n^m$$

where a is the Earth's mean radius, λ is longitude East, r is radial distance from the Earth's centre, g_n^m, h_n^m are variables whose estimated values for the epoch 1965.0 are given in the following table, and X_n^m, Y_n^m, P_n^m are numerical coefficients which are functions of North co-latitude (θ) and the integers m, n.

The surface spherical harmonics currently used in Geomagnetism are sometimes termed 'normalized'. They are, however, not fully 'normalized' in the sense of the theory of real functions. They are such that the mean square value of $P_n^m \cos m\lambda$ or $P_n^m \sin m\lambda$ taken over a sphere is $1/(2n+1)$. To avoid confusion, they should be referred to as being 'Schmidt quasi-normalized'. The coefficients may be derived as follows:

$$P_n^m(c) = \left(\frac{\delta_m (n-m)! (1 - c^2)^m}{(n+m)!} \right)^{1/2} \frac{d^m}{dc^m} P_n(c)$$

$$n X_n^m = \frac{d}{d\theta} P_n^m$$

$$n Y_n^m = m \operatorname{cosec} \theta P_n^m$$

where $c = \cos \theta$

$\delta_m = 1$ for $m = 0$; 2 for $m \geqslant 1$

and $P_n(c)$ is a Legendre polynomial in c; given by

$$P_n(c) = \frac{1}{2^n n!} \frac{d^n}{dc^n} (c^2 - 1)^n$$

For practical use, numerical values may be more conveniently computed from the initial values

$$P_0^0 = 1 \qquad P_1^1 = \sin \theta \qquad X_1^1 = \cos \theta$$

and the recurrence relationships:

$$P_m^m = \left(\frac{2m-1}{2m} \right)^{1/2} P_{m-1}^{m-1} \sin \theta, \quad \text{where } m > 1$$

$$m X_m^m = \left(\frac{2m-1}{2m} \right)^{1/2} \{ (m-1) X_{m-1}^{m-1} \sin \theta + P_{m-1}^{m-1} \cos \theta \}, \quad \text{where } m > 1$$

$$P_n^m = P_{n-1}^m (2n-1)(n^2 - m^2)^{-1/2} \cos \theta - P_{n-2}^m \left(\frac{(n-1)^2 - m^2}{n^2 - m^2} \right)^{1/2}$$

$$n X_n^m = (2n-1)(n^2 - m^2)^{-1/2} \{ (n-1) X_{n-1}^m \cos \theta - P_{n-1}^m \sin \theta \}$$

$$- \left(\frac{(n-1)^2 - m^2}{n^2 - m^2} \right)^{1/2} (n-2) X_{n-2}^m$$

$$n Y_n^m = m P_n^m \operatorname{cosec} \theta$$

Note here that m in P_n^m, etc., is an affix, not exponent.

A Fortran subroutine and equivalent Algol procedure for computing these formulae for specified values of g_n^m, h_n^m are given in *Rep. No. 71/1, Inst. geol. Sci.*, 1971 (HMSO), where also is shown how to amend the results to accord with the IAU ellipsoid.

The values of g_n^m, h_n^m change with time and are therefore given in the following table for a particular epoch (1965.0), together with the rate of change.

Coefficients of the international geomagnetic reference field

Epoch 1965.0.
Coefficients are Schmidt quasi-normalized.

n	m	g/nT	h/nT	$\dot{g}$/(nT yr^{-1})	$\dot{h}$/(nT yr^{-1})	n	m	g/nT	h/nT	$\dot{g}$/(nT yr^{-1})	$\dot{h}$/(nT yr^{-1})
1	0	− 30 339		15.3		6	2	4	106	1.1	− 0.4
1	1	− 2 123	5758	8.7	− 2.3	6	3	− 229	68	1.9	2.0
2	0	− 1 654		− 24.4		6	4	3	− 32	− 0.4	− 1.1
2	1	2 994	− 2006	0.3	− 11.8	6	5	− 4	− 10	− 0.4	0.1
2	2	1 567	130	− 1.6	− 16.7	6	6	− 112	− 13	− 0.2	0.9
3	0	1 297		0.2		7	0	71		− 0.5	
3	1	− 2 036	− 403	− 10.8	4.2	7	1	− 54	− 57	− 0.3	− 1.1
3	2	1 289	242	0.7	0.7	7	2	0	− 27	− 0.7	0.3
3	3	843	− 176	− 3.8	− 7.7	7	3	12	− 8	− 0.5	0.4
4	0	958		− 0.7		7	4	− 25	9	0.3	0.2
4	1	805	149	0.2	− 0.1	7	5	− 9	23	0.0	0.4
4	2	492	− 280	− 3.0	1.6	7	6	13	− 19	− 0.2	0.2
4	3	− 392	8	− 0.1	2.9	7	7	− 2	− 17	− 0.6	0.3
4	4	256	− 265	− 2.1	− 4.2	8	0	10		0.1	
5	0	− 223		1.9		8	1	9	3	0.4	0.1
5	1	357	16	1.1	2.3	8	2	− 3	− 13	0.6	− 0.2
5	2	246	125	2.9	1.7	8	3	− 12	5	0.0	− 0.3
5	3	− 26	− 123	0.6	− 2.4	8	4	− 4	− 17	0.0	− 0.2
5	4	− 161	− 107	0.0	0.8	8	5	7	4	− 0.1	− 0.3
5	5	− 51	77	1.3	− 0.3	8	6	− 5	22	0.3	− 0.4
6	0	47		− 0.1		8	7	12	− 3	− 0.3	− 0.3
6	1	60	− 14	− 0.3	− 0.9	8	8	6	− 16	− 0.5	− 0.3

Magnetic elements for London at different epochs

Values from 1850 onwards are for Greenwich; for 1580 the D observation was by Burrows and the I value is based on Norman (1576) and Gilbert (1600).

Year	Declination		Inclination		H/µT	Z/µT
	deg	min	deg	min		
1580.	11	17 E	72	0		
1660.	0	0	73	15		
1720.	13	0 W	74	40†		
1815.	24	27 W†	70	30		
1850.	22	24 W	68	47		
1875.	19	21 W	67	42	17.97	43.83
1907.	16	0 W	66	56	18.55†	43.57
1913.	15	15 W	66	50‡	18.53	43.32
1929.	12	23 W	66	53	18.38	43.05‡
1937.	10	58 W	66	59	18.34‡	43.17
1946.	9	38 W	67	1†	18.39	43.37
1970.	7	7 W	66	31	19.00	43.72

† Maximum.   ‡ Minimum.

The following table contains the mean values of D, H and Z observed during the year 1965 and their annual rate of change for a selection of observatories.

Geomagnetic data, 1965

Place	Lat. deg	Long. east deg	D (east) deg	Annual change min yr^{-1}	H nT	Annual change nT yr^{-1}	Z nT	Annual change nT yr^{-1}
Resolute Bay	+74.7	265.1	−79.9	+60	809	−10	58 180	+35
Bear Island	+74.5	19.2	+2.7	+2	9 245	+6	52 350	+20
Point Barrow	+71.3	203.3	+25.9	−1	9 642	+15	56 312	+5
Tromso	+69.7	18.9	+0.1	0	11 251	+12	51 094	+24
College	+64.9	212.2	+28.5	−2	12 849	+16	55 313	−3
Srednikan	+62.4	152.3	−10.3	+1	16 755	+12	54 043	0
Lerwick (UK)	+60.1	358.8	−9.4	+3	14 656	+20	47 403	+26
Krasnaya	+55.5	37.3	+7.9	0	17 182	+6	48 340	+13
Eskdalemuir (UK) . . .	+55.3	356.8	−10.0	+4	16 907	+25	45 440	+18
Patrony	+52.2	104.4	−2.2	−2	19 539	+7	57 288	−28
Hartland (UK)	+51.0	355.5	−9.5	+5	18 872	+29	43 540	+11
Memanbetsu	+43.9	144.2	−8.2	0	26 572	−4	41 501	−22
Fredericksburg	+38.2	282.6	−7.0	−5	19 511	+65	52 869	−55
San Miguel	+37.8	334.3	−15.0	+7	24 981	+60	38 495	−48
Kakioka	+36.2	140.2	−6.4	0	30 195	−6	34 772	−19
Tucson	+32.2	249.2	+13.5	−2	25 879	−12	43 639	−37
Quetta	+30.2	67.0	+1.3	−1	33 072	+15	33 725	−15
Honolulu	+21.3	202.0	+11.6	0	28 035	−22	22 460	−15
Alibag	+18.6	72.9	−0.9	0	38 793	−7	17 573	−18
San Juan	+18.1	293.9	−7.7	−7	27 550	0	33 438	−152
M'Bour	+14.4	343.4	−13.9	+7	31 695	+15	9 060	−115
Muntinlupa	+14.4	121.1	+0.1	−2	39 047	0	9 696	−2
Bangui	+4.4	18.6	−4.5	+4	32 276	+4	−8 254	−51
Lwiro	−2.2	28.8	−3.8	+4	29 804	−25	−17 019	−34
Apia	−13.8	188.2	+12.4	+2	34 765	−17	−20 325	+9
Mauritius	−20.9	57.6	−17.5	−5	22 220	−5	−30 495	0
Vassouras	−22.4	316.3	−17.2	−9	22 154	−67	−9 823	−78
Gnangara	−31.8	116.0	−2.9	0	23 906	−8	−53 496	0
Hermanus	−34.4	19.2	−24.2	0	12 526	−81	−27 139	+96
Toolangi	−37.5	145.5	+10.5	+2	22 541	−29	−56 370	+5
Macquarie	−54.5	159.0	+26.5	+10	13 152	−26	−64 214	+33
Mawson	−67.6	62.9	−61.2	−12	18 356	+4	−49 368	+98

	Lat.	Long.	
North magnetic dip-pole	+76.0	259.0 E	(1970)
South magnetic dip-pole	−66.5	139.0	(1970)

B.R.L.

1.9.6 The atmosphere

Variation of pressure, density and temperature with altitude

Values given in the table are based on the standard atmosphere of the International Civil Aviation Organization (ICAO). They are representative of average atmospheric conditions in temperate latitudes.

Altitude	Pressure	Density	Temperature	Altitude	Pressure	Density	Temperature
m	kN m^{-2}	kg m^{-3}	°C	m	kN m^{-2}	kg m^{-3}	°C
−250	104.4	1.25	17	6 000	47.2	0.66	−24
0	101.3	1.22	15	7 000	41.1	0.59	−30
250	98.4	1.20	13	8 000	35.6	0.53	−37
500	95.5	1.17	12	9 000	30.7	0.47	−44
750	92.6	1.14	10	10 000	26.4	0.41	−50
1000	89.9	1.11	8	12 000	19.3	0.31	−56
1500	84.6	1.06	5	14 000	14.1	0.23	−56
2000	79.5	1.00	2	16 000	10.3	0.17	−56
2500	74.7	0.96	−1	18 000	7.5	0.12	−56
3000	70.1	0.91	−4	20 000	5.5	0.088	−56
3500	65.8	0.86	−8	22 000	4.0	0.064	−54
4000	61.6	0.82	−11	24 000	2.9	0.046	−52
4500	57.7	0.78	−14	26 000	2.2	0.034	−50
5000	54.0	0.74	−18	28 000	1.6	0.025	−48
				30 000	1.2	0.018	−46

Reference

Manual of ICAO Standard Atmosphere, International Civil Aviation Organization Doc. 7488/2, 1964 (ICAO, Montreal, Canada).

K.W.T.E.

Pressure, mean molecular weight and temperature in the upper atmosphere

Altitude/km		Mean molecular weight	Pressure/(N m^{-2})	Temp./K
30		28.97	1.20×10^{3}	230.1
40		28.97	2.94×10^{2}	250.5
50		28.97	8.10×10^{1}	271.0
60		28.97	2.19×10^{1}	243.3
70		28.97	5.12×10^{0}	216.6
80		28.96	9.75×10^{-1}	186.0
90		28.94	1.63×10^{-1}	186.0
100		28.30	3.10×10^{-2}	208.1
120		27.01	2.73×10^{-3}	355.7
150		25.17	4.70×10^{-4}	652
200	Night . . .	22.7	9.5×10^{-5}	890
	Day	23.0	1.2×10^{-4}	1100
300	Night . . .	18.6	8.9×10^{-6}	980
	Day	19.7	1.8×10^{-5}	1360
500	Night . . .	15.0	2.8×10^{-7}	1000
	Day	16.3	1.3×10^{-6}	1420
800	Night . . .	6.5	1.1×10^{-8}	1000
	Day	12.1	6.6×10^{-8}	1430

The above values are a mean over all conditions. Considerable variation occurs with latitude, season and (above 100 km) with solar activity. Above 100 km the composition of the atmosphere changes due to dissociation and diffusive separation, the main constituents in addition to N_2 and O_2 being atomic oxygen and helium.

Reference

COSPAR International Reference Atmosphere, 1965 (North Holland Publishing Co.).

J.T.H.

1.9.7 Physical properties of sea water

The properties of sea water are a function of temperature, salinity (i.e. total dissolved solids in g kg^{-1}) and pressure. For the sources of the data see M. Hill (ed.), *The Sea*, vol. 1, chap. 1 and vol. 4, pt. 1, chap, 18 (Wiley) and Cox *et al.*, *Deep Sea Res.*, 1970, **17**, 679. For chemical composition see p. 153.

Density ρ of sea water at atmospheric pressure

Temperature °C	Salinity/(g kg^{-1})				
	20	25	30	35	40
	$(\rho - 1)/(\text{kg m}^{-3})$†				
0	16.04	20.06	24.08	28.10	32.14
5	15.84	19.78	23.73	27.68	31.64
10	15.31	19.18	23.07	26.96	30.86
15	14.48	18.30	22.13	25.97	29.82
20	13.39	17.17	20.96	24.75	28.56
25	12.07	15.82	19.57	23.34	27.12

† In oceanographical work these data are always expressed in terms of $1000(S-1)$ where S is the density relative to water at 4 °C. Values for this can be obtained by adding 0.03 to the values in the table.

Density ρ of sea water at 0 °C and salinity 35 g kg^{-1}

Pressure/(MN m^{-2})	0	20	40	60	80	100
$(\rho-1)/(\text{kg m}^{-3})$	28.10	37.44	46.37	54.92	63.12	71.02

Mechanical and thermal properties of sea water at salinity 35 g kg^{-1} and atmospheric pressure (unless otherwise stated)

Property	0 °C	20 °C
Dynamic viscosity	1.88×10^{-3} Pa s	1.08×10^{-3} Pa s
Kinematic viscosity, v	1.83×10^{-6} m^2 s^{-1}	1.05×10^{-6} m^2 s^{-1}
Thermal conductivity	0.563 W m^{-1} K^{-1}	0.596 W m^{-1} K^{-1}
Thermal diffusivity, κ	1.37×10^{-7} m^2 s^{-1}	1.46×10^{-7} m^2 s^{-1}
Prandtl number, v/κ	13.4	7.2
Specific heat capacity, C_p	3985 J kg^{-1} K^{-1}	3993 J kg^{-1} K^{-1}
Thermal expansion coefficient		
Pressure = 0.1 MN m^{-2}	52×10^{-6} K^{-1}	250×10^{-6} K^{-1}
Pressure = 100 MN m^{-2}	244×10^{-6} K^{-1}	325×10^{-6} K^{-1}
Ratio of specific heat capacities, C_p/C_v	1.0004	1.0106
Velocity of sound†	1449 m s^{-1}	1522 m s^{-1}
Compressibility	4.65×10^{-10} m^2 N^{-1}	4.28×10^{-10} m^2 N^{-1}
Freezing point		-1.910 °C
Boiling point		100.56 °C

† See also p. 67.

Electrical conductivity γ of sea water at atmospheric pressure

Temperature °C	Salinity/(g kg^{-1})				
	20	25	30	35	40
	γ/(S m^{-1})				
0	1.745	2.137	2.523	2.906	3.285
5	2.015	2.466	2.909	3.346	3.778
10	2.300	2.811	3.313	3.808	4.297
15	2.595	3.170	3.735	4.290	4.837
20	2.901	3.542	4.171	4.788	5.397
25	3.217	3.926	4.621	5.302	5.974

At a depth of 4000 m the bottom water (about 0 °C, salinity 35 g kg^{-1}) has a conductivity 6% greater than the surface water. The mean conductivity of the oceans (excluding the shallow seas) is 3.27 S m^{-1}.

E.C.B.

1.9.8 The geological time scale

Times of start (before present) and durations of geological periods are given in millions of years (*Quart. J. Geol. Soc. Lond.*, vol. 120S); uncertainties are about 2 Myr for the Tertiary rising to 10 Myr for the Palaeozoic.

Period	Start	Duration
Pleistocene	2	2
Pliocene	7	5
Miocene	26	19
Oligocene	38	12
Eocene	54	16
Palaeocene	65	11
Cretaceous	136	71
Jurassic	192	56
Triassic	225	33
Permian	280	55
Carboniferous	345	65
Devonian	395	50
Silurian	435	40
Ordovician	500	65
Cambrian	570	70
Precambrian	4600	4000

E.C.B.

1.10 Miscellaneous engineering data

1.10.1 Screw threads

The International Organization for Standardization (ISO) has now published recommendations for parallel screw threads which recognize only the ISO metric thread and the ISO inch thread. The ISO inch thread is the same as the Unified screw thread previously standardized by the USA, Britain and Canada. The ISO metric thread has the same form of thread as that of the Unified thread, namely, a symmetrical vee-form having a 60° included angle between the flanks.

Prior to the issue of the ISO Recommendations it was customary for bolts and set screws made in Britain to have a thread of Whitworth form (55° included angle) for diameters $\frac{1}{4}$ inch and above, and for sizes less than $\frac{1}{4}$ inch diameter the British Association (BA) thread ($47\frac{1}{2}$° included angle) was most commonly used. The Unified thread was also extensively used in the motor vehicle and aeronautical industries.

In 1965, following the Government announcement of the decision to adopt the metric system of measurement, the British Standards Institution decided to recommend British industry to adopt ISO metric threads for new designs in engineering; this means that BA threads and threads of Whitworth form will eventually become obsolete, but it may be many years before this occurs. It is likely also that Unified screw threads will continue to be used for many years in certain industries. Meanwhile ISO metric threads will come into increasing use in Britain.

The major diameter of a screw thread is the diameter over the crests of an external thread or to the roots of an internal thread.

The pitch is the distance between adjoining crests (say) of the same thread, measured parallel to the axis of the screw. It is specified by the reciprocal of the number of turns per inch (t.p.i.) for Whitworth and Unified screws.

For any one thread form various combinations of major diameter and pitch have been standardized to form screw-thread series; in the BA system major diameters and associated pitches are designated by means of numbers. The more common series are given in the table below.

Micrometer screws are generally made with 40 threads to the inch or, in the case of Metric screws, with a pitch of 0.5 mm.

The number designations and nominal sizes of 'woodscrews' are as follows: No. 0 (0.060 inch diameter), No. 1 (0.070), No. 2 (0.082), No. 3 (0.094), each succeeding number adding 0.014 inch to the diameter of the screw for sizes up to No. 10 (0.192): this applies to all lengths. (See BS 1210: 1963.)

Unified (ISO inch) threads				ISO metric threads	
Standard series				Standard series	
Unified Coarse (UNC)		Unified Fine (UNF)		Coarse	
Major diameter/in	t.p.i.	Major diameter/in	t.p.i.	Major diameter/mm	Pitch/mm
—	—	(No. 0) 0.0600	80	1.6	0.35
(No. 1) 0.0730	64	(No. 1) 0.0730	72	1.8	0.35
(No. 2) 0.0860	56	(No. 2) 0.0860	64	2	0.40
(No. 3) 0.0990	48	(No. 3) 0.0990	56	2.2	0.45
(No. 4) 0.1120	40	(No. 4) 0.1120	48	2.5	0.45
(No. 5) 0.1250	40	(No. 5) 0.1250	44	3	0.50
(No. 6) 0.1380	32	(No. 6) 0.1380	40	3.5	0.60
(No. 8) 0.1640	32	(No. 8) 0.1640	36	4	0.70
(No. 10) 0.1900	24	(No. 10) 0.1900	32	4.5	0.75
(No. 12) 0.2160	24	(No. 12) 0.2160	28	5	0.80
$\frac{1}{4}$	20	$\frac{1}{4}$	28	6	1.00
$\frac{5}{16}$	18	$\frac{5}{16}$	24	7	1.00
$\frac{3}{8}$	16	$\frac{3}{8}$	24	8	1.25
$\frac{7}{16}$	14	$\frac{7}{16}$	20	10	1.50
$\frac{1}{2}$	13	$\frac{1}{2}$	20	12	1.75
$\frac{9}{16}$	12	$\frac{9}{16}$	18	14	2.00
$\frac{5}{8}$	11	$\frac{5}{8}$	18	16	2.00
$\frac{3}{4}$	10	$\frac{3}{4}$	16	18	2.50
$\frac{7}{8}$	9	$\frac{7}{8}$	14	20	2.50
1	8	1	12	22	2.50
				24	3.00

Whitworth form threads				BA threads		
Standard series				Standard series		
British Standard Whitworth (BSW)		British Standard Fine (BSF)		BA		
Major diameter/in	t.p.i.	Major diameter/in	t.p.i.	No.	Major diameter/mm	Pitch/mm
$\frac{1}{4}$	20	$\frac{1}{4}$	26	0	6.0	1.00
—	—	$\frac{9}{32}$	26	1	5.3	0.90
$\frac{5}{16}$	18	$\frac{5}{16}$	22	2	4.7	0.81
$\frac{3}{8}$	16	$\frac{3}{8}$	20	3	4.1	0.73
$\frac{7}{16}$	14	$\frac{7}{16}$	18	4	3.6	0.66
$\frac{1}{2}$	12	$\frac{1}{2}$	16	5	3.2	0.59
$\frac{9}{16}$	12	$\frac{9}{16}$	16	6	2.8	0.53
$\frac{5}{8}$	11	$\frac{5}{8}$	14	7	2.5	0.48
$\frac{11}{16}$	11	$\frac{11}{16}$	14	8	2.2	0.43
$\frac{3}{4}$	10	$\frac{3}{4}$	12	9	1.9	0.39
$\frac{7}{8}$	9	$\frac{7}{8}$	11	10	1.7	0.35
1	8	1	10			

Screw thread standards

Standards are published in Great Britain, for the more common screw threads, by the British Standards Institution, 2 Park Street, London, W1A 2BS. These standards include details of tolerances for different grades of workmanship as well as recommended combinations of major diameter and pitch.

Unified (ISO inch) screw threads   BS 1580
Parts 1 and 2, 1962: Diameters $\frac{1}{4}$ inch and larger
Part 3, 1965: Diameters below $\frac{1}{4}$ inch
BS 1580 gives information relating to the following standard series:
    Unified Coarse threads (UNC)
    Unified Fine threads (UNF)

ISO metric threads   BS 3643
Part 1, 1963: Thread data and standard thread series
Part 2, 1966: Limits and tolerances for coarse pitch series threads
Part 3, 1967: Limits and tolerances for fine pitch threads

Whitworth form screw threads
BS 84: 1956, gives information relating to the following standard series:
    British Standard Whitworth threads (BSW)
    British Standard Fine threads (BSF)
BS 2779: 1956: Fastening threads of BSP sizes (parallel threads)

British Association (BA) threads   BS 93: 1951

British Standard Pipe (BSP) threads (taper)   BS 21: 1957

L.W.N.

1.10.2 Standard Wire Gauge

The sizes of wires are ordinarily expressed by an arbitrary series of numbers. There are, unfortunately, four or five independent systems of numbering, so that the wire gauge used must be specified. The wire gauge most commonly used in Great Britain is the Standard Wire Gauge (SWG). The sizes are given in the following table from 10 SWG upwards, together with the metric equivalents (1 in = 25.4 mm). For details of other wire gauges, see BS 3737: 1964.

Size	Diameter		Size	Diameter		Size	Diameter	
SWG	in	mm	SWG	in	mm	SWG	in	mm
10	0.128	3.25	23	0.024	0.610	37	0.0068	0.173
11	0.116	2.95	24	0.022	0.559	38	0.0060	0.152
12	0.104	2.64	25	0.020	0.508	39	0.0052	0.132
13	0.092	2.34	26	0.018	0.457	40	0.0048	0.122
14	0.080	2.03	27	0.0164	0.417	41	0.0044	0.112
15	0.072	1.83	28	0.0148	0.376	42	0.0040	0.102
16	0.064	1.63	29	0.0136	0.345	43	0.0036	0.091
17	0.056	1.42	30	0.0124	0.315	44	0.0032	0.081
18	0.048	1.22	31	0.0116	0.295	45	0.0028	0.071
19	0.040	1.02	32	0.0108	0.274	46	0.0024	0.061
20	0.036	0.914	33	0.0100	0.254	47	0.0020	0.051
21	0.032	0.813	34	0.0092	0.234	48	0.0016	0.041
22	0.028	0.711	35	0.0084	0.213	49	0.0012	0.030
			36	0.0076	0.193	50	0.0010	0.025

L.W.N.

1.10.3 Fuses

Tinned copper wire is the fusible element most often used in rewirable fuses. The gauge of wire for a given current-rating depends on the design of the fuse: the following table gives average values. The rated current is the largest current that can be carried indefinitely without deterioration. The fusing current will usually be between 1.5 and 2.0 times the rated current.

Rated current/A	5	15	30	60	100
Dia. of tinned copper fuse-wire/in	0.0084	0.020	0.036	0.056	0.080

Open-wire and semi-enclosed rewirable fuses have been largely superseded by cartridge fuses, in which the fusible element is contained within a sealed container (the cartridge). Standard cartridge fuse-links, on which the current rating is invariably marked, are sold in a variety of shapes and sizes for use in domestic consumers' control units, in plugs, and in other equipment. They are not intended to be rewired, and when blown they should be replaced complete by a new one of the same type and appropriate rating.

A.F.

Chemistry

2.1 The elements

2.1.1 The periodic table of the elements with atomic numbers

Period	IA	IIA	IIIA	IVA	VA	VIA	VIIA	VIII			IB	IIB	IIIB	IVB	VB	VIB	VIIB	
1	1 H																	2 He
2	3 Li	4 Be											5 B	6 C	7 N	8 O	9 F	10 Ne
3	11 Na	12 Mg											13 Al	14 Si	15 P	16 S	17 Cl	18 Ar
4	19 K	20 Ca	21 Sc	22 Ti	23 V	24 Cr	25 Mn	26 Fe	27 Co	28 Ni	29 Cu	30 Zn	31 Ga	32 Ge	33 As	34 Se	35 Br	36 Kr
5	37 Rb	38 Sr	39 Y	40 Zr	41 Nb	42 Mo	43 Tc	44 Ru	45 Rh	46 Pd	47 Ag	48 Cd	49 In	50 Sn	51 Sb	52 Te	53 I	54 Xe
6	55 Cs	56 Ba	57 La	72 Hf	73 Ta	74 W	75 Re	76 Os	77 Ir	78 Pt	79 Au	80 Hg	81 Tl	82 Pb	83 Bi	84 Po	85 At	86 Rn
7	87 Fr	88 Ra	89 Ac	104 —														

Lanthanides: 58 Ce, 59 Pr, 60 Nd, 61 Pm, 62 Sm, 63 Eu, 64 Gd, 65 Tb, 66 Dy, 67 Ho, 68 Er, 69 Tm, 70 Yb, 71 Lu

Actinides: 90 Th, 91 Pa, 92 U, 93 Np, 94 Pu, 95 Am, 96 Cm, 97 Bk, 98 Cf, 99 Es, 100 Fm, 101 Md, 102 No, 103 Lr

E.F.G.H.

2.1.2 Properties of the elements

Symbols, atomic weights, atomic numbers, densities, melting points and boiling points of the elements

The values of relative atomic masses $(A_r(E))$ of the element E given are based on the atomic mass of $^{12}C = 12$ (i.e. $A_r(^{12}C) = 12$) and apply to elements as they exist in materials of terrestrial origin and to certain artificial elements. When used with due regard to footnotes they are considered reliable to ± 1 in the last digit, or to ± 3 if that digit is subscript. The values listed are those recommended by the International Union of Pure and Applied Chemistry Commission on Atomic Weights (1971). The values in italics are those for the longest-lived isotopes selected by Dr B. Rose: for a list of isotopes see p. 286. Melting points and boiling points are for a pressure of 101.325 kN m^{-2}. Values for the vapour pressures of the elements are given on pp. 174–178, 181–182.

Element	Symbol	Atomic No.	Atomic weight	Density $\rho/(\text{kg m}^{-3})$	Melting point $\theta/°C$	Boiling point $\theta/°C$
Actinium .	Ac	89	_227_	10 060	1230	3200
Aluminium .	Al	13	26.981 54[a]	2 698	660.37†	2350
Americium .	Am	95	_243_	13 670	990	2600
Antimony .	Sb	51	121.7$_5$	6 692	630.74†	1640
Argon . .	Ar	18	39.948[b, c, d, g]	1656/−233 °C	−189.4	−185.9
Arsenic . .	As	33	74.921 6[a]	5 776	613 solid sublimes	
Astatine . .	At	85	_210_		300	350
Barium . .	Ba	56	137.34	3 594	710	1640
Berkelium .	Bk	97	_247_	14 790	986	
Beryllium .	Be	4	9.012 18[a]	1 846	1285	2470
Bismuth . .	Bi	83	208.980 4[a]	9 803	271.442†	1650
Boron . .	B	5	10.81[c, d, e]	2 466	2030	3700
Bromine . .	Br	35	79.904[c]	3 120	−7.1	58.9
Cadmium .	Cd	48	112.40	8 647	321.108†	770
Caesium . .	Cs	55	132.905 4[a]	1 900	28.6	686
Calcium . .	Ca	20	40.08[g]	1 530	840	1490
Californium .	Cf	98	_251_			
Carbon . .	C	6	12.011[b, d]	2 266	3700 solid sublimes	
Cerium . .	Ce	58	140.12	6 711	800	3000
Chlorine .	Cl	17	35.453[c]	2030/−160 °C	−101	−34.0
Chromium .	Cr	24	51.996[c]	7 194	1860	2600
Cobalt . .	Co	27	58.933 2[a]	8 800	1494†	2900
Copper . .	Cu	29	63.545[c, d]	8 933	1084.5†	2580
Curium . .	Cm	96	_247_	13 300	1340	
Dysprosium .	Dy	66	162.5$_0$	8 531	1410	2600
Einsteinium .	Es	99	_254_			
Erbium . .	Er	68	167.2$_6$	9 044	1520	2600
Europium .	Eu	63	151.96	5 248	820	1450
Fermium .	Fm	100	_257_			
Fluorine . .	F	9	18.998 40[a]	1140/−200 °C	−219.6	−188.1
Francium .	Fr	87	_223_		30	650
Gadolinium .	Gd	64	157.2$_5$	7 870	1310	3000
Gallium . .	Ga	31	69.72	5 905	29.8	2070
Germanium .	Ge	32	72.5$_9$	5 323	959	2850
Gold . . .	Au	79	196.966 5[a]	19 281	1064.43†	2850
Hafnium .	Hf	72	178.4$_9$	13 276	2230	5300
Helium . .	He	2	4.002 60[b, c]	120/4.22 K	3.5 K‡	4.22 K
Holmium .	Ho	67	164.930 4[a]	8 797	1470	2300
Hydrogen .	H	1	1.007 9[b, d]	89/−266.8 °C	−259.2	−252.87†

Properties (*contd*)

Element	Symbol	Atomic No.	Atomic weight	Density $\rho/(\text{kg m}^{-3})$	Melting point $\theta/^\circ\text{C}$	Boiling point $\theta/^\circ\text{C}$
Indium . .	In	49	114.82	7 290	$156.634\dagger$	2050
Iodine . .	I	53	$126.904\ 5^a$	4 953	113.6	184
Iridium . .	Ir	77	192.2_2	22 550	$2447\dagger$	4550
Iron . . .	Fe	26	55.84_7	7 873	1540	2760
Krypton . .	Kr	36	83.80	$3000/-188\ ^\circ\text{C}$	-157.3	-153.4
Lanthanum .	La	57	$138.905_5{}^b$	6 174	920	3450
Lawrencium	Lr	103	*256*			
Lead . . .	Pb	82	$207.2^{d,\,g}$	11 343	$327.502\dagger$	1760
Lithium . .	Li	3	$6.94_1{}^{c,\,d,\,e,\,g}$	533	180	1360
Lutetium .	Lu	71	174.97	9 842	1700	3400
Magnesium .	Mg	12	$24.305^{c,\,g}$	1 738	650	1100
Manganese .	Mn	25	$54.938\ 0^a$	7 473	1250	2120
Mendelevium	Md	101	*257*			
Mercury .	Hg	80	200.5_9	13 546	$-38.862\dagger$	$356.66\dagger$
Molybdenum	Mo	42	95.9_4	10 222	2620	4830
Neodymium .	Nd	60	144.2_4	7 000	1024	3100
Neon . .	Ne	10	$20.179_9{}^c$	$1442/-268\ ^\circ\text{C}$	-248.6	$-246.048\dagger$
Neptunium .	Np	93	$237.048\ 2^f$	20 450	640	
Nickel . .	Ni	28	58.7_1	8 907	$1455\dagger$	2150
Niobium .	Nb	41	$92.906\ 4^a$	8 578	2425	5000
Nitrogen .	N	7	$14.006\ 7^{b,\,c}$	$1035/-268.8\ ^\circ\text{C}$	-210	$-195.802\dagger$
Nobelium .	No	102	*255*			
Osmium . .	Os	76	190.2	22 580	3030	5000
Oxygen . .	O	8	$15.9994^{b,\,c,\,d}$	$1460/-252.7\ ^\circ\text{C}$	-218.8	$-182.962\dagger$
Palladium .	Pd	46	106.4	11 995	$1554\dagger$	3000
Phosphorus .	P	15	$30.973\ 76^a$	1820 (yellow)	44.2 (yellow)	280.4
Platinum .	Pt	78	195.0_9	21 450	$1772\dagger$	3720
Plutonium .	Pu	94	*244*	19 814	640	3200
Polonium .	Po	84	*209*	9 400	254	960
Potassium .	K	19	39.09_8	862	63.2	777
Praseodymium	Pr	59	$140.907\ 7^a$	6 779	935	3000
Promethium .	Pm	61			1000	2700
Protactinium .	Pa	91	$231.035\ 9^f$	15 370	1200	4000
Radium . .	Ra	88	$226.025\ 4^{f,\,g}$	5 000	700	1500
Radon . .	Rn	86	*222*	4400 (liquid $-62\ ^\circ\text{C}$)	-71	-62
Rhenium .	Re	75	186.2	21 023	3180	5600
Rhodium .	Rh	45	$102.905\ 5^a$	12 420	$1963\dagger$	3700
Rubidium .	Rb	37	$85.467_8{}^c$	1 533	38.8	705
Ruthenium .	Ru	44	101.0_7	12 360	2310	4100
Samarium .	Sm	62	150.4	7 536	1060	1600
Scandium .	Sc	21	$44.955\ 9^a$	2 992	1540	2800
Selenium .	Se	34	78.9_6	4 808	220	685
Silicon . .	Si	14	$28.08_6{}^d$	2 329	1410	2620
Silver . .	Ag	47	107.868^c	10 500	$961.93\dagger$	2160
Sodium . .	Na	11	$22.989\ 77^a$	966	97.8	900
Strontium .	Sr	38	87.62^g	2 583	770	1380
Sulphur . .	S	16	32.06^d	2 086	115	$444.674\dagger$
Tantalum .	Ta	73	$180.947_9{}^b$	16 670	3000	5400
Technetium .	Tc	43	*97*	11 496	2200	4600
Tellurium .	Te	52	127.6_0	6 247	450	990
Terbium .	Tb	65	$158.925\ 4^a$	8 267	1360	2500

Properties (*contd*)

Element	Symbol	Atomic No.	Atomic weight	Density $\rho/(\mathrm{kg\ m^{-3}})$	Melting point $\theta/°\mathrm{C}$	Boiling point $\theta/°\mathrm{C}$
Thallium .	Tl	81	204.3$_7$	11 871	304	1460
Thorium .	Th	90	232.038 1^f	11 725	1700	4500
Thulium .	Tm	69	168.934 2^a	9 325	1550	2000
Tin . . .	Sn	50	118.6$_9$	7 285	231.9681†	2720
Titanium .	Ti	22	47.9$_0$	4 508	1670	3300
Tungsten .	W	74	183.8$_5$	19 254	3387†	5420
Uranium .	U	92	238.029$^{b, c, e}$	19 050	1135	4000
Vanadium .	V	23	50.941$_4^{b, c}$	6 090	1920	3400
Xenon . .	Xe	54	131.30	3560/(-185 °C)	-111.9	-108.1
Ytterbium .	Yb	70	173.04	6 966	824	1500
Yttrium . .	Y	39	88.905 9^a	4 475	1510	3300
Zinc . . .	Zn	30	65.38	7 135	419.58†	913
Zirconium .	Zr	40	91.22	6 507	1850	4400

Notes:

[a] Element with only one stable nuclide.

[b] Element with one predominant isotope (about 99 to 100% abundance); errors in abundance determinations have a corresponding small effect on the confidence in the value of $A_r(\mathrm{E})$.

[c] Element for which the value of $A_r(\mathrm{E})$ derives its reliability from calibrated measurements (i.e. from comparisons with synthetic mixtures of known isotopic composition).

[d] Element for which known variations in isotopic abundance in terrestrial material prevent a more precise atomic weight being given; $A_r(\mathrm{E})$ values should be applicable to any 'normal' material.

[e] Element for which values of A_r may be found in commercially available products that differ from the tabulated value of $A_r(\mathrm{E})$ because of inadvertent or undisclosed changes of isotopic composition.

[f] Element for which the value of A_r is that of the most commonly available long-lived nuclide.

[g] Element for which geological specimens are known in which the element has an anomalous isotopic composition.

† Primary and secondary fixed points on the International Practical Temperature Scale of 1968.

‡ This and other values marked K are in kelvins.

E.F.G.H.

2.1.3 Abundances of the elements

This table gives an indication of the abundances of the elements in nature. The abundances stated to be in 'atoms' are relative to the number of silicon atoms taken as 10^6. The values for sea water and for crustal rocks are from Turekian, *McGraw-Hill Encyclopedia of Science and Technology*, 1970, **4**, 627; those for stony meteorites are mean values for chondrites from Urey, *Rev. Geophys.*, 1964, **2**, 1; those for major elements in iron meteorites are from Brown, *Rev. Mod. Phys.*, 1949, **21**, 625 and those for trace elements are from Smales *et al.*, in *Origin and Distribution of the Elements*, 1968, ed. Ahrens (Pergamon, London). The values for trace elements in iron meteorites are exceedingly variable and the results given are mean values for medium octahedrites. Values for the sun are from Muller, in *Origin and Distribution of the Elements, loc. cit.*, but the values listed may be incorrect by 50% because of the normalization to silicon taken as 10^6 atoms; the values for the solar system are from Cameron in the same publication.

The composition of the atmosphere is from *Atmospheres of the Earth and Planets*, 1949, ed. Kuiper (Chicago Univ. Press) with corrections from Glueckauf and Kitt, *Proc. Roy. Soc.*, 1956, **234A**, 557.

Abundances (*contd*)

At. No.	Element	Sea water $\mu g\ dm^{-3}$	Crustal rocks p.p.m.	Meteorites Stony atoms	Meteorites Iron p.p.m.	Sun atoms	Solar system atoms
1	H	1.1×10^8	—	—	—	3.2×10^{10}	2.6×10^{10}
2	He	0.0072	—	—	—	4.8×10^9	2.1×10^9
3	Li	1.7×10^2	20	53	—	<0.25	45
4	Be	0.0006	2.0	0.69	—	6.9	0.69
5	B	4.4×10^3	7.0	6.2	—	1.3×10^2	6.2
6	C	2.8×10^4	—	—	1.1×10^3	1.0×10^7	1.4×10^7
7	N	1.6×10^4	20	—	—	3.6×10^6	2.4×10^6
8	O	8.8×10^8	4.5×10^5	—	—	2.1×10^7	2.4×10^7
9	F	1.3×10^3	4.6×10^2	6.7×10^2	—	—	3.6×10^3
10	Ne	0.12	—	—	—	—	2.4×10^6
11	Na	1.1×10^7	2.3×10^4	4.6×10^4	—	6.3×10^4	6.3×10^4
12	Mg	1.3×10^6	2.8×10^4	9.4×10^5	3.2×10^2	7.2×10^5	1.0×10^6
13	Al	1	8.0×10^4	6.5×10^4	40	5.0×10^4	8.5×10^4
14	Si	2.9×10^3	2.7×10^5	1.0×10^6	40	1.0×10^6	1.0×10^6
15	P	88	1.0×10^3	5.3×10^3	2.2×10^3	6.9×10^3	1.3×10^4
16	S	9.0×10^5	3.0×10^2	1.1×10^5	3.6×10^2	6.3×10^5	5.1×10^5
17	Cl	1.9×10^7	1.9×10^2	7.0×10^2	—	—	2.0×10^3
18	Ar	4.5×10^2	—	—	—	—	2.3×10^5
19	K	3.9×10^5	1.7×10^4	3.6×10^3	—	1.6×10^3	3.2×10^3
20	Ca	4.1×10^5	5.1×10^4	4.9×10^4	5.0×10^2	3.5×10^4	7.4×10^4
21	Sc	<0.004	22	30	—	22	33
22	Ti	1	8.6×10^3	2.6×10^3	1.0×10^2	1.7×10^3	2.3×10^3
23	V	1.9	1.7×10^2	2.0×10^2	6	4.7×10^2	9.0×10^2
24	Cr	0.2	96	8.6×10^3	15	3.6×10^3	1.2×10^4
25	Mn	1.9	1.0×10^3	7.4×10^3	3.0×10^2	2.2×10^3	8.8×10^3
26	Fe	3.4	5.8×10^4	6.9×10^5	9.1×10^5	2.0×10^5	8.9×10^5
27	Co	0.39	28	1.4×10^3	6.3×10^3	1.6×10^3	2.3×10^3
28	Ni	6.6	72	3.9×10^4	6.7×10^4	1.9×10^4	4.6×10^4
29	Cu	23	58	2.2×10^2	1.3×10^2	8.9×10^2	9.2×10^2
30	Zn	11	82	1.3×10^2	28	1.0×10^2	1.5×10^3
31	Ga	0.03	17	12	80	17	46
32	Ge	0.06	1.3	19	37	9.8	1.3×10^2
33	As	2.6	2.0	4.6	11	—	7.2
34	Se	0.090	0.05	18	3	—	70
35	Br	6.7×10^4	4.0	—	1	—	21
36	Kr	0.21	—	—	—	—	64
37	Rb	1.2×10^2	70	4.9	—	9.5	6.0
38	Sr	8.1×10^3	4.5×10^2	20	—	33	58
39	Y	0.0013	35	3.4	—	50	4.6
40	Zr	0.026	1.4×10^2	46	8	14	30
41	Nb	0.015	20	—	0.2	6.3	1.2
42	Mo	10	1.2	2.5	7.3	6.3	2.5
43	Tc	Unstable with short period					
44	Ru	—	—	1.4	11	2.1	1.6
45	Rh	—	—	0.27	4.1	0.74	0.33
46	Pd	—	0.003	0.94	3.8	1.2	1.5
47	Ag	0.28	0.08	0.09	0.035	0.18	0.5
48	Cd	0.11	0.18	0.079	8	1.1	2.1
49	In	—	0.2	0.0011	0.010	0.89	0.22
50	Sn	0.81	1.5	0.56	77	—	4.2

Abundances (*contd*)

At. No.	Element	Sea water $\mu g\ dm^{-3}$	Crustal rocks p.p.m.	Meteorites		Sun atoms	Solar system atoms
				Stony atoms	Iron p.p.m.		
51	Sb	0.33	0.2	0.11	0.34	—	0.38
52	Te	—	—	0.68	—	—	6.8
53	I	64	0.5	0.051	0.6	—	1.4
54	Xe	0.047	—	—	—	—	7.1
55	Cs	0.30	1.6	0.13	—	—	0.37
56	Ba	21	3.8×10^2	4.6	—	4.0	4.7
57	La	0.003 4	50	0.39	—	3.4	0.36
58	Ce	0.001 2	83	1.2	—	1.9	1.2
59	Pr	0.000 64	13	0.14	—	0.89	0.17
60	Nd	0.002 8	44	0.64	—	2.7	0.77
61	Pm	Unstable with short period					
62	Sm	0.000 45	7.7	0.23	—	1.3	0.23
63	Eu	0.000 13	2.2	0.082	—	0.29	0.091
64	Gd	0.000 70	6.3	0.34	—	0.43	0.34
65	Tb	0.001 4	1.0	0.051	—	—	0.052
66	Dy	0.000 91	8.5	0.33	—	0.32	0.36
67	Ho	0.000 22	1.6	0.076	—	—	0.090
68	Er	0.008 7	3.6	0.23	—	—	0.22
69	Tm	0.000 17	0.52	0.031	—	—	0.035
70	Yb	0.000 82	3.4	0.18	—	1.1	0.21
71	Lu	0.000 15	0.8	0.031	—	—	0.035
72	Hf	<0.008	4	0.16	—	—	0.16
73	Ta	<0.002 5	2.4	0.020	0.06	—	0.022
74	W	<0.001	1.0	0.13	8.1	—	0.16
75	Re	—	0.000 4	0.046	0.85	—	0.005
76	Os	—	0.000 2	0.066	7.6	—	0.71
77	Ir	—	0.000 2	0.34	3.0	—	0.43
78	Pt	—	—	0.92	19	—	1.1
79	Au	0.011	0.002	0.13	1.8	—	0.20
80	Hg	0.15	0.02	0.023	—	—	0.75
81	Tl	—	0.47	0.000 74	—	—	0.18
82	Pb	0.03	10	0.14	60	1.3	2.9
83	Bi	0.02	0.004	0.002 0	0.5	—	0.16
84–89	Po–Ac	Unstable with short period					
90	Th	0.001 5	5.8	0.027	0.04	—	0.034
91	Pa	Unstable with short period					
92	U	3.3	1.6	0.008 8	0.007	—	0.023
93–102	Np–No	Unstable with short period					

E.C.B.

2.1.4 Composition of the Earth's atmosphere

(Parts in 10^6 of dry air by volume)

N_2	O_2	A	CO_2	Ne	He	CH_4	Kr	N_2O	H_2	O_3	Xe
780 900	209 500	9300	300	18	5.2	1.5	1.14	0.5	0.5	0.4	0.086

E.C.B.

2.2 Properties of inorganic compounds

Formula (including water of crystallization), melting and boiling temperatures, density and solubility

The properties, including melting θ_m, and boiling θ_b temperatures, are for the substances as formulated and at a pressure of 101.325 kN m^{-2} (1 atm). Loss of water at a specific temperature $\theta/$°C is denoted by -1, -2, etc., H$_2$O/θ, and loss of oxygen similarly. Densities ρ are for 20 °C; values ρ/θ are for temperature $\theta/$°C. Qualitative solubilities are for 0 °C (cold) and 100 °C (hot). Quantitative solubilities are given as S/θ where S is 100 times the mass of solute soluble per unit mass of water at $\theta/$°C. For gases the values relate to a total pressure of 101.325 kN m^{-2}. For some gases the value of the absorption coefficient (A) is given: the volume of gas, measured at 0 °C and 101.325 kN m^{-2} which will dissolve in unit volume of water at the stated temperature, under a partial pressure of 101.325 kN m^{-2} of the gas.

Abbreviations used are: a., acid; abs., absolute; acet., acetone; ac., acetic; al., alcohol; alk., alkali; aq., aqueous; aq. reg., aqua regia; bz., benzene; c., cold; chl., chloroform; conc., concentrated; d., decomposes; dil., dilute; eth., ether; h., hot; i., insoluble; meth., methyl; min., mineral; pyr., pyridine; s., soluble; sl., slightly; solv., solvents; sb., sublimes; v., very.

Compound	$\theta_m/$°C	$\theta_b/$°C	$\rho/$(kg m^{-3})	Solubility		
				Cold water	Hot water	Other solvents
Aluminium						
Bromide, AlBr$_3$ . . .	97.5	263	3010	d.	d.	s. al., CS$_2$, acet.
Chloride, AlCl$_3$ . . .	sb. 178	—	2440	d.	d.	s. al., eth., CCl$_4$; sl. s. bz.
Fluoride, AlF$_3$ . . .	sb. 1291	—	3070	0.56/25	s.	i. a., alk., al., acet.
Nitrate, Al(NO$_3$)$_3$.9H$_2$O	73.5	d. 150	—	63.7/25	v. s. d.	s. a., alk., al.
Nitride, AlN	2200 (in N$_2$)	d.	3260	d.	d.	d. a., alk.
Oxide, α-Al$_2$O$_3$ . . .	2015	2980	3970	i.	i.	v. sl. s. a., alk.
Sulphate, Al$_2$(SO$_4$)$_3$.	d. 770	—	2710	31.3/0	98.1/100	s. dil. a.; sl. s. al.
Sulphide, Al$_2$S$_3$	1100	sb. 1500 (in N$_2$)	2020	d.	—	s. a., alk.; i. acet.
Ammonium						
Ammonia, NH$_3$ . . .	-77.7	-34.35	0.77	89.5/0	7.4/100	s. al., eth., org. solv.
Acetate, NH$_4$C$_2$H$_3$O$_2$.	114	d.	1170	148/4	d.	s. al.; sl. s. acet.
Azide, NH$_4$N$_3$ . . .	160	sb. 134	1346	16/0	37/40	s. al.; i. eth., bz.
Bromide, NH$_4$Br . .	sb. 452	—	2429	60/0	146/100	s. al., eth., acet.
Carbonate, (NH$_4$)$_2$CO$_3$	d. 58	—	—	100/15	d.	i. al., CS$_2$, NH$_3$
Chloride, NH$_4$Cl . .	sb. 340	520	1527	29.4/0	77.3/100	s. al., NH$_3$
Chloroplatinate, (NH$_4$)$_2$PtCl$_6$ . . .	d.	—	3065	0.29/0	1.25/100	sl. s. al.; i. eth.
Chromate, (NH$_4$)$_2$CrO$_4$	d. 180	—	1910	25/0	d.	sl. s NH$_3$, acet.; i. al.
Dichromate, (NH$_4$)$_2$Cr$_2$O$_7$ . .	d. 170	—	2150	18.2/0	149/100	s. al.; i. acet.
Dihydrogen phosphate, (NH$_4$)H$_2$PO$_4$ . . .	190	—	1803	22.7/0	173/100	i. acet.
Fluoride, NH$_4$F . . .	sb.	—	1009/25	100/0	d.	s. al.; i. NH$_3$
Formate, NH$_4$CHO$_2$.	116	d. 180	1280	102/0	531/80	s. al.; NH$_3$
Hydrogen phosphate, (NH$_4$)$_2$HPO$_4$ . . .	d. 155	—	1619	43/0	106/70	i. al. acet.
Iodate, NH$_4$IO$_3$ . .	d. 150	—	3309/21	2.7/16	14.5/101	—
Iodide, NH$_4$I . . .	sb. 551	—	2514/25	154/0	250/100	v. s. al., acet., NH$_3$
Molybdate,(NH$_4$)$_2$MoO$_4$	d.	—	2276/25	s.d.	d.	s. a.; i. al., NH$_3$
Nitrate, NH$_4$NO$_3$ . .	170	d. 210	1725/25	118/0	871/100	s. al., acet., NH$_3$
Nitrite, NH$_4$NO$_2$ . .	d.	—	1690	127/1	d.	s. al.; i. eth.
Oxalate, (NH$_4$)$_2$C$_2$O$_4$.H$_2$O .	d.	—	1500	2.54/0	33.2/100	i. NH$_3$
Sulphate, (NH$_4$)$_2$SO$_4$.	d. 235	—	1769/50	71/0	104/100	i. al., acet., NH$_3$
Tartrate, (NH$_4$)$_2$C$_2$H$_4$O$_6$ . .	d.	—	1601	45/0	85/60	sl. s. al.
Thiocyanate, NH$_4$CNS .	149.6	d. 170	1305	128/0	v. s.	s. al., acet., NH$_3$
Vanadate, meta-, NH$_4$VO$_3$	d. 200	—	2326	0.5/15	6.95/96, d.	i. al., eth., NH$_4$Cl

Inorganic compounds (*contd*)

Compound	$\theta_m/°C$	$\theta_b/°C$	$\rho/(kg\ m^{-3})$	Solubility		
				Cold water	Hot water	Other solvents
Antimony						
Bromide	96.6	280	4148/23	d.	d.	s. HCl, CS_2, al., NH_3
Chloride, tri-, $SbCl_3$.	73.2	221	3140/25	602/0	v. s.	s. HCl, a., CS_2
,, , penta-,						
$SbCl_5$	2.8	140	2336	d.	d.	s. HCl, chl.
Fluoride, tri-, SbF_3 . .	290	376	4379/21	385/0	564/30	i. NH_3
Fluoride, penta-, SbF_5 .	7	149	2990/23	s.	—	s. KF
Hydride, SbH_3 . . .	−88	−17	2260/−25	0.2/0 (A)	—	v. s. al., CS_2
Iodide, SbI_3	−170.5	400	4917/17	6/25	d.	s. HCl, KI, al., acet. CS_2
Oxide, tri-, Sb_2O_3 . .	655	1425	5250	v. sl. s.	sl. s.	s. HCl, KOH
,, , penta-, Sb_2O_5 .	−O/380	−2O/930	3800	i.	i.	s. HCl, KOH
III Oxychloride, SbOCl	d.	—	—	i.	i.	s. HCl, acet., CS_2
III Sulphate $Sb_2(SO_4)_3$	d.	—	3625/4	i.	d.	s. a.
Sulphide, tri-, Sb_2S_3 .	547	—	4640	i.	d.	s. alk., HCl, K_2S
,, , penta-, Sb_2S_5	300	d.	4120	i.	i.	s. alk., HCl
Arsenic						
Bromide, $AsBr_3$ . .	32.8	221	3540/25	d.	d.	s. HCl, CS_2
Chloride, $AsCl_3$ . . .	−8.5	130.2	2163	d.	d.	s. HCl, al., eth.
Fluoride, tri-, AsF_3 .	−8.5	−63	2666/0	d.	d.	s. al. eth., bz.
Fluoride, penta-, AsF_5 .	−80.3	−52.6	7710	s.	—	s. alk., al., eth.
Hydride, AsH_3 . . .	−116.3	−62.5	2695	20/0	—	sl. s. alk.; s. chl.
Iodide, AsI_3 . . .	146	403	4390/13	6/25	d.	s. al., eth., chl., bz.
Oxide, tri-, As_2O_3 . .	315	457	3738	1.5/15	11.5/100	s. alk., HCl; i. eth.
,, , penta-, As_2O_5 .	d. 315	—	4320	59.5/0	77/100	s. alk., a.
Sulphide, tri-, As_2S_3 .	300	707	3430	i.	sl. s.	s. K_2S, $NaHCO_3$
,, , penta-, As_2S_5	sb. d. 500	—	—	i.	d.	s. alk., HNO_3
Barium						
Acetate, $Ba(C_3H_3O_2)_2$	—	—	2468	58.8/0	75/100	i. al.
Bromate,						
$Ba(BrO_3)_2.H_2O$ . .	−H_2O/170	d. 260	3990/18	0.3/0	5.7/100	i. al.; s. acet.
Bromide, $BaBr_2.2H_2O$.	−H_2O/75	−2H_2O/120	3584/24	151/20	204/100	s. al.
,, , $BaBr_2$.	847	d.	4781/24	98/0	149/100	v. s. al.; v. sl. s. acet.
Carbonate, $BaCO_3$ . .	850	d.	4430	1.6×10^{-3}/9	6×10^{-3}/100	s. a., NH_4Cl; i. al.
Chlorate,						
$Ba(ClO_3)_2.H_2O$ . .	−H_2O/120	−O/250	3180	27.4/15	111.2/100	sl. s. al., acet.
Chlorate, $Ba(ClO_3)_2$.	414	—	—	20/0	100/100	—
Chloride, $BaCl_2.2H_2O$.	−2H_2O/113	—	3097/24	31.6/0	58.8/100	sl. s. HCl, HNO_3
,, , $BaCl_2$ . .	963	1560	3917	30.5/0	59/100	v. sl. s. al.
Fluoride, BaF_2 . .	1280	2137	4890	0.12/25	sl. s.	s. a., NH_4Cl
Hydride, BaH_2 . . .	d. 675	—	4210/0	d.	sl. s.	d. a.
Hydroxide,						
$Ba(OH)_2.8H_2O$ . .	78	−8H_2O/700	2180/16	1.7/0	101.4/80	sl. s. al.; i. acet.
Iodate, $Ba(IO_3)_2$ . .	d.	—	4998	8×10^{-3}/0	197/100	s. HNO_3, HCl
Iodide, $BaI_2.2H_2O$ . .	−H_2O/100	−2H_2O/540	5150	—	269/100	s. al., acet.
,, , BaI_2 . . .	d. 740	—	5150/25	170/0	—	v. s. al.
Nitrate, $Ba(NO_3)_2$ . .	592	d.	3240/23	4.95/0	33/100	sl. s. meth. al.; i. al.
Nitrite, $Ba(NO_2)_2.H_2O$	d. 115	—	3173	63/20	110/80	—
,, , $Ba(NO_2)_2$.	d. 217	—	3230/23	47/1	300/100	sl. s. al.; i. acet.
Oxide, BaO	1923	2000	5720	3.5/20	91/100	s. dil. a.
Peroxide	450	−O/800	4960	v. sl. s.	d.	s. dil. a.; i. acet.
Sulphate, $BaSO_4$ . .	1580	—	4500/15	1×10^{-3}/0	4×10^{-3}/100	sl. s. H_2SO_4
Sulphide, BaS . . .	1200	—	4250/15	d.	d.	i. al.
Beryllium						
Bromide, $BeBr_2$ . . .	488	520	3465/25	s.	v. s.	s. al. eth.; i. bz.
Chloride, $BeCl_2$ . . .	410	492	1899/25	v. s.	v. s. d.	v. s. al., eth., bz.; i. acet.
Fluoride, BeF_2 . . .	sb. 800	—	1986/25	v. s.	v. s.	s. H_2SO_4, al.
Nitrate,						
$Be(NO_3)_2.3H_2O$ . .	60	142	1557	v. s.	v. s.	v. s. al.
Oxide, BeO	2540	4120	3010	2×10^{-5}/30	—	s. conc. H_2SO_4
Sulphate, $BeSO_4.4H_2O$	−2H_2O/100	−4H_2O/400	1713/10.5	42.5/25	100/100	i. al., acet.

Inorganic compounds (*contd*)

Compound	$\theta_m/°C$	$\theta_b/°C$	$\rho/(\text{kg m}^{-3})$	Solubility Cold water	Hot water	Other solvents
Bismuth						
Bromide, $BiBr_3$ . . .	218	461	5720/25	d.	d.	s. HCl, eth.; i. al.
Chloride, $BiCl_3$ . . .	232	441	4750/25	d.	d.	s. a., al., eth.
Fluoride, BiF_3 . . .	730	—	5320	i.	—	s. a., acet.; i. al.
Iodide, BiI_3	403	500	5778/15	i.	d.	s. HCl, HI, abs. a.
Nitrate, $Bi(NO_3)_3.5H_2O$	d.	—	2830	d.	d.	v. s. HNO_3; s. a., acet.
Oxide, Bi_2O_3 . . .	820	1890	8900	i.	i.	s. a.
Oxychloride, BiOCl .	—	—	7720/15	i.	i.	s. a.
Boron						
Chloride, BCl_3 . . .	− 107	12.4	1349/11	d.	d.	d. al.
Fluoride, BF_3 . . .	− 126.7	− 99	2.99	1.06/0(A)	d.	s. conc. H_2SO_4
Hydride, B_2H_6 . . .	− 165.5	− 92.5	447/ − 11	sl. s. d.	d.	s. NH_3
,, , B_4H_{10} . .	− 120.8	16	560/ − 35	sl. s. d.	d.	d. al.; s. bz.
Nitride, BN	3000	—	2250	i.	d.	sl. s. hot a.
Oxide, B_2O_3. . . .	460	1860	2460	1.1/0	20/103	s. a. al.
Boric acid, H_3BO_3 . .	d. 185	—	1435/15	2.7/0	27.7/100	s. al.; sl. s. acet.
Cadmium						
Acetate, $Cd(C_3H_3O_2)_2$.	256	d.	2341	v. s.	—	s. meth. al.
Bromide, $CdBr_2$. . .	567	863	5192/25	57/10	162/104	s. al.; sl. s. eth. acet.
Chloride, $CdCl_2$. . .	568	960	4047/25	140/20	150/100	s. al.; i. acet., eth.
Fluoride, CdF_2 . . .	1110	1758	6640	4.35/25	—	s. a., HF; i. al.
Iodide, CdI_2. . . .	387	796	5670/30	80/0	125/100	s. a., al., eth.; s. sl. acet.
Nitrate, $Cd(NO_3)_3.4H_2O$.	59.4	132	2455/17	s.	s.	s. al., eth., ac., NH_3
Nitrate, $Cd(NO_3)_2$ . .	350	—	—	109/0	682/100	v. s. a.
Oxide, CdO	—	sb. 1559	8150	i.	i.	s. a.; i. alk.
Sulphate, $CdSO_4$ . .	1000	—	4691	75	60/100	i. al., acet.
Caesium						
Carbonate, Cs_2CO_3 .	d. 610	—	—	260/15	v. s.	s. al., eth.
Chloride, CsCl . . .	645	1300	3988	161/0	260/89	v. s. al.
Fluoride, CsF . . .	682	1251	4115	340/18	—	i. al.
Hydroxide, CsOH . .	272	—	3675	385/15	—	s. al.
Nitrate, $CsNO_3$ . . .	417	d.	3685	9.3/0	188/100	s. acet.; sl. s. al.
Calcium						
Acetate, $Ca(C_2H_3O_2)_2$.	d.	—	—	37.4/0	29.7/100	sl. s. al.
Bromide, $CaBr_2$ . . .	sb. 730 d.	810	3353/25	125/0	286/100	s. al., acet.
Carbide, CaC_2 . . .	—	2300	2200	d.	d.	—
Carbonate, $CaCO_3$ (aragonite) . . .	—	d. 825	2930	i.	i.	s. a., NH_4Cl
Carbonate, $CaCO_3$ (calcite)	—	d. 899	2710/18	i.	i.	s. a., NH_4Cl
Chloride, $CaCl_2.6H_2O$	− 4H_2O/30	− 6H_2O/200	1710/25	279/0	536/20	s. al.
,, , $CaCl_2$. . .	782	2000	2150/25	59.5/0	159/100	s. al., acet.
Fluoride, CaF_2 . . .	1360	2500	3180	$1.7 \times 10^{-3}/26$	—	sl. s. a.
Hydride, CaH_2 . . .	816 (in H_2)	d. 600	1900	d.	d.	d. a.; i. bz.
Hydroxide, $Ca(OH)_2$.	− H_2O/580	d.	2240	0.19/0	0.077/100	s. NH_4Cl aq., a.; i. al.
Iodide, CaI_2. . . .	740	1100	3956/25	182/0	426/100	s. al., acet.
Nitrate, $CaNO_3.4H_2O$.	40	d. 132	1820	266/0	—	s. al., acet.
,, , $CaNO_3$ . .	561	—	2504/18	102/0	376/100	s. acet.
Oxide, CaO	2600	3000	3300	0.13/10, d.	0.03/120, d.	s. a.
Phosphate, ortho-, $Ca_3(PO_4)_2$ . . .	1670	—	3140	0.03/25	d.	s. a.; i. al.
Silicate, meta-, $CaSiO_3$.	1540	—	2905	i.	—	s. HCl
Sulphate, $CaSO_4.2H_2O$	− 1½H_2O/130	− 2H_2O/163	2320	0.241/0	0.222/100	s. a. NH_4 salts.
,, , $CaSO_4$ . .	—	—	2960	—	0.16/100	—
Sulphide, CaS . . .	d.	—	2500	sl. s. d.	—	d. a.
Tungstate, $CaWO_3$ . .	—	—	6060	i.	i.	s. NH_4Cl; i. a., al.
Carbon						
Oxide, mon-, CO . .	− 205.06	− 191.50	1.25	$3.5 \times 10^{-2}/0(A)$	$1.5 \times 10^{-2}/60(A)$	s. al., bz.
,, , di-, CO_2 . .	—	sb. − 78.48	1.98	1.68/0(A)	0.36/60(A)	s. a., alk., al., acet.
Cyanogen, C_2N_2 . .	− 27.84	− 20.7	2.34	4.5/20(A)	—	s. al., eth.

Inorganic compounds (*contd*)

Compound	$\theta_m/°C$	$\theta_b/°C$	$\rho/(\text{kg m}^{-3})$	Solubility		
				Cold water	Hot water	Other solvents
Cerium						
Ammonium nitrate (III)						
$(NH_4)_2Ce(NO_3)_6$.	—	—	—	143/25	232/90	sl. s. HNO_3; s. al.
(III) Chloride, $CeCl_3$.	848	1727	3920/0	100/0	d.	s. al., acet.
(III) Fluoride, CeF_3 .	1460	2300	6160	i.	—	s. al., acet.
(IV) ,, , $CeF_4.H_2O$	650	d.	4770	i.	—	s. a.
(III) Nitrate,						
$Ce(NO_3)_3.6H_2O$.	$-3H_2O/150$	d. 200	—	v. s.	v. s.	s. al.; acet.
(III) Oxide, Ce_2O_3 .	1687	—	6860	i.	i.	s. H_2SO_4; i. HCl
(IV) ,, , CeO_2 . .	2600	—	7132/23	i.	i.	i. dil. a.; sol. H_2SO_4
(III) Sulphate,						
$Ce_2(SO_4)_3$ . . .	d. 920	—	3912	10.1/0	2.25/100	—
(III) Sulphide, Ce_2S_3 .	d. 2100	—	5020/11	i.	d.	s. dil. a.
Chlorine						
Oxide, mon-, Cl_2O . .	-20	—	3.89	2/0(A)	—	s. alk., H_2SO_4
,, , di-, ClO_2 .	-60	—	3.09	20/4(A)	d.	s. alk., H_2SO_4
,, , hepta-, Cl_2O_7 .	-91.5	82	—	s. d.	—	s. bz.
Perchloric acid, $HClO_4$.	-112		1764/22	s.	s.	
Chromium						
Ammonium sulphate(III)						
$Cr_2(SO_4)_3(NH_4)_2SO_4$.						
$24H_2O$.	94	—	1750	21.2/25	32.8/40	s. al., dil. a.
(II) Chloride, $CrCl_2$.	824	—	2878/25	v. s.	v. s.	sl. s. al.; i. eth.
(III) Chloride, $CrCl_3$.	1150	sb. 1300	2760/15	i.	sl. s.	i. a., acet., al.
(III) Nitrate,						
$Cr(NO_3)_3.9H_2O$.	60	d. 100	—	s.	s.	s. a., alk., al., acet.
Oxide, sesqui-, Cr_2O_3 .	2435	4000	5210	i.	i.	i. a., alk.
,, , tri-, CrO_3 .	196	d.	2700	165/15	207/108	s. H_2SO_4, eth., al.
Oxychloride, CrO_2Cl_2 .	-95	117	1911	d.	d.	s. eth., ac. a.; d. al.
Potassium sulphate,						
$Cr_2(SO_4)_3.H_2SO_4$.						
$24H_2O$	89	d.	1830	24.4/25	s.	i. al.
Cobalt						
(II) Chloride, $CoCl_2$.	730 (in HCl)	1050	3356/25	44/0	105/96	s. al. acet.
(II) Hydroxide, $Co(OH)_2$.	d.	—	3597/15	$3 \times 10^{-4}/0$	—	s. a., NH_4 salts.
Nitrate,						
$Co(NO_3)_2.6H_2O$.	$-3H_2O/55$	d.	1870/25	134/0	335/90	s. al., acet.; sl. s. NH_3
(II) Oxide, CoO . .	1935	d.	6450	i.	i.	s. a. NH_3 aq.; i. al.
(III) ,, , Co_3O_4 . .	d. 895	—	6070	i.	i.	s. H_2SO_4; i. HCl, HNO_3
Sulphate, $CoSO_4.7H_2O$	96.8	$-7H_2O/420$	1948/25	33/0	67/70	s. al., meth. al.
Copper						
(I) Acetate,						
$Cu(C_2H_3O_2)_2.H_2O$.	115	d. 240	1882	7.2/0	20/100	s. al. eth.
(I) Bromide, CuBr . .	492	1345	4980	v. sl. s.	d.	s. HCl, HNO_3; i. acet.
(II) ,, , $CuBr_2$ . .	498	—	4770/25	v. s.	—	s. al., acet.; i. bz.
(I) Chloride, CuCl . .	430	1490	4140	$6 \times 10^{-3}/0$	—	s. HCl, NH_4OH
(II) ,, , $CuCl_2$ . .	620	d. 990	3386/25	71/0	108/100	s. al.
(II) Fluoride,						
$CuF_2.2H_2O$ . . .	d.	—	2930/25	sl. s.	d.	s. al., HCl; i. acet.
(I) Iodide, CuI . .	605	1290	5620	$8 \times 10^{-4}/18$	—	s. HCl, KI; i. a., al.
(II) Nitrate,						
$Cu(NO_3)_2.3H_2O$.	114.5	d. 170	2320/25	138/0	1270/100	v. s. al.
(I) Oxide, Cu_2O . .	1235	$-O/1800$	6000	i.	i.	s. HCl; sl. s. HNO_3; i. al.
(II) ,, , CuO . . .	1326	—	6400	i.	i.	s. a., NH_4Cl
(II) Sulphate,						
$CuSO_4.5H_2O$ . .	$-4H_2O/110$	$-5H_2O/150$	2284	31.6/0	203.3/100	sl. s. meth. al.; i. al.

Inorganic compounds (*contd*)

Compound	$\theta_m/°C$	$\theta_b/°C$	$\rho/(kg\ m^{-3})$	Solubility		
				Cold water	Hot water	Other solvents
Dysprosium						
Oxide, Dy_2O_3 . . .	2340	—	7810/27	—	—	s. a.
Oxide, $Dy(SO_4)_3.8H_2O$	$-8H_2O/360$	—	—	5.07/20	3.34/40	—
Erbium						
Oxide, Er_2O_3 . . .	—	—	8640	i.	—	sl. s. a.
Sulphate,						
$\quad Er_2(SO_4)_3.8H_2O$.	$-8H_2O/400$	—	3217	16/20	6.5/40	—
Europium						
Oxide, Eu_2O_3 . . .	—	—	7420	—	—	—
Sulphate,						
$\quad Eu_2(SO_4)_3.8H_2O$.	$-8H_2O/375$	—	—	2.6/20	1.9/40	—
Gadolinium						
Nitrate,						
$\quad Gd(NO_3)_3.6H_2O$.	91	—	2332	v. s.	v. s.	s. al.
Oxide, Gd_2O_3 . . .	—	—	7407/15	v. sl. s.	—	s. a.
Sulphate,						
$\quad Gd_2(SO_4)_3.8H_2O$.	—	—	3010/14.6	3.3/20	2.5/40	—
Gallium						
Bromide, $GaBr_3$ . . .	122	279	3690/25	s.	s.	sl. s. NH_3
Chloride, $GaCl_3$ . . .	77.5	200	2470/25	v. s.	v. s.	s. bz., CCl_4, CS_2
Nitrate, $Ga(NO_3)_3$ . .	d. 110	d. 200	—	v. s.	v. s.	s. abs. al.; i. eth.
Oxide, Ga_2O_3 . . .	1900	—	6440	i.	i.	s. alk.
Sulphate, $Ga_2(SO_4)_3$.	—	—	—	v. s.	v. s.	s. al.; i. eth.
Germanium						
Chloride, $GeCl_4$ . . .	-49.5	84	1844/30	d.	d.	v. s. dil. HCl.; s. al., eth.
Oxide, di-, GeO_2 . .	1086	—	6239	i.	—	s. a., alk.
Sulphide, GeS . . .	530	sb. 430	3310	0.24/20	i.	s. alk., HCl.; sl. s. NH_4OH
Gold						
(III) Chloride, $AuCl_3$.	d. 254	sb. 265	3900	68/20	v. s.	s. al. eth.; sl. s. NH_3 aq.
Hafnium						
Chloride, $HfCl_4$ . . .	sb. 319	—	—	d.	d.	s. al., eth.
Oxide, di-, HfO_2 . .	2812	—	9680/20	i.	i.	—
Hydrogen						
Bromide, HBr . . .	-86.82	-66.73	3.50	221/0	130/100	s. al.
Chloride, HCl . . .	-114.22	-85.05	1.64	82.3/0	56.1/60	s. al., eth., bz.
Cyanide, HCN . . .	-13.24	25.70	0.901	v. s.	v. s.	v. s. al.; s. eth.
Fluoride, HF . . .	-93.07	19.9	0.991/19.5	v. s.	v. s.	—
Iodide, HI	-50.80	-35.36	5.66/0	0.43/0(A)	v. s.	s. al.
Oxide-deuterium, D_2O.	3.8	101.42	1105.6	v. s.	v. s.	—
Peroxide, H_2O_2 . . .	-0.41	150.2	1460/0	v. s.	—	s. al., eth.
Phosphide, H_3P . . .	-133.78	-87.74	1.53	0.26/17(A)	i.	s. al., eth.
Selenide, H_2Se . . .	-65.73	-41.3	3.61	3.77/4(A)	2.7/22.5(A)	s. CS_2, $COCl_2$
Sulphide, H_2S . . .	-85.53	-60.34	1.54	4.53/0(A)	1.18/60(A)	sl. s. al.; s. CS_2
Telluride, H_2Te. . .	-51	-2.3	5.81	s.	—	s. al., alk.
Indium						
Chloride, tri-, $InCl_3$.	sb. 300	—	3460/25	v. s.	v. s.	sl. s. al., eth.
Iodide, mono-, InI . .	351	712	5310	—	s. d.	s. dil. a.; i. al.
Oxide, sesqui-, In_2O_3 .	—	—	7179	i.	—	—
Sulphide, sesqui-, In_2S_3	1100	d.	4900	i.	—	d. a.; sl. s. Na_2S

Inorganic compounds (*contd*)

Compound	$\theta_m/°C$	$\theta_b/°C$	$\rho/(kg\ m^{-3})$	Solubility		
				Cold water	Hot water	Other solvents
Iodine						
Bromide, mono-, IBr .	—	d. 116	4416/0	s. d.	—	s. al., eth., chl., CS_2
Chloride, mono-, ICl .	27.3	97.4	3182/0	d.	—	s. HCl, al., eth., CS_2
,, , tri-, ICl_3 . .	—	d. 77	3177/15	s. d.	—	s. ac., a., al., bz., eth., CCl_4
Iodic acid, HIO_3 . .	d.	—	4630	286/0	576/101	v. s. al.; sl. s. HNO_3
Oxide, penta, I_2O_5 . .	d. 300	—	4799/25	187/13	v. s.	sl. s. dil. a.; i. abs. al., eth. chl.
Iridium						
Chloride, tri-, $IrCl_3$.	d. 763	—	5300	i.	—	i. a., alk.
Fluoride, hexa-, IrF_6 .	—	53	6000	d.	d.	—
Oxide, di-, IrO_2 . . .	d. 1100	—	3150	i.	i.	i. a., alk.
Iron						
Ammonium sulphate (III) $Fe(NH_4)(SO_4)_2 . 12H_2O$	40	$-12H_2O/230$	1710	124/25	400/100	s. dil. a.; i. al.
Ammonium sulphate (II) $Fe(NH_4)_2(SO_4)_2 . 6H_2O$	d. 100	—	1860	26.9/20	73/80	i. al.
Carbonyl, penta, $Fe(CO)_5$	-21	103	1457	i.	—	s. H_2SO_4., alk., al., eth.
(III) Chloride, $FeCl_3 . 6H_2O$ . .	37	280	—	92/20	v. s.	s. al. eth.
(III) Chloride, $FeCl_3$	307	d. 319	—	74/0	536/100	v. s. al., eth., acet.
(II) Chloride, $FeCl_2$	677	sb.	—	51/1.5	106/100	v. s. al.; s. acet.; i. eth.
Dicyclopentadiene, $Fe(C_5H_5)_2$ . .	174	249	—	i.	—	s. org. solv.
(III) Nitrate, $Fe(NO_3)_3 . 9H_2O$.	50	d. 125	1684	s.	v. s.	s. HCl, H_2SO_4, al., acet.
(III) Oxide, Fe_2O_3 . .	1565	—	5240	i.	i.	s. HCl
(III) Oxide, (magnetite), Fe_3O_4	d. 1538	—	5180	i.	i.	sl. s. a.
(II) Oxide, $Fe_{0.95}O$ . .	1420	—	5700	i.	i.	s. a.
(II) Phosphide, FeP .	1365	—	6070	i.	i.	—
(II) Sulphate, $FeSO_4 . 7H_2O$ . . .	$-6H_2O/90$	$-7H_2O/300$	1898	15.7/0	52.1/55	i. al.
(II) Sulphide, FeS . .	1196	d.	4740	$6 \times 10^{-4}/18$	d.	s. a. d.; i. NH_3
Lanthanum						
Chloride, $LaCl_3$ . . .	860	>1000	3842/25	v. s.	d.	v. s. al., pyr.; i. eth.
Nitrate, $La(NO_3)_3 . 6H_2O$.	40	d. 126	—	151/25	v. s.	v. s. al.; s. pyr.
Oxide, La_2O_3 . . .	2320	4200	6510/15	v. sl. s.	d.	s. a.; i. acet.
Sulphate, $La_2(SO_4)_3 . 9H_2O$	d.	—	2821	3.8/0	0.87/100	sl. s. HCl; i. acet.
Lead						
Acetate, $Pb(C_2H_3O_2)_2$.	280	—	3250	19.8/0	221/50	v. sl. s. al.
Carbonate, $PbCO_3$ . .	d. 315	—	6600	$3 \times 10^{-3}/0$	d.	s. a., alk.; i. al.
Chloride, $PbCl_2$ . . .	501	950	5850	0.67/0	3.2/100	sl. s. dil. HCl, NH_3; i. al.
Chromate, $PbCrO_4$ . .	844	d.	6120/15	i.	i.	s. a. alk.
Fluoride, PbF_2 . .	855	1290	8240	0.06/9	0.07/27	s. HNO_3; i. acet.
Iodide, PbI_2	402	954	6160	0.04/0	0.42/100	s. alk., KI.; i. al.
Nitrate, $Pb(NO_3)_2$ . .	d. 470	—	4530	37.6/0	128/100	s. aq. al., NH_3, alk.
Oxide, mono-, PbO . .	—	—	8000	$2 \times 10^{-3}/20$	—	s. HNO_3, alk.
,, (red-lead), Pb_3O_4	d. 500	—	9100	i.	i.	s. h. HCl, ac. a.
,, , di-, PbO_2 . .	d. 290	—	9375	i.	i.	s. dil. HCl; sl. s. ac. a.
Sulphate, $PbSO_4$ . .	1170	—	6200	$3 \times 10^{-3}/0$	$6 \times 10^{-3}/50$	s. a., NH_4 salts
Sulphide, PbS . . .	1114	—	7500	$5 \times 10^{-5}/25$	i.	s. a.; i. alk.
Tetraethyl, $Pb(C_2H_5)_4$.	-136	d. 200	1659/11	i.	—	s. bz., al. eth.

Inorganic compounds (*contd*)

Compound	$\theta_m/°C$	$\theta_b/°C$	$\rho/(\text{kg m}^{-3})$	Solubility		
				Cold water	Hot water	Other solvents
Lithium						
Aluminium hydride, LiAlH$_4$	d. 125	—	917	d.	d.	s. eth.
Borohydride LiBH$_4$ . .	d. 279	—	660	d.	d.	s. eth., s. al. (d.)
Bromide, LiBr . . .	547	1265	3464/25	143/0	256/100	s. al., eth.
Carbonate, Li$_2$CO$_3$ . .	723	d. 1310	2110	1.54/0	0.72/100	s. a.; i. acet., al.
Chloride, LiCl . . .	614	1382	2068/25	63.7/0	130/95	s. al., acet.
Fluoride, LiF . . .	842	1676	2635	0.12/0	0.135/35	s. HF.; i. al. acet.
Hydride, LiH . . .	680	—	820	d.	d.	i. eth.
Iodide, LiI	450	1180	3494	151/0	461/100	v. s. al. NH$_3$
Nitrate, LiNO$_3$ . . .	264	d. 600	2380	53.4/0.1	234/100	s. al., acet.
Oxide, Li$_2$O	>1700	—	2013/25	d.	d.	—
Phosphate, Li$_3$PO$_4$ . .	837	—	2537/18	0.02/0	0.03/25	s. a.; i. acet.
Sulphate, Li$_2$SO$_4$ . .	860	—	2221	26/0	23/100	i. acet.
Magnesium						
Bromide, MgBr$_2$ . .	700	1230	3720/25	98/0	126/100	s. HBr., al., acet.; sl. s. NH$_3$
Carbonate, MgCO$_3$.	d. 350	$-$CO$_2$/900	2958	0.04/0	0.011/100	s. a.
Chloride, MgCl$_2$.6H$_2$O .	d. 117	d.	1569	281/0	367/100	v. s. al.
,, , MgCl$_2$ . . .	708	1412	2320	53/0	70/100	v. s. al.
Fluoride, MgF$_2$ . . .	1266	2239	2970	$8 \times 10^{-3}/18$	—	s. HNO$_3$; i. al.
Nitrate, Mg(NO$_3$)$_2$.6H$_2$O .	89	$-$5H$_2$O/330	1636/25	223/0	v. s.	s. al.
Oxide, MgO . . .	2800	3600	3580/25	$6 \times 10^{-4}/0$	$8 \times 10^{-3}/30$	s. a.; i. al.
Perchlorate, Mg(ClO$_4$)$_2$	d. 251	—	2210/18	99.6/25	v. s.	s. al., sl. s. eth.
Phosphate, Mg$_2$(PO$_4$)$_2$.22H$_2$O .	$-$18H$_2$O/100	d. 200	1640/15	v. sl. s.	—	s. a.; i. NH$_4$ salts
Sulphate, MgSO$_4$.	d. 1124	—	2660	22/0	68.3/100	s. al., eth.; i. acet.
(II) Carbonate, MnCO$_3$	d.	—	3125	$6.5 \times 10^{-3}/25$	—	s. dil. a.; i. al.
(II) Chloride, MnCl$_2$.	650	1190	2977/25	63.4/0	110/100	s. al.; i. eth.
(II) Nitrate, Mn(NO$_3$)$_2$.4H$_2$O .	26	129	1820	426/0	v. s.	v. s. al.
(II) Oxide, MnO . .	—	—	5450	i.	i.	s. a.
(III) Oxide, Mn$_2$O$_3$.	$-$O/1080	—	4500	i.	i.	s. a.
,, , di-, MnO$_2$.	$-$O/535	—	5026	i.	i.	s. HCl; i. HNO$_3$
(II) Sulphate, MnSO$_4$.	700	d. 850	3250	53/0	70/70	s. al.; i. eth.
(II) Sulphide, MnS . .	d.	—	3990	$5 \times 10^{-4}/18$	—	s. dil. a.
Mercury						
(II) Bromide, HgBr$_2$.	236	322	6109/25	0.62/25	4/100	s. al.; sl. s. eth.
(I) ,, , HgBr .	sb. 345	—	7307	$4 \times 10^{-6}/25$	—	s. a.; i. al., acet.
(II) Chloride, HgCl$_2$.	276	302	5440/25	3.61/0	61/100	v. s. al., eth.; s. ac. a.
(I) ,, , HgCl .	sb. 400	—	7150	$1.4 \times 10^{-4}/0$	$1 \times 10^{-3}/43$	s. aq. reg.; sl. s. HNO$_3$, HCl
(II) Cyanide, Hg(CN)$_2$	d.	—	3996	12.5/15	33/100	s. al., NH$_3$; i. bz.
(II) Iodide, HgI$_2$ . .	259	354	6360/25	$4 \times 10^{-3}/17.5$	sl. s.	s. al., eth.
(II) Nitrate, Hg(NO$_3$)$_2$.$\frac{1}{2}$H$_2$O .	79	d.	4390	v. s.	d.	s. HNO$_3$, acet.; i. al.
(II) Oxide, HgO . .	d. 500	—	11100/4	$5 \times 10^{-3}/25$	0.04/100	s. a.; i. al., eth.
(II) Sulphate, HgSO$_4$.	d.	—	6470	d.	—	s. a.; i. al., acet.
(I) ,, , Hg$_2$SO$_4$.	d.	d.	7560	0.05/25	0.09/100	s. H$_2$SO$_4$, HNO$_3$
Molybdenum						
Fluoride, hexa-, MoF$_6$.	17	35	2550/18	s. d.	d.	s. NH$_4$OH, alk.
Oxide, tri-, MoO$_3$ . .	795	sb. 1155	4692/21	0.106/18	2.06/100	s. HF, H$_2$SO$_4$, NH$_4$OH
Silicide, di-, MoSi$_2$ . .	2030	—	6310/21	—	—	—
Sulphide, di-, MoS$_2$.	1185	—	4800/14	i.	i.	s. aq. reg., H$_2$SO$_4$
Neodymium						
Chloride, NdCl$_3$ . .	784	1600	4134/25	97/10	140/100	s. al.; i. eth., chl.
Oxide, Nd$_2$O$_3$ . . .	1900	—	7240	i.	i.	s. HCl
Sulphate, Nd$_2$(SO$_4$)$_3$.8H$_2$O .	1176	—	2850	16/0	1.5/100	—

Inorganic compounds (contd)

Compound	$\theta_m/°C$	$\theta_b/°C$	$\rho/(kg\ m^{-3})$	Solubility		
				Cold water	Hot water	Other solvents
Nickel						
Carbonyl, $Ni(CO)_4$ . .	-25	42.4	1320/17	0.018/9.8	—	s. HNO_3, aq. reg., al., eth.
Chloride, $NiCl_2$ . . .	1001	sb. 973	3550	53/0	88/100	s. al., NH_4OH
Fluoride, NiF_2 . . .	sb. 1000 (in HF)	—	4630	2.55/0	2.49/100	i. a., al., eth.
Nitrate, $NiNO_3 . 6H_2O$.	56.7	136.7	2050	240/0	v. s.	s. al., NH_4OH
Oxide, NiO	1990	—	6670	i.	i.	s. a., NH_4OH
Sulphate, $NiSO_4 . 6H_2O$	$-6H_2O/103$	—	2070	63/0	340/100	v. s. al., NH_4OH
Niobium						
Chloride, $NbCl_5$ . .	205	254	2750	d.	—	s. al., HCl, CCl_4
Fluoride, NbF_5 . . .	73	236	3293	d.	—	s. HNO_3, al.; sl. s. chl.
Oxide, mon-, NbO . .	—	—	7300	i.	i.	s. HCl, H_2SO_4; i. HNO_3
,, , pent-, Nb_2O_5 .	1460	—	4470	i.	i.	s. HF, alk.
Nitrogen						
Chloride, tri-, NCl_3 . .	-40	71	1653	i.	d.	s. chl., bz., CCl_4, CS_2
Oxide, (ous), N_2O . .	-90.86	-88.48	1.98	1.3/0(A)	0.57/25(A)	s. al., eth., H_2SO_4
,, , (ic), NO . .	-163.65	-151.77	1.34	$7.1 \times 10^{-2}/0$(A)	$2.9 \times 10^{-2}/60$(A)	s. aq. $FeSO_4$, CS_2; sl. s. H_2SO_4
,, , tri-, N_2O_3 .	-111	d. 3.5	1447/2	s.	d.	s. alk., a., eth.
,, , di-, NO_2 . .	-11.20	21.15	1449/20	s. d.	—	s. alk., CS_2, chl.
,, , penta-, N_2O_5 .	30	d. 47	1642/18	s.	d.	s. chl.
Oxychloride, NOCl .	-64.5	-5.5	2.99	d.	d.	s. fuming H_2SO_4
Hydrazine, N_2H_4 . .	1.5	113.5	1.01/15	v. s.	v. s.	s. al.
Hydrazine sulphate, $N_2H_4 . H_2SO_4$	254	—	1370	2.87/20	14.0/80	i. al.
Hydroxylamine, NH_2OH	33.1	—	1200	s.	d.	s. a. al.; v. sl. s. eth.
Hydroxylamine hydrochloride, $NH_2OH . HCl$ . .	151	d.	1670	83/17	v. s.	s. al.; i. eth.
Osmium						
Oxide, tetr-, OsO_4 . .	39.5	130	4906/22	5.1/0	6.2/25	v. s. CCl_4; s. al. eth.
Oxygen						
Fluoride, di-, OF_2 . .	-224	144.9	—	sl. s. d.	—	sl. s. a., alk.
Ozone, O_3	-192.7	-110.51	2.14	0.49/0(A)	—	s. alk. solns., oils
Palladium						
Chloride, $PdCl_2$ . . .	d. 500	d.	4000/18	s.	s.	s. HCl, acet.
Nitrate, $Pd(NO_3)_2$ . .	d.	—	—	s. d.	—	s. HNO_3
Oxide, PdO	d. 750	—	8700	i.	i.	sl. s. h. a.
Phosphorus						
Bromide, tri-, PBr_3 . .	-40	173	2852/15	d.	d.	s. eth., chl., CCl_4
Chloride, tri-, PCl_3 . .	-112	76	1574/21	d.	d.	s. eth., chl., CS_2, CCl_4
,, , penta-, PCl_5 .	d. 167	sb. 162	1600	d.	d.	s. CCl_4, CS_2
Fluoride, tri-, PF_3 . .	-151.5	-101.5	3.91	d.	d.	s. al.
,, , penta-, PF_5 .	-83	-75	5.81	d.	d.	d. alk.
Oxide, tri-, P_2O_3 . .	23.8	174 (in N_2)	2135/21	d.	d.	s. CS_2, eth., chl.
,, , penta-, P_2O_5 .	582	sb. 300	2390	d.	d.	s. H_2SO_4; i. acet.
Oxychloride, $POCl_3$.	2	105	1675	d.	d.	d. al.
Sulphide, penta-, P_2S_5 .	288	514	2030	d.	d.	s. alk.; sl. s. CS_2
Phosphoric acid, ortho-, H_3PO_4	42.4	$-\frac{1}{2}H_2O/213$	1834/18	548/0	v. s.	s. al.
Platinum						
Chloride, tetra-, $PtCl_4$.	d. 370 (in Cl_2)	—	4303/25	v. s.	v. s.	s. acet., sl. s. al.; i. acet.

Inorganic compounds (*contd*)

Compound	$\theta_m/°C$	$\theta_b/°C$	$\rho/(kg\ m^{-3})$	Solubility		
				Cold water	Hot water	Other solvents
Potassium						
Aluminium sulphate, $KAl(SO_4)_2.12H_2O$.	$-9H_2O/65$	$-12H_2O/200$	1757	11.4/20	v. s.	s. dil. a.; i. al.
Antimonyl tartrate, $KSbC_4H_4O_7.\frac{1}{2}H_2O$.	$-\frac{1}{2}H_2O/100$	—	2607	5.3/8.7	35.7/100	i. al.
Bromate, $KBrO_3$ . .	d. 370	—	3270/18	3.1/0	49.7/100	sl. s. al.
Bromide, KBr . .	730	1435	2750/25	53.5/0	104/100	sl. s. al. eth.
Carbonate, K_2CO_3 . .	891	d.	2428	105.5/0	156/100	i. al., acet.
Chlorate, $KClO_3$.	356	d. 400	2320	3.3/0	57/100	sl. s. al., alk.
Chloride, KCl . .	776	sb. 1500	1984	28.1/0	56.3/100	sl. s. al., alk.
Chromate, K_2CrO_4 .	968.3	—	2732/18	59.2/0	79.2/100	i. al.
Citrate, $K_3C_6H_5O_7.H_2O$	d. 230	—	1980	162/0	194/100	sl. s. al.
Cyanide, KCN . . .	634.5	—	1520/16	50/0	100/100	s. meth. al.; sl. s. al.
Dichromate, $K_2Cr_2O_7$.	398	d. 500	2676/25	4.7/0	102/100	i. al.
Dihydrogen phosphate, KH_2PO_4	252.6	—	2338	14.8/0	80.6/90	i. al.
Ferricyanide, $K_3Fe(CN)_6$ . . .	d.	—	1850/25	30.1/0	86.6/100	s. acet.; i. al.
Ferrocyanide, $K_4Fe(CN)_6.3H_2O$.	$-3H_2O/70$	d.	1850/17	27.8/12	90.6/96.3	s. acet.; i. al.
Fluoride, KF . .	846	1505	2480	92.3/18	v. s.	s. HF.; i. al.
Fluoroberyllate, K_2BeF_4	—	—	—	2/20	5.3/100	—
Hydrogen carbonate, $KHCO_3$	d. 100–200	—	2170	22.6/0	74.9/100	i. al.
Hydrogen phthalate, $KHC_4H_4O_6$ . .	—	—	1636	10.2/25	33/100	—
Hydrogen sulphate $KHSO_4$	214	d.	2322	36/0	117/100	i. al.
Hydrogen tartrate, $KHC_4H_4O_6$ . .	—	—	1954	0.3/0	6.6/100	s. a.; i. al.
Hydroxide, KOH . .	360	1322	2044	96/0	175/100	v. s. al.; i. eth.
Iodate, KIO_3 . . .	560	d.	3930/32	4.6/0	30.9/100	s. KI.; i. al.
Iodide, KI	686	1330	3130	128/0	208/100	v. s. al.; s. NH_3; sl. s. eth.
Metabisulphite, $K_2S_2O_5$	d. 190	—	2340	28/0	120/94	sl. s. al.
Nitrate, KNO_3 . . .	334	d. 400	2109/16	13.3/0	246/100	i. al., eth.
Nitrite, KNO_2 . . .	440	d.	1915	280/0	395/100	v. s. NH_3; sl. s. al.
Oxalate, $K_2C_2O_4.H_2O$	$-H_2O/100$	—	2127/4	25.6/0	66.8/90	—
Oxide, mon-, K_2O . .	d. 350	—	2320/0	v. s. d.	v. s. d.	s. al., eth.
,, , super-, KO_2 . .	380	d.	2140	v. s. d.		d. al.
Periodate, KIO_4 . .	$-O/300$	—	3618/15	0.17/0.2	7.0/97	v. sl. s. KOH
Permanganate, $KMnO_4$.	d. <240	—	2703	4.2/0	24.5/100	v. sl. s. acet.; s. H_2SO_4
Persulphate, $K_2S_2O_8$.	d. <100	—	2477	1.65/0	10.9/40	i. al.
Sulphate, K_2SO_4 . .	1069	1689	2662	7.4/0	24.1/100	i. al., acet.
Thiocyanate, $KCNS$.	173	d. 500	1886/14	177/0	644/99	s. al., acet.
Praseodymium						
Chloride, $PrCl_3$ . .	786	1700	4020/25	104/14	v. s.	s. al.; i. eth., chl.
Sulphate, $Pr_2(SO_4)_3.8H_2O$.	—	—	2827/13	17.4/20	sl. s.	—
Radium						
Bromide, $RaBr_2$ . .	728	sb. 900	5790	s.	s.	s. al.
Sulphate, $RaSO_4$ . .	—	—	—	$2\times10^{-4}/25$	$5\times10^{-6}/45$	i. al.
Rhenium						
Fluoride, hexa-, ReF_6 .	19	48	6157	s. d.	s. d.	—
Oxide, hepta-, Re_2O_7 .	297	sb. 250	6103	v. s.	s. d.	v. s. al.; s. a., alk.
Rhodium						
Chloride, $RhCl_3$ . .	d. 450–500	—		i.	i.	i. aq. reg.
Oxide, Rh_2O_3 . . .	d. 1100–1150	—	8200	i.	i.	i. aq. reg., alk.

Inorganic compounds (*contd*)

Compound	$\theta_m/°C$	$\theta_b/°C$	$\rho/(kg\ m^{-3})$	Solubility		
				Cold water	Hot water	Other solvents
Rubidium						
Carbonate, Rb_2CO_3 .	d. 740	—	—	450/20	s.	sl. s. abs. al.
Chloride, RbCl . .	715	1390	2800	77/0	137/100	sl. s. al., NH_3
Fluoride, RbF . .	775	1410	3557	130/18	—	s. dil. HF; i. al. eth.
Nitrate, $RbNO_3$. .	310	—	3110	19.5/0	452/100	v. s. HNO_3; s. acet.
Sulphate, Rb_2SO_4 . .	1074	—	3613	37.6/0	84.8/100	—
Ruthenium						
Chloride, tri-, $RuCl_3$.	d. 500	—	3110	i.	d.	sl. s. abs. al.; i. a.
Oxide, di-, RuO_2 . .	d.	—	6970	i.	i.	i. a.
„ , tetr-, RuO_4 .	26	—	3290/21	2.0/0	2.2/74	s. a., alk., al.
Samarium						
Chloride, $SmCl_3$ . .	678	d.	4460/18	92/10	100/50	v. s. al.; s. pyr.
Oxide, Sm_2O_3 . . .	—	—	8347	i.	i.	v. s. a., hot HF
Sulphate, $Sm_2(SO_4)_3.8H_2O$.	$-8H_2O/450$	—	2930	2.7/20	2.0/40	—
Scandium						
Chloride, $ScCl_3$ . .	939	sb. 800–850	2390/25	v. s.	v. s.	i. al.
Oxide, Sc_2O_3 . . .	—	—	3864	i.	i.	s. h. a.; sl. s. c. a.
Selenium						
Chloride, mono-, Se_2Cl_2 .	-85	d. 130	2770/25	d.	d.	s. CS_2, CCl_4, bz.
„ , tetra-, $SeCl_4$.	sb. 170–196	d. 288	3800/360	d.	d.	—
Oxide, di-, SeO_2 . .	sb. 316	—	3950	38.4/14	82.5/65	s. al., acet., bz., ac. a.
Oxychloride, $SeOCl_2$.	8.5	176.4	2420/22	d.	d.	s. CS_2, CCl_4, bz.
Selenic acid, H_2SeO_4 .	60	d.	2950	426/0	v. s.	s. H_2SO_4; i. NH_3
Selenous acid, H_2SeO_3 .	d. 70/$-H_2O$	—	3004/15	90/0	372/100	v. s. al.
Silicon						
Chloride, $SiCl_4$ . . .	-70	57.6	1483	d.	d.	d. conc. H_2SO_4, al.
Fluoride, SiF_4 . . .	-90.2	-86	4.69	d.	d.	s. abs. al., eth., HF
Hydride, SiH_4 . . .	-185	-112	1.44	i.	—	i. al.; d. KOH
Oxide (quartz), SiO_2 . .	1610	2230	2650	i.	i.	s. HF; v. sl. s. alk.
Silver						
Bromide, AgBr . .	432	d. >1300	6473/25	$3.5 \times 10^{-5}/20$	$3.7 \times 10^{-4}/100$	s. KCN, $Na_2S_2O_3$
Chloride, AgCl . .	455	1550	5560	$7 \times 10^{-5}/0$	$2 \times 10^{-3}/100$	s. NH_4OH, KCN, $Na_2S_2O_3$
Chromate, Ag_2CrO_4 .	—	—	5625	$1 \times 10^{-4}/0$	$5 \times 10^{-3}/50$	s. a., NH_4OH, KCN
Cyanide, AgCN . .	d. 320	—	3950	$2 \times 10^{-5}/20$	—	s. NH_4OH, KCN, HNO_3
Fluoride, AgF . .	435	1160	5852/16	182/16	205/108	s. NH_4OH
Iodide, AgI	558	1506	6010/15	$3 \times 10^{-6}/20$	$1.4 \times 10^{-5}/80$	s. KCN; sl. s. NH_4OH
Nitrate, $AgNO_3$. .	212	d. 444	4352/19	122/0	952/100	s. eth.; v. sl. s. abs. al.
Oxide, Ag_2O . . .	d. 300	—	7143/17	$1.7 \times 10^{-4}/20$	$5.5 \times 10^{-3}/80$	s. a., KCN, NH_4OH; i. al.
Sulphate, Ag_2SO_4 . .	652	d. 1085	5450/29	0.57/0	1.41/100	s. a., NH_4OH
Sodium						
Acetate, $NaC_2H_3O_2$.	324	—	1528	119/0	170/100	sl. s. al.
Aluminium fluoride, Na_3AlF_6 . . .	1000	—	2900	sl. s.	—	d. alk.; i. HCl
Arsenate, $Na_3AsO_4.12H_2O$.	86.3	—	1760	39/16	—	s. al.
Borate, tetra-, $Na_2B_4O_7.10H_2O$.	$-8H_2O/60$	$-10H_2O/320$	1730	8/25	170/100	i. al.
Borate, tetra-, $Na_2B_4O_7$	741	d. 1575	2367	1.5/0	50/100	i. al.
Bromide, NaBr . .	755	1390	3203/25	82/0	121/100	sl. s. al.
Carbonate, $Na_2CO_3.10H_2O$ . .	$-H_2O/34$	—	1440/15	7/0	45.5/100	i. al.

Inorganic compounds (*contd*)

Compound	$\theta_m/°C$	$\theta_b/°C$	$\rho/(kg\ m^{-3})$	Solubility		
				Cold water	Hot water	Other solvents
Sodium (*contd*)						
Carbonate, Na_2CO_3 .	851	d.	2532	7/0	46/100	sl. s. abs. al.; i. acet.
Chloride, NaCl . . .	801	1413	2165/25	35.7/0	39.8/100	sl. s. al.
Citrate, $Na_3C_6H_5O_7.2H_2O$.	$-2H_2O/150$	—	—	72/25	167/100	i. al.
Fluoride, NaF . . .	988	1695	2558/41	3.7/0	4.9/100	s. HF; v. sl. s. al.
Hydrogen carbonate, $NaHCO_3$	$-CO_2/270$	—	2159	7/0	23/100	sl. s. al.
Hydrogen sulphate, $NaHSO_4$	>315	d.	2435/13	28/25	100/100	sl. s. al.
Hydroxide, NaOH . .	318	1390	2130	42/0	347/100	v. s. al.; i. acet.
Iodide, NaI . . .	651	1304	3667/25	160/0	302/100	v. s. al.
Molybdate, Na_2MoO_4 .	687	—	3280/18	44/0	80/100	—
Nitrate, $NaNO_3$ . . .	307	d. 380	2261	73/0	180/100	v. s. NH_3; sl. s. al.
Nitrite, $NaNO_2$ . . .	271	d. 320	2168/0	71/0	163/100	s. meth. al.; sl. s. al., eth.
Nitroprusside, $Na_2Fe(CN)_5NO.2H_2O$	—	—	1720	40/16	—	s. al.
Oxalate, $Na_2C_2O_4$ . .	d. 250–270	—	2340	2.7/0	6.3/100	i. al., eth.
Oxide, mon-, Na_2O .	sb. 1275	—	2270	d.	d.	d. al.
,, per-, Na_2O_2 .	d. 460	d. 657	2805	s.	d.	d. al.; s. dil. a.
Selenate, Na_2SeO_4 . .	—	—	3213/17	84/35	73/100	s. H_2SO_4; i. al.
Sulphate, Na_2SO_4 . .	890	—	2680	4.3/0.7	40.3/100	s. H_2SO_4; i. al.
Sulphide, mono-, Na_2S .	1180	—	1856/14	18/18	57/90	sl. s. al.; i. eth.
Sulphite, $Na_2SO_3.7H_2O$.	$-7H_2O/150$	d.	1539/15	33/0	196/40	sl. s. al.
Thiosulphate, $Na_2S_2O_3.5H_2O$ . .	d. 48	$-5H_2O/100$	1729/17	79/0	302/60	s. NH_3; v. sl. s. al.
Tungstate, Na_2WO_4 .	698	—	4179	—	—	—
Strontium						
Bromide, $SrBr_2.6H_2O$.	$-6H_2O/>180$	—	2386/25	200/0	v. s.	—
,, , $SrBr_2$ . . .	643	d.	4216/24	85/0	223/100	s. al.
Carbonate, $SrCO_3$ . . .	$-CO_2/1340$	—	3700	$8.2 \times 10^{-5}/9$	0.065/100	s. a.; sl. s. NH_4 salts
Chloride, $SrCl_2$ . .	873	1250	3052	44/0	101/100	v. sl. s. acet. abs. al.
Fluoride, SrF_2 . . .	1400	2480	4240	0.011/0	0.012/27	s. h. HCl; i. HF
Nitrate, $Sr(NO_3)_2$ . .	570	—	2986	39.3/0	99/100	s. NH_3; sl. s. acet.
Sulphur						
Chloride, mono-, S_2Cl_2 .	-80	136	1678	d.	d.	s. CS_2, eth., bz.
,, , di-, SCl_2 . .	-78	d. 59	1621/15	d.	d.	s. bz., CCl_4; d. al.
Fluoride, hexa-, SF_6 .	-51	64	6.602	v. sl. s.	sl. s.	—
Oxide, di-, SO_2 . . .	-75.48	-10.02	2.927	22.8/0	0.58/90	sl. s. al.
,, , tri-, SO_3 . .	16.85	43.3	1920	d.	d.	s. al., H_2SO_4, ac. a.
Sulphuric acid, H_2SO_4 .	10.35	338	1841	∞	∞	d. al.
Sulphonyl chloride SO_2Cl_2	-54.1	69.1	1667	d.	d.	s. ac. a., bz.
Thionyl chloride, $SOCl_2$	-104.5	75.7	1660/10	d.	d.	s. bz., chl.; d. a., alk.
Tantalum						
Chloride, $TaCl_5$ . . .	216	242	3680/27	d.	—	s. abs. al.
Oxide, Ta_2O_5 . . .	1800	—	8200	i.	i.	s. fused $KHSO_4$; i. al.
Tellurium						
Bromide, $TeBr_4$ . . .	380	d. 421	4310/15	sl. s. d.	d.	s. min. a.
Chloride, $TeCl_4$ . . .	224	380	3260/18	s. d.	s. d.	s. dil. HCl, bz., al.; i. CS_2
Oxide, TeO_2 . . .	d. 395	—	6210	i.	i.	s. a., alk.; i. NH_4OH
Terbium						
Sulphate, $Tb_2(SO_4)_3.8H_2O$	$-8H_2O/360$	—	—	3.6/20	2.5/20	—

Inorganic compounds (*contd*)

Compound	$\theta_m/°C$	$\theta_b/°C$	$\rho/(\text{kg m}^{-3})$	Solubility		
				Cold water	Hot water	Other solvents
Thallium						
Chloride, TlCl . . .	430	720	7004/30	0.17/0	2.4/100	sl. s. HCl; i. al.
Nitrate, TlNO$_3$ . . .	206	430	—	3.9/0	4.1/100	s. acet.; i. al.
Oxide, Tl$_2$O	300	−O/1865	9520/16	v. s. d.	v. s. d.	s. a., al.
Sulphate, Tl$_2$SO$_4$ . .	632	d.	6770	2.7/0	19.1/100	—
Thorium						
Chloride, ThCl$_4$ . . .	770	d. 928	4590	v. s.	v. s.	s. KCl, al., eth.
Oxide, di-, ThO$_2$ . .	3050	4400	9860	i.	i.	i. alk.
Sulphate,						
Th(SO$_4$)$_2$.9H$_2$O . .	−9H$_2$O/400	—	2770	0.7/0	6.8/55	s. HCl, HNO$_3$
Tin						
(IV) Chloride, SnCl$_4$.	−33	114.1	2226	s.	d.	s. CS$_2$, eth.
(II) ,, , SnCl$_2$.	246	652	3950/25	84/0	270/15 d.	s. al., eth.
(IV) Oxide, SnO$_2$ . .	1127	sb. 1800−1900	6950	i.	i.	s. fus. alk.
(II) ,, , SnO . .	d. 1080	—	6446/0	i.	i.	sl. s. NH$_4$Cl
(II) Sulphate, SnSO$_4$.	d.	—	—	33/25	—	s. H$_2$SO$_4$
(IV) Sulphide, SnS$_2$.	d. 600	—	4500	i.	—	s. alk.; i. HCl
(II) ,, , SnS .	882	1230	5220/25	i.	—	s. alk., HCl
Titanium						
Chloride, tetra-,TiCl$_4$.	−25	136.4	1726	d.	d.	s. dil. HCl, al.
Oxide, di-, TiO$_2$ . .	1825	—	4170	i.	i.	s. H$_2$SO$_4$, alk.
Tungsten						
Carbide, WC . . .	2870	6000	15 630/18	i.	i.	s. HNO$_3$ + HF
Chloride, WCl$_6$ . . .	275	347	3520/25	—	d.	v. s. CS$_2$; s. al. bz. CCl$_4$
Oxide, tri-, WO$_3$ . .	1473	—	7160	i.	i.	s. h. alk., HF; i. a.
Uranium						
Fluoride, hexa-, UF$_6$.	65	56	4680/21	s.	—	s. CCl$_4$, chl.; i. CS$_2$
Oxide, di-, UO$_2$ . . .	2500	—	10 960	i.	i.	s. HNO$_3$, conc. H$_2$SO$_4$
,, , tri-, UO$_3$ . .	d.	—	7290	i.	i.	s. HNO$_3$, H$_2$SO$_4$
,, , U$_3$O$_8$ . . .	d.	—	8300	i.	i.	s. HNO$_3$, H$_2$SO$_4$
Uranyl acetate,						
UO$_2$(C$_2$H$_3$O$_2$)$_2$.2H$_2$O	−2H$_2$O/110	—	2893/15	7.7/15	d.	v. s. al.
Uranyl chloride, UO$_2$Cl$_2$	578	d.	—	320/18	v. s.	s. al., eth.
Uranyl nitrate,						
UO$_2$(NO$_3$)$_2$.6H$_2$O .	d. 100	118	2807/13	170/0	v. s.	v. s. al., eth., ac. a.
Uranyl sulphate,						
UO$_2$SO$_4$.3H$_2$O . .	d. 100	—	3280/16	19/13	22.2/100	s. H$_2$SO$_4$
Vanadium						
Chloride, tetra-, VCl$_4$.	d.	—	3000/18	s. d.	—	s. abs. al., eth., chl.
Oxide, penta-, V$_2$O$_5$.	690	d. 1750	3357/18	0.8/20	—	s. a., alk.
Xenon						
Fluoride, di-, XeF$_2$ . .	120−140	—	4320	d.	d.	v. s. HF
,, , tetra-, XeF$_4$.	100	—	4070	d.	d.	s. HF, C$_7$F$_{16}$
Ytterbium						
Chloride, YbCl$_3$.6H$_2$O	−6H$_2$O/180	—	2757	v. s.	v. s.	s. al.
Oxide, Yb$_2$O$_3$ . . .	—	—	9170	i.	i.	s. h. dil. a.; i. HF
Sulphate,						
Yb$_2$(SO$_4$)$_3$.8H$_2$O .	—	—	3286	36/25	21/40	—
Yttrium						
Nitrate, Y(NO$_3$)$_3$.6H$_2$O	−3H$_2$O/100	—	2680	135/22	v. s.	v. s. al., eth., HNO$_3$
Oxide, Y$_2$O$_3$ . . .	2410	—	5010	i.	i.	s. a.; i. alk., HF
Sulphate,						
Y$_2$(SO$_4$)$_3$.8H$_2$O . .	−8H$_2$O/120	d. 700	2558	9.8/20	2.6/95	v. sl. s. H$_2$SO$_4$

Inorganic compounds (*contd*)

Compound	$\theta_m/°C$	$\theta_b/°C$	$\rho/(kg\ m^{-3})$	Solubility		
				Cold water	Hot water	Other solvents
Zinc						
Acetate,						
$\quad Zn(C_2H_3O_2)_2.2H_2O$	d. 200	—	1840	31/20	67/100	s. al.
Carbonate, $ZnCO_3$ . .	$-CO_2/300$	—	4398	0.001/15	—	s. a., alk.; i. acet.
Chloride, $ZnCl_2$ . . .	283	732	2910/25	432/25	615/100	v. s. al., eth.
Nitrate,						
$\quad Zn(NO_3)_2.6H_2O$.	$-6H_2O/105-131$	—	2065/14	184/20	v. s.	v. s. al.
Oxide, ZnO	1975	—	5606	i.	—	s. a., alk.
Sulphate,						
$\quad ZnSO_4.7H_2O$ . .	$-7H_2O/280$	—	1957/25	97/20	663/100	sl. s. al.
Zirconium						
Chloride, tetra-, $ZrCl_4$.	sb. 331	—	2803/15	s.	d.	s. HCl, al., eth.
Oxide, ZrO_2	2715	—	5600	i.	i.	s. H_2SO_4, HF
Sulphate,						
$\quad Zr(SO_4)_2.4H_2O$ . .	$-3H_2O/135-150$	—	3220/16	v. s.	v. s.	s. H_2SO_4; i. al.
Zirconyl chloride,						
$\quad ZrOCl_2.8H_2O$ . .	$-6H_2O/150$	$-8H_2O/210$	—	s.	d.	s. al., eth.; sl. s. HCl

J.H.S.G.

2.3 Properties of organic compounds

Formula, melting point, boiling point, density and refractive index

Hydrated forms are indicated thus: Alloxan $+4H_2O$ and the molecules of water are included in the formula.

Melting, θ_m, and boiling, θ_b, points are for a pressure of 101.325 kN m^{-2} unless other conditions are indicated, e.g. 235/2.4 is the boiling point at 2.4 kN m^{-2}. For boiling points at other pressures see pp. 178–183.

Densities and refractive indices (D lines) are for 20 °C unless some other temperature is indicated, e.g. 788/16 is the density at 16 °C; 1.6048/98.8 is the refractive index at 98.8 °C.

Abbreviations: d., decomposes; exp., explodes; sub., sublimes.

Organic compounds

Substance	Formula	Melting point $\theta_m/°C$	Boiling point $\theta_b/°C$	Density $\rho/(kg\ m^{-3})$	Refractive index
Acenaphthene	$C_{12}H_{10}$	95.4	277	1200	1.6048/98.8
Acetaldehyde	C_2H_4O	-121	20	788/16	1.3316
Acetamide	C_2H_5ON	82	222	1159	1.4274/78
Acetanilide	C_8H_9ON	114	304	1210/4	
Acetic acid	$C_2H_4O_2$	16.7	118	1049	1.3718
Acetic anhydride . . .	$C_4H_6O_3$	-73	140	1085/15	1.3904
Acetone	C_3H_6O	-95	56	787/25	1.3602/15
Acetonitrile	C_2H_3N	-41	82	783	1.3460
Acetophenone	C_8H_8O	20.5	202	1033/15	1.5342/19
Acetyl chloride . . .	C_2H_3OCl	-112	51	1105	1.3898
Acetylene	C_2H_2	-81.5	sub. -84	610/-80	
Adipic acid	$C_6H_{10}O_4$	149	238	1366	
Alloxan $+4H_2O$ . . .	$C_4H_{10}O_8N_2$	170 d.			

Organic compounds (*contd*)

Substance	Formula	Melting point $\theta_m/°C$	Boiling point $\theta_b/°C$	Density $\rho/(\text{kg m}^{-3})$	Refractive index
Allyl alcohol	C_3H_6O	-129	97	857/15	1.4135
4-Aminobenzoic acid . .	$C_7H_7O_2N$	186			
2-Aminopyridine . . .	$C_5H_6N_2$	56	204		
Aniline.	C_6H_7N	-6	184	1026/15	1.5863
Anilinium hydrochloride .	C_6H_8NCl	198	245	1222/4	
Anisole.	C_7H_8O	-37.5	156	993/25	1.5150/22
Anthracene . . .	$C_{14}H_{10}$	217	342	1243	
Anthraquinone . . .	$C_{14}H_8O$	286	379	1419/4	
Azobenzene	$C_{12}H_{10}N_2$	68	293		
Benzaldehyde . . .	C_7H_6O	-26	179	1050/15	1.5463/17.6
Benzene . . .	C_6H_6	5.5	80.1	879	1.5011
Benzoic acid . . .	$C_7H_6O_2$	122	249	1266/15	1.5397/15
Benzoic anhydride . .	$C_{14}H_{10}O_3$	42	360	1199/15	1.5767/15
Benzoin	$C_{14}H_{12}O_2$	137	343		
Benzonitrile . . .	C_7H_5N	-13	191	1009/15	1.5289
Benzophenone (α) . .	$C_{13}H_{10}O$	49	306	1085/50	
p-Benzoquinone . .	$C_6H_4O_2$	115.7			
Benzoyl chloride . . .	C_7H_5OCl	-1	197	1212	1.5537
Benzoyl peroxide . .	$C_{14}H_{10}O_4$	104	exp.		
Benzyl alcohol. . .	C_7H_8O	-15.3	205	1049/15	1.5396
Benzyl benzoate . .	$C_{14}H_{12}O_2$	21	323	1114/18	1.5681/21
Benzyl chloride . .	C_7H_7Cl	-43	179	1098	1.5415/15
Benzyl cinnamate . . .	$C_{16}H_{14}O_2$	39	350		
Biphenyl	$C_{12}H_{10}$	70	255	1180/0	1.5852/79
($\pm$)-Borneol . . .	$C_{10}H_{18}O$	208	sub.	1010	
Bromobenzene . . .	C_6H_5Br	-31	156	1498/15	1.5625/15
Bromoethane	C_2H_5Br	-125.5	38.4	1456	1.4239
Bromoform	$CHBr_3$	8	151	2900/15	1.6005/15
Bromomethane . . .	CH_3Br	-93	3.6	1.732/0	
1-Bromonaphthalene . .	$C_{10}H_7Br$	6.2	281	1489/16.5	1.6601/16.5
Butane.	C_4H_{10}	-138.4	-0.5		
iso-Butane	C_4H_{10}	-159.6	-11.7		
Butan-1-ol	$C_4H_{10}O$	-79.9	118	810	1.3993
Butan-2-ol	$C_4H_{10}O$	-114.7	100	808	1.3949/25
Butanone	C_4H_8O	-86.4	79.6	805	1.3814/15
Butyric acid . . .	$C_4H_8O_2$	-8	164	958	1.3991
iso-Butyric acid . . .	$C_4H_8O_2$	-47	155	950	
($\pm$)-Camphene . . .	$C_{10}H_{16}$	50	161	879	1.4402/80
(+)-Camphor. . . .	$C_{10}H_{16}O$	176	208	992/10	
Catechol	$C_6H_6O_2$	105	246	1344	
Carbon disulphide . .	CS_2	-111	46.5	1293/0	1.6276
Carbon tetrabromide. .	CBr_4	90	190	2911/99.5	
Carbon tetrachloride. .	CCl_4	-23	76.7	1632/0	1.4607
Chloral hydrate . . .	$C_2H_3O_2Cl_3$	51.7	96	1908	
Chloroacetic acid . . .	$C_2H_3O_2Cl$	63	190	1390/75	1.4297/65
Chlorobenzene . . .	C_6H_5Cl	-45	132	1107	1.5248
1-Chlorobutane . . .	C_4H_9Cl	-123	78	907/0	1.4015
Chlorodifluoromethane .	$CHClF_2$	-160	-40.8		
1-Chloro-2,3-epoxy propane	C_3H_5OCl	-25.6	118	1180	1.4420/11.6
Chloroethane	C_2H_5Cl	-142.5	12.3	921/0	
Chloroform	$CHCl_3$	-63.5	61	1.499/15	1.4467

Organic compounds (*contd*)

Substance	Formula	Melting point $\theta_m/°C$	Boiling point $\theta_b/°C$	Density $\rho/(kg\ m^{-3})$	Refractive index
Chloromethane . . .	CH_3Cl	-93	-24.0	$991/-25$	
1-Chloro-2,4,6- trinitrobenzene . . .	$C_6H_2O_6N_3Cl$	83	d.	1797	
Chloresterol	$C_{27}H_{46}O$	148.5	360 d.	1067	
Cineole	$C_{10}H_{18}O$	1.5	176	927	1.4584/18
trans-Cinnamic acid . .	$C_9H_8O_2$	133	300	1247	
Cinnamyl alcohol . . .	$C_9H_{10}O$	33	250	1044	1.5819
Citric acid	$C_6H_8O_7$	153	d.	1542/18	
o-Cresol	C_7H_8O	31.0	191.0	1051	1.5372/40
m-Cresol	C_7H_8O	12.0	202.2	1035	1.5406
p-Cresol	C_7H_8O	34.7	202.0	1035	1.5316
Cumene	C_9H_{12}	-96.5	152	862	1.4909
Cyclohexane	C_6H_{12}	6.6	80.7	779	1.4262
Cyclohexanol	$C_6H_{12}O$	24	161	962	1.4656/22
Cyclohexanone . . .	$C_6H_{10}O$	-31.2	156	948	1.4507
Cyclohexene	C_6H_{10}	-103.7	83.2	811	1.4467
p-Cymene	$C_{10}H_{14}$	-71.6	177	877	1.5006
cis-Decahydronaphthlene	$C_{10}H_{18}$	-43	196	896	1.4811
trans-Decahydro- naphthlene . . .	$C_{10}H_{18}$	-31.5	187	870/18	1.4697/18
1,2-Diaminoethane . .	$C_2H_8N_2$	8.5	117	902/15	1.4540/26.1
1,2-Dibromoethane . .	$C_2H_4Br_2$	10	132	2179	1.5379
Dibromomethane . . .	CH_2Br_2	-52.7	99	2810/15	
Dibutyl phthlate . . .	$C_{16}H_{22}O_4$		340	1047	
Dichlorodifluoromethane	CCl_2F_2		-29.8		
1,2-Dichloroethane . .	$C_2H_4Cl_2$	-42	82.4	1252	1.4443
Dichloromethane . . .	CH_2Cl_2	-96.8	40.7	1335/15	1.4237
Diethanolamine . . .	$C_4H_{11}O_2N$	28	268	1097	1.4776
Diethylamine	$C_4H_{11}N$	-50	55.5	711/18	1.3873/18
Diethyl ether	$C_4H_{10}O$	-116	34.4	714/20	1.3538
Diethyl oxalate . . .	$C_6H_{10}O_4$	-40.6	186	1079	1.4101
Diethyl sulphate . . .	$C_4H_{10}O_4S$	-24.5	209	1180/18	1.4010/18
Dimethylamine . . .	C_2H_7N	-96	7.4	680/0	1.350/17
N,N-Dimethylaniline .	$C_8H_{11}N$	2.5	193	956	1.5582
Dimethyl carbonate . .	$C_3H_6O_3$	0.5	90	1069	1.3687
Dimethyl sulphate . .	$C_2H_6O_4S$	-27	188	1335/15	1.3874
1,4-Dioxan	$C_4H_8O_2$	11	101	1034	1.4224
Diphenylamine . . .	$C_{12}H_{11}N$	54	302	1159	
1,2-Diphenylethane . .	$C_{14}H_{14}$	52	284	995	
Ethane	C_2H_6	-183.3	-88.6		
Ethanediol	$C_2H_6O_2$	-15.6	197	1116	1.4274
Ethanethiol	C_2H_6S	-121	35	832/25	1.4351
Ethanol	C_2H_6O	-117	78.3	789	1.3610/20.5
Ethanolamine . . .	C_2H_7ON	10.5	171	1022	1.4539
2-Ethoxyethanol . . .	$C_4H_{10}O_2$		135.1	931	
Ethyl acetate	$C_4H_8O_2$	-83.6	77.1	925	1.3701/25
Ethyl acetoacetate . .	$C_6H_{10}O_3$	-45	181	1028	1.4209/16
Ethylamine	C_2H_7N	-80.6	16.6	706/0	
Ethylbenzene	C_8H_{10}	-94	136	867	1.4959
Ethylbenzoate	$C_9H_{10}O_2$	-34	213	1051/15	1.5068/17.3
Ethylene	C_2H_4	-169.2	-103.7		
Ethylene oxide . . .	C_2H_4O	-111.3	10.7	877/7	1.3597/7

Organic compounds (*contd*)

Substance	Formula	Melting point $\theta_m/°C$	Boiling point $\theta_b/°C$	Density $\rho/(\text{kg m}^{-3})$	Refractive index
Ethyl formate	$C_3H_6O_2$	-79	54.3	917	1.3598
Ethyl nitrate	$C_2H_5O_3N$	-112	87.5	1109	1.3853
Ethyl nitrite	$C_2H_5O_2N$		17	900/15	
Ethyl salicylate . . .	$C_9H_{10}O_3$	1.3	232	1131	1.5226
Eugenol	$C_{10}H_{12}O_2$	10.3	254	1062/25	1.5439/19
Fluorescin	$C_{20}H_{12}O_5$	314 d.			
Fluorobenzene . . .	C_6H_5F	-41.2	84.7	1024	1.4677
Formaldehyde	CH_2O	-92	-19.1	815/-20	
Formamide	CH_3ON	2.5	111	1133	1.4472
Formic acid	CH_2O_2	8.4	100.6	1220	1.3714
Fructose	$C_6H_{12}O_6$	102		1598	
Fumaric acid	$C_4H_4O_4$	300	sub.	1635	
Furan	C_4H_4O		32	964/0	1.4216
Furfuraldehyde . . .	$C_5H_4O_2$	-36.5	162	1159	1.5261
Furfuryl alcohol . . .	$C_5H_6O_2$		170	1128/23	1.4852
Glucose	$C_6H_{12}O_6$	142		1544/25	
Glycerol	$C_3H_8O_3$	20	290	1260/17.5	1.4730
Glycerol trioleate . . .	$C_{57}H_{104}O_6$	-4	235/2.4	899/50	1.4561/60
Glycerol tripalmitate . .	$C_{51}H_{98}O_6$	65.5	315	875/70	1.4381/80
Glycerol tristearate . .	$C_{57}H_{110}O_6$	72		856/90	1.4385/80
Glycine	$C_2H_5O_2N$	236 d.	289 d.		
Guaiacol	$C_7H_8O_2$	28.2	205	1129/21.4	
Heptane	C_7H_{16}	-90.6	98.4	684	1.3877
Hexachloroethane . .	C_2Cl_6	187 (sealed tube)		2091	
Hexamine	$C_6H_{12}N_4$	263			
Hexane	C_6H_{14}	-95.3	68.7	659	1.3749
Hippuric acid	$C_9H_9O_3N$	187	d.	1371	
Indene	C_9H_8	-2	183	996	1.5766
Iodoethane	C_2H_5I	-105	72.4	1913/30	1.5168/15
Iodoform	CHI_3	119	sub. 210 exp.	4008	
Iodomethane	CH_3I	-66.5	42.4	2.251/30	1.5293/21
Isoprene	C_5H_8	-120	32.6	680	1.4194
Lactic acid	$C_3H_6O_3$	25	d.	1249	1.4414
Lactose+H_2O . . .	$C_{12}H_{24}O_{12}$	(H_2O lost at 130 °C) 230 d.		1525	
Maleic acid	$C_4H_4O_4$	130	d.	1592	
Maleic anhydride . . .	$C_4H_2O_3$	60	202	934	
Malonic acid	$C_3H_4O_4$	135.6	d.	1631/15	
Maltose+H_2O . . .	$C_{12}H_{24}O_{12}$	(H_2O lost at 100 °C)		1540	
$(-)$-Menthol	$C_{10}H_{20}O$	Four forms 44, 35, 33, 31	212	903/15	
Mesitylene	C_9H_{12}	-44.7	165	865	1.4994
Metaldehyde	$(C_2H_4O)_n$	246.2 (sealed tube)	sub.		
Methane	CH_4	-182.5	-161.5		
Methanol	CH_4O	-93.9	64.5	791	1.3276/25
Methyl acetate . . .	$C_3H_6O_2$	-98.7	57.8	928	1.3593/20
Methylamine	CH_5N	-92.5	-6.3	699/-10.8	
N-Methylaniline . . .	C_7H_9N	-57	196	989	1.5702/21.2

Organic compounds (*contd*)

Substance	Formula	Melting point $\theta_m/°C$	Boiling point $\theta_b/°C$	Density $\rho/(kg\ m^{-3})$	Refractive index
Methyl anthranilate . .	$C_8H_9O_2N$	24	135.5/2.0	1168/18.6	
Methyl benzoate . . .	$C_8H_8O_2$	−12.3	200	1094/15	1.5205/15
2-Methylbutane . . .	C_5H_{12}	−160.0	27.9	619	1.3538
2-Methyl-butan-1-ol . .	$C_5H_{12}O$	−117	131	816	
2-Methyl-butan-2-ol . .	$C_5H_{12}O$	−12	102	809	1.4045
Methyl formate . . .	$C_2H_4O_2$	−90	32.0	987/15	1.344
2 Methyl furan . . .	C_5H_6O		63	916	
Methyl methacrylate . .	$C_5H_8O_2$	−50	100	936	1.413
2-Methyl propan-1-ol .	$C_4H_{10}O$	−108	108	817	1.3968/17.5
2-Methyl propan-2-ol .	$C_4H_{10}O$	25.5	83	789	1.3878
Methyl salicylate . .	$C_8H_8O_3$	−8.6	223	1179/25	1.538/18.1
Morpholine	C_4H_9ON	−4.9	128	999	1.4545
Naphthalene	$C_{10}H_8$	80.2	218	1145	1.5822/100
1-Naphthol	$C_{10}H_8O$	95	283	1099/99	1.6206/98.7
2-Naphthol	$C_{10}H_8O$	122	288	1272	
1-Naphthylamine . . .	$C_{10}H_9N$	50	301	1120/25	1.6703/51
2-Naphthylamine . . .	$C_{10}H_9N$	113	306	1061/98	1.6493/98
(−)-Nicotine	$C_{10}H_{14}N_2$	< −80	247	1010	1.5280
Nitrobenzene	$C_6H_5O_2N$	5.9	211	1173/25	1.5530
Nitroethane	$C_2H_5O_2N$	−90	114	1050	1.3916
Nitromethane	CH_3O_2N	−29	101	1137	1.3818
1-Nitropropane . . .	$C_3H_7O_2N$	−108	132	1001	1.4015
2-Nitropropane . . .	$C_3H_7O_2N$	−93	120	990	1.3941
Octane	C_8H_{18}	−56.8	125.7	703	1.3974
Octane-1-ol	$C_8H_{18}O$	−15.4	195	827	1.4292
Oleic acid	$C_{18}H_{34}O_2$	16	361	898	1.4582
Oxalic acid	$C_2H_2O_4$	189.5			
Palmitic acid	$C_{16}H_{32}O_2$	63	351	853/62	1.4339/60
Paraldehyde	$C_6H_{12}O_3$	12.6	128	994	1.4049
Pentane	C_5H_{12}	−129.7	36	626	1.3575
Pentan-1-ol	$C_5H_{12}O$	−79	138	815	1.414/13
Pentan-2-ol	$C_5H_{12}O$		120	810	1.4053
Phenanthrene	$C_{14}H_{10}$	100	340	1170	1.6567/129
Phenol	C_6H_6O	41.0	181.9	1073	1.5425/40.6
Phosgene	$COCl_2$	−104	8.3		
Phthalic acid	$C_8H_6O_4$	d. > 191	—	1593	
Phthalic anhydride . .	$C_8H_4O_3$	131.6	285	1527/4	
Phthalimide	$C_8H_5O_2N$	238	sub.		
2-Picoline	C_6H_7N	−66.7	129.4	944	1.5010
3-Picoline	C_6H_7N	−18.2	144.1	957	1.5068
4-Picoline	C_6H_7N	3.7	145.4	955	1.5058
Picric acid	$C_6H_3O_7N_3$	122.5	exp. > 300	1763	
α-Pinene	$C_{10}H_{16}$	−50	155	861	1.4685/15
Piperidine	$C_5H_{11}N$	−9	106	861	1.4530
Propane	C_3H_8	−187.7	−42.1		
Propan-1-ol	C_3H_8O	−127	97.1	804	1.3850
Propan-2-ol	C_3H_8O	−89.5	82.2	786	1.3776
Propene	C_3H_6	−185.3	−47.7		
Propyl acetate	$C_5H_{10}O_2$	−92.5	101.8	887	1.3844
Pyridine	C_5H_5N	−42	115.3	983	1.5102
Pyrogallol	$C_6H_6O_3$	133	309		
Quinhydrone	$C_{12}H_{10}O_4$	171	sub.	1401	

Organic compounds (*contd*)

Substance	Formula	Melting point $\theta_m/°C$	Boiling point $\theta_b/°C$	Density $\rho/(\text{kg m}^{-3})$	Refractive index
Quinol	$C_6H_6O_2$	170.5	286	1358	
Quinoline	C_9H_7N	−15	238	1095	1.6269
iso-Quinoline	C_9H_7N	25	241	1099	1.6223/25
Resorcinol	$C_6H_6O_2$	111	277	1285/15	
Salicylic acid . . .	$C_7H_6O_3$	159	255	1443	
Stearic acid	$C_{18}H_{36}O_2$	71.5	375	941	1.4335/70
Styrene	C_8H_8	−30.6	145 d.	906	1.5469
Succinic acid	$C_4H_6O_4$	185	235 gives anhydride	1564/15	
Succinic anhydride . .	$C_4H_4O_3$	119	261	1234	
Sucrose	$C_{12}H_{22}O_4$	184 d.		1588/15	
meso-Tartaric acid . . .	$C_4H_6O_6$	140		1666	
(±)-Tartaric acid + H_2O	$C_4H_8O_7$	loses H_2O at 100 °C		1697	
(+)-Tartaric acid . .	$C_4H_6O_6$	170		1760	
(−)-Tartaric acid . .	$C_4H_6O_6$	170		1760	
1,2,3,4-Tetrahydro- naphthalene . . .	$C_{10}H_{12}$	−35	207.2		1.5453/17
Thiophen	C_4H_4S	−38.3	84.4	1064	1.5287
Thiourea	CH_4N_2S	180	d.	1405	
Thymol	$C_{10}H_{14}O$	51.5	232	969	
Toluene	C_7H_8	−95.0	110.6	867	1.4969
o-Toluidine	C_7H_9N	−15.5	200	1004	1.5688
m-Toluidine	C_7H_9N	−43.6	203	987/25	1.5686
p-Toluidine	C_7H_9N	44.5	200	961/50	1.5532/59.1
Trichloroethylene . . .	C_2HCl_3	−73	86.7	1460/15	1.4782
Trichlorofluoromethane .	CCl_3F	−111	23.7	1.494/17	
Triethanolamine . . .	$C_6H_{15}O_3N$	21.2	277	1124	1.4852
2,2,4-Trimethylpentane .	C_8H_{18}	−107.4	99.2	692	1.3915
2,4,6-Trinitrotoluene . .	$C_7H_5O_6N_3$	80.8		1654	
Triphenylmethane . .	$C_{19}H_{16}$	94	200/1.3		
Tri-*o*-tolylphosphate . .	$C_{21}H_{21}O_4P$		410 d.		
Tri-*p*-tolylphosphate . .	$C_{21}H_{21}O_4P$	77			
Urea	CH_4ON_2	132		1335	
Uric acid	$C_5H_4O_3N_4$	d.		1893	
Valeric acid	$C_5H_{10}O_2$	−34.5	186	942	1.4086
iso-Valeric acid . . .	$C_5H_{10}O_2$	−37.6	175	937/15	1.4018/22.4
Vanilin	$C_8H_6O_3$	80	285		
o-Xylene	C_8H_{10}	−25.2	144.4	880	1.5054
m-Xylene	C_8H_{10}	−47.9	139.1	864	1.4972
p-Xylene	C_8H_{10}	13.3	138.3	861	1.4958

E.F.G.H.

2.4 Vapour pressures and boiling points

The plot of the logarithm of vapour pressure ($\log p$) against the reciprocal of absolute temperature ($1/T$) for a single substance is usually nearly linear and this relation may be used for interpolation and, with caution, for extrapolation. The formula to obtain a value of $\log p_2$ for a temperature T_2 from values of $\log p_1$ and $\log p_3$ for temperatures T_1 and T_3 is

$$\log p_2 = \log p_1 + (\log p_3 - \log p_1)(T_2 - T_1)T_3/(T_3 - T_1)T_2$$

2.4.1 Vapour pressure of water at temperatures between 0 and 360 °C

Pressures are tabulated in kN m^{-2} for values of °C (IPTS-1968); for example at 14 °C the pressure is 1.5983 kN m^{-2} and at 76 °C it is 40.206 kN m^{-2}.

θ/°C	0	1	2	3	4	5	6	7	8	9
0	0.6107	0.6566	0.7055	0.7576	0.8130	0.8720	0.9348	1.0015	1.0724	1.1477
10	1.2276	1.3123	1.4022	1.4974	1.5983	1.7051	1.8180	1.9375	2.0639	2.1974
20	2.3384	2.4872	2.6443	2.8099	2.9846	3.1686	3.3625	3.5666	3.7814	4.0074
30	4.2451	4.4949	4.7574	5.0332	5.3226	5.6264	5.9451	6.2793	6.6296	6.9967

θ/°C	0	2	4	6	8	10	12	14	16	18
40	7.3812	8.2053	9.1075	10.094	11.171	12.345	13.623	15.013	16.522	18.160
60	19.933	21.852	23.926	26.164	28.578	31.177	33.974	36.980	40.206	43.667
80	47.375	51.344	55.587	60.121	64.960	70.120	75.617	81.468	87.691	94.304
100	101.325	108.77	116.67	125.03	133.88	143.25	153.14	163.59	174.61	186.24
120	198.49	211.39	224.97	239.26	254.27	270.03	286.58	303.95	322.16	341.23
140	361.21	382.12	404.00	426.87	450.77	475.74	501.79	528.98	557.34	586.89
160	617.68	649.74	683.12	717.84	753.95	791.48	830.48	870.98	913.03	956.66
180	1001.9	1048.9	1097.5	1147.9	1200.1	1254.1	1310.1	1368.0	1427.8	1489.7
200	1553.6	1619.7	1687.9	1758.4	1831.1	1906.1	1983.5	2063.4	2145.6	2230.5

θ/°C	220	240	260	280	300	320	340	360
Pressure/(kN m^{-2})	2317.8	3344.7	4689.3	6412.7	8583.0	11 279	14 593	18 656

I.J.L.

2.4.2 Boiling point of water at pressures from 90 to 106.9 kN m^{-2}

The manner of reading the table is indicated by the following examples: for a pressure of 94.3 kN m^{-2} the temperature is 97.999 °C and for a pressure of 94.4 kN m^{-2} the temperature is 98.028 °C.

p/kN m^{-2}	0	0.1	0.2	0.3	0.4	0.5	0.6	0.7	0.8	0.9
90	96.712	96.743	96.773	96.803	96.834	96.864	96.895	96.925	96.955	96.986
91	97.016	97.046	97.076	97.106	97.136	97.167	97.197	97.227	97.257	97.287
92	97.317	97.347	97.377	97.407	97.436	97.466	97.496	97.526	97.556	97.585
93	97.615	97.645	97.674	97.704	97.734	97.763	97.793	97.822	97.852	97.881
94	97.911	97.940	97.969	97.999	98.028	98.057	98.087	98.116	98.145	98.174
95	98.204	98.233	98.262	98.291	98.320	98.349	98.378	98.407	98.436	98.465
96	98.494	98.523	98.552	98.581	98.610	98.638	98.667	98.696	98.725	98.753
97	98.782	98.811	98.839	98.868	98.897	98.925	98.954	98.982	99.011	99.039
98	99.068	99.096	99.124	99.153	99.181	99.209	99.238	99.266	99.294	99.323
99	99.351	99.379	99.407	99.435	99.463	99.491	99.519	99.547	99.576	99.604
100	99.631	99.659	99.687	99.715	99.743	99.771	99.799	99.827	99.854	99.882
101	99.910	99.938	99.965	99.993	100.021	100.048	100.076	100.104	100.131	100.159
102	100.186	100.214	100.241	100.269	100.296	100.323	100.351	100.378	100.406	100.433
103	100.460	100.487	100.515	100.542	100.569	100.596	100.623	100.651	100.678	100.705
104	100.732	100.759	100.786	100.813	100.840	100.867	100.894	100.921	100.948	100.975
105	101.001	101.028	101.055	101.082	101.109	101.135	101.162	101.189	101.216	101.242
106	101.269	101.296	101.322	101.349	101.375	101.402	101.428	101.455	101.481	101.508

I.J.L.

2.4.3 Vapour pressure at 25 °C of liquids important in high vacuum technique

(Mainly from values given by G. Burrows. *J. Soc. Chem. Ind.*, 1946, **65**, 360.)

Substance	Vapour pressure/(N m^{-2})
Mercury	2.24×10^{-1}
Dibutyl phthalate	2.5×10^{-3}
Apiezon A oil.	2.7×10^{-4}
Apiezon B oil	5.6×10^{-5}
Tritolyl phosphate	4.1×10^{-5}

D.A.

2.4.4 Vapour pressures from 0.2 to 101.325 kN m^{-2}

Kelvin temperatures are listed for pressures p from 0.2 to 101.325 kN m^{-2}. Values of dT/dp in the right-hand column are those applicable at the normal boiling point, i.e. when $p = 101.325$ kN m^{-2}. The letter 's' indicates a solid, 'd' decomposition and 'p' polymerization.

Elements and inorganic compounds

Element or compound	Pressure, $p/(\text{kN m}^{-2})$								dT/dp
	0.2	1	2	5	10	20	50	101.325	$\text{K}(\text{kN m}^{-2})^{-1}$
Aluminium	1780	1940	2020	2130	2230	2330	2490	2620	2.0
Aluminium borohydride	s	226	236	251	264	278	299	318	0.28
Aluminium bromide .	356 s	385	400	422	441	463	497	529	0.50
Aluminium chloride .	377 s	394 s	402 s	413 s	422 s	431 s	444 s	454 s	0.14
Aluminium fluoride .	1245 s	1306 s	1334 s	1375 s	1407 s	1441 s	1490 s	1530 s	0.58
Aluminium oxide . .	2460	2630	2710	2820	2910	3000	3140	3260	2
Ammonia	166 s	178.6 s	184.7 s	193.5 s	201.8 s	211.6	226.5	239.7	0.198
Ammonium azide . .	307 s	328 s	338 s	352 s	364 s	376 s	393 s	407 s	0.21
Ammonium bromide .	483 s	519 s	537 s	562 s	584 s	606 s	640.0 s	668.9 s	0.421
Ammonium chloride .	445 s	478 s	494 s	517.2 s	536.4 s	557.1 s	587.4 s	613.3 s	0.379
Ammonium cyanide .	226 s	241.5 s	249.0 s	259.8 s	268.7 s	278.4 s	292.6 s	304.8 s	0.179
Ammonium iodide. .	488 s	526 s	544 s	570 s	592 s	615.2 s	649.6 s	679.1 s	0.432
Antimony	1064	1201	1272	1380	1474	1580	1750	1910	2.4
Antimony pentachloride	302	329	343	363	380	d	—	—	
Antimony tribromide .	375	408	425	448	468	489	521	548	0.40
Antimony trichloride .	326 s	354 s	367	389	408	430	462	492	0.44
Antimony triiodide .	450 s	488	508	539	565	594	638	674	0.53
Antimony trioxide . .	792	911	976	1081	1180	1300	1504	1700	3.0
Argon	55.0 s	60.7 s	63.6 s	67.8 s	71.4 s	75.4 s	81.4 s	87.3	0.093
Arsenic	656	701	723	754	780	808	849	883	0.50
Arsenic hydride . .	133 s	146 s	152 s	162.2	171.2	181.3	196.8	211.1	0.219
Arsenic pentafluoride .	157 s	168 s	173 s	180.4 s	186.4 s	192.9	207.5	220.3	0.190
Arsenic tribromide. .	322	352	368	390	410	431	464	493	0.43
Arsenic trichloride. .	267	291	304	322	337	354	381	404	0.34
Arsenic trifluoride . .	s	s	s	270	281	294	313	329	0.24
Arsenic trioxide . .	493 s	526 s	543 s	570 s	592	627	682	730	0.73
Barium	1164	1292	1358	1455	1539	1633	1778	1910	2.0
Beryllium chloride . .	584 s	618 s	634 s	656 s	674 s	696	728	754	0.39
Bismuth.	1225	1350	1412	1503	1581	1668	1800	1918	1.8
Bismuth tribromide .	503	546	567	598	623	651	694	734	0.62
Bismuth trichloride .	478	528	550	581	606	633	674	714	0.65

Elements and inorganic compounds (*contd*)

Element or compound	Pressure, $p/(kN\ m^{-2})$								dT/dp
	0.2	1	2	5	10	20	50	101.325	$K(kN\ m^{-2})^{-1}$
Boron tribromide . .	237	259	270	286	300	316	341.2	364.8	0.366
Boron trichloride . .	186	203	211	224	236	248	268.1	285.8	0.267
Boron trifluoride . .	120 s	130 s	134 s	141 s	147	154	164	172	0.13
Bromine . . .	227 s	244.1 s	252.1 s	263.6 s	275.7	290.0	312.0	332.0	0.301
Bromine pentafluoride	s	226.7	236.4	250.6	262.6	276.0	296.2	314.0	0.266
Cadmium	683	748	781	829	870	915	983	1043	0.90
Cadmium chloride . .	s	913	951	1006	1051	1100	1174	1241	1.0
Cadmium fluoride . .	1409	1525	1582	1664	1734	1811	1926	2027	1.5
Cadmium iodide . .	704	772	805	854	895	941	1010	1070	0.89
Cadmium oxide . .	1298	1402	1453	1530	1590	1655	1750	1835	1.2
Caesium	559	624	657	708	752	802	883	859	1.2
Caesium bromide . .	1043	1141	1189	1260	1320	1387	1487	1575	1.3
Caesium chloride . .	1040	1138	1187	1258	1317	1384	1484	1575	1.4
Caesium fluoride . .	1006	1099	1146	1215	1273	1339	1438	1526	1.3
Caesium iodide . .	1033	1128	1175	1245	1304	1369	1467	1555	1.3
Calcium	1106	1221	1279	1366	1441	1524	1651	1764	1.7
Carbon dioxide . .	140 s	151.2 s	156.4 s	163.8 s	170.0 s	176.7 s	186.4 s	194.7 s	0.120
Carbon monoxide . .	52 s	57 s	59.3 s	62.8 s	66.0 s	69.7	76.0	81.7	0.086
Carbon selenosulphide	231	253	264	280	294	310	335	358	0.34
Carbonyl selenide . .	158	175	183	196	206	218	236	252	0.22
Carbonyl sulphide . .	144	157	164	174	183	193	209	223	0.21
Chlorine	158 s	170 s	176.3	187.4	196.9	207.7	224.2	239.1	0.224
Chlorine dioxide . .	s	s	218.4	229.2	238.8	249.9	267.4	284.0	0.254
Chlorine fluoride . .	s	129	134	143	150	159	171	183	0.17
Chlorine monoxide . .	182	198.2	206.5	218.7	229.3	241.1	259.1	275.3	0.242
Chlorine trifluoride . .	193	209.4	217.6	229.8	240.1	251.7	269.3	284.9	0.234
Chromium	2020	2180	2260	2370	2470	2580	2730	2870	2.0
Chromium carbonyl .	314	337	348	363	376	389	408	424	0.23
Chromium oxychloride	260	282	294	310	325	341	366	390	0.37
Cobalt (II) chloride .	946 s	1001 s	1032	1086	1133	1184	1260	1327	1.0
Copper	1940	2110	2195	2320	2425	2540	2710	2855	2.2
Copper (I) bromide .	867	970	1025	1110	1188	1283	1448	1630	3.1
Copper (I) chloride .	842	952	1011	1104	1191	1298	1498	1765	5.5
Copper (I) iodide . .	s	910	965	1054	1138	1241	1422	1611	3.0
Cyanogen	180 s	194 s	201 s	210 s	218 s	227 s	240 s	252	0.22
Cyanogen bromide . .	240 s	259 s	268 s	280 s	291 s	302 s	318 s	332 s	0.20
Cyanogen chloride . .	200 s	216 s	224 s	235 s	244 s	254 s	269	288	0.17
Cyanogen fluoride . .	140 s	152 s	158 s	166 s	173 s	180 s	191 s	200 s	0.14
Cyanogen iodide . .	304 s	326 s	337 s	352 s	364 s	378 s	397 s	413 s	0.23
Deuterium oxide . .	s	282.8	293.2	308.4	321.1	335.2	356.1	374.6	0.273
Fluorine	53	58	61	65.3	68.9	73.0	79.3	85.0	0.086
Fluorine monoxide .	80	89	93	93	104	110	120	128	0.13
Gallium	1655	1790	1855	1950	2025	2110	2235	2340	1.6
Gallium trichloride .	320 s	346 s	358	379	397	417	447	473	0.38
Germanium bromide .	298 s	324	338	360	378	399	432	462	0.46
Germanium chloride .	233	254	264	280	294	310	335	357	0.34
Germanium hydride .	113	125	132	141	149	158	172	184	0.18
Gold	2100	2290	2390	2520	2640	2770	2960	3120	3.0
Helium	1.3	1.66	1.85	2.17	2.48	2.87	3.54	4.22	0.010
Hydrogen	10 s	11.4 s	12.2 s	13.4 s	14.5	16.0	18.2	20.3	0.033
Hydrogen bromide . .	137 s	149.3 s	155.3 s	164.1 s	171.6 s	179.9 s	193.3	206.4	0.200
Hydrogen chloride . .	123 s	134.8 s	140.4 s	148.6 s	155.5 s	163.6	176.5	188.1	0.174
Hydrogen cyanide . .	206 s	222 s	230 s	242 s	251 s	260.8	276.1	289.2	0.191
Hydrogen fluoride . .	183 s	202.0	212.1	226.9	239.4	253.3	274.2	292.7	0.275
Hydrogen iodide . .	159 s	172.4 s	179.2 s	189.3 s	197.9 s	207.6 s	222.2 s	237.6	0.240
Hydrogen peroxide . .	295	318.5	330.3	347.6	362.2	378.2	402.3	423.3	0.312
Hydrogen selenide . .	158 s	170 s	177 s	186 s	194 s	203 s	218	231	0.20

Elements and inorganic compounds (*contd*)

Element or compound	Pressure, $p/(\text{kN m}^{-2})$								dT/dp $\text{K}(\text{kN m}^{-2})^{-1}$
	0.2	1	2	5	10	20	50	101.325	
Hydrogen sulphide .	142 s	154.3 s	160.4 s	169.5 s	177.2 s	185.7 s	199.8	212.8	0.195
Hydrogen telluride .	180 s	195 s	203 s	213 s	227	238	255	271	0.24
Hydroxylamine . .	s	316.9	325.2	336.9	346.5	356.6	371.1	383.2	0.174
Iodine	318 s	341.8 s	353.1 s	369.3 s	382.7 s	400.8	430.6	457.5	0.403
Iodine heptafluoride .	188 s	205 s	214 s	226 s	236 s	247 s	264 s	279 s	0.22
Iodine pentafluoride .	263 s	280 s	289	305	318	333	356	376	0.30
Iron	2110	2290	2380	2500	2610	2720	2890	3030	2.1
Iron (III) chloride . .	475 s	503 s	514 s	529 s	541 s	553 s	570 s	584 s	0.22
Iron (II) chloride . .	s	959	995	1048	1095	1148	1229	1300	1.06
Iron pentacarbonyl .	s	273	285	302	317	334	358	378	0.30
Krypton	77 s	84.3 s	88.1 s	93.8 s	98.6 s	103.9 s	112.0 s	119.7	0.127
Lead	1285	1420	1485	1585	1670	1760	1905	2030	1.9
Lead bromide . . .	801	870	904	955	999	1049	1123	1188	0.96
Lead chloride . . .	836	908	943	994	1038	1086	1160	1228	1.02
Lead fluoride . . .	s	1160	1205	1270	1329	1392	1485	1568	1.21
Lead iodide . . .	766	832	864	913	955	1000	1077	1146	1.06
Lead oxide	1240	1340	1388	1458	1515	1580	1670	1750	1.14
Lead sulphide . . .	1145 s	1227 s	1265 s	1320 s	1363 s	1411	1480	1540	0.92
Lithium	1041	1146	1199	1277	1343	1416	1528	1630	1.48
Lithium bromide . .	1043	1142	1191	1262	1323	1391	1493	1585	1.37
Lithium chloride . .	1080	1185	1237	1312	1376	1447	1555	1660	1.55
Lithium fluoride . .	1347	1463	1520	1600	1670	1745	1860	1955	1.47
Lithium iodide . . .	1016	1098	1139	1197	1246	1300	1378	1446	0.99
Magnesium	891	979	1023	1087	1142	1203	1295	1376	1.21
Magnesium chloride .	1075	1183	1237	1317	1386	1463	1585	1695	1.68
Manganese	1560	1710	1780	1890	1980	2080	2240	2390	2.3
Manganese (II) chloride	1010	1100	1144	1210	1266	1328	1422	1505	1.25
Mercury	408	448.8	468.8	498.4	523.4	551.2	592.9	629.8	0.555
Mercury (II) bromide .	417 s	448 s	463 s	485 s	502 s	524	561	593	0.49
Mercury (II) chloride .	415	446	461	482	500	519	546	570	0.35
Mercury (II) iodide .	437 s	469 s	485 s	509 s	525	551	591	627	0.54
Molybdenum . . .	3450	3760	3910	4130	4320	4520	4830	5100	4.0
Molybdenum hexafluoride . . .	212 s	229 s	237 s	250 s	260 s	272 s	289 s	309	0.21
Molybdenum trioxide .	1020 s	1073 s	1095	1155	1206	1263	1350	1428	1.17
Neon	16.3 s	18.1 s	19.0 s	20.4 s	21.6 s	22.9 s	24.9	27.1	0.033
Nickel	1715 s	1850	1920	2010	2090	2180	2310	2420	1.6
Nickel chloride . . .	972	1030	1058	1096	1127	1160	1206	1245	0.56
Nickel carbonyl . .	s	s	s	249	261	276	297	316	0.26
Nitric oxide . . .	87 s	93.8 s	96.9 s	100.1	104.3	108.9	115.6	121.4	0.085
Nitrogen	48.1 s	53.0 s	55.4 s	59.0 s	62.1 s	65.8	71.8	77.4	0.084
Nitrogen dioxide . .	218 s	232.1 s	238.9 s	248.6 s	256.6 s	266.6	285.4	302.2	0.251
Nitrogen trifluoride .	91	100.4	104.9	111.8	117.7	124.4	134.9	144.1	0.142
Nitrogen pentoxide .	239 s	253 s	260 s	269 s	277 s	285 s	297 s	307 s	0.140
Nitrosyl chloride . .	179 s	194.4 s	201.9 s	212.6	223.0	234.6	252.1	267.8	0.234
Nitrosyl fluoride . .	146	157	163	171	179	187.5	200.9	213.2	0.188
Nitrous oxide . . .	132 s	142.1 s	147.0 s	154.2 s	160.2 s	166.6 s	176.0 s	184.7	0.165
Nitryl fluoride . . .	s	145.0	150.9	159.6	167.2	175.7	188.8	200.8	0.180
Osmium tetroxide (white)	273 s	294.5 s	305.0 s	321.0	336.8	354.1	380.1	402.8	0.337
Osmium tetroxide (yellow)	281 s	300.7 s	310.2 s	321.0	336.8	354.1	380.1	402.8	0.337
Oxygen	55.4	61.3	64.3	68.8	72.7	77.1	83.9	90.2	0.094
Ozone	105	114.8	119.7	127.1	133.5	140.7	151.8	161.8	0.151
Phosphonium bromide	233 s	249 s	256 s	268 s	277 s	286 s	300 s	312 d	0.16
Phosphonium chloride	185 s	197 s	203 s	211 s	218 s	226 s	237 s	246 s	0.13
Phosphonium iodide .	252 s	269 s	277 s	289 s	298 s	309 s	323 s	335 s	0.17

Elements and inorganic compounds (*contd*)

Element or compound	Pressure, $p/(kN\,m^{-2})$								dT/dp
	0.2	1	2	5	10	20	50	101.325	$K(kN\,m^{-2})^{-1}$
Phosphorus (yellow) .	365	398	414	439	460	484	520.5	553.6	0.497
Phosphorus (violet) .	542 s	576 s	592 s	614 s	633 s	652 s	680 s	704 s	0.34
Phosphorus (black) .	572 s	605 s	620 s	642 s	660 s	678 s	705 s	727 s	0.32
Phosphorus hydride .	117 s	128.2 s	133.9 s	142.6	150.4	159.3	172.9	185.4	0.188
Phosphorus oxychloride	s	s	281.1	298.3	313.1	329.8	355.4	378.7	0.350
Phosphorus pentachloride . .	326 s	347 s	358 s	372 s	384.6 s	397.7 s	416.5 s	432.4 s	0.232
Phosphorus pentoxide (stable) form. . .	662 s	705 s	725 s	753 s	776 s	801 s	837 s	878	0.80
Phosphorus thiobromide	328	352	364	380	394	408	429	447	0.25
Phosphorus thiochloride	260	284	297	315	330	347	374	397	0.35
Phosphorus tribromide	288	316	329	350	367.7	387.7	418.5	446.4	0.419
Phosphorus trichloride	226	247	258	274	288	303	327	349.3	0.329
Phosphorus trioxide .	296	322.0	335.2	354.9	372.0	391.3	421.2	448.5	0.412
Platinum	2860	3090	3200	3360	3490	3630	3830	3990	2.4
Potassium	633	704	740	794	841	894	975	1050	1.11
Potassium bromide .	1101	1218	1268	1336	1392	1453	1550	1660	2.05
Potassium chloride. .	1118	1221	1272	1347	1411	1482	1590	1680	1.4
Potassium fluoride .	1183	1292	1346	1424	1491	1565	1675	1775	1.5
Potassium hydroxide .	1015	1117	1168	1244	1310	1384	1498	1600	1.6
Potassium iodide .	1041	1141	1190	1264	1326	1396	1503	1600	1.4
Radon	139 s	152 s	158 s	168 s	176 s	184 s	198 s	211	0.21
Rhenium heptoxide .	494	518	529	545	557	570	589	604	0.21
Rubidium	583	649	684	735	779	829	907	978	1.07
Rubidium bromide .	1077	1177	1227	1300	1361	1430	1535	1625	1.4
Rubidium chloride .	1088	1190	1241	1316	1380	1451	1560	1655	1.4
Rubidium fluoride . .	1209	1275	1310	1366	1418	1482	1590	1685	1.3
Rubidium iodide .	1043	1140	1188	1259	1320	1387	1489	1580	1.3
Selenium . . .	636	695	724	767	803	843.5	904.4	958.4	0.800
Selenium dioxide . .	469 s	498 s	512 s	532 s	548 s	565 s	590 s	610 s	0.30
Selenium hexafluoride .	160 s	172 s	179 s	188 s	195.4 s	203.8 s	216.1 s	226.9 s	0.159
Selenium oxychloride .	321	342	352	368.5	382.7	399.0	424.7	448.7	0.365
Selenium tetrachloride	353 s	376 s	386 s	402 s	414.5 s	428.2 s	447.9 d	464.5 d	0.242
Silane	96	107.8	113.6	122.1	129.4	137.7	150.3	161.7	0.172
Silicon	1980	2150	2240	2360	2470	2580	2750	2890	2.0
Silicon dioxide . . .	1850	1985	2050	2140	2210	2290	2410	2500	1.3
Silicon tetrachloride .	214	234.5	244.7	259.8	272.8	287.4	309.8	329.9	0.303
Silicon tetrafluoride .	131 s	141 s	145 s	152 s	158 s	163 s	172 s	178 s	0.10
Silver	1640	1790	1865	1970	2060	2160	2310	2440	1.9
Silver chloride . . .	1211	1326	1383	1468	1540	1620	1735	1840	1.5
Silver iodide . . .	1119	1234	1292	1378	1452	1535	1665	1780	1.8
Sodium	729	807	846	904	954	1010	1096	1173	1.16
Sodium bromide . .	1103	1206	1257	1331	1394	1464	1570	1665	1.5
Sodium chloride . .	1163	1270	1323	1400	1466	1540	1645	1740	1.4
Sodium cyanide . .	1117	1234	1293	1381	1457	1540	1665	1770	1.6
Sodium fluoride . .	1377	1493	1550	1630	1700	1775	1885	1980	1.4
Sodium hydroxide . .	1037	1148	1203	1285	1355	1433	1550	1655	1.5
Sodium iodide . . .	1062	1158	1205	1274	1332	1396	1493	1580	1.3
Strontium	1040 s	1148	1203	1285	1354	1432	1550	1660	1.6
Sulphur.	462	505.7	528.0	561.3	590.1	622.5	672.4	717.8	0.682
Sulphur dioxide . .	180 s	193.1 s	199.3 s	211.2	221.0	231.9	248.4	263.2	0.221
Sulphur hexafluoride .	144 s	156.6 s	162.6 s	171.4 s	178.8 s	186.9 s	198.8 s	209.2 s	0.153
Sulphuric acid . . .	427	461	477	501	521	543	576	604 d	0.41
Sulphur monochloride .	270	294	306	325	340	358	386	410.8	0.377
Sulphur trioxide (α) .	237 s	252.8 s	260.3 s	271.0 s	279.8 s	289.2 s	304.6	317.9	0.195
Sulphur trioxide (β) .	243 s	257.9 s	265.1 s	275.3 s	283.5 s	292.3 s	304.8 s	317.9	0.195
Sulphur trioxide (γ) .	262 s	276.1 s	282.4 s	291.3 s	298.4 s	305.8 s	316.2 s	324.7 s	0.122

Elements and inorganic compounds (*contd*)

Element or compound	Pressure, $p/(kN\ m^{-2})$								dT/dp
	0.2	1	2	5	10	20	50	101.325	$K(kN\ m^{-2})^{-1}$
Sulphuryl chloride . .	226	246	257	272	285	299.7	322.2	342.5	0.305
Tantalum fluoride . .	s	s	s	s	s	420	464	503	0.59
Tellurium	806	888	929	990	1042	1100	1188	1267	1.18
Tellurium tetrachloride	s	510	527	552	572	596	631	662	0.46
Tellurium hexafluoride	164	178	184	194	202	211	223.7	234.9	0.166
Thallium . . .	1124	1235	1290	1371	1440	1515	1630	1730	1.5
Thallium bromide . .	s	782	816	867	909	955	1027	1093	1.0
Thallium chloride . .	713	777	810	858	899	946	1017	1081	1.0
Thallium iodide . .	726	793	826	875	917	964	1034	1097	0.94
Thionyl bromide . .	277	301	313	331	346	363	390	413	0.36
Thionyl chloride . .	224	246	257	274	288	303.5	327.4	348.8	0.319
Tin	1930	2120	2210	2350	2470	2600	2800	2990	2.8
Tin (IV) bromide . .	309	339.7	354.8	376.9	395.6	416.5	448.5	477.9	0.448
Tin (IV) chloride . .	256	278.5	290.0	306.9	321.5	337.8	363.1	386.2	0.350
Tin (II) chloride . .	604	660	688	730	765	805	866	922	0.84
Tin (IV) hydride . .	136	151	159	170	180	190	207	221	0.21
Tin (IV) iodide . .	403 s	446	466	497	523	552	595	634	0.57
Titanium tetrachloride	265	289.2	301.7	320.4	336.7	355.1	383.6	409.2	0.382
Tungsten	4230	4530	4680	4870	5040	5220	5470	5690	3.1
Tungsten hexafluoride	205 s	220 s	228 s	239 s	248 s	258 s	272 s	291	0.18
Uranium hexafluoride .	238 s	255.8 s	264.3 s	296.6 s	286.9 s	298.1 s	314.6 s	328.8 s	0.209
Vanadyl trichloride .	255	280.5	293.2	312.1	328.3	346.7	374.8	400.4	0.384
Water	260 s	280.1	290.6	306.0	319.0	333.2	354.5	373.15	0.276
Xenon	107 s	117.3 s	122.5 s	130.1 s	136.6 s	143.8 s	154.7 s	165.0	0.171
Zinc	780	854	891	945	991	1042	1118	1186	1.01
Zinc chloride . . .	714	770	797	836	869	905	958	1006	0.70
Zinc fluoride . . .	1220	1323	1375	1450	1515	1585	1690	1785	1.4
Zirconium tetrabromide	488 s	518 s	532 s	552 s	568 s	584 s	609 s	630 s	0.32
Zirconium tetrachloride	470 s	498 s	511 s	530 s	545 s	562 s	585 s	604 s	0.28
Zirconium tetraiodide .	545 s	579 s	594 s	616 s	634 s	653 s	681 s	704 s	0.34

Organic compounds

Compound	Pressure, $p/(kN\ m^{-2})$								dT/dp
	0.2	1.0	2.0	5.0	10	20	50	101.325	$K(kN\ m^{-2})^{-1}$
Acenaphthene . . .	s	397.3	414.4	439.7	461.2	485.0	520.3	550.7	0.444
Acetaldehyde . . .	193	210.3	219.2	232.4	243.8	256.5	276.0	293.5	0.263
Acetamide	s	372.8	386.8	407.2	424.3	443.1	470.9	495.2	0.360
Acetic acid	s	s	297.9	314.7	329.0	345.0	369.3	391.0	0.325
Acetic anhydride . .	280	304.3	316.2	333.9	348.9	365.6	390.7	412.8	0.326
Acetone	217	236.7	246.7	261.6	274.2	288.4	310.0	329.3	0.289
Acetonitrile	s	252.4	263.3	279.6	293.8	309.7	333.8	354.9	0.311
Acetophenone . . .	316	345.2	359.5	380.7	398.7	418.7	448.9	475.6	0.395
Acetylene	126 s	142.3 s	147.8 s	155.8 s	162.4 s	169.7 s	180.2 s	189.3 s	0.172
Adipic acid	440	472.6	488.4	511.3	530.5	551.6	583.0	610.7	0.414
Allyl alcohol	258	279.5	289.9	305.1	317.7	331.5	351.9	369.7	0.263
Aniline	313	337.3	349.7	368.7	385.3	404.2	433.0	457.6	0.355
Anisole	284	310.1	323.1	342.4	358.9	377.2	404.7	428.7	0.352
Anthracene	s	s	s	s	512.8	539.7	580.6	615.2	0.495
Benzaldehyde . . .	305	330.1	342.8	361.7	378.0	396.5	425.4	452.2	0.409
Benzene	s	s	s	279.5	293.1	308.4	331.9	353.2	0.320
Benzoic acid	s	399.4	413.7	434.4	451.6	470.5	498.3	522.4	0.354
Benzoic anhydride . .	426	463.6	482.6	510.4	534.0	560.1	599.1	633.2	0.504

Organic compounds (*contd*)

Compound	Pressure, $p/(\text{kN m}^{-2})$								dT/dp $\text{K}(\text{kN m}^{-2})^{-1}$
	0.2	1.0	2.0	5.0	10	20	50	101.325	
Benzoin	417	453.7	472.0	498.8	521.5	546.5	583.8	616.2	0.476
Benzonitrile	307	335.7	349.9	371.0	388.8	408.6	438.1	463.8	0.378
Benzophenone . . .	389	424.1	441.4	466.8	488.3	512.1	547.6	578.6	0.458
Benzoyl chloride . .	312	340.2	354.4	375.5	393.5	413.5	443.7	470.4	0.396
Benzyl alcohol . .	336	360.8	373.1	391.6	407.6	425.7	453.4	477.9	0.362
Benzyl chloride . . .	301	328.5	342.1	362.3	379.4	398.5	427.2	452.6	0.375
Benzyl cinnamate . .	453	486.6	503.0	527.0	546.8	568.1	598.5	623.2	0.352
Biphenyl	351	383.7	400.1	424.3	444.8	467.3	500.3	528.1	0.401
Bromobenzene . . .	282	308.1	321.1	340.2	356.6	375.0	403.3	429.4	0.398
Bromoethane . . .	203	221.8	231.3	245.4	257.6	271.4	292.5	311.5	0.284
Bromoform	s	302.1	315.3	335.0	351.9	370.7	399.0	423.7	0.363
Bromomethane . . .	180	197.0	205.4	218.0	228.8	241.1	259.9	276.7	0.252
1-Bromonaphthalene .	365	399.6	416.6	441.7	463.0	486.6	522.4	554.3	0.478
Buta-1,3-diene . . .	167	189.9	198.2	210.7	221.2	233.3	258.1	268.7	0.255
Butane	175	192.0	200.4	213.1	224.0	236.4	255.4	272.7	0.259
Butan-1-ol	281	300.6	310.4	325.0	337.4	351.3	372.2	390.9	0.278
Butan-2-ol	269	287.6	296.8	310.4	322.1	335.1	354.9	372.7	0.266
Butanone	234	255.0	265.4	280.9	294.3	309.2	332.1	352.7	0.309
Butyric acid	306	330.2	342.0	359.4	374.1	390.4	414.9	436.4	0.320
*iso*Butyric acid . .	294	319.2	331.6	349.6	364.7	381.2	405.9	427.7	0.326
Camphene	s	s	327.8	347.2	363.8	382.2	409.8	433.7	0.348
(+)-Camphor . . .	320 s	349.8 s	364.5 s	385.9 s	403.8 s	423.4 s	452.4 s	480.4	0.430
Carbon disulphide . .	204	224.3	234.4	249.5	262.4	276.9	299.2	319.6	0.310
Carbon tetrabromide .	s	s	s	368.0	385.1	404.3	434.2	462.7	0.447
Carbon tetrachloride .	s	s	258.8	274.9	288.7	304.3	328.2	349.8	0.325
Catechol	s	385.4	400.3	422.4	441.0	461.6	492.3	518.7	0.387
Chloracetic acid . . .	s	348.7	361.8	380.8	396.8	414.3	440.2	462.7	0.330
Chloral hydrate . . .	s	s	s	311.4	323.0	335.6	353.9	369.4 d	0.226
Chlorobenzene . . .	266	290.5	302.9	321.2	336.9	354.4	381.2	405.4	0.362
Chlorodifluoromethane .	153	166.7	173.5	183.8	192.7	202.8	218.4	232.4	0.208
1-Chloro-2,3-epoxypropane . .	262	285.1	296.8	314.0	328.6	344.9	369.5	391.1	0.319
Chloroethane . . .	187	203.9	212.5	225.2	236.2	248.7	268.0	285.4	0.263
Chloroform	220	239.6	249.7	264.8	277.7	292.3	314.6	334.4	0.296
Chloromethane . . .	s	s	185.1	196.2	205.9	216.9	233.9	249.1	0.228
Cineole (Eucalyptol) .	294	321.6	335.4	355.9	373.4	393.0	422.7	449.2	0.394
trans-Cinnamic acid .	407	439.8	455.7	478.5	497.3	517.7	547.5	573.2	0.379
Cinnamyl alcohol . .	353	384.3	399.9	422.7	442.1	463.9	495.3	523.2	0.413
o-Cresol	320	345.0	357.7	376.7	392.9	411.2	439.1	464.2	0.377
m-Cresol	330	355.6	368.5	387.7	404.0	422.4	450.3	475.4	0.376
p-Cresol	331	356.4	369.1	388.0	404.2	422.4	450.2	475.1	0.374
Cumene	282	306.5	319.1	337.8	354.0	372.2	400.2	425.6	0.381
Cyclohexane	s	s	s	s	292.4	308.0	332.1	353.8	0.326
Cyclohexanol . . .	299	324.0	336.4	354.7	370.2	387.2	412.5	434.2	0.317
Cyclohexanone . . .	280	306.6	319.8	339.4	356.2	375.0	403.5	428.8	0.376
p-Cymene	298	324.2	337.5	357.4	374.6	393.9	423.5	450.3	0.403
cis-Decahydro-naphthalene . . .	308	335.5	349.5	370.5	388.6	409.1	440.5	468.9	0.428
trans-Decahydro-naphthalene . . .	301	328.5	342.3	363.1	381.0	401.2	432.2	460.4	0.424
1,2-Diaminoethane . .	s	289.9	301.0	317.3	331.2	346.6	369.9	390.4	0.304
1,2-Dibromoethane . .	s	286.2	299.8	319.7	336.4	354.6	381.5	404.7	0.341
Dibromomethane . .	243	266.3	277.8	294.4	309.2	325.2	349.7	371.7	0.332
Dibutyl phthalate . .	429	464.7	481.9	506.9	527.7	550.5	584.1	613.2	0.429
Dichlorodifluoro-methane	158	172.4	179.8	190.8	200.3	211.1	227.9	243.4	0.233
1,2-Dichloroethane . .	234	255.0	265.8	282.1	296.1	311.8	335.3	355.5	0.293

Organic compounds (contd)

Compound	Pressure, $p/(kN\ m^{-2})$								dT/dp
	0.2	1.0	2.0	5.0	10	20	50	101.325	$K(kN\ m^{-2})^{-1}$
Dichloromethane . .	207	226.1	235.5	249.6	261.6	275.2	295.7	313.8	0.270
Diethylamine	s	236.2	246.1	260.8	273.4	287.7	309.4	328.6	0.284
Diethyl ether . . .	202	220.5	229.7	243.5	255.3	268.6	289.1	307.6	0.278
Diethyl oxalate . .	326	352.0	364.4	382.4	397.4	413.8	437.9	458.9	0.309
Diethyl sulphate . .	326	354.9	369.1	389.9	407.5	427.1	456.5	482.2 d	0.381
Dimethylamine . .	189	205.4	213.5	225.6	235.9	247.4	264.9	280.5	0.233
Dimethylaniline . .	309	337.4	351.7	372.9	390.9	410.8	440.5	466.3	0.377
1,4-Dioxan . . .	s	s	s	297.1	311.7	328.2	352.9	374.3	0.310
Diphenylamine . .	389	423.7	440.8	465.8	486.8	510.1	544.8	575.2	0.449
1,2-Diphenylethane .	368	402.3	419.5	444.8	466.3	490.2	525.9	557.2	0.464
Ethane	116	127.9	133.8	142.7	150.3	159.0	172.4	184.6	0.182
Ethanediol . . .	332	358.5	371.6	390.5	406.2	423.4	448.7	470.5	0.321
Ethanethiol . . .	200	219.1	228.4	242.5	254.6	268.3	289.4	308.1	0.279
Ethanol	248	267.0	276.6	290.6	302.3	315.3	334.6	351.5	0.249
Ethyl acetate . . .	234	255.3	265.7	281.1	294.1	308.7	330.7	350.2	0.291
Ethyl acetoacetate .	308	334.9	348.3	367.9	384.4	402.7	430.0	454.0	0.357
Ethylamine . . .	194	211.5	220.1	232.7	243.5	255.4	273.5	289.7	0.244
Ethylbenzene . . .	266	293.9	306.3	324.6	340.3	357.8	384.8	409.4	0.372
Ethyl benzoate . .	324	353.2	368.0	389.7	408.2	428.7	459.5	486.6	0.401
Ethylene	107	117.8	123.1	131.2	138.2	146.1	158.3	169.5	0.167
Ethylene oxide . .	187	204.1	212.6	225.2	236.1	248.2	266.9	283.8	0.255
Ethyl formate . .	217	236.3	246.0	260.5	273.0	287.0	308.4	327.4	0.284
Ethyl salicylate . .	341	371.2	386.1	408.1	426.8	447.3	478.0	504.7	0.393
Eugenol . . .	358	389.6	405.1	427.7	446.8	468.0	499.4	526.7	0.401
Fluorobenzene . .	236	256.7	267.4	283.4	297.1	312.6	336.4	357.9	0.323
Formaldehyde . .	s	182.6	190.2	201.6	211.4	222.4	239.1	254.1	0.224
Formic acid . . .	s	s	s	295.7	310.5	326.9	351.7	373.8	0.330
Furfuryl alcohol . .	311	336.1	348.7	367.1	382.4	399.1	423.1	443.2	0.289
Glycerol	405	435.1	449.1	469.3	486.2	504.9	534.2	563.2	0.473
Heptane	244	266.5	277.7	294.3	308.5	324.6	349.2	371.5	0.338
Hexane	223	243.9	254.2	269.7	283.0	297.9	321.0	341.9	0.314
Indene	303	329.3	342.6	362.5	379.7	399.0	428.8	455.7	0.406
Iodoethane . . .	223	244.6	255.3	271.2	284.8	300.2	323.9	345.5	0.327
Iodomethane . . .	s	223.4	233.1	247.8	260.5	274.8	296.6	315.5	0.278
Isoprene	197	216.0	225.4	239.5	251.8	265.6	286.8	305.7	0.281
Maleic anhydride . .	s	345.4	361.1	383.2	401.3	420.8	449.7	475.2	0.383
(−)-Menthol . . .	336	363.7	377.5	397.8	414.8	433.5	461.2	485.2	0.354
Mesitylene . . .	289	315.1	328.3	347.8	364.5	383.3	411.9	437.9	0.391
Methane	s	s	s	s	s	95.1	103.7	111.5	0.119
Methanol . . .	234	253.0	262.6	276.6	288.4	301.4	320.7	337.7	0.251
Methyl acetate . .	220	239.9	249.7	264.3	276.7	290.7	311.9	330.9	0.284
Methylamine . . .	s	196.1	203.9	215.4	225.1	235.9	252.2	266.8	0.219
Methylaniline . . .	316	343.6	357.4	377.8	395.0	414.0	442.9	468.7	0.387
Methyl benzoate . .	318	345.2	358.8	379.0	396.3	415.8	445.7	472.7	0.404
3-Methylbutan-1-ol .	290	309.8	320.0	335.1	348.0	362.5	384.5	404.2	0.295
Methyl formate . .	203	220.9	230.0	243.5	255.1	268.0	287.7	305.1	0.258
2-Methylpropan-1-ol .	274	293.5	303.0	317.2	329.2	342.6	362.9	381.0	0.270
2-Methylpropan-2-ol .	s	s	s	s	307.8	320.1	338.8	355.6	0.251
Methyl salicylate . .	334	362.4	376.7	397.8	415.9	436.2	467.6	496.4	0.439
Naphthalene . . .	s	353.9	368.4	390.2	408.9	429.9	462.1	491.1	0.436
1-Naphthol . . .	374	407.9	424.5	449.1	469.8	492.7	526.6	555.7	0.426
2-Naphthol . . .	s	410.6	427.6	452.6	473.7	497.0	531.5	561.2	0.436
1-Naphthylamine . .	385	420.1	437.4	462.6	483.9	507.5	542.8	574.0	0.464
2-Naphthylamine . .	389	424.0	441.3	466.8	488.3	512.1	547.8	579.3	0.467
Nicotine	342	373.8	389.8	413.6	434.0	456.6	490.7	520.5	0.439
Nitrobenzene . . .	324	352.1	366.3	387.2	405.1	425.3	456.0	483.8	0.416
Nitroethane . . .	257	281.0	292.7	309.9	324.5	340.7	365.2	387.2	0.328

Organic compounds (*contd*)

Compound	Pressure, $p/(\text{kN m}^{-2})$								dT/dp $\text{K}(\text{kN m}^{-2})^{-1}$
	0.2	1.0	2.0	5.0	10	20	50	101.325	
Nitromethane . . .	249	271.3	282.5	299.2	313.4	329.4	353.4	374.4	0.309
1-Nitropropane . . .	269	293.4	305.6	323.5	338.8	355.8	381.6	404.8	0.347
2-Nitropropane . . .	260	283.9	295.9	313.6	328.7	345.4	370.8	393.5	0.338
Octane	264	287.7	299.4	316.9	332.0	349.1	375.2	398.8	0.355
Octan-1-ol . . .	334	357.3	368.8	386.2	401.2	418.2	444.4	468.3	0.360
Oleic acid . . .	457	489.8	505.9	529.7	549.9	572.2	605.4	634.2 d	0.422
Palmitic acid . .	449	478.1	493.0	515.5	535.0	557.3	592.0	624.0	0.485
Pentane	200	218.9	228.5	242.8	255.0	268.8	290.0	309.2	0.289
Pentan-1-ol . . .	296	315.9	326.2	341.4	354.4	369.0	391.2	411.2	0.299
Pentan-2-ol . . .	280	301.0	311.3	326.2	338.7	352.5	373.5	392.9	0.296
Phenanthrene . .	400	438.5	457.9	486.7	511.2	538.3	578.6	613.4	0.509
Phenol	316	340.7	352.9	371.2	386.8	404.3	431.0	455.0	0.359
Phosgene. . . .	184	200.6	209.1	221.8	232.7	245.1	264.2	281.4	0.258
Phthalic anhydride . .	s	400.2	417.4	443.3	465.6	490.4	527.1	557.7	0.439
Pinene	278	305.0	318.5	338.5	355.7	374.7	403.3	428.2	0.366
Piperidine . . .	s	274.8	285.8	302.3	316.4	332.4	357.0	379.4	0.336
Propane	147	161.7	169.0	179.9	189.3	199.9	216.3	231.1	0.220
Propan-1-ol . . .	265	284.0	293.6	307.7	319.6	332.9	352.8	370.3	0.261
Propan-2-ol . . .	255	273.6	282.7	296.1	307.4	320.0	338.8	355.4	0.247
Propene	144	158.2	165.2	175.8	184.9	195.2	211.1	225.5	0.217
Propyl acetate . . .	251	273.6	284.6	300.8	314.6	330.0	353.6	375.0	0.322
Pyridine	258	280.8	292.3	309.4	324.1	340.5	365.7	388.4	0.340
Pyrogallol . . .	s	434.1	451.1	475.6	496.1	518.6	552.2	581.7 d	0.441
Quinol	s	s	s	463.3	481.9	502.4	532.9	559.4	0.391
*iso*Quinoline . .	344	374.5	390.0	412.7	432.0	453.4	485.5	513.7	0.417
Resorcinol . . .	388	419.4	434.6	456.6	475.0	495.1	524.5	549.7	0.369
Salicylic acid . . .	s	s	s	442.4	459.6	478.2	505.2	528.2	0.336
Stearic acid . . .	467	496.2	511.4	534.5	554.8	578.0	614.5	648.3 d	0.516
Styrene	272	298.6	311.8	331.3	347.9	366.2 p	393.8 p	418.2 p	0.363
Succinic anhydride . .	s	395.3	411.0	434.2	453.9	475.6	507.4	534.2	0.388
1,2,3,4-Tetrahydro-naphthalene	318	346.2	360.6	382.0	400.4	421.0	452.4	480.4	0.417
Thiophen . . .	237	258.0	268.6	284.3	297.7	312.8	336.1	357.5	0.327
Thymol	344	374.2	389.1	411.0	429.5	449.7	479.5	505.0	0.371
Toluene	252	274.9	286.5	303.7	318.4	335.0	360.5	383.8	0.350
o-Toluidine . . .	323	349.2	362.4	381.9	398.7	417.6	446.7	472.9	0.394
m-Toluidine . . .	320	349.3	363.6	384.7	402.5	422.2	451.4	476.5	0.368
p-Toluidine . . .	321	349.2	362.9	382.9	399.9	418.8	447.5	473.6	0.394
Trichloroethylene . .	234	256.2	267.3	283.8	298.0	314.0	338.2	359.8	0.322
Trichlorofluoromethane	193	210.5	219.5	233.0	244.8	258.0	278.5	296.8	0.274
Trimethylamine. . .	180	196.0	204.4	217.0	228.0	240.4	259.4	276.0	0.246
2,2,4-Trimethylpentane	242	264.2	275.6	292.6	307.2	323.7	349.2	372.3	0.349
Valeric acid . . .	324	348.6	360.8	378.7	393.9	410.8	436.2	458.7	0.334
*iso*Valeric acid . .	314	339.4	351.9	370.1	385.3	401.9	426.6	448.3	0.322
Vanillin	388	420.6	436.7	460.0	479.4	500.5	531.6	558.2	0.391
o-Xylene	275	300.1	312.7	331.3	347.3	365.2	392.7	417.6	0.375
m-Xylene	272	296.5	308.8	327.1	342.9	360.6	387.7	412.3	0.370
p-Xylene	271	295.4	307.8	326.2	342.0	359.7	386.9	411.5	0.370

D.A.

2.4.5 Vapour pressures from 0.2 to 6 MN m^{-2}

Kelvin temperatures are listed for pressures p from 0.2 to 6 MN m^{-2}. Values of the critical temperatures and pressures are given on pp. 183–186. The letter 's' indicates a solid.

Elements and inorganic compounds

Element or compound	Pressure, $p/(MN\ m^{-2})$							
	0.2	0.5	1	2	3	4	5	6
Ammonia	254.2	277.2	298.0	322.5	338.8	351.5	361.9	370.9
Argon	94.2	105.7	116.6	129.7	138.7	145.7		
Bromine	354	389	421	459	485	506	523	537
Carbon dioxide	203.3 s	216.0 s	233.0	253.6	267.6	278.4	287.4	295.1
Carbon monoxide	88.0	98.8	108.9	121.2	129.5			
Chlorine	255	283	308	339	360	376	390	401
Cyanogen	268	294	317	345	364	379	391	401
Hydrogen	22.8	27.1	31.2					
Hydrogen bromide	221	244	265	291	309	322	334	343
Hydrogen chloride	201	222	241	264	279	291	301	309
Hydrogen cyanide	318	350	378	410	430	445		
Hydrogen fluoride	312	344	371	402	421	436	448	457
Hydrogen iodide	255	282	307	338	359	376	390	402
Hydrogen sulphide	227	251	273	299	316	330	341	350
Krypton	129.2	144.7	159.2	176.8	188.8	198.2	206.0	
Neon	29.6	33.7	37.5	42.2				
Nitric oxide	128	138	148	159	166	171	175	179
Nitrogen	83.6	94.0	103.8	115.6	123.6			
Nitrogen dioxide	310	332	352	373	387	396	404	411
Nitrous oxide	197	217	235	257	272	283	293	301
Oxygen	97.2	108.8	119.6	132.8	141.7	148.7	154.3	154.8
Silicon tetrafluoride . . .	185	201	217	238	252			
Sulphur dioxide	279	306	330	358	377	392	404	414
Sulphur hexafluoride . . .	219.6	243.4	265.2	291.1	308.4			
Sulphur trioxide	333	358	381	411	431	447	460	471
Tin (IV) chloride	414	456	495	543	574			
Xenon	177.8	198.9	218.6	242.5	258.8	271.6	282.2	

Organic compounds

Compound	Pressure, $p/(MN\ m^{-2})$							
	0.2	0.5	1	2	3	4	5	6
Acetic acid	415	452	485	524	549	568	584	
Acetic anhydride . .	435	467	494	525	546	561		
Acetone	351	385	416	453	478	497		
Acetylene	201	221	240	263	278	289	299	307
Aniline	485	527	566	614	647	673	694	
Benzene	377	416	452	495	524	546		
Bromobenzene . . .	459	505	547	598	633	659		
Bromoethane . . .	333	367	399	437	461	479	492	502
Buta-1,3-diene . . .	289	319	348	388	410	422		
Butane	292	323	352	388	411			
Carbon disulphide . .	342	377	409	448	474	495	512	528
Carbon tetrachloride .	375	414	450	494	524	546		
Chlorobenzene . . .	429	481	521	558	581	602		
Chlorodifluoromethane	248	273	297	324	343	358		

Organic compounds (*contd*)

Compound	Pressure, $p/(\text{MN m}^{-2})$							
	0.2	0.5	1	2	3	4	5	6
Chloroethane . . .	305	337	366	399	421	439	456	
Chloroform	356	393	425	463	489	509	528	
Chloromethane . . .	267	294	320	350	370	386	399	410
Dichlorodifluoromethane	261	289	315	347	368	383		
Diethyl ether . . .	328	362	393	430	454			
Ethane	198	220	241	266	283	296		
Ethanol	370	398	424	454	474	489	502	513
Ethyl acetate . . .	373	409	442	482	508			
Ethylene	182	202	220	243	259	271	281	
Hexane	365	403	438	480	507			
Methane	121	135	149	165	177	186		
Methanol	356	385	410	439	458	473	485	495
Methyl acetate . . .	352	386	417	453	478	496		
Pentane	331	366	398	436	462			
Phenol	480	521	557	600	630	653	674	692
Phosgene	299	330	359	393	415	433	447	
Toluene	410	452	490	536	566	590		

D.A.

2.5 Critical constants and second virial coefficients of gases

The critical pressure P_c, critical molar volume V_c, and critical temperature T_c, are the values of the pressure P, molar volume V_m, and thermodynamic temperature T, at which the densities of coexisting liquid and gaseous phases just become identical. At the critical point

$$(\partial P/\partial V)_{T=T_c} = 0, \qquad (\partial^2 P/\partial V^2)_{T=T_c} = 0$$

At temperatures above the critical temperature a gas cannot be liquefied.

The state of a gas can be accurately expressed by an equation of the form:

$$z \equiv PV_m/RT = 1 + B/V_m + C/V_m^2 + \ldots \tag{1}$$

where R is the molar gas constant ($R = 8.3143 \pm 0.0004 \text{ J K}^{-1} \text{ mol}^{-1}$) and $B, C, \ldots$, are the second, third, $\ldots$, virial coefficients, which are functions of temperature only. As a rough rule neglect of the third and higher virial coefficients in equation (1) leads to errors in z of less than x per cent provided that P/P_c is less than about $0.2x^{1/2}$ at $T/T_c = 0.6$, or $0.6x^{1/2}$ at $T/T_c = 1$, or $1.3x^{1/2}$ at $T/T_c = 2$, or $3x^{1/2}$ at $T/T_c = 4$.

For Ar, Kr, Xe, CH_4, N_2, and CO accurately, and with useful but lower accuracy for other substances, the dimensionless quantity $z_c = P_c V_c/RT_c$ is about 0.29, and the second virial coefficient B, which is usually a negative quantity for gases at ordinary temperatures, is given by the formula:

$$B/V_c = 0.430 - 0.866(T_c/T) - 0.694(T_c/T)^2, \qquad (0.55 \leqslant T/T_c \leqslant 6) \tag{2}$$

A better estimate of B for a substance with a value of z_c appreciably different from 0.29 can be obtained by the method described in Lewis and Randall's *Thermodynamics*, 2nd edn., 1961, p. 608, rev. by Pitzer and Brewer (McGraw-Hill, New York).

Critical constants and second virial coefficients of gases

Substance	$\dfrac{P_c}{MN\ m^{-2}}$	$\dfrac{V_c}{cm^3\ mol^{-1}}$	$\dfrac{T_c}{K}$	$\dfrac{P_c V_c}{R T_c}$	$\dfrac{-B}{cm^3\ mol^{-1}}\left(\dfrac{T}{K}\right)$
Acetaldehyde . . .	—	—	461	—	1400 (290); 510 (380); 270 (470)
Acetic acid	5.79	171	594.5	0.200	— —
Acetic anhydride . .	4.7	—	569		— —
Acetone	4.70	213	508.1	0.237	1950 (300); 1250 (350); 700 (400)
Acetonitrile	4.83	173	548	0.183	5000 (315); 2300 (360); 1450 (405)
Acetylene	6.14	113	308.4	0.271	191 (273); 158 (293); 133 (313)
Allene (propadiene) .	—	—	393		710 (223); 395 (293); 265 (353)
Ammonia	11.30	72.5	405.4	0.243	345 (273); 100 (423); 45 (573)
Aniline	5.30	274	698.9	0.250	
Argon	4.86	75.2	150.7	0.292	225 (90); 65 (173); 22 (273); −16 (673)
Benzene.	4.90	254	562.2	0.266	1460 (300); 970 (350); 700 (400)
Bromine.	10.3	135	584	0.286	
Bromobenzene . . .	4.52	323	670	0.262	— —
Bromoethane . . .	6.23	215	503.8	0.320	— —
Bromomethane . . .	—	—	464	—	560 (290); 390 (335); 280 (380)
Buta-1,3-diene . . .	4.33	221	425	0.271	
Butane	3.80	255	425.2	0.274	705 (300); 370 (400); 225 (500)
*iso*Butane	3.65	263	408.0	0.283	710 (290); 320 (400); 230 (473); 137 (573)
Butanone	4.15	267	536	0.249	1975 (315); 1560 (340); 1140 (370)
But-1-ene	4.02	238	419.6	0.274	740 (280); 440 (350); 295 (420)
*iso*Butene	4.00	240	417.8	0.276	1200 (240); 810 (273); 510 (333)
cis-But-2-ene . . .	4.20	234	435.4	0.271	1050 (260); 700 (300); 500 (340)
trans-But-2-ene . . .	4.10	238	428.5	0.274	1250 (240); 850 (280); 600 (320)
Carbon dioxide. . .	7.38	94.0	304.2	0.274	300 (210); 148 (273); 70 (373); 29 (473)
Carbon disulphide . .	7.9	173	552	0.298	850 (290); 535 (360); 375 (430)
Carbon monoxide . .	3.50	93.1	133	0.295	230 (91); 14 (273); −1 (348); −10 (423)
Carbon tetrachloride .	4.56	276	556.4	0.272	1440 (315); 1260 (330); 1100 (345)
Carbon tetrafluoride .	3.74	140	227.5	0.277	111 (273); 44 (373); 10 (473); −17 (623)
Chlorine	7.71	124	417	0.276	— —
Chlorobenzene . . .	4.52	308	632.4	0.265	
Chlorodifluoromethane	4.91	165	369.4	0.264	— —
Chloroethane . . .	5.24	—	460.4	—	570 (330); 450 (365); 350 (400)
Chloroform	5.5	240	536.4	0.296	1050 (315); 810 (355); 640 (395)
Chloromethane. . .	6.68	143	416.2	0.276	750 (240); 310 (340); 160 (440)
Cyanogen	6.0	—	400	—	350 (308); 215 (373); 145 (423)
Cyclohexane . . .	4.02	308	553.4	0.270	1550 (310); 1050 (335); 700 (400)
Cyclohexene . . .	—	—	560.4	—	— —
Cyclopentane . . .	4.52	260	511.6	0.276	
Cyclopropane . . .	5.49	—	397.8	—	355 (310); 255 (360); 195 (410)
Deuterium	1.650	60.3	38.26	0.312	−13 (273); −15 (423)
Deuterium oxide . .	21.88	54.9	644.2	0.224	
Dichlorodifluoromethane	4.12	218	385.0	0.280	630 (240); 480 (275); 350 (310)
Dichlorofluoromethane	5.17	197	451.6	0.271	730 (250); 430 (350); 270 (450)
Diethylamine . . .	3.71	301	496	0.271	1220 (320); 930 (360); 710 (400)
Diethyl ether . . .	3.64	280	466.7	0.263	1060 (310); 670 (360); 500 (400)
Dimethylamine. . .	5.31	—	437.6	—	560 (320); 430 (360); 320 (400)
Dimethyl ether . . .	5.35	190	400.0	0.306	575 (273); 460 (303); 380 (333)
1,4-Dioxan	5.20	245	585	0.262	— —
Ethane	4.88	148	305.4	0.284	220 (273); 115 (373); 65 (473)
Ethanol	6.14	167	513.9	0.240	2000 (330); 1000 (365); 500 (400)
Ethyl acetate . . .	3.83	286	523.2	0.252	1570 (330); 1110 (365); 870 (400)
Ethylamine	5.62	—	456	—	760 (300); 520 (350); 360 (400)

Critical constants and second virial coefficients of gases (*contd*)

Substance	$\dfrac{P_c}{\text{MN m}^{-2}}$	$\dfrac{V_c}{\text{cm}^3 \text{ mol}^{-1}}$	$\dfrac{T_c}{K}$	$\dfrac{P_c V_c}{R T_c}$	$\dfrac{-B}{\text{cm}^3 \text{ mol}^{-1}}\left(\dfrac{T}{K}\right)$
Ethylene . . .	5.12	127.4	282.4	0.278	305 (200); 134 (300); 43 (473)
Ethyl formate . . .	4.74	229	508.4	0.257	1000 (330); 740 (365); 570 (400)
Ethyl methyl ether . .	4.40	221	437.8	0.267	—
Fluorine	—	—	—		385 (80); 98 (150); 26 (250)
Fluorobenzene . . .	4.55	269	560.1	0.263	380 (548); 340 (573); 280 (623)
Fluoromethane . . .	5.88	113	318	0.251	260 (273); 140 (348); 85 (423)
Helium	0.229	58	5.2	0.307	110 (3); −3 (30); −12 (273); −9 (1273)
Heptane	2.74	426	540.4	0.260	1780 (350); 1420 (380); 1160 (410)
Hept-1-ene	—	—	537.2	—	1790 (340); 1360 (375); 1080 (410)
Hexane	3.03	368	507.5	0.264	1860 (300); 1180 (355); 820 (410)
Hex-1-ene	—	—	504.0	—	1420 (320); 1000 (365); 750 (410)
Hydrogen	1.294	65.5	32.99	0.309	150 (20); −3 (123); −14 (273); −16 (473)
Hydrogen bromide . .	8.51	—	363	—	— —
Hydrogen chloride . .	8.26	87	324.6	0.266	— —
Hydrogen deuteride . .	1.484	62.8	35.91	0.312	— —
Hydrogen iodide . .	8.2	—	423	—	— —
Hydrogen sulphide . .	9.01	97.9	373.6	0.284	— —
Iodobenzene . . .	4.52	351	721	0.265	— —
Iodomethane . . .	—	—	528	—	600 (323); 490 (353); 420 (383)
Krypton	5.50	92.3	209.4	0.292	310 (120); 63 (273); 0 (573); −17 (873)
Methane	4.62	98.7	190.6	0.288	270 (125); 54 (273); 11 (423); −7 (573)
Methanol	8.09	118	512.6	0.224	1450 (320); 900 (365); 500 (410)
Methyl acetate . . .	4.69	228	506.8	0.254	1200 (330); 900 (360); 750 (390)
Methylamine . . .	7.46	—	430.1	—	510 (300); 330 (350); 240 (400)
2-Methylbutane . .	3.33	308	460.6	0.268	1220 (290); 625 (380); 380 (470)
Methylcyclohexane . .	3.48	344	572.2	0.252	— —
Methylcyclopentane . .	3.79	319	532.8	0.273	1450 (304); 1260 (326); 1120 (345)
Methyl formate . . .	6.00	172	487.2	0.255	830 (320); 575 (360); 430 (400)
2-Methylpentane . .	3.03	367	497.6	0.269	— —
Naphthalene . . .	—	—	748.0	—	— —
Neon	2.73	41.7	44.4	0.308	21 (65); −6 (173); −11 (273); −14 (673)
Neopentane . . .	3.20	303	433.8	0.269	700 (330); 460 (400); 215 (548)
Nitric oxide . . .	6.5	—	180	—	175 (130); 51 (210); 23 (290); 9 (370)
Nitrogen	3.39	90.1	126.2	0.291	200 (90); 80 (143); 10 (273); −24 (673)
Nitrogen dioxide . .	10.1	82	431	0.231	
Nitrous oxide . . .	7.24	96.7	309.6	0.272	270 (220); 142 (290); 86 (360); 57 (430)
Octafluorocyclobutane	2.78	325	388.4	0.280	435 (373); 196 (498); 82 (623)
Octane	2.49	490	568.6	0.258	2140 (370); 1840 (390); 1600 (410)
Oct-1-ene	—	—	566.6	—	2160 (360); 1780 (385); 1490 (410)
Oxygen	5.08	78	154.8	0.308	245 (90); 80 (157); 22 (273); 3 (373)
Pentane	3.37	311	469.6	0.268	1180 (300); 585 (400); 350 (500)
Pent-1-ene	4.0	—	464.7	—	980 (310); 675 (360); 500 (410)
Phenol	6.1	—	694.3	—	1450 (428); 1340 (438); 1140 (454)
Phosgene	5.7	190	453	0.288	— —
Phosphine	6.5	—	324.5	—	450 (190); 310 (230); 200 (273)
Propane	4.25	200	369.8	0.277	380 (300); 205 (400); 125 (500)
Propan-1-ol	5.17	219	536.8	0.254	890 (378); 760 (393); 600 (423)
Propene	4.62	181	365.0	0.276	665 (220); 230 (360); 110 (500)
Propyne	5.62	—	402	—	— —
Pyridine	5.63	—	620.0	—	1260 (350); 960 (390); 700 (430)
Radon	6.3	—	377	—	— —

Critical constants and second virial coefficients of gases (*contd*)

Substance	$\dfrac{P_c}{MN\,m^{-2}}$	$\dfrac{V_c}{cm^3\,mol^{-1}}$	$\dfrac{T_c}{K}$	$\dfrac{P_cV_c}{RT_c}$	$\dfrac{-B}{cm^3\,mol^{-1}}\left(\dfrac{T}{K}\right)$
Silane	4.84	—	269.6	—	— —
Silicon tetrachloride .	3.76	—	506	—	— —
Silicon tetrafluoride .	3.72	—	259.0	—	145 (293); 110 (323); 85 (353)
Stannic chloride . .	3.75	351	591.8	0.268	— —
Sulphur dioxide . .	7.88	122	430.6	0.269	560 (270); 240 (370); 130 (470)
Sulphur hexafluoride .	3.77	195	318.8	0.277	335 (273); 140 (398); 52 (523)
Sulphur trioxide . .	8.49	126	491.4	0.262	— —
Tetramethylsilane . .	—	—	—	—	955 (323); 735 (363); 580 (403)
Toluene	4.11	320	591.8	0.267	1630 (350); 1200 (390); 900 (430)
Trichlorofluoromethane	4.38	248	471.2	0.277	1080 (250); 580 (350); 330 (450)
Triethylamine . . .	3.0	—	532	—	1540 (330); 1220 (365); 980 (400)
Trimethylamine . .	4.07	254	433.2	0.287	660 (313); 530 (343); 440 (373)
2,2,4-Trimethylpentane	2.57	476	543.8	0.271	— —
Water	22.12	59.1	647.3	0.243	450 (373); 196 (473); 72 (673); 25 (973)
Xenon	5.88	119	289.7	0.290	150 (273); 25 (573); −8 (873)
m-Xylene	3.55	376	616.5	0.261	2070 (380); 1610 (405); 1280 (430)
o-Xylene	3.73	368	630.5	0.262	2040 (380); 1630 (405); 1320 (430)
p-Xylene	3.51	378	616.2	0.259	2050 (380); 1600 (405); 1270 (430)

D.A.
M.L.M.

2.6 Properties of solutions

2.6.1 Solubilities of gases in water

Air in water

A kilogram of water saturated with air at a pressure of 101.325 kN m^{-2} contains the following volumes of dissolved oxygen, etc., in cm^3 at 0 °C and 101.325 kN m^{-2}.

Gas	*Temperature of water, $\theta/°C$*						
	0	5	10	15	20	25	30
Oxygen	10.19	8.9	7.9	7.0	6.4	5.8	5.3
Nitrogen, argon, etc.	19.0	16.8	15.0	13.5	12.3	11.3	10.4
Sum of above	29.2	25.7	22.9	20.5	18.7	17.1	15.7
% of oxygen in dissolved air (by vol.)	34.9	34.7	34.5	34.2	34.0	33.8	33.6

Gases in water

S indicates the number of m^3 of gas measured at 0 °C and 101.325 kN m^{-2} which dissolve in 1 m^3 of water at the temperature stated, and when the pressure of the gas plus that of the water vapour is 101.325 kN m^{-2}.

A indicates the same quantity except that the gas itself is at the uniform pressure of 101.325 kN m^{-2} when in equilibrium with water. (For other values see pp. 155–167.)

Gases in water (contd)

Gas		Temperature of water, $\theta/°C$							
		0	10	15	20	30	40	50	60
Ammonia	S	1130	870	770	680	530	400	290	200
Argon	A	0.054	0.041	0.035	0.032	0.028	0.025	0.024	0.023
Carbon dioxide	A	1.676	1.163	0.988	0.848	0.652	0.518	0.424	0.360
Carbon monoxide	A	0.035	0.028	0.025	0.023	0.020	0.018	0.016	0.015
Chlorine	S	4.61	3.09	2.63	2.26	1.77	1.41	1.20	1.01
Helium	A	0.0098	0.0091	0.0089	0.0086	0.0084	0.0084	0.0086	0.0090
Hydrogen	A	0.0214	0.0195	0.0188	0.0182	0.0170	0.0164	0.0161	0.0160
Hydrogen sulphide	A	4.53	3.28	2.86	2.51	1.97	1.62	1.37	1.18
Hydrochloric acid	S	512	475	458	442	412	385	362	339
Nitrogen	A	0.0230	0.0183	0.0165	0.0152	0.0133	0.0119	0.0108	0.0100
Nitrous oxide	A	—	0.88	0.74	0.63	—	—	—	—
Nitric oxide	A	0.071	0.055	0.049	0.046	0.039	0.034	0.031	0.029
Oxygen	A	0.047	0.037	0.033	0.030	0.026	0.022	0.020	0.019
Sulphur oxide	S	79.8	56.6	47.3	39.4	27.2	18.8	—	—

Values of A for 20 °C for other rare gases are: Ne, 0.0101; Kr, 0.0594; Xe, 0.126.

E.F.G.H.

2.6.2 Solubilities of solids in water

Solubilities are expressed as the number of g of *anhydrous* substance which when dissolved in 100 g of water make a saturated solution at the temperature stated.

The formula given is that of the solid phase which is in equilibrium with the solution. (For more complete data see Seidell's *Solubilities*, 1958 (New York).) For other solutions, see pp. 155–167.

Substance	Temperature of water/°C							
	0	10	15	20	40	60	80	100
Ammonium chloride, NH_4Cl	29.4	33.3	35.2	37.2	45.8	55.2	65.6	77.3
Barium chloride, $BaCl_2 . 2H_2O$	31.6	33.3	34.4	35.7	40.7	46.4	52.4	58.8
Barium hydroxide, $Ba(OH)_2 . 8H_2O$	1.67	2.48	3.23	3.89	8.22	20.94	101.4	—
Cadmium sulphate, $3CdSO_4 . 8H_2O$	76.5	76.0	76.3	76.6	78.5	83.7	69.7†	60.77†
Calcium hydroxide, $Ca(OH)_2$	0.185	0.176	0.170	0.165	0.141	0.116	0.094	0.077
Copper sulphate, $CuSO_4 . 5H_2O$	14.3	17.4	18.8	20.7	28.5	40.0	55.0	75.4
Lithium carbonate, Li_2CO_3	1.54	1.43	1.38	1.33	1.17	1.01	0.85	0.72
Mercuric chloride, $HgCl_2$	3.6	4.8	5.6	6.5	10.3	16.3	30.0	61.3
Potassium chloride, KCl	28.07	31.23	32.8	34.2	40.0	45.8	51.3	56.3
Potassium bromide, KBr	53.5	59.5	62.5	65.2	75.5	85.5	95.0	104
Potassium iodide, KI	127.5	136	140	144	160	176	192	208
Potassium hydroxide, $KOH . 2H_2O$	97.0	103	107	112	138‡	—	—	178‡
Potassium nitrate, KNO_3	13.3	20.9	25.8	31.6	63.9	110	169	246
Silver nitrate, $AgNO_3$	122	170	196	222	376	525	669	952
Sodium carbonate, $Na_2CO_3 . 10H_2O$	7.0	12.5	16.4	21.5	48.5§	46.4§	45.8§	45.5§
Sodium chloride, $NaCl$	35.7	35.8	35.9	36.0	36.6	37.3	38.4	39.8

Solubilities of solids in water (contd)

Substance	Temperature of water/°C							
	0	10	15	20	40	60	80	100
Sodium sulphate, $Na_2SO_4.10H_2O$	5.0	9.0	13.4	19.4	48.8‖	45.3‖	43.7‖	42.5‖
Strontium chloride, $SrCl_2.6H_2O$	43.5	47.7		52.9	65.3	81.8	90.5**	100.8**
Succinic acid, $(CH_2)_2(COOH)_2$	2.80	4.51	5.7	6.9	16.2	35.8	70.8	125
Sugar (Cane), $C_{12}H_{22}O_{11}$ . .	179.2	190.5	197.0	203.9	238.1	287.3	362.1	487.2

† Solid phase becomes $CdSO_4.H_2O$ at 74 °C.
‡ Becomes $KOH.\frac{3}{2}H_2O$ at 32.5 °C and $KOH.H_2O$ at 50 °C.
§ Becomes $Na_2CO_3.H_2O$ at 35 °C.
‖ Becomes Na_2SO_4 at 32.38 °C.
** Becomes $SrCl_2.2H_2O$ at 70 °C.

E.F.G.H.

2.6.3 Densities of aqueous solutions

The values listed are $\rho(kg\ m^{-3})$ for a temperature of 20 °C unless otherwise stated and the indicated % is the number of g of substance in 100 g of aqueous solution.

Density of ethyl alcohol aqueous solutions

As an example of the values listed the density of a 24% solution is 963.1 kg m^{-3}.

%	0	1	2	3	4	5	6	7	8	9
0	998.2	996.4	994.5	992.8	991.0	989.4	987.8	986.3	984.8	983.3
10	981.9	980.5	979.1	977.8	976.4	975.1	973.9	972.6	971.3	970.0
20	968.6	967.3	965.9	964.3	963.1	961.7	960.2	958.7	957.1	955.5
30	953.8	952.1	950.4	948.6	946.8	944.9	943.1	941.1	939.2	937.2
40	935.2	933.1	931.1	929.0	926.9	924.7	922.6	920.4	918.2	916.0
50	913.8	911.6	909.4	907.1	904.9	902.6	900.3	898.0	895.7	893.4
60	891.1	888.8	886.5	884.2	881.8	879.5	877.1	874.8	872.4	870.0
70	867.7	865.3	862.9	860.5	858.1	855.6	853.2	850.8	848.4	845.9
80	843.4	841.0	838.5	836.0	833.5	831.0	828.4	825.8	823.2	820.6
90	818.0	815.3	812.6	809.8	807.1	804.2	801.4	798.5	795.5	792.4
100	789.3									

For other temperatures, interpolate from the above and the following:

At 15 °C: 0%, 999.13; 10%, 983.0; 20%, 970.7; 30%, 956.8; 40%, 938.8; 50%, 917.8; 60%, 895.2; 70%, 871.9; 80%, 847.8; 90%, 822.3; 100%, 793.5.

At 25 °C: 0%, 997.1; 10%, 980.4; 20%, 966.4; 30%, 950.7; 40%, 931.5; 50%, 909.9; 60%, 887.0; 70%, 863.4; 80%, 839.1; 90%, 813.6; 100%, 785.1.

Density of hydrochloric acid aqueous solutions

$\rho/(kg\ m^{-3})$	·% HCl	$\rho/(kg\ m^{-3})$	% HCl	$\rho/(kg\ m^{-3})$	% HCl	$\rho/(kg\ m^{-3})$	% HCl	$\rho/(kg\ m^{-3})$	% HCl
1010	2.40	1050	10.57	1090	18.65	1130	26.60	1170	34.53
1020	4.44	1060	12.60	1100	20.65	1140	28.56	1180	36.58
1030	6.48	1070	14.63	1110	22.65	1150	30.54	1190	38.63
1040	8.53	1080	16.64	1120	24.63	1160	32.54		

Density of nitric acid aqueous solutions

$\rho/(kg\ m^{-3})$	% HNO$_3$	$\rho/(kg\ m^{-3})$	% HNO$_3$	$\rho/(kg\ m^{-3})$	% HNO$_3$	$\rho/(kg\ m^{-3})$	% HNO$_3$	$\rho/(kg\ m^{-3})$	% HNO$_3$
1010	2.162	1120	20.80	1240	39.02	1360	58.74	1480	90.00
1020	3.983	1140	23.94	1260	42.14	1380	62.64	1490	93.40
1040	7.536	1160	27.00	1280	45.27	1400	66.90	1495	95.90
1060	10.98	1180	30.00	1300	48.38	1420	71.57	1500	97.76
1080	14.32	1200	32.98	1320	51.68	1440	76.66	1505	98.88
1100	17.67	1220	35.92	1340	55.13	1460	82.33	1510	99.60

Density of sulphuric acid aqueous solutions

$\rho/(kg\ m^{-3})$	% H$_2$SO$_4$	$\rho/(kg\ m^{-3})$	% H$_2$SO$_4$	$\rho/(kg\ m^{-3})$	% H$_2$SO$_4$	$\rho/(kg\ m^{-3})$	% H$_2$SO$_4$	$\rho/(kg\ m^{-3})$	% H$_2$SO$_4$
1010	1.731	1260	35.01	1520	62.00	1780	85.16	1836.5	97.25
1020	3.242	1280	37.36	1540	63.81	1800	87.69	1836.0	98.06
1040	6.221	1300	39.68	1560	65.59	1810	89.23	1835.5	98.38
1060	9.129	1320	41.94	1580	67.35	1820	91.11	1835.0	98.65
1080	11.97	1340	44.17	1600	69.09	1822	91.54	1834.5	98.87
1100	14.73	1360	46.33	1620	70.82	1824	92.00	1834.0	99.04
1120	17.43	1380	48.44	1640	72.53	1826	92.51	1833.5	99.21
1140	20.08	1400	50.50	1660	74.22	1828	93.03	1833.0	99.35
1160	22.67	1420	52.52	1680	75.91	1830	93.64	1832.5	99.50
1180	25.22	1440	54.49	1700	77.63	1832	94.32	1832.0	99.63
1200	27.72	1460	56.41	1720	79.37	1834	95.17	1831.5	99.76
1220	30.18	1480	58.30	1740	81.16	1835	95.72	1831.0	99.88
1240	32.61	1500	60.16	1760	83.06	1836	96.56	1830.5	100

Density of ammonia aqueous solutions

$\rho/(kg\ m^{-3})$	% NH$_3$	$\rho/(kg\ m^{-3})$	% NH$_3$	$\rho/(kg\ m^{-3})$	% NH$_3$	$\rho/(kg\ m^{-3})$	% NH$_3$	$\rho/(kg\ m^{-3})$	% NH$_3$
994	0.98	966	7.77	938	15.47	910	24.03	882	33.16
990	1.89	962	8.82	934	16.66	906	25.34	878	34.43
986	2.83	958	9.87	930	17.85	902	26.67	874	35.71
982	3.79	954	10.95	926	19.06	898	28.00	870	37.00
978	4.77	950	12.03	922	20.28	894	29.33	866	38.29
974	5.75	946	13.15	918	21.51	890	30.63		
970	6.76	942	14.29	914	22.76	886	31.89		

Density of sodium carbonate aqueous solutions

$\rho/(kg\ m^{-3})$	% Na$_2$CO$_3$	$\rho/(kg\ m^{-3})$	% Na$_2$CO$_3$	$\rho/(kg\ m^{-3})$	% Na$_2$CO$_3$	$\rho/(kg\ m^{-3})$	% Na$_2$CO$_3$	$\rho/(kg\ m^{-3})$	% Na$_2$CO$_3$
1010	1.135	1040	4.019	1070	6.895	1100	9.728	1130	12.51
1020	2.096	1050	4.980	1080	7.848	1110	10.66	1140	13.43
1030	3.058	1060	5.942	1090	8.788	1120	11.59	1145	13.88

Density of sodium hydroxide aqueous solutions

$\rho/(kg\ m^{-3})$	% NaOH	$\rho/(kg\ m^{-3})$	% NaOH	$\rho/(kg\ m^{-3})$	% NaOH	$\rho/(kg\ m^{-3})$	% NaOH	$\rho/(kg\ m^{-3})$	% NaOH
1020	1.938	1130	11.92	1240	21.90	1350	32.10	1460	43.11
1030	2.838	1140	12.82	1250	22.82	1360	33.07	1470	44.16
1040	3.746	1150	13.73	1260	23.73	1370	34.04	1480	45.22
1050	4.655	1160	14.63	1270	24.65	1380	35.02	1490	46.28
1060	5.564	1170	15.54	1280	25.56	1390	36.00	1500	47.32
1070	6.473	1180	16.44	1290	26.48	1400	37.00	1510	48.37
1080	7.378	1190	17.35	1300	27.41	1410	37.99	1520	49.44
1090	8.282	1200	18.26	1310	28.34	1420	39.00		
1100	9.191	1210	19.17	1320	29.27	1430	40.00		
1110	10.10	1220	20.08	1330	30.20	1440	41.03		
1120	11.01	1230	20.99	1340	31.15	1450	42.06		

Density of phosphoric acid aqueous solutions

$\rho/(kg\ m^{-3})$	% H_3PO_4	$\rho/(kg\ m^{-3})$	% H_3PO_4	$\rho/(kg\ m^{-3})$	% H_3PO_4	$\rho/(kg\ m^{-3})$	% H_3PO_4	$\rho/(kg\ m^{-3})$	% H_3PO_4
1010	2.15	1090	16.26	1240	38.16	1400	57.23	1720	87.72
1020	4.00	1100	17.87	1260	40.77	1440	61.43	1760	91.17
1030	5.84	1120	21.02	1280	43.33	1480	65.49	1780	92.83
1040	7.64	1140	24.07	1300	45.83	1520	69.41	1800	94.48
1050	9.43	1160	27.04	1320	48.21	1560	73.21	1820	96.08
1060	11.18	1180	29.93	1340	50.57	1600	76.94	1840	97.68
1070	12.91	1200	32.75	1360	52.84	1640	80.63	1860	99.23
1080	14.60	1220	35.53	1380	55.11	1680	84.20		

Densities of some other aqueous solutions

Values of $\rho/(kg\ m^{-3})$ at 20 °C excepting where a different temperature is given. The indicated % is the number of g of anhydrous substance in 100 g of solution.

Substance	5%	10%	15%	20%	25%	Substance	5%	10%	15%	20%
$BaCl_2$.	1043	1092	1145	1203	1266	NH_4Cl .	1014	1029	1043	1057
$CdSO_4$.	1049	1102	1161	1224	1295	$CuSO_4$.	1051	1107	1167	1230
$FeCl_3$.	1041	1085	1132	1182	1234	$MgSO_4$.	1050	1103	1160	1220
$MgCl_2$.	1040	1082	1125	1171	1218	KCl . .	1030	1063	1098	1133
NaA . .	1024	1050	1075	1102	1130	$K_2Cr_2O_7$.	1034	1070	—	—
NaCl .	1034	1071	1109	1148	1189	$K_3Fe(CN)_6$	1026	1054	1083	1113
$NaNO_3$.	1032	1067	1104	1143	1183	$K_4Fe(CN)_6$	1033	1068	1105	—
$SrCl_2$.	1044	1093	1145	1201	1260	KNO_3 . .	1030	1063	1097	1133
$ZnSO_4$.	1051	1107	1168	—	—	K_2SO_4 .	1039	1082	—	—

Densities of some other aqueous solutions (*contd*)

Substance	5%	10%	15%	20%	25%	30%	35%	40%	45%	50%
NH_4NO_3 . .	1019	1040	1061	1083	1105	1128	1151	1175	1200	1226
$CaCl_2$ . . .	1040	1084	1129	1178	1228	1282	1337	1396	—	—
H_2O_2† (18 °C)	1017	1035	1054	1073	1092	1112	1133	1154	1175	1197
PbA_2 (18 °C)	1037	1077	1120	1166	1217	1271	1330	1399	—	—
LiCl . . .	1027	1056	1085	1115	1146	1179	—	—	—	—
$HClO_4$‡ (15 °C)	1029	1060	1093	1128	1166	1207	1251	1299	1352	1410
KBr . . .	1035	1074	1116	1160	1208	1259	1315	1375	—	—
K_2CO_3 . .	1044	1090	1139	1190	1243	1298	1355	1414	1476	1540
KI . . .	1036	1076	1119	1166	1216	1271	1331	1396	1467	1546
$AgNO_3$ . .	1042	1088	1139	1194	1255	1321	1393	1474	1565	1668
$Na_2S_2O_3$ . .	1040	1083	1127	1174	1223	1274	1327	1383	—	—
Sugar§ . .	1018	1038	1059	1081	1104	1127	1151	1176	1203	1230

† H_2O_2: 60%, 1242; 70%, 1290; 80%, 1341; 90%, 1393; 100%, 1447.
‡ $HClO_4$: 60%, 1539; 70%, 1674.
§ Sugar: 60%, 1287; 70%, 1347.

E.F.G.H.

2.7 Properties of chemical bonds

2.7.1 Dipole moments and dipole lengths

Most of the following data were calculated from an extensive collection given by A. L. McClellan, *Tables of Experimental Dipole Moments*, San Francisco, Copyright by W. H. Freeman and Co., 1963. The values listed represent average values for some common chemicals. The conditions under which the measurements were made are indicated by the following letters, b = substance in benzene solution, g = substance as a gas, l = substance as a liquid.

The SI unit of dipole moment, p, is the coulomb metre (C m) and the dipole length, l_p is equal to p/e where e is the charge of a proton. The unit previously in use for dipole moments was the Debye, symbol D; $1D = 3.335\ 640 \times 10^{-30}$ C m.

Substance	$10^{30}p/(\text{C m})$	l_p/pm	Substance	$10^{30}p/(\text{C m})$	l_p/pm
Acetaldehyde	8.3b	52	Benzonitrile	10.7l	67
Acetamide	12.3b	77	Benzophenone	9.0b	56
Acetic acid	3.3–5.0b	21–31	Bromobenzene	5.0l	31
Acetone	10.0l	62	Bromoethane	6.0b	37
Acetonitrile	11.7b	73	Bromoform	3.3b	21
Acetophenone	9.7b	60	Carbon dioxide	0 g	0
Acetylacetone	9.3b	58	Carbon disulphide . . .	0 l	0
Acetyl chloride	8.0b	50	Carbon tetrahalides . . .	0 l	0
Acrolein	9.7b	60	Carbonyl sulphide . . .	2.4g	15
Acrylonitrile	11.7b	73	Chloroacetic acid	7.7b	48
Ammonia	3.0l	19	Chlorobenzene	5.2l	32
Aniline	5.2b	32	Chloroethane	6.7b	42
Anisole	4.0l	25	Chloroform	3.7b	23
Benzamide	12.2b	76	Cyclohexanol	6.0b	37
Benzene	0 l	0	1,2-Dibromoethane . . .	4.0b	25

Dipole moments and dipole lengths (*contd*)

Substance	$10^{30}p/(\text{C m})$	l_p/pm	Substance	$10^{30}p/(\text{C m})$	l_p/pm
Dichloromethane	6.0b	37	Methyl benzoate	6.2b	39
Diethylamine	3.7l	23	Methyl ether	4.2b	26
Dimethylaniline	5.2b	32	Morpholine	5.0b	31
Dimethylformamide . . .	12.7l	79	Nitric oxide	0.5g	3
Dimethylsulphoxide . . .	13.0b	81	Nitrobenzene	13.3b	83
Dioxan	1.3b	8	Nitrogen dioxide	1.3g	8
Ethanol	5.7b	35	Nitromethane	10.3b	65
Ethyl acetate	6.2b	39	Oxygen	0 g	0
Ethylamine	4.3b	27	Ozone	2.2g	14
Ethylbenzene	1.2l	7	Phenol	5.3b	33
Ethylene carbonate . . .	16.0b	100	Propylene oxide	6.5b	41
Ethylene glycol	6.7b	42	Pyridine	7.7l	48
Ethylene oxide	6.3l	40	Pyrrole	6.0b	37
Ethyl ether	4.2b	26	Quinoline	7.0l	44
Formamide	11.3b	71	Styrene	0.3l	2
Formic acid	5.0b	31	Sulphur hexafluoride . . .	0 g	0
Furan	2.2b	14	Tetrahydrofuran	5.7b	35
Hexane	0 l	0	Thiophen	1.6l	10
Hydrazine	10.0l	62	Toluene	1.3l	8
Hydrogen cyanide . . .	9.8g	61	Triethylamine	2.8l	18
Hydrogen sulphide . . .	3.2g	20	s-Trioxane	7.3b	46
Iodoethane	5.8b	36	Water	6.7–10.0l	42–62
Methanol	5.5b	34	Water	6.2g	39

E.F.G.H.

2.7.2 Bond lengths and dissociation energies of diatomic molecules

The tabulated values are for the internuclear separation r_e in the absence of vibration; those marked * are for the vibrational ground state. They are taken from the compilations: *Tables of Interatomic Distances and Configuration of Molecules and Ions*, Special Publication No 11; Supplement 1956–1959, Special Publication No 18, Chemical Society, London, 1958, 1965. Bond Dissociation Energies, *D*, are taken from A. G. Gaydon, *Dissociation Energies and Spectra of Diatomic Molecules*, 3rd edn, 1968 (Chapman and Hall, London).

Bond	r_e/pm	D/kJ	Bond	r_e/pm	D/kJ
H—H	74.130	432.00	C—H	112.0	335
H—D	74.140	435.39	N—H	103.8	310
D—D	74.143	439.53	P—H	143.3*	340
C—C	131.2	603	O—H	97.1	425
N—N	109.76	942	F—H	91.7	564
P—P	189.3	485	Cl—H	127.5	428
O—O	120.741	493.6	Br—H	140.8	362
S—S	188.7	423	I—H	160.0	295
F—F	141.8*	155	C—N	117.7	745
Cl—Cl	198.8	239	C—O	113.1	1069
Br—Br	228.4	190	N—O	115.02	627
I—I	266.7	147	Cl—F	162.8	247

J.H.S.G.

2.7.3 Bond lengths and angles in polyatomic molecules

The values are selected from those given in *Tables of Interatomic Distances and Configuration of Molecules and Ions*, Special Publication No. 11; Supplement 1956–1959, Special Publication No 18, Chemical Society, London 1958, 1965.

Molecule	Bond	length/pm	Bond	angle/deg.	Molecule	Bond	length/pm	Bond	angle/deg.
CO_2 . .	C—O	115.98	O—C—O	180	C_2H_2 . .	C—H	109.3	H—C—H	180
CS_2 . .	C—S	155.30	S—C—S	180		C—C	176.6		
CSe_2 .	C—Se	198	Se—C—Se	180	C_2H_4 . .	C—H	108.4	H—C—H	115.5
SO_2 . .	S—O	143.21	O—S—O	119.5		C—C	133.2		
SO_3 . .	S—O	143	O—S—O	120	C_2H_6 . .	C—H	109.3	H—C—H	109.75
H_2O . .	O—H	95.8	H—O—H	104.45		C—C	153.4		
H_2O_2 .	O—O	148	O—O—H	100	C_6H_6 . .	C—H	108.4	H—C—C	120
ClO_2 .	Cl—O	149	O—Cl—O	118.5		C—C	139.7	C—C—C	120
H_2S . .	S—H	134.55	H—S—H	93.3	CH_3OH .	C—H	142.8	C—O—H	109
NH_3 . .	N—H	100.8	H—N—H	107.3		C—O	109.5		
PH_3 . .	P—H	143.7	H—P—H	93.3		O—H	96.0		
AsH_3 .	As—H	151.9	H—As—H	91.83	$(CH_3)_2O$.	C—H	109.4	C—O—C	111.5
$AsCl_3$.	As—Cl	216.1	Cl—As—Cl	98.4		C—O	141.6		
$SbCl_3$.	Sb—Cl	235.2	Cl—Sb—Cl	99.5	$(CH_3)_3N$.	C—H	109	H—C—H	107.1
$BiCl_3$.	Bi—Cl	248	Cl—Bi—Cl	100		C—N	147.2	C—N—C	108.7
SiH_3F .	Si—H	146.0	H—Si—H	109.3	C_2H_5Cl .	C—C	159.5	H—C—H	110
	Si—F	159.5				C—Cl	177.9	C—C—Cl	110.5
GeH_3Cl .	Ge—H	152	H—Ge—H	110.9	$(CH_3)_2CO$	C—C	151.5	C—C—C	116.22
$POCl_3$.	P—Cl	199	Cl—P—Cl	103.5		C—O	121.5	C—C—O	121.9
CH_4 . .	C—H	109.3	H—C—H	109.5	CH_3SH .	C—S	181.9	C—S—H	96.5
CCl_4 . .	C—Cl	176.6	Cl—C—Cl	109.5	$(CH_3)_3As$.	C—As	195.9	C—As—C	96

J.H.S.G.

2.7.4 Bond dissociation energies in polyatomic molecules

The values have been selected from those listed by T. L. Cottrell, *The Strengths of Chemical Bonds*, 2nd edn, 1958 (Butterworths, London); V. I. Vedeneyev, L. V. Gurvich, V. N. Kondrat'yev, V. A. Medvedev and Y. L. Frankevich, *Bond Energies, Ionization Potentials and Electron Affinities*, 1966 (Edward Arnold, London); and the references cited in J. A. Kerr and A. F. Trotman-Dickenson, *Handbook of Chemistry and Physics*, 50th edn, 1969/70 (The Chemical Rubber Co., Cleveland, Ohio).

X—Y	D(X—Y)/kJ	X—Y	D(X—Y)/kJ	X—Y	D(X—Y)/kJ
CH—H	448	NH_2—H	427	CH_3—NH_2	335
CH_2—H	439	OH—H	492	C_6H_5—NH_2	418
CH_3—H	435	SH—H	377	CH_3—OH	377
C_2H_5—H	435	HC≡CH	962	C_6H_5—OH	469
tC_3H_7—H	395	H_2C=CH_2	699	CH_3—OCH_3	335
C_6H_5—H	469	F_2C=CF_2	319	C_6H_5—OCH_3	423
$C_6H_5CH_2$—H	356	CH_3—CH_3	368	tC_3H_7—OCH_3	347
CCl_3—H	401	CF_3—CF_3	406	CH_3—SH	368
CF_3—H	444	CH_3—CF_3	418	CH_3—F	452
CH_3NH—H	385	CH_3—CHO	314	tC_3H_7—F	439
$(CH_3)_2N$—H	360	CH_3—CN	431	C_6H_5—F	523
C_6H_5NH—H	335	NC—CN	607	CF_3—F	540

Bond dissociation energies in polyatomic molecules (*contd*)

$X-Y$	$D(X-Y)/kJ$	$X-Y$	$D(X-Y)/kJ$	$X-Y$	$D(X-Y)/kJ$
CCl_3-F	444	C_6H_5-Br	335	O_2N-NO_2	57
CH_3-Cl	352	CF_3-Br	293	$CO=O$	531
tC_3H_7-Cl	339	CCl_3-Br	226	$HO-OH$	213
CF_3-Cl	356	CH_3-I	234	CH_3O-OCH_3	151
CCl_3-Cl	305	tC_3H_7-I	222	$F_3P=O$	544
CH_3-Br	293	C_6H_5-I	272	$Cl_3P=O$	510
tC_3H_7-Br	285	O_2N-NO	40	$Br_3P=O$	498

J.H.S.G.

2.7.5 Ionic radii

Empirical values for ionic radii are used for predicting interatomic distances and packing in crystals, even when the forces are far from being purely ionic. The ions are regarded as hard spheres, whose size is, to a first approximation, constant in all environments. To a second approximation, it has long been recognized that there is a dependence on *coordination number*, and, more recently, for transition-metal ions, a dependence also on *spin state*. Separate values are listed accordingly in the table.

The *coordination number* of an atom is the number of its first-nearest neighbours of opposite sign. It may sometimes be ambiguous when the neighbours are not all at exactly the same distance; judgement and experience, rather than formal rules, must then decide which of them are to be counted as first-nearest. Different decisions on this point will lead to small differences in estimates of the mean cation–anion distance within the polyhedron (the figure whose vertices are the first-neighbour anions round the cation).

The *spin state* of a transition-metal ion refers to its spin angular momentum. Normally this depends on the ground state of the free atom, and is given by Hund's rule; this is the *high-spin state*. In a particular crystal structure, however, ions may be situated in strong ligand fields which reduce the spin angular momentum; this is the *low-spin state*.

In using the table, no *a priori* knowledge of the ionic character of the bonds, i.e. of the degree of polarization or covalent character, is needed. An estimate of the degree of polarization is given by the *Pauling electrostatic valence*, whose value (measured in v.u.) is the valency of the cation divided by its coordination number. The radii tabulated for the different coordination numbers have been so derived as to allow for this, and will predict the mean cation–anion distance within a polyhedron to an accuracy, generally, of about ±3 pm. Individual distances may, however, show a rather larger scatter. The tabulated values also predict the shortest anion–anion polyhedron edges and packing distances for polyhedra where the Pauling valence is not greater than 1 v.u.; for greater Pauling valences, the polyhedron edges are considerably shorter than the anion diameter.

The radii tabulated are taken, with some simplification, from those listed by R. D. Shannon and C. T. Prewitt, *Acta Cryst.*, 1969, **B25**, 925, with corrections *ibid.*, 1970, **B26**, 1046. Methods of derivation and references to earlier work will be found there. For the sort of corrections to be made for more precise work, see, for example, W. H. Baur, *Trans. Am. Cryst. Assoc.*, 1970, **6**, 129.

Ionic Radii, in pm

Cation	Coordination no.				Cation	Coordination no.				Cation	Coordination no.			
	4	6	8	12		4	6	8	12		4	6	8	12
Li^+	59	74			Pb^{2+}	94	118	129	149	Si^{4+}	26	40		
Na^+		102	116							Ge^{4+}	40	54		
K^+		138	151	160	B^{3+}	12				Sn^{4+}		69		
Rb^+		149	160	173	Al^{3+}	39	53			Pb^{4+}		78	94	
Cs^+		170	182	188	Ga^{3+}	47	62							
					In^{3+}		79	92		Ti^{4+}		60		
Ag^+		115	130		Tl^{3+}		88	100		Zr^{4+}		72	84	
Tl^+		150	160	176						Hf^{4+}		71	83	
					Sc^{3+}		73	87		Th^{4+}		100	106	
Be^{2+}	27				Y^{3+}		90	101		U^{4+}		97	100	
Mg^{2+}	58	72			La^{3+}		106	118	132					
Ca^{2+}		100	112	135						P^{5+}	17	35		
Sr^{2+}		116	125	144	Cr^{3+}		61			As^{5+}	34	50		
Ba^{2+}		136	142	160	Mn^{3+} LS		58			Sb^{5+}		61		
					Mn^{3+} HS		65			Bi^{5+}		74		
Zn^{2+}	60	75			Fe^{3+} LS		55			V^{5+}	36	54		
Cd^{2+}	80	95	107	131	Fe^{3+} HS	49	65			Nb^{5+}	32	64		
					Co^{3+} LS		53			Ta^{5+}		64	69	
Mn^{2+} LS		67			Co^{3+} HS		61							
Mn^{2+} HS		82								S^{6+}	12			
Fe^{2+} LS		61			Bi^{3+}		102	111						
Fe^{2+} HS	63	78								Cr^{6+}	30	52		
Co^{2+} LS		57			Pu^{3+}		100			Mo^{6+}	42	60		
Co^{2+} HS		74								W^{6+}	41	58		
Ni^{2+}		70								U^{6+}	48	75		
Cu^{2+}		73												

Anion	Coordination no.			Anion	Coordination no.		
	2	4	6		2	4	6
F^-	128	131	133	O^{2-}	135	138	140
Cl^-		181†		S^{2-}		184†	
Br^-		196†					
I^-		220†					

LS = low-spin state; HS = high-spin state.
† Independent of coordination number.

H.D.M.

2.7.6 Crystal structures

A perfect infinite crystal possesses a *lattice*, an infinite set of points generated by three non-parallel vectors, such that each point is identical in itself and its surroundings.

With each lattice point may be associated a number of atoms. If their coordinates relative to the lattice point are given, together with the lengths and directions of the lattice vectors chosen to define the axes of reference, the complete structure is defined.

Position coordinates x, y, z are commonly expressed as fractions of the *lattice parameters a, b, c*; the parallelepiped defined by the lattice vectors $\mathbf{a}$, $\mathbf{b}$, $\mathbf{c}$, is the *unit cell*.

Symmetry elements may be present imposing certain relations (i) between lattice parameters, (ii) between position coordinates of different atoms. Axes of reference are generally chosen in accordance with the symmetry. As a result of (i), and with a conventional choice of axes, crystals are classified into systems as follows:

Cubic:	$a = b = c,$	$\alpha = \beta = \gamma = 90°$
Tetragonal:	$a = b \neq c,$	$\alpha = \beta = \gamma = 90°$
Orthorhombic:	$a \neq b \neq c,$	$\alpha = \beta = \gamma = 90°$
Hexagonal:	$a = b \neq c,$	$\alpha = \beta = 90°, \gamma = 120°$
Monoclinic:	$a \neq b \neq c,$	$\alpha = \gamma = 90°, \beta \neq 90°$
Triclinic:	$a \neq b \neq c,$	$\alpha \neq \beta \neq \gamma \neq 90°$

where α is the angle between b and c, and similarly for β and γ. Accidental equality of unit cell edges, and special values of interaxial angles not required by the symmetry, are disregarded in making this classification.

Atoms may be in *general positions*, or in *special positions* on point-symmetry elements. In the latter case, some or all of the position coordinates x, y, z are simple fractions; in the former, they are variable parameters whose values may change with temperature, pressure, or composition. It often happens, however, that atoms in general positions have, accidentally, parameters which are to a good approximation simple fractions.

If the translation vectors chosen as axes of reference generate all the points of the lattice, it is said to be *primitive* (P); otherwise it is *centred*. For centred lattices, if there is an atom at x, y, z, all translation-repeats of it within one unit cell can be derived by adding to x, y, z, components which are fractions of the lattice parameters. The kind of centring is summarized in the lattice centring operator: if, for example, this is written $(0, 0, 0; 0, \frac{1}{2}, \frac{1}{2}) +$ it means that for every point at $0+x$, $0+y, 0+z$, there is another identical point at $0+x, \frac{1}{2}+y, \frac{1}{2}+z$. The complete set of possibilities is as follows:

For any symmetry except hexagonal:

One-face-centred lattice A [2 points],	$(0, 0, 0;\ \ 0, \frac{1}{2}, \frac{1}{2}) +$
Body-centred lattice I [2 points],	$(0, 0, 0;\ \ \frac{1}{2}, \frac{1}{2}, \frac{1}{2}) +$
All-face-centred lattice F [4 points]	$(0, 0, 0;\ \ 0, \frac{1}{2}, \frac{1}{2};\ \ \frac{1}{2}, 0, \frac{1}{2};\ \ \frac{1}{2}, \frac{1}{2}, 0) +$

For hexagonal symmetry only:

Rhombohedrally centred lattice R [3 points],	$(0, 0, 0;\ \ \frac{2}{3}, \frac{1}{3}, \frac{1}{3};\ \ \frac{1}{3}, \frac{2}{3}, \frac{2}{3}) +$

There are expressions similar to A for different choices of axes, giving B- and C-face-centring. A- or B-face-centring is impossible in the tetragonal system; A-, B- or C- in the cubic. The same lattice could be named either P or C in the tetragonal and monoclinic systems, according to the choice of axes; similarly for I and F. A rhombohedrally centred lattice rotated through 180° from the conventional (obverse) orientation listed above is obtained by the operations $(0, 0, 0;\ \ \frac{1}{3}, \frac{2}{3}, \frac{1}{3};\ \ \frac{2}{3}, \frac{1}{3}, \frac{2}{3}) + .$

To describe a structure fully, we specify its system, its independent lattice parameters (thus choosing our axes of reference), its lattice type, and the position parameters of a set of atoms so chosen that (a) no two are separated by a lattice vector, (b) no atom not excluded by (a) is omitted. This description is complete and general, applicable to crystals of any system and any degree of complexity. Note that, if a letter or number specifying a position parameter is negative, it is conventional to write a bar over it instead of a minus sign in front of it: thus $\bar{x}, \bar{x}, \frac{1}{4}$ stands for $-x, -x, \frac{1}{4}$.

If the space group is known, the description can be abbreviated by listing position parameters for the *asymmetric unit* only, i.e. the set of atoms not related to each other by any symmetry element. This kind of abbreviation will not be used for the structures described below (see *International Tables for X-ray Crystallography*, Vol. 1, for details of its use).

Some simple structures

The structures described below are particularly simple and important types. Most of them are cubic, and most of them have all their atoms in special positions. The structure type is commonly named after the particular material described, except where otherwise specified. Lattice parameters and variable position parameters (if any) apply to this material. Other isostructural (isomorphous) materials (of which examples are given in the table) have slightly different numerical values for these parameters. A short qualitative comment is added to the formal description, which, however, is sufficient in itself to allow scale diagrams to be drawn and interatomic distances calculated.

Note that atoms are not separately listed if their coordinates can be derived from those listed by the use of lattice centring operators, as explained above. As a check, the total number of atoms of each kind in the unit cell is indicated.

1. *Copper ('Monatomic face-centred cubic')*
   Cubic; all-face-centred lattice F
   $a = 361.5$ pm
   4 Cu at 0, 0, 0
        Each atom has 12 equidistant neighbours, in directions parallel to the face diagonal of the cube, at distances $a/\sqrt{2}$. The structure is a *cubic close packing* of equal spheres.
        An alternative description uses hexagonal axes of reference, with c_H along the diagonal of the cube and a_H along a face diagonal perpendicular to it; this unit cell is rhombohedrally centred and contains 3 Cu.

2. *Iron ('Monatomic body-centred cubic')*
   Cubic; body-centred lattice I
   $a = 286.6$ pm
   2 Fe at 0, 0, 0
        Each atom has 8 equidistant neighbours in directions parallel to the cube body diagonal, at distances $\sqrt{(3)}a/2$.

3. *Magnesium ('Hexagonal close-packed')*
   Hexagonal; primitive lattice P
   $a = 320.9$ pm, $c = 521.0$ pm
   2 Mg at $\pm(\frac{2}{3}, \frac{1}{3}, \frac{1}{4})$
   With an alternative choice of origin, Mg atoms are at 0, 0, 0 and $\frac{2}{3}, \frac{1}{3}, \frac{1}{2}$.
        Each atom has 6 equidistant neighbours in its own plane at a distance a, and two sets of 3 above and below it. If c/a has the ideal value of 1.633, the whole array is in *hexagonal close packing*. The unit cell contains two close-packed layers, not related by a lattice vector (in contrast to cubic close packing, with three layers related by lattice vectors).

4. *Diamond*
   Cubic; all-face-centred lattice F
   $a = 356$ pm
   8 C at $\pm(\frac{1}{8}, \frac{1}{8}, \frac{1}{8})$
   With an alternative choice of origin, C atoms are at 0, 0, 0 and $\frac{1}{4}, \frac{1}{4}, \frac{1}{4}$.
        Each atom has 4 equidistant neighbours in directions parallel to the body diagonals of the cube, at distances $\sqrt{(3)}a/4$; but adjacent atoms have all their nearest-neighbour bonds oppositely directed. The structure could be derived by taking 8 body-centred cells, of side $a/2$, and leaving out half the atoms systematically. The C—C bonds are at the tetrahedral angle of $109\frac{1}{2}°$.

5. *Rock Salt*, NaCl
   Cubic; all face-centred lattice F
   $a = 563$ pm
   4 Na at 0, 0, 0
   4 Cl at $\frac{1}{2}$, 0, 0
   With an alternative choice of origin, the position coordinates of Na and Cl can be interchanged.

Each atom has 6 neighbours of the opposite kind, in directions parallel to the cube edges, at a distance $a/2$.

6. *Caesium Chloride*, CsCl
   Cubic; primitive lattice P
   $a = 411$ pm
   1 Cs at 0, 0, 0
   1 Cl at $\frac{1}{2}, \frac{1}{2}, \frac{1}{2}$
   With an alternative choice of origin, the position coordinates of Cs and Cl can be interchanged.

   Each atom has 8 neighbours of the opposite kind, in directions parallel to the cube body diagonal, at a distance of $\sqrt{(3)}a/2$.

   The structure is closely related to that of iron, to which it would reduce if Cs and Cl were replaced by indistinguishable atoms.

7. *Fluorite*, CaF_2
   Cubic; all-face-centred lattice F
   $a = 545$ pm
   4 Ca at 0, 0, 0
   8 F at $\pm\left(\frac{1}{4}, \frac{1}{4}, \frac{1}{4}\right)$
   Each Ca atom has 8 equidistant F neighbours, at the corners of a cube; each F atom has 4 equidistant Ca neighbours, at the vertices of a regular tetrahedron. The Ca–F distance is $\sqrt{(3)}a/4$.

8. *Zinc blende*, ZnS
   Cubic; all-face-centred lattice F
   $a = 542$ pm
   4 Zn at 0, 0, 0
   4 S at $\frac{1}{4}, \frac{1}{4}, \frac{1}{4}$
   Each Zn atom is surrounded by 4 equidistant S atoms, at the corners of a regular tetrahedron, at a distance $\sqrt{(3)}a/4$; similarly, each S atom is surrounded tetrahedrally by 4 Zn atoms. All the Zn—S bonds lying parallel to a given body diagonal of the cube point in the same direction.

   The structure is closely related to that of diamond, to which it would reduce if Zn and S were replaced by indistinguishable atoms.

   Alternatively, the structure may be described using hexagonal axes of reference, chosen as for copper (q.v.); the unit cell is rhombohedrally centred, and contains 3 atoms of each kind, with Zn at 0, 0, 0 and S at 0, 0, $\frac{1}{4}$ (or by reversing the sense of the c-axis and changing the origin these coordinates may be interchanged).

   Unlike any of the structures previously described, this has no centre of symmetry.

9. *Wurtzite*, ZnS
   Hexagonal; primitive lattice P
   $a = 382$ pm, $c = 626$ pm
   2 Zn at $\frac{2}{3}, \frac{1}{3}, 0$; $\frac{1}{3}, \frac{2}{3}, \frac{1}{2}$
   2 S at $\frac{2}{3}, \frac{1}{3}, u$; $\frac{1}{3}, \frac{2}{3}, \frac{1}{2}+u$; with $u \simeq \frac{3}{8}$
   As in the zinc-blende structure, each Zn atom is surrounded by 4 S atoms at the corners of a tetrahedron, and each S similarly by 4 Zn atoms. The tetrahedra are, however, only regular if c/a and u have the ideal values of 1.633 and 0.375 respectively. The Zn—S bonds parallel to the c-axis all point in the same direction; in contrast to the zinc-blende structure, this is a unique axis, and the symmetry is therefore *polar*.

   The relation between the structures of wurtzite and zinc blende is the same as that between the structures of magnesium and copper, i.e. between the operations of hexagonal close packing and cubic close packing.

   Intermediates between the wurtzite and zinc-blende structure, sometimes of considerable complexity ('polytypes'), are found in a number of materials, including ZnS itself and SiC (carborundum).

10. *Rutile*, TiO_2

Tetragonal; primitive lattice P

$a = 459.4$ pm, $c = 296.2$ pm

2 Ti at o, o, o and $\frac{1}{2}, \frac{1}{2}, \frac{1}{2}$

4 O at $\pm (u, u, o)$ and $\pm (\frac{1}{2}+u, \frac{1}{2}-u, \frac{1}{2})$, with $u = 0.31$

Each Ti atom has as neighbours 6 O atoms, forming a nearly regular octahedron. Edges of the octahedra parallel to face-diagonals of the square base are shared with similar octahedra, forming chains parallel to the c-axis; octahedra in neighbouring chains share corners, making each O atom neighbour to 3 Ti atoms.

11. *Ideal perovskite: example*, $SrTiO_3$

('Perovskite' is the mineral name of $CaTiO_3$, whose actual structure, though it approximates to that of $SrTiO_3$, is more complicated and of lower symmetry.)

Cubic; primitive lattice P

$a = 390.5$ pm

1 Ti at o, o, o

1 Sr at $\frac{1}{2}, \frac{1}{2}, \frac{1}{2}$

3 O at $\frac{1}{2}, o, o$; $o, \frac{1}{2}, o$; $o, o, \frac{1}{2}$

Each Ti atom has 6 neighbouring O atoms, at a distance $a/2$, forming a regular octahedron. Each octahedron shares each corner with one similar octahedron to form a three-dimensional framework, with cavities holding the larger Sr atoms. Each Sr atom has 12 neighbouring O atoms at a distance $\sqrt{(2)a/2}$. Each O atom is linked to 2 Ti atoms and 4 Sr atoms.

The ideal perovskite structure imposes a particular relation between ionic radii as a condition of cation–anion contact for both cations, and only occurs when this is nearly satisfied, as in $SrTiO_3$. With moderate misfit, related structures of lower symmetry are found, collectively described as 'structures of the perovskite family'. Many materials possessing such structures at room temperature have a high-temperature form with the ideal perovskite structure.

12. *Calcite*, $CaCO_3$

Rhombohedrally centred hexagonal lattice, R

$a_H = 499.0$ pm, $c_H = 1700$ pm

6 Ca at o, o, o and o, o, $\frac{1}{2}$

6 C at o, o, $\frac{1}{4}$ and o, o, $\frac{3}{4}$

18 O at $x, o, \frac{1}{4}$; $o, x, \frac{1}{4}$; $\bar{x}, \bar{x}, \frac{1}{4}$; $\bar{x}, o, \frac{3}{4}$; $o, \bar{x}, \frac{3}{4}$; $x, x, \frac{3}{4}$; with $x = 0.257$

Each Ca atom has 6 equidistant O atoms at the vertices of an octahedron. Each O atom is shared between two octahedra, and also forms one corner of an equilateral triangle, perpendicular to the c-axis, with the carbon atom at its centre. The edges of the CO_3 triangle are shorter than any of the octahedron edges. CO_3 groups in successive layers perpendicular to the c-axis are rotated through 180°.

An alternative description uses the edges of the primitive rhombohedron as axes of reference; then $a_r = 636$ pm, $\alpha = 46°$ 05′.

If the CO_3 groups were replaced by cylindrical discs, the unit cell would be half the height; the primitive rhombohedron would have $a'_r = 404$ pm, $\alpha' = 76°$ 10′.

References

(1) For a full account of crystallographic symmetry and notation, see *International Tables for X-ray Crystallography*, vol. I (ed. N. F. M. Henry and K. Lonsdale), Kynoch Press (Birmingham, England) for the International Union of Crystallography.

(2) For a classified reference book of crystal structures, with lattice parameters and position parameters, see R. W. G. Wyckoff, *Crystal Structures*, 2nd edn (Interscience, New York).

(3) For a description and discussion of numerous important structures and their symmetry, see H. D. Megaw, *Crystal Structures—A Working Approach* (Saunders, Philadelphia, Pa.).

Table of Materials

Name of structure type	*Some representative examples†*
1. Copper (A1)‡	Cu, Ag, Au, Al, Pt, Ni, Pb
2. Iron (A2)	Fe, Mo, W, Li, Na, K, Ba
3. Magnesium (A3)	Mg, Be, Gd, Rh, Zr; Zn§, Cd§
4. Diamond (A4)	C (diamond), Si, Ge, Sn (grey)
5. Rock salt (sodium chloride) (B1)	NaCl, and most alkali halides except Cs compounds; MgO, CaO, SrO, FeO, CaS, CaSe, AsSn, ThC
6. Caesium chloride (B2)	CsCl, CsBr, CsI, TlCl; many ordered intermetallic compounds such as AgMg, AuMg, BaCd
7. Fluorite (calcium fluoride) (C1)	CaF_2, BaF_2, UO_2, ThO_2, K_2O, UN_2, $AuAl_2$
8. Zinc blende (B3)	ZnS, ZnSe, BeS, CdS, GaAs, BN
9. Wurtzite (B4)	ZnS, ZnSe, ZnO, BeO, CdS, GaN
10. Rutile (C4)	TiO_2, SnO_2, MnO_2, CrO_2, MgF_2, FeF_2, MnF_2
11. Ideal perovskite	$SrTiO_3$, $KTaO_3$, $BaZrO_3$, $BaSnO_3$, $DyMnO_3$, $KMgF_3$, $KFeF_3$; also high-temperature forms of $BaTiO_3$, $KNbO_3$, $NaNbO_3$, $LaAlO_3$; also, by omission of the 12-coordinated cation, ReO_3, RhF_3, MoF_3
12. Calcite	$CaCO_3$, $MgCO_3$, $FeCO_3$, $NaNO_3$

† Unless otherwise specified, the structures listed are those found at room temperature.

‡ The type names (A1), (A2), etc. are those given to the structures in Vol. 1 of *Strukturbericht* (1920) and are still sometimes found in the literature.

§ These elements have axial ratios c/a much greater than the ideal value characteristic of close packing.

H.D.M.

2.8 Nuclear magnetic resonance and Mössbauer effect

2.8.1 Nuclear moments and magnetic resonance

The table includes all naturally occurring isotopes having non-zero spin (excluding ^{180}Ta). The magnetic moment, μ, is given in units of the nuclear magneton μ_N. The majority of values are derived from nuclear magnetic resonance (n.m.r.) measurements. The resonant frequency is given by

$$\nu = \mu H/hI$$

where H is the magnetic field at the nucleus, h is Planck's constant and I is the nuclear spin. Where the ordinary nuclear magnetic resonance of an isotope has not been observed, the nuclear moments listed were obtained from atomic beam resonance, electron spin resonance, optical spectroscopy measurements or electron–nuclear double resonance (ENDOR). In these circumstances, an approximate n.m.r. frequency is derived from the nuclear moment and spin values.

Nuclear magnetic resonance frequencies can be determined with considerable precision: the values tabulated are for an externally applied field of 1 tesla (10 kilogauss). Owing to chemical shift and shielding effects, the field at the nucleus differs slightly from the applied field and thus the n.m.r. frequencies are dependent to some extent on the chemical compound used. Occasionally, nuclear moment values obtained by ENDOR are given in preference to those from n.m.r. because of uncertainties in the chemical shifts of the compounds studied.

Determination of the true value of a nuclear magnetic moment requires a correction for shielding. This may be up to 1% for heavy elements and usually limits the accuracy of the nuclear moment values. In the table, the last decimal place given is significant.

Nuclear electric quadrupole moments are conveniently written $e.Q$, where e is the charge on the electron and the parameter Q is in units of 10^{-28} m^2 (barns). As there is no generally satisfactory method of estimating shielding effects for quadrupole interactions, corrections for these effects have not been applied. Errors of up to 30% in the tabulated values are possible from this source. Nuclei with spin $\frac{1}{2}$ have no quadrupole moment.

Where the ordinary nuclear magnetic resonance of an isotope has not been observed the techniques used to determine the moments are indicated in the table as follows:

at.—atomic beam magnetic resonance
el.—electron spin resonance
op.—optical spectroscopy
†—electron–nuclear double resonance.

The symbol aq. in the 'State' column of the table indicates that the measurement was carried out with the substance in aqueous solution. The sign of μ/μ_N and of $Q/10^{-28}$ m^2 is uncertain for those nuclides for which no sign is given.

Nuclide	I	μ/μ_N	$Q/10^{-28}$ m^2	ν/MHz	Compound	State
^{0}n	$\frac{1}{2}$	$-1.913\ 2$	—	29.1670	at.	
^{1}H	$\frac{1}{2}$	$+2.792\ 85$	—	42.5760	H_2O	liquid
^{2}H	1	$+0.857\ 45$	$+0.002\ 74$	6.5357	H_2O	liquid
^{3}H	$\frac{1}{2}$	$+2.978\ 96$	—	45.4131	H_2O	liquid
^{3}He	$\frac{1}{2}$	$-2.127\ 65$	—	32.4338	He	gas
^{6}Li	1	$+0.822\ 04$	$-0.000\ 80$	6.2655	LiCl	aq.
^{7}Li	$\frac{3}{2}$	$+3.256\ 3$	-0.045	16.5465	LiCl	aq.
^{9}Be	$\frac{3}{2}$	$-1.177\ 4$	0.02	5.9827	$BeCl_2$	aq.
^{10}B	3	$+1.800\ 5$	$+0.06$	4.574	$Na_2B_2O_4$	aq.
^{11}B	$\frac{3}{2}$	$+2.688\ 4$	$+0.04$	13.6595	$K_2B_2O_4$	aq.
^{13}C	$\frac{1}{2}$	$+0.702\ 3$	—	10.7054	CH_3I	liquid
^{14}N	1	$+0.403\ 56$	$+0.07$	3.076	HNO_3	aq.
^{15}N	$\frac{1}{2}$	$-0.283\ 05$	—	4.315	HNO_3	aq.
^{17}O	$\frac{5}{2}$	-1.893	-0.004	5.7719	H_2O	liquid
^{19}F	$\frac{1}{2}$	$+2.628\ 35$	—	40.0543	HF	aq.
^{21}Ne	$\frac{3}{2}$	$-0.661\ 7$	$+0.09$	3.3611	at.	—
^{23}Na	$\frac{3}{2}$	$+2.217\ 4$	$+0.1$	11.2621	NaCl	aq.
^{25}Mg	$\frac{5}{2}$	$-0.855\ 3$	$+0.2$	2.606	$MgCl_2$	aq.
^{27}Al	$\frac{5}{2}$	$+3.641\ 2$	$+0.15$	11.0940	$AlCl_3$	aq.
^{29}Si	$\frac{1}{2}$	$-0.555\ 2$	—	8.458	cobalt glass	solid
^{31}P	$\frac{1}{2}$	$+1.131\ 7$	—	17.238	H_3PO_4	aq.
^{33}S	$\frac{3}{2}$	$+0.643\ 4$	-0.06	3.266	CS_2	liquid
^{35}Cl	$\frac{3}{2}$	$+0.821\ 8$	-0.078	4.171	$CaCl_2$	aq.
^{36}Cl	2	$+1.285\ 3$	-0.017	4.893	HCl	aq.
^{37}Cl	$\frac{3}{2}$	$+0.684\ 1$	-0.062	3.472	$CaCl_2$	aq.
^{39}K	$\frac{3}{2}$	$+0.391\ 4$	$+0.1$	1.9864	HCOOK	aq.
^{40}K	4	-1.298	-0.006	2.470	at.	—
^{41}K	$\frac{3}{2}$	$+0.214\ 9$	0.1	1.0903	HCOOK	aq.
^{43}Ca	$\frac{7}{2}$	-1.317	—	2.8654	$CaCl_2$	aq.
^{45}Sc	$\frac{7}{2}$	$+4.756$	-0.22	10.3434	$ScCl_3$	aq.
^{47}Ti	$\frac{5}{2}$	-0.788	—	2.3997	$TiCl_4$	liquid
^{49}Ti	$\frac{7}{2}$	-1.104	—	2.4004	$TiCl_4$	liquid
^{50}V	6	$+3.345$	0.01	4.243	$NaVO_3$	aq.
^{51}V	$\frac{7}{2}$	$+5.147$	0.007	11.1922	$NaVO_3$	aq.
^{53}Cr	$\frac{3}{2}$	-0.474	-0.03	2.4063	Na_2CrO_4	aq.

Nuclear moments and magnetic resonance (*contd*)

Nuclide	I	μ/μ_N	$Q/10^{-28}\ m^2$	ν/MHz	Compound	State
^{55}Mn	$\frac{5}{2}$	+3.450†	+0.3	10.5542	$KMnO_4$	aq.
^{57}Fe	$\frac{1}{2}$	+0.090 5†	—	1.38	—	—
^{59}Co	$\frac{7}{2}$	+4.627	+0.5	10.072	CoSi	solid
^{61}Ni	$\frac{3}{2}$	−0.748†	—	3.8048	$Ni(CO)_4$	liquid
^{63}Cu	$\frac{3}{2}$	+2.226	−0.16	11.285	CuCl	solid
^{65}Cu	$\frac{3}{2}$	+2.386	−0.15	12.090	CuCl	solid
^{67}Zn	$\frac{5}{2}$	+0.876	+0.16	2.663	$ZnSO_4$	aq.
^{69}Ga	$\frac{3}{2}$	+2.016	+0.19	10.2188	$GaCl_3$	aq.
^{71}Ga	$\frac{3}{2}$	+2.562	+0.12	12.9840	$GaCl_3$	aq.
^{73}Ge	$\frac{9}{2}$	−0.879	−0.2	1.485	$GeCl_4$	liquid
^{75}As	$\frac{3}{2}$	+1.439	+0.3	7.292	Na_3AsO_4	aq.
^{77}Se	$\frac{1}{2}$	+0.534	—	8.118	H_2Se	aq.
^{79}Br	$\frac{3}{2}$	+2.106	+0.33	10.669	NaBr	aq.
^{81}Br	$\frac{3}{2}$	+2.269	+0.28	11.498	NaBr	aq.
^{83}Kr	$\frac{9}{2}$	−0.970	+0.15	1.6380	Kr	gas
^{85}Rb	$\frac{5}{2}$	+1.352	+0.29	4.111	RbCl	aq.
^{87}Rb	$\frac{3}{2}$	+2.750	+0.14	13.932	RbCl	aq.
^{87}Sr	$\frac{9}{2}$	−1.093	—	1.8451	$SrBr_2$	aq.
^{89}Y	$\frac{1}{2}$	−0.1373	—	2.0860	YCl_3	aq.
^{91}Zr	$\frac{5}{2}$	−1.303	—	3.9578	$(NH_4)_2ZrF_6$	aq.
^{93}Nb	$\frac{9}{2}$	+6.17	−0.2	10.4048	$Nb_2O_5 + HF$	aq.
^{95}Mo	$\frac{5}{2}$	−0.913	0.12	2.774	K_2MoO_4	aq.
^{97}Mo	$\frac{5}{2}$	−0.932	1.1	2.832	K_2MoO_4	aq.
^{99}Tc	$\frac{9}{2}$	+5.68	+0.3	9.5832	NH_4TcO_4	aq.
^{99}Ru	$\frac{5}{2}$	−0.6	—	1.9	op.	—
^{101}Ru	$\frac{5}{2}$	−0.7	—	2.1	op.	—
^{103}Rh	$\frac{1}{2}$	−0.0883	—	1.3434	Rh	metal
^{105}Pd	$\frac{5}{2}$	−0.642	—	1.8921	Pd	metal
^{107}Ag	$\frac{1}{2}$	−0.1135	—	1.7230	$AgNO_3$	aq.
^{109}Ag	$\frac{1}{2}$	−0.1305	—	1.9808	$AgNO_3$	aq.
^{111}Cd	$\frac{1}{2}$	−0.5950	—	9.0283	$CdCl_2$	aq.
^{113}Cd	$\frac{1}{2}$	−0.6225	—	9.4441	$CdCl_2$	aq.
^{113}In	$\frac{9}{2}$	+5.524	+0.75	9.312	$In(NO_3)_3$	aq.
^{115}In	$\frac{9}{2}$	+5.536	+0.76	9.331	$In(NO_3)_3$	aq.
^{115}Sn	$\frac{1}{2}$	−0.918	—	13.922	$SnCl_2$	aq.
^{117}Sn	$\frac{1}{2}$	−1.000	—	15.168	$SnCl_2$	aq.
^{119}Sn	$\frac{1}{2}$	−1.046	—	15.868	$SnCl_2$	aq.
^{121}Sb	$\frac{5}{2}$	+3.360	−0.5	10.192	$NaSbF_6$	aq.
^{123}Sb	$\frac{7}{2}$	+2.547	−0.7	5.519	$NaSbF_6$	aq.
^{123}Te	$\frac{1}{2}$	−0.736	—	11.159	$TeO_2 + HCl$	aq.
^{125}Te	$\frac{1}{2}$	−0.887	—	13.453	$TeO_2 + HCl$	aq.
^{127}I	$\frac{5}{2}$	+2.809	−0.82	8.517	KI	aq.
^{129}I	$\frac{7}{2}$	+2.618	−0.49	5.669	$I + N_2H_4$	aq.
^{129}Xe	$\frac{1}{2}$	−0.7770	—	11.7779	Xe	gas
^{131}Xe	$\frac{3}{2}$	+0.6907	−0.12	3.4902	Xe	gas
^{133}Cs	$\frac{7}{2}$	+2.579	−0.003	5.5846	CsCl	aq.
^{135}Ba	$\frac{3}{2}$	+0.8372	—	4.2295	$BaCl_2$	aq.
^{137}Ba	$\frac{3}{2}$	+0.9366	—	4.7314	$BaCl_2$	aq.

Nuclear moments and magnetic resonance (*contd*)

Nuclide	I	μ/μ_N	$Q/10^{-28}$ m^2	v/MHz	Compound	State
^{138}La	5	+3.706	0.9	5.617	LaCl$_3$	aq.
^{139}La	$\frac{7}{2}$	+2.778	+0.27	6.014	LaCl$_3$	aq.
^{141}Pr	$\frac{5}{2}$	+4.28	−0.06	13.0	at.	—
^{143}Nd	$\frac{7}{2}$	−1.064	−0.48	2.303	at.	—
^{145}Nd	$\frac{7}{2}$	−0.653	−0.25	1.414	at.	—
^{147}Sm	$\frac{7}{2}$	−0.796	−0.29	1.72	at.	—
^{149}Sm	$\frac{7}{2}$	−0.643	+0.06	1.39	at.	—
^{151}Eu	$\frac{5}{2}$	+3.419	—	10.35	at.	—
^{153}Eu	$\frac{5}{2}$	+1.507	—	4.56	at.	—
^{155}Gd	$\frac{3}{2}$	−0.28	—	1.4	el.	—
^{157}Gd	$\frac{3}{2}$	−0.38	—	1.9	el.	—
^{159}Tb	$\frac{3}{2}$	1.90	+1.32	9.7	el.	—
^{161}Dy	$\frac{5}{2}$	−0.455	+1.35	1.39	el.	—
^{163}Dy	$\frac{5}{2}$	+0.635	+1.62	1.94	el.	—
^{165}Ho	$\frac{7}{2}$	+4.01	+2.82	8.7	at.	—
^{167}Er	$\frac{7}{2}$	−0.564	+2.82	1.22	at.	—
^{169}Tm	$\frac{1}{2}$	−0.229	—	3.47	at.	—
^{171}Yb	$\frac{1}{2}$	+0.4926	—	7.456	YbCl$_2$	solid (4 K)
^{173}Yb	$\frac{5}{2}$	−0.677	+2.4	2.050	el.	—
^{175}Lu	$\frac{7}{2}$	+2.23	+5.6	4.8189	LuSb	solid
^{176}Lu	7	3.14	7.0	3.39	op.	—
^{177}Hf	$\frac{7}{2}$	+0.61	+3.0	1.3	op.	—
^{179}Hf	$\frac{9}{2}$	−0.47	+3.0	0.8	op.	—
^{181}Ta	$\frac{7}{2}$	+2.35	+4.0	5.096	KTaO$_3$	solid
^{183}W	$\frac{1}{2}$	+0.117	—	1.7715	WF$_6$	liquid
^{185}Re	$\frac{5}{2}$	+3.17	+2.8	9.586	NaReO$_4$	aq.
^{187}Re	$\frac{5}{2}$	+3.20	+2.6	9.684	NaReO$_4$	aq.
^{187}Os	$\frac{1}{2}$	+0.066	—	1.0	op.	—
^{189}Os	$\frac{3}{2}$	+0.656	+2.0	3.304	OsO$_4$	liquid
^{191}Ir	$\frac{3}{2}$	+0.2	+1.0	0.9	op.	—
^{193}Ir	$\frac{3}{2}$	+0.2	+1.0	0.9	op.	—
^{195}Pt	$\frac{1}{2}$	+0.602	—	9.1523	H$_2$PtCl$_6$	aq.
^{197}Au	$\frac{3}{2}$	+0.145†	+0.6	0.7412	Au	metal
^{199}Hg	$\frac{1}{2}$	+0.504	—	7.612	HgNO$_3$	aq.
^{201}Hg	$\frac{3}{2}$	−0.558	+0.5	2.8781	Hg	metal
^{203}Tl	$\frac{1}{2}$	+1.612	—	24.332	Tl(C$_2$H$_3$O$_2$)$_2$	aq.
^{205}Tl	$\frac{1}{2}$	+1.627	—	24.567	Tl(C$_2$H$_3$O$_2$)$_2$	aq.
^{207}Pb	$\frac{1}{2}$	+0.589	—	8.898	Pb(C$_2$H$_3$O$_2$)$_2$	aq.
^{209}Bi	$\frac{9}{2}$	+4.08	−0.4	6.842	Bi(NO$_3$)$_3$	aq.
^{235}U	$\frac{7}{2}$	0.34	4.0	0.75	el.	—

References

(1) A. Abragam, *The Principles of Nuclear Magnetism*, 1961 (Oxford University Press).
(2) B. Bleaney, 'Nuclear moments of the lanthanons', *Quantum Electronics*, 1964, **3**, 595–611 (Columbia University Press).
(3) *N.M.R. Table*, 5th edn, 1965 (Varian Associates).

L.E.D.

2.8.2 Proton chemical shifts of organic compounds

(Selected from data compiled by Dr G. V. D. Tiers, Minnesota Mining and Manufacturing Co.)

The chemical shift scale used here is defined by $10^6(H_{TMS}-H)/H_{TMS}$, where H and H_{TMS} are the magnetic field strengths corresponding to resonance for a particular nucleus in the sample and for the protons in tetramethylsilane respectively. Thus the chemical shift is dimensionless, and is expressed in parts per million (p.p.m.) with that of tetramethylsilane defined as zero.

The values tabulated were obtained from measurements on dilute solutions in carbon tetrachloride containing a little tetramethylsilane as reference substance.

The compounds are listed in order of increasing value of the chemical shift. The chemical formula of each compound is shown and the particular hydrogen nucleus having the value of chemical shift listed is indicated by bold type or by the numbering of the hydrogen nucleus given in brackets.

Compound	Formula	Shift, p.p.m.
Tetramethylsilane	$(\mathbf{CH_3})_4Si$	0.00
Tetramethyltin	$(\mathbf{CH_3})_4Sn$	0.06
Cyclopropane	$C_3\mathbf{H_6}$	0.22
s-Butylbenzene	$C_6H_5CH(CH_3)CH_2\mathbf{CH_3}$	0.80
Isopentane	$(CH_3)_2CHCH_2\mathbf{CH_3}$	0.86
Octyl bromide	$\mathbf{CH_3}(CH_2)_7Br$	0.88
Isobutane	$(\mathbf{CH_3})_3CH$	0.89
Hexadecane	$\mathbf{CH_3}(CH_2)_{14}CH_3$	0.89
Heptane	$\mathbf{CH_3}(CH_2)_5CH_3$	0.89
Hexane	$\mathbf{CH_3}(CH_2)_4CH_3$	0.90
Hexyl fluoride	$\mathbf{CH_3}(CH_2)_5F$	0.90
Methylcyclohexane	$C_6H_{11}\mathbf{CH_3}$	0.92
Butylamine	$\mathbf{CH_3}(CH_2)_3NH_2$	0.93
Neopentane	$(\mathbf{CH_3})_4C$	0.94
Triethylamine	$(\mathbf{CH_3}CH_2)_3N$	0.95
Butyl iodide	$\mathbf{CH_3}(CH_2)_3I$	0.96
Butyl bromide	$\mathbf{CH_3}(CH_2)_3Br$	0.96
Butyl chloride	$\mathbf{CH_3}(CH_2)_3Cl$	0.96
s-Butyl chloride	$\mathbf{CH_3}CHClCH_2CH_3$	0.98
Isobutyl iodide	$(\mathbf{CH_3})_2CHCH_2I$	0.98
Dipropyl sulphide	$(\mathbf{CH_3}CH_2CH_2)_2S$	1.00
Isobutyl bromide	$(\mathbf{CH_3})_2CHCH_2Br$	1.04
3-Pentanone	$(\mathbf{CH_3}CH_2)_2CO$	1.04
Propyl iodide	$\mathbf{CH_3}CH_2CH_2I$	1.04
Propyl bromide	$\mathbf{CH_3}CH_2CH_2Br$	1.04
Propyl chloride	$\mathbf{CH_3}CH_2CH_2Cl$	1.04
t-Butylamine	$(\mathbf{CH_3})_3CNH_2$	1.08
2-Methylpropanal	$(\mathbf{CH_3})_2CHCHO$	1.12
Diethyl ether	$(\mathbf{CH_3}CH_2)_2O$	1.16
Ethanol	$\mathbf{CH_3}CH_2OH$	1.17
t-Butyl hydroperoxide	$(\mathbf{CH_3})_3CO_2H$	1.19
Triethoxymethane	$(\mathbf{CH_3}CH_2O)_2CH$	1.19
Ethylbenzene	$C_6H_5CH_2\mathbf{CH_3}$	1.20
Ethyl acetate	$CH_3COOCH_2\mathbf{CH_3}$	1.21
t-Butyl alcohol	$(\mathbf{CH_3})_3COH$	1.22
s-Butylbenzene	$C_6H_5CH(\mathbf{CH_3})CH_2CH_3$	1.23
Cumene	$C_6H_5CH(\mathbf{CH_3})_2$	1.23
Pivalic acid	$(\mathbf{CH_3})_3CCOOH$	1.23
Hexadecane	$CH_3(\mathbf{CH_2})_{14}CH_3$	1.25
Heptane	$CH_3(\mathbf{CH_2})_5CH_3$	1.25
Hexane	$CH_3(\mathbf{CH_2})_4CH_3$	1.25

Proton chemical shifts of organic compounds (*contd*)

Compound	Formula	Shift, p.p.m.
1,2-Epoxypropane	$CH_3\overline{CHCH_2O}$	1.28
Methylcyclobutane	$C_4H_7CH_3$	1.30
Ethyl fumarate	$(CHCOOCH_2CH_3)_2$	1.31
Ethyl maleate	$(CHCOOCH_2CH_3)_2$	1.31
Ethanethiol	CH_3CH_2SH	1.31
t-Butylbenzene	$C_6H_5C(CH_3)_3$	1.32
p-Diethoxybenzene	$C_6H_4(OCH_2CH_3)_2$	1.36
t-Butanethiol	$(CH_3)_3CSH$	1.39
Ethyl isothiocyanate	CH_3CH_2NCS	1.40
Decahydronaphthalene	$C_{10}H_8$	1.41
Cyclohexane	C_6H_{12}	1.44
Piperidine	$(CH_2)_5NH$ (3,4,5)	1.49
Cyclopentane	C_5H_{10}	1.50
Cycloheptane	C_7H_{14}	1.53
Isopropyl chloride	$(CH_3)_2CHCl$	1.54
Methylcyclohexane	$C_6H_{11}CH_3$	1.54
Bromocyclohexane	$C_6H_{11}Br$ (3,4,5)	1.54
Chlorocyclohexane	$C_6H_{11}Cl$ (3,4,5)	1.55
Isobutane	$(CH_3)_3CH$	1.56
Isobutyl iodide	$(CH_3)_2CHCH_2I$	1.56
Dipropyl sulphide	$(CH_3CH_2CH_2)_2S$	1.56
2-Nitropropane	$(CH_3)_2CHNO_2$	1.57
Isobutyl chloride	$(CH_3)_2CHCH_2Cl$	1.58
Tetrahydropyran	$C_5H_{10}O$ (3,4,5)	1.58
t-Butyl chloride	$(CH_3)_3CCl$	1.58
1-Methylcyclohexene	$C_6H_9CH_3$	1.60
1-Methylcyclohexene	$C_6H_9CH_3$ (4,5)	1.60
Pyrrolidine	C_4H_9N (3,4)	1.62
2-Methylbut-2-ene	$(CH_3)_2C{=}CH(CH_3)$	1.63
Ethyl bromide	CH_3CH_2Br	1.65
Butyl chloride	$CH_3CH_2CH_2CH_2Cl$	1.66
Cycloheptanone	$C_7H_{12}O$ (3,4,5,6)	1.66
Butyl iodide	$CH_3CH_2CH_2CH_2I$	1.67
s-Butylbenzene	$C_6H_5CH(CH_3)CH_2CH_3$	1.70
Butyl bromide	$CH_3CH_2CH_2CH_2Br$	1.70
Isobutene	$CH_2{=}C(CH_3)_2$	1.70
2-Bromo-octane	$C_6H_{13}CHBrCH_3$	1.71
Isopropyl bromide	$(CH_3)_2CHBr$	1.71
t-Butyl bromide	$(CH_3)_3CBr$	1.76
Tetrahydrofuran	C_4H_8O (3,4)	1.79
Isobutyl bromide	$(CH_3)_2CHCH_2Br$	1.80
Propyl chloride	$CH_3CH_2CH_2Cl$	1.83
1-Methylcyclohexene	$C_6H_9CH_3$ (3,6)	1.86
Propyl iodide	$CH_3CH_2CH_2I$	1.86
Cyclohexanone	$C_6H_{10}O$ (3,4,5)	1.87
Chlorocyclohexane	$C_6H_{11}Cl$ (2,6)	1.87
Isopropyl iodide	$(CH_3)_2CHI$	1.88
Propyl bromide	$CH_3CH_2CH_2Br$	1.89
Bromocyclohexane	$C_6H_{11}Br$ (2,6)	1.93
Bromocyclopentane	C_5H_9Br (2,3,4,5)	1.95
Ethyl acetate	$CH_3COOCH_2CH_3$	1.95
Cyclohexene	C_6H_{10} (3,6)	1.96
1,4-Dichlorobutane	$ClCH_2CH_2CH_2CH_2Cl$	1.96

Proton chemical shifts of organic compounds (*contd*)

Compound	Formula	Shift, p.p.m.
Acetonitrile	CH_3CN	1.97
Methyl acetate	CH_3COOCH_3	2.01
2-Butenal	$CH_3CH{=}CHCHO$ (*trans*)	2.02
Cyclopentanone	C_5H_8O (2,3,4,5)	2.02
1,4-Dibromobutane	$BrCH_2CH_2CH_2CH_2Br$	2.03
Acetic acid	CH_3COOH	2.07
Dimethyl sulphide	$(CH_3)_2S$	2.08
Acetone	$(CH_3)_2CO$	2.09
Trimethylamine	$(CH_3)_3N$	2.12
1,2,4,5-Tetramethylbenzene	$(CH_3)_4C_6H_2$	2.14
2-Pyrrolidone	C_4H_7NO (4)	2.14
Methyl iodide	CH_3I	2.16
Acetaldehyde	CH_3CHO	2.17
N,N-Dimethylbenzylamine	$C_6H_5CH_2N(CH_3)_2$	2.17
Acetic anhydride	$(CH_3CO)_2O$	2.20
Mesitylene	$(CH_3)_3C_6H_3$	2.22
Butanedione	$CH_3COCOCH_3$	2.23
2-Pyrrolidone	C_4H_7NO (3)	2.23
o-Xylene	$C_6H_4(CH_3)_2$	2.23
2,5-Dimethylfuran	$C_4H_2O(CH_3)_2$	2.24
Cyclohexanone	$C_6H_{10}O$ (2,6)	2.25
p-Methoxytoluene	$CH_3OC_6H_4CH_3$	2.25
p-Xylene	$C_6H_4(CH_3)_2$	2.27
m-Xylene	$C_6H_4(CH_3)_2$	2.27
1,2-Epoxypropane	$CH_3\overline{CHCH_2}O$	2.29
p-Cresol	$CH_3C_6H_4OH$	2.30
p-Fluorotoluene	$CH_3C_6H_4F$	2.30
3-Bromopropyne	$HC{\equiv}CCH_2Br$	2.33
1,1-Dimethylhydrazine	$(CH_3)_2NNH_2$	2.33
Toluene	$C_6H_5CH_3$	2.34
1,3-Dibromopropane	$BrCH_2CH_2CH_2Br$	2.36
3-Pentanone	$(CH_3CH_2)_2CO$	2.39
3-Chloropropyne	$HC{\equiv}CCH_2Cl$	2.40
Cycloheptanone	$C_7H_{12}O$ (2,7)	2.42
Triethylamine	$(CH_3CH_2)_3N$	2.42
Ethanethiol	CH_3CH_2SH	2.44
Dimethyl sulphoxide	$(CH_3)_2SO$	2.50
Acetophenone	$C_6H_5COCH_3$	2.55
Ethylbenzene	$C_6H_5CH_2CH_3$	2.62
1,2,3,4-Diepoxybutane	$(\overline{CHCH_2}O)_2$	2.63
Butylamine	$C_3H_7CH_2NH_2$	2.63
1,2-Epoxypropane	$CH_3\overline{CHCH_2}O$	2.66
Piperidine	$(CH_2)_5NH$ (2,6)	2.69
1,2,3,4-Diepoxybutane	$(\overline{CHCH_2}O)_2$	2.72
Pyrrolidine	C_4H_9N (2,5)	2.74
Acetyl bromide	CH_3COBr	2.81
Cumene	$C_6H_5CH(CH_3)_2$	2.87
Isopropylamine	$(CH_3)_2CHNH_2$	2.87
Dibenzyl	$C_6H_5CH_2CH_2C_6H_5$	2.87
1,3-Cyclopentadiene	C_5H_6 (5)	2.90
Ethynylbenzene	$C_6H_5C{\equiv}CH$	2.93

Proton chemical shifts of organic compounds (*contd*)

Compound	Formula	Shift, p.p.m.
Methyl chloride	CH_3Cl	3.05
p-Dimethylaminobenzaldehyde	$(CH_3)_2NC_6H_4CHO$	3.05
Isobutyl iodide	$(CH_3)_2CHCH_2I$	3.07
Ethyl iodide	CH_3CH_2I	3.15
Butyl iodide	$C_3H_7CH_2I$	3.19
Propyl iodide	$C_2H_5CH_2I$	3.20
Dimethyl ether	$(CH_3)_2O$	3.24
Isobutyl bromide	$(CH_3)_2CHCH_2Br$	3.26
Dimethoxymethane	$(CH_3O)_2CH_2$	3.28
N,N-Dimethylbenzylamine	$C_6H_5CH_2N(CH_3)_2$	3.32
Acenaphthene	$C_{10}H_6(CH_2)_2$	3.34
Octyl bromide	$C_7H_{15}CH_2Br$	3.35
Isobutyl chloride	$(CH_3)_2CHCH_2Cl$	3.35
Diethyl ether	$(CH_3CH_2)_2O$	3.36
Propyl bromide	$C_2H_5CH_2Br$	3.36
Ethyl bromide	CH_3CH_2Br	3.36
2-Pyrrolidone	C_4H_7NO (5)	3.37
Methanol	CH_3OH	3.38
1,4-Dibromobutane	$BrCH_2CH_2CH_2CH_2Br$	3.42
Propyl chloride	$C_2H_5CH_2Cl$	3.45
Butyl chloride.	$C_3H_7CH_2Cl$	3.46
Triethoxymethane	$(CH_3CH_2O)_3CH$	3.50
Trimethyl borate	$(CH_3O)_3B$	3.53
1,1,1-Trifluoro-2-iodoethane	CF_3CH_2I	3.56
Tetrahydropyran	$C_5H_{10}O$ (2,6)	3.56
1,4-Dichlorobutane	$ClCH_2CH_2CH_2CH_2Cl$	3.57
1,4,-Dioxan	$C_4H_8O_2$	3.57
Dimethyl sulphite	$(CH_3)_2SO_3$	3.58
1,3-Dibromopropane	$BrCH_2CH_2CH_2Br$	3.58
1,1,1,2,2,3,3-Heptafluoro-4-iodobutane . .	$C_3F_7CH_2I$	3.58
Ethanol	CH_3CH_2OH	3.59
Ethyl isothiocyanate	CH_3CH_2NCS	3.61
1,2-Dibromoethane	$BrCH_2CH_2Br$	3.63
Tetrahydrofuran	C_4H_8O (2,5)	3.63
Methanesulphonyl chloride	CH_3SO_2Cl	3.64
Methyl acetate	CH_3COOCH_3	3.65
1,2-Dichloroethane	$ClCH_2CH_2Cl$	3.69
Methyl formate	$HCOOCH_3$	3.69
p-Methoxytoluene	$CH_3OC_6H_4CH_3$	3.73
Anisole	$C_6H_5OCH_3$	3.73
Dimethyl carbonate	$CO(OCH_3)_2$	3.77
Dimethyl phthalate	$C_6H_4(COOCH_3)_2$	3.82
3-Bromopropyne	$HC{\equiv}CCH_2Br$	3.82
2,2,2-Trifluoroethanol	CF_3CH_2OH	3.85
p-Methoxybenzaldehyde	$CH_3OC_6H_4CHO$	3.86
Chlorocyclohexane	$C_6H_{11}Cl$ (1)	3.89
Methyl benzoate	$C_6H_5COOCH_3$	3.90
Methylene iodide	CH_2I_2	3.90
s-Butyl chloride	$C_2H_5CHClCH_3$	3.90
Diphenylmethane	$(C_6H_5)_2CH_2$	3.92
p-Diethoxybenzene	$C_6H_4(OCH_2CH_3)_2$	3.92
Dimethyl sulphate	$(CH_3)_2SO_4$	3.94
2-Methyl-3-chloropropene	$CH_2{=}CCH_3CH_2Cl$	3.96
1,1,2-Trichloroethane	$CHCl_2CH_2Cl$	3.97

Proton chemical shifts of organic compounds (*contd*)

Compound	Formula	Shift, p.p.m.
2-Bromo-octane	$CH_3(CH_2)_5CHBrCH_3$	4.03
Ethyl acetate	$CH_3COOCH_2CH_3$	4.05
Ferrocene	$C_{10}H_{10}Fe$	4.05
Chlorocyanomethane	$ClCH_2CN$	4.07
3-Chloropropyne	$HC{\equiv}CCH_2Cl$	4.09
Bromocyclohexane	$C_6H_{11}Br$ (1)	4.11
Isopropyl chloride	$(CH_3)_2CHCl$	4.13
Prop-1-yne-3-ol	$HC{\equiv}CCH_2OH$	4.18
Phenylsilane	$C_6H_5SiH_3$	4.19
Isopropyl bromide	$(CH_3)_2CHBr$	4.20
Isopropyl iodide	$(CH_3)_2CHI$	4.24
Methyl fluoride	CH_3F	4.26
Nitromethane	CH_3NO_2	4.28
Hexyl fluoride	$C_5H_{11}CH_2F$	4.35
Bromocyclopentane	C_5H_9Br (1)	4.39
Benzyl alcohol	$C_6H_5CH_2OH$	4.40
α,α'-Dibromo-*p*-xylene	$C_6H_4(CH_2Br)_2$	4.41
Benzyl bromide	$C_6H_5CH_2Br$	4.43
2-Nitropropane	$(CH_3)_2CHNO_2$	4.44
Dimethoxymethane	$(CH_3O)_2CH_2$	4.49
Dibenzyl ether	$(C_6H_5CH_2)_2O$	4.49
α,α'-Dichloro-*p*-xylene	$C_6H_4(CH_2Cl)_2$	4.50
Benzyl chloride	$C_6H_5CH_2Cl$	4.50
Isobutene	$CH_2{=}C(CH_3)_2$	4.60
Diphenylsilane	$(C_6H_5)_2SiH_2$	4.86
2-Methyl-3-chloroprop-1-ene	$\begin{array}{c} ClCH_2 \qquad\quad H \\ \diagdown\qquad\diagup \\ C{=}C \\ \diagup\qquad\diagdown \\ CH_3 \qquad\quad H \end{array}$	4.90
Methylene bromide	CH_2Br_2	4.94
Triethoxyethane	$(CH_3CH_2O)_3CH$	4.96
Chloroiodomethane	CH_2ClI	4.99
Trioxane	C_3H_6O	5.00
2-Methyl-3-chloroprop-1-ene	$\begin{array}{c} ClCH_2 \qquad\quad H \\ \diagdown\qquad\diagup \\ C{=}C \\ \diagup\qquad\diagdown \\ CH_3 \qquad\quad H \end{array}$	5.02
Isopropenylbenzene	$\begin{array}{c} C_6H_5 \qquad\quad H \\ \diagdown\qquad\diagup \\ C{=}C \\ \diagup\qquad\diagdown \\ CH_3 \qquad\quad H \end{array}$	5.02
Bromochloromethane	CH_2BrCl	5.16
2-Methylbut-2-ene	$(CH_3)_2C{=}CHCH_3$	5.21
Isopropenylbenzene	$\begin{array}{c} C_6H_5 \qquad\quad H \\ \diagdown\qquad\diagup \\ C{=}C \\ \diagup\qquad\diagdown \\ CH_3 \qquad\quad H \end{array}$	5.28
1-Methylcyclohexene	$C_6H_9CH_3$ (2)	5.30
Methylene chloride	CH_2Cl_2	5.33
1,1-Diphenylethylene	$(C_6H_5)_2C{=}CH_2$	5.40
Triphenylsilane	$(C_6H_5)_3SiH$	5.43

Proton chemical shifts of organic compounds (*contd*)

Compound	Formula	Shift, p.p.m.
Cyclohexene	C_6H_{10} (1,2)	5.57
Cyclo-octatetraene	C_8H_8	5.69
2,5-Dimethylfuran	$C_4H_2O(CH_3)_2$ (3,4)	5.70
1,1,2-Trichloroethane	$CHCl_2CH_2Cl$	5.74
Methyl dichloroacetate	$CHCl_2COOCH_3$	5.89
1,1-Difluoro-1,2,2-trichloroethane . . .	$CF_2ClCHCl_2$	5.92
1,1,2,2-Tetrachloroethane	$CHCl_2CHCl_2$	5.94
1,1,2,2-Tetrabromoethane	$CHBr_2CHBr_2$	6.03
2-Butenal	$CH_3CH{=}CHCHO$ (*trans*)	6.05
1,1,1,2,2-Pentachloroethane	CCl_3CHCl_2	6.05
Pyrrole	C_4H_5N (3,4)	6.09
Furan	C_4H_4O (3,4)	6.30
Cinnamic acid	$C_6H_5CH{=}CHCOOH$ (*trans*)	6.39
1,3-Cyclopentadiene	C_5H_6 (1,2,3,4)	6.42
1,1,2-Trichloroethylene	$CCl_2{=}CHCl$	6.44
Aniline	$C_6H_5NH_2$	6.59
Pyrrole	C_4H_5N (2,5)	6.60
Benzylidene chloride	$C_6H_5CHCl_2$	6.61
1,2-Dibromoethylene	$BrCH{=}CHBr$ (*trans*)	6.62
p-Dimethylaminobenzaldehyde	$(CH_3)_2NC_6H_4CHO$ (3,5)	6.63
Cinnamaldehyde	$C_6H_5CH{=}CHCHO$ (*trans*)	6.64
Mesitylene	$(CH_3)_3C_6H_3$	6.64
1,4-Diethoxybenzene	$C_6H_4(OCH_2CH_3)_2$	6.67
β-Bromostyrene	$C_6H_5CH{=}CHBr$ (*trans*)	6.71
1,2,4,5-Tetramethylbenzene	$(CH_3)_4C_6H_2$	6.74
p-Cresol	$CH_3C_6H_4OH$	6.75
p-Benzoquinone	$C_6H_4O_2$	6.76
1,2,4-Trimethylbenzene	$C_6H_3(CH_3)_3$	6.82
Tribromomethane	$CHBr_3$	6.82
2-Butenal	$CH_3CH{=}CHCHO$ (*trans*)	6.85
p-Acetoxytoluene	$CH_3COOC_6H_4CH_3$ (3,5)	6.87
m-Xylene	$C_6H_4(CH_3)_2$	6.89
Anisaldehyde	$CH_3OC_6H_4CHO$ (3,5)	6.91
p-Fluorotoluene	$C_6H_4FCH_3$ (3,5)	6.92
p-Xylene	$C_6H_4(CH_3)_2$	6.95
Stilbene	$C_6H_5CH{=}CHC_6H_5$ (*trans*)	6.99
1,2-Dibromoethylene	$BrCH{=}CHBr$ (*cis*)	7.00
o-Xylene	$C_6H_4(CH_3)_2$	7.01
p-Acetoxytoluene	$CH_3COOC_6H_4CH_3$ (2,6)	7.03
p-Fluorotoluene	$C_6H_4FCH_3$ (2,6)	7.04
Thiophen	C_4H_4S (3,4)	7.04
β-Bromostyrene	$C_6H_5CH{=}CHBr$ (*trans*)	7.05
Dibromochloromethane	$CHClBr_2$	7.06
Toluene	$C_6H_5CH_3$	7.09
Dibenzyl	$C_6H_5CH_2CH_2C_6H_5$	7.11
Ethylbenzene	$C_6H_5CH_2CH_3$	7.11
Cumene	$C_6H_5CH(CH_3)_2$	7.12
Diphenylmethane	$(C_6H_5)_2CH_2$	7.14
Benzyl alcohol	$C_6H_5CH_2OH$	7.17
t-Butylbenzene	$C_6H_5C(CH_3)_3$	7.19
Thiophen	C_4H_4S (2,5)	7.19
Bromodichloromethane	$CHCl_2Br$	7.20
3,5-Dibromotoluene	$C_6H_3Br_2CH_3$ (2,6)	7.21
β-Bromostyrene	$C_6H_5CH{=}CHBr$ (*trans*)	7.23

Proton chemical shifts of organic compounds (*contd*)

Compound	Formula	Shift, p.p.m.
1,1-Diphenylethylene	$(C_6H_5)_2C=CH_2$	7.24
Dibenzyl ether	$(C_6H_5CH_2)_2O$	7.24
p-Dichlorobenzene	$C_6H_4Cl_2$	7.24
Chlorobenzene	C_6H_5Cl	7.24
Chloroform	$CHCl_3$	7.25
Bromobenzene	C_6H_5Br	7.26
Benzyl bromide	$C_6H_5CH_2Br$	7.26
Benzene	C_6H_6	7.27
Stilbene	$C_6H_5CH=CHC_6H_5$ (*trans*)	7.27
Benzyl chloride	$C_6H_5CH_2Cl$	7.28
Ethynylbenzene	$C_6H_5C\equiv CH$	7.32
α,α'-Dichloro-*p*-xylene	$C_6H_4(CH_2Cl)_2$	7.33
p-Dibromobenzene	$C_6H_4Br_2$	7.34
Phenylsilane	$C_6H_5SiH_3$	7.37
Naphthalene	$C_{10}H_8$ (2,3,6,7)	7.37
Cinnamaldehyde	$C_6H_5CH=CHCHO$ (*trans*)	7.38
Furan	C_4H_4O (2,5)	7.38
p-Di-iodobenzene	$C_6H_4I_2$	7.38
4,4'-Dichlorobiphenyl	$ClC_6H_4C_6H_4Cl$	7.41
3,5-Dibromotoluene	$C_6H_3Br_2CH_3$ (4)	7.43
Cinnamaldehyde	$C_6H_5CH=CHCHO$ (*trans*)	7.46
Acetophenone	$C_6H_5COCH_3$	7.48
p-Chlorobenzaldehyde	ClC_6H_4CHO (3,5)	7.48
α-Trifluorotoluene	$C_6H_5CF_3$	7.51
(Perfluoropropyl)benzene	$C_6H_5C_3F_7$	7.53
Benzonitrile	C_6H_5CN	7.54
p-Bromoacetophenone	$BrC_6H_4COCH_3$ (3,5)	7.57
p-Terphenyl	$C_6H_5C_6H_4C_6H_5$	7.60
Methyl phthalate	$C_6H_4(COOCH_3)_2$	7.62
p-Dimethylaminobenzaldehyde	$(CH_3)_2NC_6H_4CHO$ (2,6)	7.63
Benzaldehyde	C_6H_5CHO	7.63
Cinnamic acid	$C_6H_5CH=CHCOOH$ (*trans*)	7.72
Naphthalene	$C_{10}H_8$ (1,4,5,8)	7.73
Anisaldehyde	$CH_3OC_6H_4CHO$ (2,6)	7.75
p-Bromoacetophenone	$BrC_6H_4COCH_3$ (2,6)	7.77
α,α'-Trifluoro-*p*-xylene	$C_6H_4(CF_3)_2$	7.77
p-Chlorobenzaldehyde	ClC_6H_4CHO (2,6)	7.78
N,*N*-Dimethylformamide	$HCON(CH_3)_2$	7.84
Terephthalaldehyde	$C_6H_4(CHO)_2$	8.02
Methyl formate	$HCOOCH_3$	8.03
p-Dinitrobenzene	$C_6H_4(NO_2)_2$	8.45
Pyridine	C_5H_5N (2,6)	8.50
2-Butenal	$CH_3CH=CHCHO$ (*trans*)	9.43
2-Methyl propanal	$(CH_3)_2CHCHO$	9.57
Cinnamaldehyde	$C_6H_5CH=CHCHO$ (*trans*)	9.63
p-Dimethylaminobenzaldehyde	$(CH_3)_2NC_6H_4CHO$	9.65
Acetaldehyde	CH_3CHO	9.72
Anisaldehyde	$CH_3OC_6H_4CHO$	9.80
p-Chlorobenzaldehyde	ClC_6H_4CHO	9.93
Benzaldehyde	C_6H_5CHO	9.96
Terephthalaldehyde	$C_6H_4(CHO)_2$	10.08

I.J.L.

2.8.3 Nuclear spin relaxation time

When a material is in thermal equilibrium, the nuclear spins are either randomly oriented or have a distribution of orientations related to the crystal structure. This equilibrium distribution of nuclear spins may be altered by various re-orienting processes, but after such a process is removed the spin orientations (and hence the nuclear magnetization of the sample) will revert to their original equilibrium distribution exponentially with time. The time constant for this reversion is known as the spin-lattice relaxation time, τ. Relaxation times are usually temperature sensitive but are not very dependent on magnetic field. In the table a magnetic field of about 1 tesla (10 kilogauss) is assumed and the temperature T, or temperature range of the observations, is listed. The relaxation times in insulating diamagnetic materials are often very much reduced if paramagnetic impurities are present.

Nuclide	Material	τ/s	T/K
^{1}H	H_2 (gas, 10 atm)	0.015	293
	H_2 (liquid, 75% ortho)	0.017	20
	H_2O (liquid)	2.8	293
	H_2O (solid)	20	263
	C_6H_6 (liquid)	15	293
	Polyethylene (solid)	0.9	295
^{3}He	^{3}He (liquid)	550	3
^{7}Li	LiF (solid)	300	293
^{19}F	LiF (solid)	120	293
^{23}Na	NaCl (solid)	12	293
^{35}Cl	NaCl (solid)	6	293
^{127}I	KI (solid)	0.04	293
^{29}Si	Si (solid)	2×10^4	293
^{69}Ga	GaAs (solid)	0.27	293
^{115}In	InSb (solid)	0.05	298
^{23}Na	Na (solid)	$4.9/T$	1–300
^{27}Al	Al (solid to liquid)	$1.82/T$	1–1000
^{63}Cu	Cu (solid)	$1.27/T$	1–300
^{195}Pt	Pt (solid)	$0.032/T$	0.5–290

Reference: A. Abragam, *The Principles of Nuclear Magnetism*, 1961 (Oxford University Press).

L.E.D.

2.8.4 The Mössbauer effect

The recoil-free resonance absorption and emission of γ-rays arising from nuclear excited states (with half-lives typically in the range 10^{-6} to 10^{-10} second) are known as the Mössbauer effect. In the optical spectrum, resonance absorption, for example the absorption of sodium light by sodium vapour, is well known, but in the γ-ray case it is not normally observed because the frequency shift and broadening of the line by the Doppler effect are much greater than the line width. However, if the emitting and absorbing atoms are tightly bound in a crystal so that the recoil energy is less than the binding energy, a fraction f of the γ-rays (known as the Lamb–Mössbauer fraction) is emitted without suffering these broadening effects.

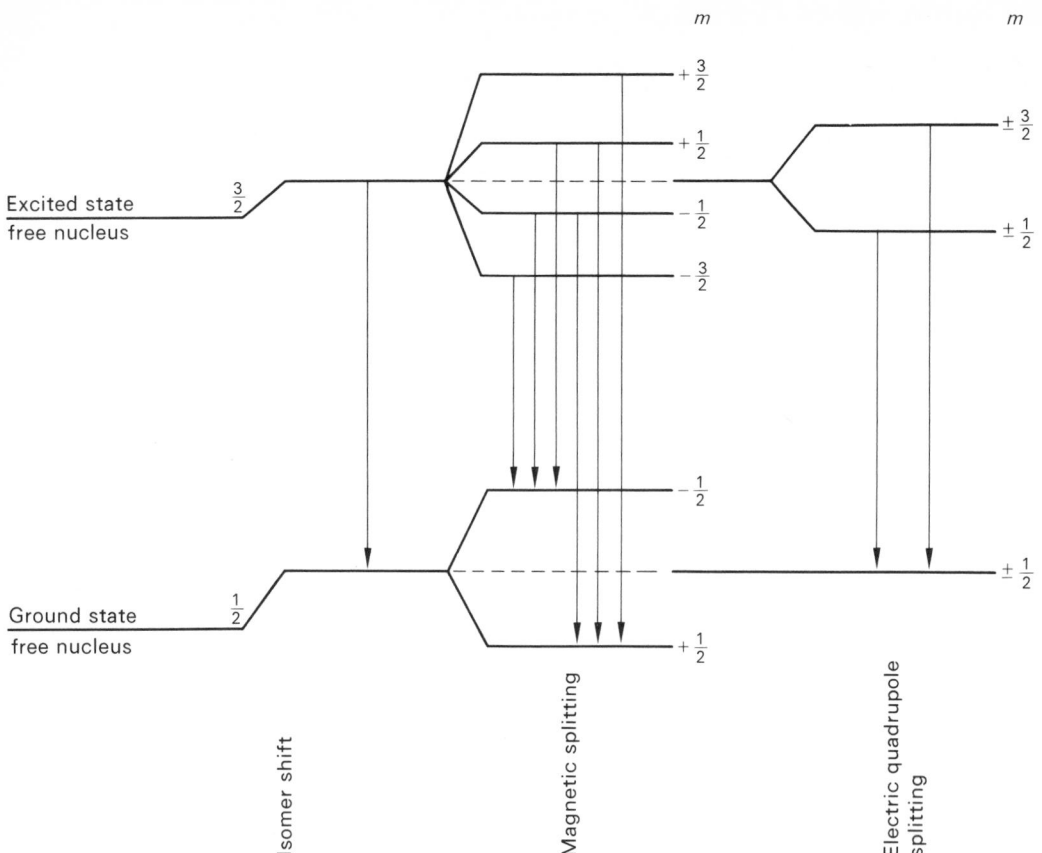

The usefulness of the Mössbauer effect results from the fact that transition energies and, hence, absorption spectra are affected by interaction between the nucleus and its surroundings—local electrons or applied magnetic fields. Mössbauer spectra are obtained by measuring the resonance absorption as a function of γ-ray energy. The variation of energy is provided by moving the source or absorber and using the Doppler effect. It is then convenient to express the effect—chemical shift or quadrupole splitting—in mm s^{-1}. The relation between Mössbauer spectrum and hyperfine interactions can be seen from the figure, which shows the allowed transitions in the case of ^{57}Fe, the most widely studied nucleus.

A change in the electronic environment of a nucleus causes changes in the energy of the ground and excited states. This results in a change in the Mössbauer spectrum known as the *chemical shift* or *isomer shift*. The shift is proportional to the electron density in the vicinity of the nucleus and can give useful information on valency states.

Interaction between the nucleus and a magnetic field can give rise to *magnetic splitting* or *hyperfine Zeeman splitting* which can be used to detect magnetic exchange interactions and to determine the magnitude and sign of the magnetic field at a selected site in a crystal. This has some similarity to the splitting observed in n.m.r., but in this case the transition is always from a magnetic sublevel of the excited state to a magnetic sublevel of the ground state, and not between levels of one state as in n.m.r.

Hyperfine interactions between nuclear quadrupoles and electric field gradients give rise to *electric quadrupole splitting* in Mössbauer spectra. Such splitting can be used to obtain information about site symmetries and field gradients in crystals.

Isotopes commonly used in Mössbauer spectroscopy

I_π and I_π^* are the spins and parities of the ground and excited states, E_γ is the γ-ray energy, $t_{1/2}$ is the half-life of the excited state, W_0 is the width of the resonance and σ_0 is the cross-section for resonance absorption.

Nuclide	I_π	I_π^*	E_γ/keV	$t_{1/2}$/ns	$W_0/(\text{mm s}^{-1})$	σ_0/mm^2
^{197}Au	$\frac{3}{2}+$	$\frac{1}{2}+$	77.3	1.8	2.0	4.4×10^{-18}
^{151}Eu	$\frac{5}{2}+$	$\frac{7}{2}+$	21.6	8.8	1.4	2.3×10^{-17}
^{57}Fe	$\frac{1}{2}-$	$\frac{3}{2}-$	14.4	97.7	0.19	2.4×10^{-16}
^{127}I	$\frac{5}{2}+$	$\frac{7}{2}+$	57.6	1.9	2.5	2.1×10^{-17}
^{121}Sb	$\frac{5}{2}+$	$\frac{7}{2}+$	37.2	3.5	2.1	2.1×10^{-17}
^{119}Sn	$\frac{1}{2}+$	$\frac{3}{2}+$	23.9	18.4	0.63	1.3×10^{-16}

To obtain the probability of absorption, one must take account of any splitting of the line by hyperfine interactions and of the factor f. When f is not nearly unity it is strongly dependent on wavelength and temperature. A typical value for ^{57}Fe γ-rays is 0.7 at room temperature, but for ^{197}Au it is about 0.1 at 4 K. Values of f much less than 0.1 make observation of the resonance difficult or impossible.

Mössbauer spectroscopy

Chemical (isomer) shifts, internal magnetic field (H_{int}) in ferromagnetic materials and quadrupole splittings for the isotopes listed in the preceding table.

Nuclide	Absorbing substance	$\dfrac{Temp}{\text{K}}$	$\dfrac{\textit{Chemical shift}}{\text{mm s}^{-1}}$	$\dfrac{H_{\text{int}}}{\text{T}}$	$\dfrac{\textit{Quadrupole splitting}}{\text{mm s}^{-1}}$
^{197}Au	Pt Au		0		0
	Fe Au dilute alloy	4.2	4.4	128	0
	Au$_2$Mn	4.2	1.5	163	1.45
	Ni Au dilute alloy	4.2	3.8	29	0
	AuCl	4.2	-1.42	0	4.65
^{151}Eu	EuF$_3$		0		⎫
	Eu	4.2	-7.6	-25.7	⎪
	EuS	4.2	-11.8	-31.4	⎪
	EuSO$_4$	0.054	0	32.4	⎬ †
		300	-14.0	0	⎪
	EuCl$_2$	0.054	0	31.5	⎪
		300	-13.0	0	⎪
	EuF$_2$	300	-15.0	0	⎪
	Eu$_2$(SO$_4$)$_3$	300	0.35	0	⎭
^{57}Fe	Fe metal	293	0	-33.0	0
		4.2		-33.8	0
	αFe$_2$O$_3$	293	0.36	-51.7	†
	Na$_2$(Fe(CN)$_3$NO)$_2$H$_2$O	293	0.257	0	1.712
	K$_4$Fe(CN)$_6$3H$_2$O	293	0.045	0	0
	FeF$_2$	298	1.36	0	2.79
		4.2	1.47	33	2.85
	FeF$_3$	298	0.49	0	-0.04
		4.2	0.56	62	-0.04

Mössbauer spectroscopy (*contd*)

Nuclide	Absorbing substance	$\dfrac{Temp}{K}$	$\dfrac{Chemical\ shift}{mm\ s^{-1}}$	$\dfrac{H_{int}}{T}$	$\dfrac{Quadrupole\ splitting}{mm\ s^{-1}}$
^{127}I	ZnTe(I)		0		
	$KICl_4$	4.2	-1.4	0	
	ICl	4.2	-0.6	0	
	KIO_4	4.2	0.7	0	$\left.\right\}$ †
	KI	4.2	0.14	0	
^{121}Sb	InSb		0		
	$KSbF_6$	77	12.3	0	
	Sb_2O_5	77	9.9	0	
	Sb_2O_3	77	-3.02	0	$\left.\right\}$ †
	SbF_3	77	-6.0	0	
^{119}Sn	βSn		0		
	SnO_2	77	-2.8	0	0.45
	SnO	77	0.0	0	1.4
	SnF_2	77	0.8	0	1.67
	SnF_4	77	-3.1	0	1.7
	$SnCl_2$	77	1.8	0	0
	$(CH_3)_4Sn$	77	-1.6	0	0
	$(CH_2)_3SnCH_2COCH_3$	77	-1.7	0	2.1
	FeSn dilute alloy	77	-1.1	-8.5	0

† When either of the spins is greater than $\frac{3}{2}$, or there is a magnetic field plus an electric field gradient, the line splitting is more complicated and cannot be represented by a single splitting parameter.

References

(1) G. K. Wertheim, *Mössbauer Effect: Principles and Applications*, 1964 (Academic Press, New York).
(2) N. N. Greenwood and T. C. Gibb, *Mössbauer Spectroscopy*, 1971 (Chapman and Hall).
(3) J. G. and V. E. Stevens, *Mössbauer Effect Data Index*, 1970 (Adam Hilger Ltd, London).

T.E.C.

2.9 Electrochemistry

2.9.1 Standard solutions for calibrating conductivity vessels

The following table is from Jones and Bradshaw, *J. Am. Chem. Soc.*, 1933, **55**, 1799. The unit of conductivity κ is $\Omega^{-1}\ cm^{-1}$. The original data have been converted from (int. ohm)$^{-1}$ cm^{-1} where 1 int. ohm = 1.000 49 ohm. It is still common practice not to do this, and conductivity results are therefore usually quoted in (int. ohm)$^{-1}$ cm^{-1} units. The experimental values of conductivity have been corrected for that of the water used.

Potassium chloride solutions

g KCl per kg solution in vacuo	$\kappa/(\Omega^{-1}\ cm^{-1})$		
	0 °C	18 °C	25 °C
71.135 2	0.065 14$_4$	0.097 79$_0$	0.111 28$_7$
7.419 13	0.007 134$_4$	0.011 161$_2$	0.012 849$_7$
0.745 263	0.000 773 2$_6$	0.001 219 9$_2$	0.001 408 0$_8$

The following equation, which gives the molar conductivity in $\Omega^{-1}\ cm^{-1}$ of aqueous potassium chloride solution at 25 °C in the range $0.01\ mol\ dm^{-3} < c < 0.10\ mol\ dm^{-3}$, is also useful for cell calibration (Ying-Chech Chiu and Fuoss, *J. Phys. Chem.*, 1968, **72**, 4123):

$$\Lambda = 149.86 - 94.83c + 25.47c\ \ln c + 220.9c - 229c^{3/2}$$

J.E.P.

2.9.2 Conductivities

Molar conductivities of electrolytes in aqueous solution at 25 °C

The unit of Λ in the table is $\Omega^{-1}\ cm^2\ mol^{-1}$. In the following table the entities to which the mole relates are given in the first column, e.g. $\frac{1}{2}MgCl_2$ (not $MgCl_2$). The values are taken by permission from Harned and Owen, *Physical Chemistry of Electrolytic Solutions*, 2nd edn, 1950, Appendix A, p. 537 (Reinhold, New York).

Electrolyte	$c/(\text{mol dm}^{-3})$							
	0	0.0005	0.001	0.005	0.01	0.02	0.05	0.1
HCl	425.95	422.53	421.15	415.60	411.80	407.04	398.89	391.13
LiCl	114.97	113.09	112.34	109.34	107.27	104.60	100.06	95.81
NaCl	126.39	124.44	123.68	120.59	118.45	115.70	111.01	106.69
KCl	149.79	147.74	146.88	143.48	141.20	138.27	133.30	128.90
NH_4Cl	149.6	—	—	—	141.21	138.26	133.26	128.69
KBr	151.8	—	—	146.02	143.36	140.41	135.61	131.32
NaI	126.88	125.30	124.19	121.19	119.17	116.64	112.73	108.73
KI	150.31	—	—	144.30	142.11	139.38	134.90	131.05
KNO_3	144.89	142.70	141.77	138.41	135.75	132.34	126.25	120.34
$KHCO_3$	117.94	116.04	115.28	112.18	110.03	107.17	—	—
$NaOOC.CH_3$	91.0	89.2	88.5	85.68	83.72	81.20	76.88	72.76
NaOH	247.7	245.5	244.6	240.7	237.9	—	—	—
$AgNO_3$	133.29	131.30	130.45	127.14	124.68	121.35	115.18	109.09
$\frac{1}{2}MgCl_2$	129.36	125.55	124.05	118.25	114.49	109.99	103.03	97.05
$\frac{1}{2}CaCl_2$	135.77	131.87	130.30	124.19	120.30	115.59	108.42	102.41
$\frac{1}{2}SrCl_2$	135.73	131.84	130.27	124.18	120.23	115.48	108.20	102.14
$\frac{1}{2}BaCl_2$	139.91	135.89	134.27	127.96	123.88	119.03	111.42	105.14
$\frac{1}{2}Na_2SO_4$	129.8	125.68	124.09	117.09	112.38	106.73	97.70	89.94
$\frac{1}{2}CuSO_4$	133.5	121.5	115.20	94.02	83.08	72.16	59.02	50.55
$\frac{1}{2}ZnSO_4$	132.7	121.3	115.47	95.44	84.87	72.20	61.17	52.61
$\frac{1}{3}LaCl_3$	145.7	139.5	136.9	127.4	121.7	115.2	106.1	99.1
$\frac{1}{3}K_3Fe(CN)_6$	174.4	166.3	163.0	150.6	—	—	—	—
$\frac{1}{4}K_4Fe(CN)_6$	184.4	—	167.16	146.02	134.76	122.76	107.64	97.82

There are extensive tables of conductivity data for both aqueous and non-aqueous solutions in Landolt–Börnstein, 6th edn, 1960, Vol. II, Pt 7, pp. 27–726 (Springer-Verlag, Berlin).

Ionic conductivities at infinite dilution at 25 °C

The unit of ionic conductance in the table is Ω^{-1} cm^2 mol^{-1}. The values in the table are taken by permission from Robinson and Stokes, *Electrolyte Solutions*, 2nd edn, 1959, p. 463 (Butterworths, London).

Cation	Λ^∞	Cation	Λ^∞	Anion	Λ^∞	Anion	Λ^∞
H^+ . . .	349.6	$\frac{1}{2}Mg^{2+}$ . .	53.0	OH^- . . .	199.1†	HCO_2^- . .	54.6
Li^+ . . .	38.7	$\frac{1}{2}Ca^{2+}$ . . .	59.5	Cl^- . . .	76.3	$CH_3CO_2^-$ . .	40.9
Na^+ . . .	50.1	$\frac{1}{2}Sr^{2+}$ . .	59.4	Br^- . . .	78.1	$\frac{1}{2}CO_3^{2-}$ . .	69.3
K^+ . . .	73.5	$\frac{1}{2}Ba^{2+}$ . .	63.6	I^-	76.8	$\frac{1}{2}C_2O_4^{2-}$ . .	74.1
NH_4^+ . . .	73.6	$\frac{1}{2}Cu^{2+}$ . .	53.6	NO_3^- . . .	71.4	$\frac{1}{2}SO_4^{2-}$ . .	80.0
Ag^+ . . .	61.9	$\frac{1}{2}Zn^{2+}$ . .	52.8	ClO_4^- . . .	67.3	$\frac{1}{3}Fe(CN)_6^{3-}$.	100.9
Tl^+ . . .	74.7	$\frac{1}{3}La^{3+}$ . . .	69.7	HCO_3^- . .	44.5	$\frac{1}{4}Fe(CN)_6^{4-}$.	110.5

† Marsh and Stokes, *Austral. J. Chem.*, 1964, **17**, 740.

J.E.P.

2.9.3 Standard half-cell EMFs at 25 °C

Each conventional half-cell process implies the real process obtained by adding the half-cell process $H^+ + e^- \rightarrow \frac{1}{2}H_2(g)$. The word 'standard' implies that each solute species is present at an activity equal to one mole kg^{-1} and each gaseous species at a partial pressure of 101.325 kN m^{-2}. All values relate to aqueous solutions at 25 °C and are taken by permission from Latimer, *The Oxidation States of the Elements and their Potentials in Aqueous Solutions*, 1952, pp. 340–345 (Prentice-Hall, New York). If the sign is reversed, the quantities may be called standard electrode potentials (IUPAC *Manual of Symbols and Terminology for Physicochemical Quantities and Units*, 1970, p. 32 (Butterworths, London)).

Process	$E^\ominus$ volt	Process	$E^\ominus$ volt
$Li \rightarrow Li^+ + e^-$	3.04	$\frac{1}{2}Fe \rightarrow \frac{1}{2}Fe^{2+} + e^-$	0.44
$K \rightarrow K^+ + e^-$	2.92	$Cr^{2+} \rightarrow Cr^{3+} + e^-$	0.41
$Rb \rightarrow Rb^+ + e^-$	2.92	$\frac{1}{2}Cd \rightarrow \frac{1}{2}Cd^{2+} + e^-$	0.40
$Cs \rightarrow Cs^+ + e^-$	2.92	$Ti^{2+} \rightarrow Ti^{3+} + e^-$	0.37
$\frac{1}{2}Ba \rightarrow \frac{1}{2}Ba^{2+} + e^-$	2.90	$\frac{1}{2}Pb + \frac{1}{2}SO_4^{2-} \rightarrow \frac{1}{2}PbSO_4 + e^-$.	0.36
$\frac{1}{2}Sr \rightarrow \frac{1}{2}Sr^{2+} + e^-$	2.89	$\frac{1}{3}In \rightarrow \frac{1}{3}In^{3+} + e^-$	0.34
$\frac{1}{2}Ca \rightarrow \frac{1}{2}Ca^{2+} + e^-$	2.87	$Tl \rightarrow Tl^+ + e^-$	0.34
$Na \rightarrow Na^+ + e^-$	2.71	$\frac{1}{2}Co \rightarrow \frac{1}{2}Co^{2+} + e^-$	0.28
$\frac{1}{3}La \rightarrow \frac{1}{3}La^{3+} + e^-$	2.52	$V^{2+} \rightarrow V^{3+} + e^-$	0.26
$\frac{1}{2}Mg \rightarrow \frac{1}{2}Mg^{2+} + e^-$	2.37	$\frac{1}{2}Ni \rightarrow \frac{1}{2}Ni^{2+} + e^-$	0.25
$\frac{1}{3}Sc \rightarrow \frac{1}{3}Sc^{3+} + e^-$	2.08	$Ag + I^- \rightarrow AgI + e^-$	0.15
$\frac{1}{4}Th \rightarrow \frac{1}{4}Th^{4+} + e^-$	1.90	$\frac{1}{2}Sn \rightarrow \frac{1}{2}Sn^{2+} + e^-$	0.14
$\frac{1}{2}Be \rightarrow \frac{1}{2}Be^{2+} + e^-$	1.85	$\frac{1}{2}Pb \rightarrow \frac{1}{2}Pb^{2+} + e^-$	0.13
$\frac{1}{3}Al \rightarrow \frac{1}{3}Al^{3+} + e^-$	1.66	$\frac{1}{2}H_2 \rightarrow H^+ + e^-$	0 exactly
$\frac{1}{2}Ti \rightarrow \frac{1}{2}Ti^{2+} + e^-$	1.63	$Ag + Br^- \rightarrow AgBr + e^-$ . . .	−0.10
$\frac{1}{2}Mn \rightarrow \frac{1}{2}Mn^{2+} + e^-$	1.18	$\frac{1}{2}Sn^{2+} \rightarrow \frac{1}{2}Sn^{4+} + e^-$ . . .	−0.15
$\frac{1}{2}Zn \rightarrow \frac{1}{2}Zn^{2+} + e^-$	0.76	$Cu^+ \rightarrow Cu^{2+} + e^-$	−0.15
$\frac{1}{3}Cr \rightarrow \frac{1}{3}Cr^{3+} + e^-$	0.74	$Ag + Cl^- \rightarrow AgCl + e^-$ . . .	−0.22
$\frac{1}{3}Ga \rightarrow \frac{1}{3}Ga^{3+} + e^-$	0.53	$Hg + Cl^- \rightarrow HgCl + e^-$ . . .	−0.27

Standard half-cell EMFs at 25 °C (contd)

Process	$E^{\ominus}$ volt	Process	$E^{\ominus}$ volt
$\frac{1}{2}Cu \rightarrow \frac{1}{2}Cu^{2+} + e^-$	-0.34	$\frac{1}{2}Hg_2^{2+} \rightarrow Hg^{2+} + e^-$	-0.92
$Fe(CN)_6^{4-} \rightarrow Fe(CN)_6^{3-} + e^-$ . .	-0.36	$Pu^{3+} \rightarrow Pu^{4+} + e^-$	-0.97
$Cu \rightarrow Cu^+ + e^-$	-0.52	$\frac{1}{2}Pd \rightarrow \frac{1}{2}Pd^{2+} + e^-$	-0.99
$I^- \rightarrow \frac{1}{2}I_2 + e^-$	-0.54	$\frac{1}{3}Au + \frac{4}{3}Cl^- \rightarrow \frac{1}{3}AuCl_4^- + e^-$ . .	-1.00
$\frac{3}{2}I^- \rightarrow \frac{1}{2}I_3^- + e^-$	-0.54	$Br^- \rightarrow \frac{1}{2}Br_2(l) + e^-$	-1.06
$Hg + \frac{1}{2}SO_4^{2-} \rightarrow \frac{1}{2}Hg_2SO_4 + e^-$ . .	-0.62	$\frac{1}{2}H_2O \rightarrow \frac{1}{4}O_2 + H^+ + e^-$ . . .	-1.23
$\frac{1}{3}Au + \frac{4}{3}SCN^- \rightarrow \frac{1}{3}(AuSCN)_4^- + e^-$	-0.66	$\frac{1}{2}Tl^+ \rightarrow \frac{1}{2}Tl^{3+} + e^-$	-1.25
$Fe^{2+} \rightarrow Fe^{3+} + e^-$	-0.77	$Cl^- \rightarrow \frac{1}{2}Cl_2 + e^-$	-1.36
$Hg \rightarrow \frac{1}{2}Hg_2^{2+} + e^-$	-0.79	$\frac{1}{3}Au \rightarrow \frac{1}{3}Au^{3+} + e^-$	-1.50
$Ag \rightarrow Ag^+ + e^-$	-0.80		

J.E.P.

2.9.4 Ionization constant of water

Values taken by permission from Harned and Owen, *Physical Chemistry of Electrolytic Solutions*, 2nd edn, 1950, p. 485 (Reinhold, New York). The unit of K_w in the table is $mol^2 (kg\ water)^{-2}$.

°C	$10^{14}K_w$	°C	$10^{14}K_w$	°C	$10^{14}K_w$
0	0.114	25	1.008	45	4.02
5	0.185	30	1.47	50	5.47
10	0.292	35	2.09	55	7.30
15	0.450	40	2.92	60	9.61
20	0.681				

J.E.P.

2.9.5 pH values

Standardization of pH

According to BS 1647: 1961 the pH of a solution X, denoted by pH(X), is related to the pH of a standard solution, denoted by pH(S) by

$$pH(X) - pH(S) = E/2.3026RTF^{-1}$$

where R denotes the gas constant, T the absolute temperature, F the Faraday constant and E the electromotive force of the cell

$$Pt, H_2 \,|\, Solution\ X \,|\, 3.5M\ KCl \,|\, Solution\ S \,|\, H_2, Pt$$

The primary standard solution S is a one-twentieth molar solution of pure potassium hydrogen phthalate, of which the pH is defined as having the value 4 exactly at 15 °C and at other temperatures t °C between 0 °C and 95 °C values defined by the formulae:

$$pH(S) = 4.000 + \frac{1}{2}\left(\frac{t-15}{100}\right)^2 \qquad \text{if } t \text{ is between 0 and 55};$$

$$pH(S) = 4.000 + \frac{1}{2}\left(\frac{t-15}{100}\right) - \frac{t-55}{500} \qquad \text{if } t \text{ is between 55 and 95}.$$

The following table is from the British Standard:

°C	$\dfrac{2.3026RT/F}{mV}$	$\dfrac{F/2.3026RT}{V^{-1}}$	$pH(S)$	°C	$\dfrac{2.3026RT/F}{mV}$	$\dfrac{F/2.3026RT}{V^{-1}}$	$pH(S)$
0	54.20	18.45	4.011	50	64.12	15.60	4.061
5	55.19	18.12	4.005	55	65.11	15.36	4.080
10	56.18	17.80	4.001	60	66.10	15.13	4.091
15	57.17	17.49	4.000	65	67.09	14.90	4.105
20	58.17	17.19	4.001	70	68.08	14.69	4.121
25	59.16	16.90	4.005	75	69.08	14.48	4.140
30	60.15	16.63	4.011	80	70.07	14.27	4.161
35	61.14	16.36	4.020	85	71.06	14.07	4.185
40	62.13	16.09	4.031	90	72.06	13.88	4.211
45	63.13	15.84	4.045	95	73.04	13.69	4.240

The following table gives solutions of which the pH, as defined in the British Standard, has been accurately determined and which are recommended for calibration of glass electrodes.

pH Values of aqueous solutions recommended for calibration of glass electrodes

Composition/(mol dm^{-3} soln)	pH values		
	12 °C	25 °C	38 °C
0.1 $KH_3(C_2O_4)_2 . 2H_2O$	—	1.48	1.50
0.01 HCl + 0.09 KCl	—	2.07	2.08
0.05 $HOOC . C_6H_4 . COOK$	4.000	4.005	4.026
(Primary Standard)			
0.1 $CH_3 . COOH + 0.1\ CH_3 . COONa$ . . .	4.65	4.64	4.65
0.01 $CH_3 . COOH + 0.01\ CH_3 . COONa$ . . .	4.71	4.70	4.72
0.025 $KH_2PO_4 + 0.025\ Na_2HPO_4 . 2H_2O$ . .	—	6.85	6.84
0.05 $Na_2B_4O_7 . 10H_2O$	—	9.18	9.07
0.025 $NaHCO_3 + 0.025\ Na_2CO_3$	—	10.00	—

Buffer solutions

The following table of pH values covering a wider range of temperatures is taken from the IUPAC *Manual of Symbols and Terminology for Physico-chemical Quantities and Units*, 1970, p. 34 (Butterworths, London). The values are based on a convention slightly different from the British Standard, but the values agree almost certainly to within better than ± 0.01.

Values of pH(S) for five standard solutions

$t/°C$	A	B	C	D	E
0 . . .		4.003	6.984	7.534	9.464
5 . . .		3.999	6.951	7.500	9.395
10 . . .		3.998	6.923	7.472	9.332
15 . . .		3.999	6.900	7.448	9.276
20 . . .		4.002	6.881	7.429	9.225
25 . . .	3.557	4.008	6.865	7.413	9.180
30 . . .	3.552	4.015	6.853	7.400	9.139
35 . . .	3.549	4.024	6.844	7.389	9.102
38 . . .	3.548	4.030	6.840	7.384	9.081
40 . . .	3.547	4.035	6.838	7.380	9.068
45 . . .	3.547	4.047	6.834	7.373	9.038
50 . . .	3.549	4.060	6.833	7.367	9.011
55 . . .	3.554	4.075	6.834		8.985
60 . . .	3.560	4.091	6.836		8.962
70 . . .	3.580	4.126	6.845		8.921
80 . . .	3.609	4.164	6.859		8.885
90 . . .	3.650	4.205	6.877		8.850
95 . . .	3.674	4.227	6.886		8.833

The compositions of the standard solutions are:

A: KH tartrate (saturated at 25 °C)
B: KH phthalate, $m = 0.05$ mol kg^{-1}
C: KH_2PO_4, $m = 0.025$ mol kg^{-1};
   Na_2HPO_4, $m = 0.025$ mol kg^{-1}
D: KH_2PO_4, $m = 0.008\,695$ mol kg^{-1};
   Na_2HPO_4, $m = 0.030\,43$ mol kg^{-1}
E: $Na_2B_4O_7$, $m = 0.01$ mol kg^{-1}

where m denotes molality and the solvent is water.

Acid-base indicators

An acid-base indicator is a conjugate acid-base pair such that the acid form and the base form have different colours. The following table is from A. I. Vogel, *Quantitative Inorganic Analysis*, 3rd edn, 1961, p. 55 (Longman, Green & Co.).

Indicator	*pH range*	*Acid*	*Base*
Brilliant cresyl blue (acid)	0.0–1.0	Red-orange	Blue
Cresol red (acid) . . .	0.2–1.8	Red	Yellow
Quinaldine red . . .	1.0–2.0	Colourless	Red
Thymol blue (acid) . .	1.2–2.8	Red	Yellow
m-Cresol purple . . .	1.2–2.8	Red	Yellow
Pentamethoxy red . . .	1.2–3.2	Red-violet	Colourless
Tropaeolin OO . . .	1.3–3.0	Red	Yellow
Meta-cresol purple . .	1.2–2.8	Red	Yellow
Bromo-phenol blue . .	3.0–4.6	Yellow	Blue
Methyl yellow	2.9–4.0	Red	Yellow
Ethyl orange	3.0–4.5	Red	Orange
Methyl orange	3.1–4.4	Red	Orange
Congo red	3.0–5.0	Violet	Red
α-Naphthyl red . . .	3.7–5.0	Red	Yellow

Acid-base indicators (*contd*)

Indicator	pH range	Acid	Base
Bromo-cresol green . .	3.8–5.4	Yellow	Blue
Methyl red	4.2–6.3	Red	Yellow
Ethyl red	4.5–6.5	Red	Orange
Propyl red	4.6–6.6	Red	Yellow
Chlorophenol red . . .	4.8–6.4	Yellow	Red
p-Nitrophenol	5.6–7.6	Colourless	Yellow
Bromocresol purple . .	5.2–6.8	Yellow	Purple
Bromophenol red . . .	5.2–6.8	Yellow	Red
Azolitmin (litmus) . .	5.0–8.0	Red	Blue
Bromo-thymol blue . .	6.0–7.6	Yellow	Blue
Neutral red	6.8–8.0	Red	Orange
Phenol red	6.8–8.4	Yellow	Red
Cresol red (base) . . .	7.2–8.8	Yellow	Red
α-Naphtholphthalein . .	7.3–8.7	Yellow	Blue
m-Cresol purple . . .	7.6–9.2	Yellow	Purple
Thymol blue (base) . .	8.0–9.6	Yellow	Blue
o-Cresolphthalein . . .	8.2–9.8	Colourless	Red
Phenolphthalein . . .	8.3–10.0	Colourless	Red
Turmeric	8.0–10.0	Yellow	Orange
Thymolphthalein . . .	8.3–10.5	Colourless	Blue
Alizarin yellow R . . .	10.1–12.0	Yellow	Orange red
Brilliant cresyl blue (base).	10.8–12.0	Blue	Yellow
Tropaeolin O	11.1–12.7	Yellow	Orange
Nitramine	10.8–13.0	Colourless	Orange-brown

J.E.P.

2.9.6 Half-wave potentials of metals

Half-wave potentials $E_{1/2}$ are referred to the saturated calomel electrode at 25 °C for which $E^{\ominus} = -0.242$ V. The subjoined data are taken by permission from Kolthoff and Lingane, *Polarography*, 2nd edn, 1952 (Interscience, New York).

Element	Supporting electrolyte	$E_{1/2}/V$	Reaction
Al	0.025M $BaCl_2$	−1.75	
Ba	Tetramethylammonium iodide in 80% ethanol .	−1.82	
Bi	Bi^{3+} in 0.5M tartrate (pH = 4.5)	−0.29	
Ca	Tetramethylammonium iodide in 80% ethanol .	−2.13	
Cd	0.1M KCl or HCl	−0.599	
	0.5M tartrate (pH = 4.5)	−0.64	
	1M NH_3 + 1M NH_4Cl	−0.81	

Half-wave potentials of metals (*contd*)

Element	Supporting electrolyte	$E_{1/2}/V$	Reaction
Co	Co^{2+} in o.1M KCl or NaCl	-1.4	
	Co^{3+} in 1M NH_3 + 1M NH_4Cl	-0.5	$Co^{3+} \rightarrow Co^{2+}$
		-1.3	$Co^{2+} \rightarrow Co$
	Co^{2+} in 1M KSCN	-1.03	
	Co^{2+} in o.1M pyridine + o.1M pyridinium chloride	-1.07	
Cr	Cr^{3+} in o.1M KCl	-0.91	$Cr^{3+} \rightarrow Cr^{2+}$
		-1.47	$Cr^{2+} \rightarrow Cr$
	CrO_4^{2-} in 1M NaOH	-0.85	$Cr^{6+} \rightarrow Cr^{3+}$
	CrO_4^{2-} in o.1M KCl	-0.3	
		-1.0	$Cr^{6+} \rightarrow Cr^{3+}$
		-1.5	$Cr^{3+} \rightarrow Cr^{2+}$
		-1.7	$Cr^{2+} \rightarrow Cr$
Cu	o.1M KCl or HCl	$+0.04$	$Cu^{2+} \rightarrow Cu$
	1M NH_3 + 1M NH_4Cl	-0.24	$Cu^{2+} \rightarrow Cu^+$
		-0.50	$Cu^+ \rightarrow Cu$
	o.1M KSCN	-0.02	$Cu^{2+} \rightarrow Cu^+$
		-0.39	$Cu^+ \rightarrow Cu$
	Cu^{2+} in 0.5M tartrate (pH = 4.5) . . .	-0.09	
Fe	Fe^{3+} in KCl, HCl, $HClO_4$, etc. . . .	0.0†	
	Fe^{2+} in o.1M KCl or 0.05M $BaCl_2$	-1.3	$Fe^{2+} \rightarrow Fe$
	Fe^{3+} and/or Fe^{2+} in 0.5M tartrate (pH = 5.8) .	-0.17‡	$Fe^{3+} \rightleftharpoons Fe^{2+}$
	Fe^{3+} and/or Fe^{2+} in 0.2M oxalate (pH = 5.25) .	-0.245‡	$Fe^{3+} \rightleftharpoons Fe^{2+}$
	Fe^{3+} in 1M KOH + 8% mannitol	-0.9	$Fe^{3+} \rightarrow Fe^{2+}$
		-1.5	$Fe^{2+} \rightarrow Fe$
K	Tetraethylammonium hydroxide in 50% ethanol	-2.10	
Li	Tetraethylammonium hydroxide in 50% ethanol	-2.31	
Mn	Mn^{2+} in 1M KCl	-1.51	$Mn^{2+} \rightarrow Mn$
	Mn^{2+} in 1M NH_3 + 1M NH_4Cl	-1.65	
	Mn^{2+} in 1M KSCN	-1.55	
Na	Tetraethylammonium hydroxide in 50% ethanol	-2.07	
Ni	1M KCl	-1.1	
	1M NH_3 + 0.2M NH_4Cl	-1.06	
	1M KSCN	-0.70	
	1M KCl + 0.5M pyridine	-0.78	
Pb	1M KNO_3 or HNO_3	-0.405	
	o.1M KCl or HCl	-0.396	
	$HPbO_2^-$ in 1M NaOH	-0.755	
	0.5M tartrate (pH = 4.5)	-0.48	
Sb	Sb^{3+} in 1M HCl	-0.15	$Sb^{3+} \rightarrow Sb$
	Sb^{3+} in 1M NaOH	-1.26	
Sn	Sn^{2+} in 1M HCl	-0.47	
	SnO_2^{2-} in 1M NaOH	-1.22	$SnO_2^{2-} \rightarrow Sn$
		-0.73§	$SnO_2^{2-} \rightarrow SnO_3^{2-}$
Ti	Ti^{4+} in o.1M HCl or HNO_3	-0.81	$Ti^{4+} \rightarrow Ti^{3+}$
Tl	Tl^+ in o.1M KCl, KNO_3, HCl or NaOH . . .	-0.460	
U	UO_2^{2+} in 0.02M HCl	-0.18	$U^{6+} \rightarrow U^{5+}$
		-0.92	$U^{5+} \rightarrow U^{3+}$
Zn	o.1M KCl or KNO_3	-0.995	
	ZnO_2^{2-} in 1M NaOH	-1.53	
	o.1M KSCN	-1.01	
	0.5M tartrate (pH = 4.5)	-1.23	

† Diffusion current of $Fe^{3+} \rightarrow Fe^{2+}$ at zero applied potential.   ‡ Reversible.   § Anodic wave.

J.E.P.

2.9.7 Osmotic coefficients

Osmotic coefficients at 25 °C in aqueous solutions used as standards for isopiestic measurements

The practical osmotic coefficient ϕ in a solution of an electrolyte consisting of v ions is given by $vr\phi = \ln (p^0/p)$, where r is the mole ratio of electrolyte to solvent, p is the partial vapour pressure of the solvent over the solution and p^0 is the vapour pressure of the pure solvent. In aqueous solutions the mole ratio r is related to the molality m, expressed in mol (kg water)$^{-1}$ by $m = 55.51r$. The values of ϕ in the following table are taken by permission from Robinson and Stokes, *Electrolyte Solutions*, 2nd edn, 1959, Appendices 8.3 and 8.5 (Butterworths, London).

m $\overline{\text{mol kg}^{-1}}$	NaCl ϕ	KCl ϕ	CaCl$_2$ ϕ	m $\overline{\text{mol kg}^{-1}}$	NaCl ϕ	KCl ϕ	CaCl$_2$ ϕ
0.1	0.9324	0.9266	0.854	3.6	1.0867	0.9531	
0.2	0.9245	0.9130	0.862	3.8	1.1013	0.9588	
0.3	0.9215	0.9063	0.876	4.0	1.1158	0.9647	2.182
0.4	0.9203	0.9017	0.894	4.2	1.1306	0.9707	
0.5	0.9209	0.8989	0.917	4.4	1.1456	0.9766	
0.6	0.9230	0.8976	0.940	4.5			2.383
0.7	0.9257	0.8970	0.963	4.6	1.1608	0.9824	
0.8	0.9288	0.8970	0.988	4.8	1.1761	0.9883	
0.9	0.9320	0.8971	1.017	5.0	1.1916		2.574
1.0	0.9355	0.8974	1.046	5.2	1.2072		
1.2	0.9428	0.8986	1.107	5.4	1.2229		
1.4	0.9513	0.9010	1.171	5.5			2.743
1.6	0.9616	0.9042	1.237	5.6	1.2389		
1.8	0.9723	0.9081	1.305	5.8	1.2548		
2.0	0.9833	0.9124	1.376	6.0	1.2706		2.891
2.2	0.9948	0.9168		6.5			3.003
2.4	1.0068	0.9214		7.0			3.081
2.5			1.568	7.5			3.127
2.6	1.0192	0.9264		8.0			3.151
2.8	1.0321	0.9315		8.5			3.165
3.0	1.0453	0.9367	1.779	9.0			3.171
3.2	1.0587	0.9421		9.5			3.171
3.4	1.0725	0.9477		10.0			3.169
3.5			1.981				

J.E.P.

2.9.8 Activity coefficients

Activity coefficients of typical strong electrolytes in aqueous solution at 25 °C

The mean activity coefficient γ of an electrolyte, consisting of v_+ cations R and v_- anions X, is related to the chemical potential μ of one mole of electrolyte by

$$\mu = \mu^\ominus + v_+ RT \ln m_R + v_- RT \ln m_X + (v_+ + v_-)RT \ln \gamma$$

where m_R and m_X denote the molalities of the cation R and anion X respectively while $\mu^\ominus$ is independent of the composition of the solution and is so chosen that $\gamma \to 1$ at infinite dilution.

In the following table of γ for typical 1–1 electrolytes at 25 °C the values at molalities 0.1 upwards are from Robinson and Stokes, *Trans. Faraday Soc.*, 1949, **45**, 612. The values at lower molalities for HCl, NaCl and KCl have been obtained from e.m.f. measurements and for the other electrolytes have been interpolated by a Debye–Hückel type of formula. All γ values are considered to be accurate to within ± 0.002.

Typical 1–1 electrolytes at 25 °C

$m/(\text{mol kg}^{-1})$	HCl	LiCl	NaCl	KCl	CsCl	LiNO$_3$	NaNO$_3$	KNO$_3$	CsNO$_3$
0.01	0.904	0.903	0.902	0.901	0.899	0.903	0.900	0.897	0.896
0.02	0.875	0.873	0.870	0.868	0.865	0.872	0.866	0.861	0.860
0.05	0.830	0.825	0.820	0.816	0.807	0.825	0.811	0.799	0.795
0.10	0.796	0.790	0.778	0.770	0.756	0.788	0.762	0.739	0.733
0.2	0.767	0.757	0.735	0.718	0.694	0.752	0.703	0.663	0.655
0.4	0.755	0.740	0.693	0.666	0.628	0.728	0.638	0.576	0.561
0.6	0.763	0.743	0.673	0.637	0.589	0.727	0.599	0.519	0.501
0.8	0.783	0.755	0.662	0.618	0.563	0.733	0.570	0.476	0.458
1.0	0.809	0.774	0.657	0.604	0.544	0.743	0.548	0.443	0.422
1.2	0.840	0.796	0.654	0.593	0.529	0.757	0.530	0.414	0.393
1.4	0.876	0.823	0.655	0.586	0.518	0.774	0.514	0.390	0.368
1.6	0.916	0.853	0.657	0.580	0.509	0.792	0.501	0.369	
1.8	0.960	0.885	0.662	0.576	0.501	0.812	0.489	0.350	
2.0	1.009	0.921	0.668	0.573	0.495	0.835	0.478	0.333	
2.5	1.147	1.026	0.688	0.569	0.484	0.896	0.455	0.297	
3.0	1.316	1.156	0.714	0.569	0.478	0.966	0.437	0.269	
3.5	1.518	1.317	0.746	0.572	0.474	1.044	0.422	0.246	
4.0	1.762	1.510	0.783	0.577	0.473	1.125	0.408		
4.5	2.04	1.741	0.826	0.583	0.473	1.215	0.396		
5.0	2.38	2.02	0.874		0.474	1.310	0.386		

The values of γ for typical 2–1 and 1–2 electrolytes in the following table are from Stokes, *Trans. Faraday Soc.*, 1948, **44**, 295 as revised by Guggenheim and Stokes, *Trans. Faraday Soc.*, 1958, **54**, 1646. They are derived from experimental values of the osmotic coefficient ϕ by use of the Gibbs–Duhem relation.

Typical 2–1 and 1–2 electrolytes at 25 °C

$m/(\text{mol kg}^{-1})$	0.1	0.2	0.3	0.4	0.5	0.6	0.7	0.8	0.9	1.0
MgCl$_2$	0.528	0.488	0.476	0.474	0.480	0.490	0.505	0.521	0.543	0.569
CaCl$_2$	0.518	0.472	0.455	0.488	0.448	0.453	0.460	0.470	0.484	0.500
SrCl$_2$	0.515	0.466	0.446	0.436	0.433	0.434	0.437	0.445	0.453	0.465
BaCl$_2$	0.508	0.451	0.426	0.411	0.403	0.397	0.397	0.397	0.398	0.401
MnCl$_2$	0.518	0.471	0.452	0.444	0.442	0.445	0.450	0.457	0.468	0.481
FeCl$_2$	0.520	0.475	0.456	0.450	0.452	0.456	0.465	0.475	0.490	0.508
CoCl$_2$	0.523	0.480	0.464	0.460	0.463	0.471	0.480	0.493	0.512	0.532
NiCl$_2$	0.523	0.480	0.464	0.461	0.465	0.472	0.483	0.497	0.516	0.537
CuCl$_2$	0.510	0.457	0.431	0.419	0.413	0.411	0.411	0.412	0.415	0.419
MgBr$_2$	0.542	0.511	0.510	0.519	0.537	0.562	0.590	0.625	0.667	0.712
CaBr$_2$	0.532	0.492	0.482	0.483	0.491	0.505	0.522	0.543	0.568	0.597
SrBr$_2$	0.527	0.484	0.469	0.466	0.468	0.474	0.485	0.498	0.516	0.536
BaBr$_2$	0.517	0.469	0.450	0.440	0.438	0.442	0.446	0.452	0.462	0.473
Mg(NO$_3$)$_2$	0.522	0.480	0.467	0.465	0.469	0.478	0.488	0.501	0.518	0.536
Ca(NO$_3$)$_2$	0.488	0.429	0.397	0.378	0.365	0.356	0.349	0.344	0.340	0.338
Sr(NO$_3$)$_2$	0.478	0.410	0.373	0.348	0.329	0.314	0.302	0.292	0.283	0.275
Co(NO$_3$)$_2$	0.521	0.474	0.455	0.448	0.448	0.451	0.458	0.468	0.480	0.493
Cu(NO$_3$)$_2$	0.512	0.461	0.440	0.430	0.427	0.428	0.432	0.438	0.446	0.456
Li$_2$SO$_4$	0.478	0.406	0.369	0.344	0.326	0.313	0.303	0.295	0.288	0.283
Na$_2$SO$_4$	0.452	0.371	0.325	0.294	0.270	0.252	0.237	0.225	0.213	0.204
K$_2$SO$_4$	0.436	0.356	0.313	0.283	0.261	0.243	0.229	—	—	—
Rb$_2$SO$_4$	0.460	0.381	0.338	0.307	0.285	0.268	0.254	0.243	0.233	0.223
Cs$_2$SO$_4$	0.464	0.389	0.344	0.317	0.296	0.279	0.267	0.256	0.246	0.239

Acidity constants in water at 25 °C

The acidity constant K of an acid-base equilibrium $A \rightleftharpoons B + H^+$ is defined as the limiting value at infinite dilution of $[B][H^+]/[A]$ and pK is an abbreviation for $-\log_{10}(K/\text{mol kg}^{-1})$. The values in the following table relate to aqueous solutions at 25 °C and are taken from the sources quoted.

Name	Formula	pK	Formula	pK	Reference
Oxalic	$H_2C_2O_4$	1.271	$HC_2O_4^-$	4.266	1
Phosphoric	H_3PO_4	2.148	$H_2PO_4^-$	7.198	1
Glycine	$^+H_3N.CH_2.CO_2H$	2.350	$^+H_3N.CH_2.CO_2^-$	9.780	1
p-Aminobenzoic . .	$^+H_3N.C_6H_4.CO_2H$	2.45	$^+H_3N.C_6H_4.CO_2^-$	4.85	1
Cyanoacetic . . .	$CH_2CN.CO_2H$	2.469			1
Malonic	$CH_2.(CO_2H)_2$	2.855	$(CO_2H)CH_2.CO_2^-$	5.696	1
Chloroacetic . . .	$CH_2Cl.CO_2H$	2.865			1
Bromoacetic . . .	$CH_2Br.CO_2H$	2.901			1
o-Phthalic	$C_6H_4.(CO_2H)_2$	2.950	$(CO_2H)C_6H_4.CO_2^-$	5.408	1
o-Hydroxybenzoic . .	$HO.C_6H_4.CO_2H$	2.996			1
Tartaric	$C_2H_4O_2.(CO_2H)_2$	3.033	$(CO_2H)C_2H_4O_2.CO_2^-$	4.366	1
Citric	$C_3H_5O.(CO_2H)_3$	3.128	$(CO_2H)_2C_3H_5O.CO_2^-$	4.761	1
Iodoacetic	$CH_2I.CO_2H$	3.174			1
2,6-Dinitrophenol . .	$(O_2N)_2.C_6H_4.OH$	3.72			2
Formic	$H.CO_2H$	3.739			1
Glycollic	$CH_2OH.CO_2H$	3.831			1
Lactic	$CH_3.CHOH.CO_2H$	3.860			1
2,4-Dinitrophenol . .	$(O_2N)_2.C_6H_4.OH$	4.09			2
Benzoic	$C_6H_5.CO_2H$	4.201			1
Succinic	$C_2H_4.(CO_2H)_2$	4.207	$(CO_2H)C_2H_4.CO_2^-$	5.638	1
Anilinium	$C_6H_5.NH_3^+$	4.60			3
Acetic	$CH_3.CO_2H$	4.756			1
Butyric	$C_3H_7.CO_2H$	4.820			1
Propionic	$C_2H_5.CO_2H$	4.874			1
2,3-Dinitrophenol . .	$(O_2N)_2.C_6H_4.OH$	4.96			5
2,5-Dinitrophenol . .	$(O_2N)_2.C_6H_4.OH$	5.21			2
3,4-Dinitrophenol . .	$(O_2N)_2.C_6H_4.OH$	5.42			2
Hydroxylammonium . .	NH_3OH^+	5.96			5
Carbonic	H_2CO_3	6.352	HCO_3^-	10.329	1
3,5-Dinitrophenol . .	$(O_2N)_2.C_6H_4.OH$	6.69			5
p-Nitrophenol . . .	$O_2N.C_6H_4.OH$	7.156			4
o-Nitrophenol . . .	$O_2N.C_6H_4.OH$	7.230			6
Triethanolammonium . .	$(C_2H_4OH)_3.NH^+$	7.762			3
Tris-hydroxymethyl-methyl-ammonium . .	$(CH_2OH)_3.CNH_3^+$	8.075			1
Hydrazinium	$N_2H_5^+$	8.11			1
m-Nitrophenol . . .	$O_2N.C_6H_4.OH$	8.355			7
Diethanolammonium . .	$(C_2H_4OH)_2.NH_2^+$	8.883			3
Phenol-sulphonate . .	$HO.C_6H_4.SO_3^-$	9.053			1
Hydrogen cyanide . . .	HCN	9.216			4
Boric	H_3BO_3	9.234			1
Ammonium	NH_4^+	9.245			1
Ethanol-ammonium . .	$C_2H_4OH.NH_3^+$	9.498			1
Trimethyl-ammonium . .	$(CH_3)_3.NH^+$	9.800			1
Phenol	C_6H_5OH	9.998			1
m-Cresol	$CH_3.C_6H_4.OH$	10.09			1
p-Cresol	$CH_3.C_6H_4.OH$	10.26			1
o-Cresol	$CH_3.C_6H_4.OH$	10.29			1
Methyl-ammonium . .	$CH_3.NH_3^+$	10.624			1
Ethyl-ammonium . . .	$C_2H_5.NH_3^+$	10.631			1

Acidity constants in water at 25 °C (*contd*)

Name	Formula	pK	Formula	pK	Reference
Triethyl-ammonium . .	$(C_2H_5)_3.NH^+$	10.715			1
Dimethyl-ammonium . .	$(CH_3)_2.NH_2^+$	10.774			1
Diethyl-ammonium . .	$(C_2H_5)_2.NH_2^+$	10.933			1
Piperidinium	$C_5H_{10}NH_2^+$	11.123			1

References: (1) Robinson and Stokes, *Electrolyte Solutions*, 2nd edn, 1959, pp. 517–535 (Butterworths, London). (2) Kortüm, Vogel and Andrussow, *Dissociation Constants of Organic Acids in Aqueous Solution*, 1961 (Butterworths, London). (3) Perrin, *Dissociation Constants of Organic Bases in Aqueous Solution*, 1965 (Butterworths, London). Perrin, *Dissociation Constants of Inorganic Acids and Bases in Aqueous Solution*, 1969 (Butterworths, London). (5) Robinson, Davis, Paabo and Bower, *J. Res. Nat. Bur. Stand.*, 1960 **64A**, 347. (6) Robinson and Peiperl, *J. Phys. Chem.*, 1963, **67**, 1723. (7) Robinson and Peiperl, *J. Phys. Chem.*, 1963, **67**, 2860.

Extensive tables of acidity constants will also be found in Landolt–Börnstein, 6th edn, 1960, Vol. II, Pt 7, pp. 839–953 (Springer-Verlag, Berlin).

J.E.P.

2.9.9 Stability constants and solubility products

Stability constants in aqueous solutions

The stability constants K_n of step-wise complex ion equilibria $ML_{n-1} + L \rightleftharpoons ML_n$ are defined as the limiting values at infinite dilution of $[ML_n]/[ML_{n-1}][L]$. pK_n is an abbreviation for $-\log_{10}(K_n/dm^3\ mol^{-1})$. Q_n is used to denote the corresponding stability quotient in a swamping concentration of a specified inert salt. The values in the following tables are selected from Sillén and Martell, *Stability Constants*, Chem. Soc. Special Publ., 1964, No. 17.

$L = NH_3$; *temperature* 30 °C.

n	Co^{2+} $-pK_n$	Ni^{2+} $-pK_n$	Cu^{2+} $-pK_n$	Zn^{2+} $-pK_n$	Cd^{2+} $-pK_n$	Hg^{2+}† $-pQ_n$	Ag^+‡ $-pK_n$
1	1.99	2.67	4.25	2.18	2.51	8.8	3.32
2	1.51	2.12	3.61	2.25	1.96	8.7	3.92
3	0.93	1.61	2.98	2.31	1.30	1.0	
4	0.64	1.07	2.24	1.96	0.79	0.8	
5	0.06	0.63	0.52				
6	−0.74	−0.09					

† pQ_n in 2M ammonium nitrate at ∼22 °C.
‡ 25 °C.

$L = $ *ethylenediaminetetraacetate ion; temperature* 20 °C.

	Mg^{2+}	Ca^{2+}	Ba^{2+}	Mn^{2+}	Fe^{2+}	Co^{2+}	Ni^{2+}	Cu^{2+}	Fe^{3+}	La^{3+}
$-pQ_1$† . .	8.69	10.70	7.76	13.58	14.33	16.21	18.56	18.79	25.1	15.1

	Zn^{2+}	Cd^{2+}	Hg^{2+}	Pb^{2+}	Ag^+					
$-pQ_1$‡ . .	16.50	16.46	21.80	18.04	7.32					

† In 0.1M potassium chloride.
‡ In 0.1M potassium nitrate.

Solubility products in water

The solubility product P of a salt M_nX_m is defined by the value extrapolated to infinite dilution of $[M]_n[X]_m$ in a saturated solution of the salt. pP is an abbreviation for $-\log_{10}(P/mol^{m+n}\,dm^{-3(m+n)})$. The values in the following table relate to aqueous solutions at 25 °C and are selected from Sillén and Martell, *Stability Constants*, Chem. Soc. Special Publ., 1964, No. 17.

	AgCl	AgBr	AgI	AgIO$_3$	AgSCN	TlCl	TlIO$_3$
pP . .	9.75	12.28	16.08	7.51	12.0	3.7	5.51

	BaSO$_4$	PbSO$_4$	PbCl$_2$	Hg$_2$Cl$_2$	Ca(IO$_3$)$_2$	La(IO$_3$)$_3$	Ag$_2$SO$_4$
pP . .	10.0	7.80	4.7	17.88	6.15	11.21	4.8

J.E.P.

2.10 Chemical thermodynamics

2.10.1 Molar heat capacities, heats of fusion and vaporization of the elements

Values for the molar heat capacities of the elements at 298.15 K are given on pp. 235–249. In the following table heat capacities are expressed in J mol^{-1} K^{-1} and heats of transition are expressed in kJ mol^{-1}. The horizontal lines indicate the positions of melting and boiling points. The elements are arranged in alphabetical order of name but are listed by chemical symbol. The tables have been compiled mainly from D. R. Stull and G. C. Sinke, *Thermodynamic Properties of the Elements*, 1956 (American Chemical Society), and K. K. Kelley and E. G. King, 'Contributions to the data on theoretical metallurgy, XIV—Entropies of the elements and inorganic compounds', *US Bureau of Mines, Bulletin 592*.

T/K	Element and formula weight							
	Al 26.98	Sb 121.75	Ar 39.948	As 74.92	Ba 137.34	Be 9.012	Bi 208.98	B 10.81
50	3.8	13.0	24.8	7.9		0.2	17.7	0.1
100	13.1	20.6	20.8	16.7		1.8	23.0	1.1
150	18.5	23.2	20.8	20.7		5.7	24.4	3.2
200	21.6	24.4	20.8	22.7		10.0	25.0	6.1
400	25.6	26.0	20.8	25.6	6.64	19.8	27.8	14.4
600	28.1	27.4	20.8	27.4	30.5	23.4	31.4	19.1
800	30.6	28.9	20.8	29.3	33.9	25.4	31.4	22.4
1000	29.3	31.4	20.8		31.4	27.3	31.4	24.7
1500	29.3	31.4	20.8		31.4	32.4	31.4	27.9
2000	29.3	18.7	20.8		28.5	31.4	21.0	30.1
2500	29.3	18.7	20.8		39.2	31.4	21.4	31.4
m.p. (T/K)	933.5	903.9	83.75		983	1558	544.6	2300
$\Delta H_m/(kJ\ mol^{-1})$	10.67	19.83	1.18		7.66			22.18
b.p. (T/K)	2673	1873	87.25		1873	2743	1823	4000
$\Delta H_v/(kJ\ mol^{-1})$	293.72	67.91	6.52		150.92			538.90
Molecular formula of gas . . .	Al	Sb$_2$	Ar		Ba		Bi	B

Molar heat capacities, heats of fusion and vaporization of the elements (contd)

T/K	Element and formula weight							
	Br_2 159.808	Cd 112.40	Cs 132.905	Ca 40.08	C 12.011	Ce 140.12	Cl_2 70.906	Cr 51.996
50	33.3	16.2	24.3	10.9	0.5	20.8	29.2	2.0
100	43.6	22.2	25.8	19.5	1.7	27.0	42.3	10.0
150	49.2	24.0	26.8	23.0	3.2	28.1	51.0	16.5
200	53.8	24.9	27.4	24.7	5.0	28.9	54.2	20.1
400	36.7	27.2	31.8	27.8	11.9	30.5	35.3	25.4
600	37.3	29.7	31.8	30.6	16.9	33.9	36.6	27.5
800	37.5	29.7	31.8	33.8	19.9	37.2	37.2	28.5
1000	37.7	29.7	20.8	39.7	21.5	40.6	37.5	31.5
1500	38.0	20.8	20.8	31.0	24.0	33.5	38.0	38.6
2000	38.2	20.8	21.3	21.0	25.3	33.5	38.3	43.6
2500	38.5	20.8	23.0	21.8	26.2	33.5	38.6	40.6
m.p. (T/K)	266.1	594.3	301.8	1113		1073	172.15	2130
$\Delta H_m/(\text{kJ mol}^{-1})$	10.54	6.07	2.13	8.66		9.20	6.41	13.81
b.p. (T/K)	331.4	1040	958	1720		3270	239.1	2900
$\Delta H_v/(\text{kJ mol}^{-1})$	30.00	99.87	65.90	149.95		313.80	20.41	348.78
Molecular formula of gas	Br_2	Cd	Cs	Ca		Ce	Cl_2	Cr

T/K	Element and formula weight							
	Co 58.93	Cu 63.54	Er 167.26	Eu 151.96	F_2 38.00	Gd 157.25	Ga 69.72	Ge 72.59
50	4.4	6.3	28.4		═══	19.4	10.3	6.2
100	14.0	16.2	24.6			28.9	18.5	13.8
150	19.5	20.5	26.0			32.7	22.0	18.5
200	22.3	22.7	27.0			36.1	23.8	20.9
400	26.6	25.1	28.7	27.6	33.0	29.3	27.8	24.9
600	30.0	26.4	30.0	29.3	35.2	30.7	27.8	26.4
800	32.0	27.7	31.3	31.0	36.3	32.1	27.8	27.4
1000	37.3	28.9	32.5	32.6	37.0	33.5	27.8	28.3
1500	40.2	31.4	35.6	33.5	37.9	37.0	27.8	29.3
2000	34.7	31.4	33.5	21.3	38.4	33.5	27.8	29.3
2500	34.7	31.4	33.5	23.5	38.8	33.5	27.8	29.3
m.p. (T/K)	1767	1351.6	1790	1100	53.55	1600	303	1232
$\Delta H_m/(\text{kJ mol}^{-1})$	15.23	13.05	17.15	10.46	5.10	15.48	5.59	31.80
b.p. (T/K)	3170	2853	2900	1700	85.05	3200	2510	3100
$\Delta H_v/(\text{kJ mol}^{-1})$	382.42	304.60	292.88	175.73	6.54	311.71	256.06	334.30
Molecular formula of gas	Co	Cu	Er	Eu	F_2	Gd	Ga	Ge

Molar heat capacities, heats of fusion and vaporization of the elements (*contd*)

T/K	Element and formula weight							
	Au 196.97	Hf 178.49	He 4.003	Ho 164.93	H$_2$ 2.0158	In 114.82	I$_2$ 253.81	Ir 192.22
50	14.3		===	24.5	===	18.6	35.8	7.3
100	21.4			39.2		23.3	45.6	17.3
150	23.5			26.6		25.0	49.6	21.5
200	24.4			26.5		25.9	51.5	23.4
400	25.9	26.0	20.8	27.9	29.2	28.9	80.3	25.6
600	26.8	27.0	20.8	29.4	29.3	29.7	37.6	26.8
800	27.8	28.0	20.8	30.8	29.6	29.7	37.8	28.0
1000	28.9	29.0	20.8	32.2	30.2	29.7	37.9	29.2
1500	29.3	31.6	20.8	35.8	32.3	29.7	38.2	32.2
2000	29.3	34.1	20.8	33.5	34.3	29.7	38.5	35.1
2500	29.3	33.5	20.8	33.5	36.0	23.9	38.8	38.1
m.p. (T/K)	1337.6	2500	3.5	1740	13.95	429.8	386.8	2720
ΔH_m/(kJ mol^{-1})	12.36	21.76	0.02	17.15	0.11	3.26	0.02	26.36
b.p. (T/K)	2930	5600	4.22	2600	20.28	2320	456	4800
ΔH_v/(kJ mol^{-1})	324.43	661.07	0.08	251.04	0.90	226.35	41.71	563.58
Molecular formula of gas	Au	Hf	He	Ho	H$_2$	In	I$_2$	Ir

T/K	Element and formula weight							
	Fe 55.85	Kr 83.30	La 138.91	Pb 207.2	Li 6.941	Mg 24.31	Mn 54.94	Hg 200.59
50	3.0	25.1	18.5	21.4	4.2	5.8		19.9
100	12.0	31.6	23.6	24.4	12.8	15.7		24.3
150	18.1	20.8	25.4	25.4	17.8	20.4		25.9
200	21.5	20.8	26.3	25.9	20.6	22.7		27.3
400	27.4	20.8	6.8	27.4	27.2	26.1	28.2	27.4
600	31.5	20.8	7.1	29.4	29.5	28.3	31.5	27.2
800	38.6	20.8	7.5	30.0	28.9	31.0	34.4	27.8
1000	57.7	20.8	7.8	29.4	28.8	33.0	38.9	20.8
1500	36.7	20.8	8.0	27.9	28.5	20.8	47.3	20.8
2000	44.3	20.8	8.0	26.3	20.8	20.8	46.0	20.8
2500	45.1	20.8	8.0	29.1	21.1	20.8	21.1	20.8
m.p. (T/K)	1813	115.9	1195	600.65	453	923	1520	234.29
ΔH_m/(kJ mol^{-1})	15.36	1.64	11.30	4.77	3.02	8.95	14.64	2.30
b.p. (T/K)	3070	119.75	3700	2023	1600	1370	2330	629.81
ΔH_v/(kJ mol^{-1})	351.04	9.03	399.57	179.41	134.68	128.66	219.74	59.15
Molecular formula of gas	Fe	Kr	Fe	Pb	Li	Mg	Mn	Hg

Molar heat capacities, heats of fusion and vaporization of the elements (*contd*)

T/K	Element and formula weight							
	Mo 95.94	Nd 144.24	Ne 20.179	Ni 58.71	Nb 92.91	N_2 28.013	Os 190.2	O_2 31.999
50	3.7	21.6	═	4.0	8.5	41.5		46.1
100	13.5	26.6		13.6	17.9	29.1		29.1
150	18.8	28.3		19.3	21.6	29.1		29.1
200	21.5	29.1		22.5	23.4	29.1		29.1
400	25.0	32.4	20.8	26.1	25.3	29.2	25.3	30.1
600	26.3	36.9	20.8	28.3	26.1	30.1	26.0	32.1
800	26.9	41.3	20.8	35.0	26.9	31.4	26.7	33.7
1000	28.0	45.8	20.8	31.1	27.7	32.7	27.5	34.9
1500	32.0	33.5	20.8	32.6	29.7	34.9	29.3	36.6
2000	35.6	33.5	20.8	36.4	31.7	36.0	31.2	37.8
2500	39.2	33.5	20.8	38.5	33.7	36.6	33.0	38.9
m.p. (T/K)	2890	1297	24.56	1728	2700	63.2	3300	54.36
$\Delta H_m/(\text{kJ mol}^{-1})$	27.61	10.88	0.34	17.61	26.78	0.72	29.29	0.44
b.p. (T/K)	4900	3370	27.10	3070	5300	77.35	5300	90.19
$\Delta H_v/(\text{kJ mol}^{-1})$	594.13	283.68	1.77	371.83	696.64	5.59	627.60	6.82
Molecular formula of gas	Mo	Nd	Ne	Ni	Nb	N_2	Os	O_2

T/K	Element and formula weight							
	Pd 106.4	P (red) 30.974	Pt 195.09	K 39.098	Pr 140.91	Ra 226.03	Rn 222	Re 186.2
50	8.2		10.7	20.9				7.9
100	17.9		19.6	24.6				18.0
150	22.2		23.0	26.0				22.3
200	24.2		24.6	27.0			═	24.3
400	26.6	22.5	26.5	31.5	28.4	29.3	20.8	25.9
600	27.7	25.7	27.5	30.1	31.0	33.5	20.8	26.9
800	28.9		28.6	29.7	33.7	37.7	20.8	28.0
1000	30.0		29.7	30.4	36.4	31.4	20.8	29.1
1500	33.0		32.4	20.8	33.5	31.4	20.8	31.8
2000	34.7		34.5	21.0	33.5	21.5	20.8	34.6
2500	34.7		34.7	21.6	33.5	24.4	20.8	37.3
m.p. (T/K)	1827		2045	336.4	1208	973	202	3453
$\Delta H_m/(\text{kJ mol}^{-1})$	16.74		19.66	2.32	10.04	8.37	2.90	33.05
b.p. (T/K)	3300	704	4100	1033	3270	1800	211	5900
$\Delta H_v/(\text{kJ mol}^{-1})$	393.30	(sublimes)	510.45	77.53	332.63	136.82	16.40	707.10
Molecular formula of gas	Pd		Pt	K	Pr	Ra	Rn	Re

Molar heat capacities, heats of fusion and vaporization of the elements (*contd*)

T/K	Rh 102.91	Rb 85.47	Ru 101.1	Sm 150.4	Sc 44.96	Se 78.96	Si 28.09	Ag 107.868
Element and formula weight								
50	4.9	23.4	3.7			11.0	2.2	11.6
100	15.1	25.5	13.5			18.2	7.4	20.2
150	20.2	26.5	18.9			21.7	11.9	23.0
200	22.6	27.4	21.6			23.3	15.5	24.3
400	26.4	31.4	24.5	28.4	25.6	27.8	22.3	25.9
600	28.2	31.4	25.7	30.7	26.5	35.1	24.4	27.1
800	29.9	31.4	27.0	33.1	27.4	35.1	25.6	28.5
1000	31.6	20.8	28.2	35.4	28.3	18.8	26.5	29.8
1500	35.9	20.8	30.1	33.5	30.6	19.2	28.6	31.4
2000	40.2	21.0	31.4	26.4	33.5	19.6	29.3	31.4
2500	41.8	21.8	31.4	25.4	33.5	20.0	29.3	20.8
m.p. (T/K)	2236	312.0	2600	1320	1770	493	1683	1235.1
ΔH_m/(kJ mol^{-1})	21.76	2.34	25.52	11.09	16.11	5.44	46.44	11.30
b.p. (T/K)	4000	983	4400	1870	3070	958	3600	2430
ΔH_v/(kJ mol^{-1})	495.39	69.20	567.77	191.63	304.80	26.32	—	255.06
Molecular formula of gas	Rh	Rb	Rn	Sm	Sc	Se$_2$		Ag

T/K	Na 22.990	Sr 87.62	S 32.06	Ta 180.95	Te 127.60	Tl 204.37	Th 232.04	Sn 118.69
Element and formula weight								
50	15.8		7.4	10.9	14.9	21.0	16.9	10.9
100	22.6		12.8	19.9	21.6	24.4	22.9	19.9
150	24.7		16.6	22.8	23.7	25.4	25.0	22.8
200	26.0		19.4	24.1	24.7	25.7	26.0	24.1
400	31.5	27.8	31.5	26.2	27.9	27.8	29.3	29.0
600	29.7	30.5	34.9	27.0	32.3	30.5	33.1	30.5
800	28.9	33.5	18.3	27.5	37.7	30.5	36.9	30.5
1000	29.0	39.7	18.5	27.9	37.7	30.5	40.7	30.5
1500	20.8	31.0	18.7	28.9	18.7	30.5	50.2	30.5
2000	20.8	21.0	18.9	29.7	18.7	22.7	46.0	30.5
2500	20.9	22.3	19.0	30.6	18.7	24.4	46.0	30.5
m.p. (T/K)	371	1043	388	3270	723	577	1970	505.1
ΔH_m/(kJ mol^{-1})	2.60	9.20	1.41	31.38	17.49	4.27	15.65	7.20
b.p. (T/K)	1160	1650	717.9	5700	1260	1730	4800	2900
ΔH_v/(kJ mol^{-1})	89.04	138.91	9.62	753.12	50.63	162.09	543.92	290.37
Molecular formula of gas	Na	Sr	S$_2$	Ta	Te$_2$	Tl	Th	Sn

Molar heat capacities, heats of fusion and vaporization of the elements (*contd*)

T/K	Element and formula weight							
	Ti 47.90	W 183.85	U 238.03	V 50.94	Xe 131.30	Y 88.91	Zn 65.38	Zr 91.22
50	4.7	5.8	15.5	4.0	25.1		11.1	9.2
100	14.3	16.0	22.2	13.5	28.2		19.2	18.6
150	19.5	20.5	24.7	19.1	33.6		22.5	22.3
200	22.3	22.5	26.1	22.1	20.8		23.8	24.1
400	26.6	25.1	29.6	26.1	20.8	25.6	26.4	26.6
600	28.6	25.8	34.5	27.5	20.8	26.4	28.4	28.8
800	30.0	26.5	41.3	28.7	20.8	27.3	31.4	30.7
1000	31.3	27.2	42.5	30.5	20.8	28.1	31.4	32.6
1500	33.7	28.9	38.3	36.6	20.8	30.3	20.8	30.4
2000	33.5	30.7	38.3	42.9	20.8	33.5	20.8	33.8
2500	33.5	32.4	38.3	39.7	20.8	33.5	20.8	33.5
m.p. (T/K)	1940	3660	1408	2190	161.3	1780	692.7	2123
ΔH_m/(kJ mol^{-1})	15.48	35.22	15.48	17.57	2.30	17.15	7.38	16.74
b.p. (T/K)	3570	5800	4300	3670	165.0	3600	1180	4670
ΔH_v/(kJ mol^{-1})	428.86	799.14	422.58	458.57	12.64	393.30	115.31	581.58
Molecular formula of gas . . .	Ti	W	U	V	Xe	Y	Zn	Zr

E.F.G.H.

2.10.2 Molar heat capacities of some solids at various temperatures

The values for the molar heat capacities (J mol^{-1} K^{-1}) were calculated from data given by K. K. Kelley, *US Bureau of Mines Bulletin 584*, 1960 or by C. E. Wicks and F. E. Block, *US Bureau of Mines Bulletin 605*, 1963. Some values for solid elements are given on pp. 235–249 and for other substances on pp. 235–250.

T/K	Substance and formula weight							
	BN 24.82	CaO 56.08	CaCO$_3$ calcite 100.09	CaAl$_2$Si$_2$O$_8$ glass 278.21	Fe$_3$O$_4$ 231.54	MgO 40.30	SiO$_2$ quartz 60.08	SiC 40.10
298 . .	24.26	42.82	81.84	206.99	151.64	37.79	44.45	26.64
400 . .	27.35	46.56	97.07	250.24	172.21	41.63	53.61	34.36
600 . .	33.41	49.73	110.47	286.62	212.55	45.24	64.39	41.33
800 . .	39.46	51.42	118.00	305.79	252.88	47.45	72.63	45.40
1000 . .	45.52	52.69	123.85	319.98	200.83	49.25	68.41	48.63
1200 . .	51.58	53.80	129.02		200.83	50.90	70.03	51.53
1500 . .		55.32			200.83	53.24	72.47	55.62
2000 . .						57.00		

Molar heat capacities of some solids at various temperatures (*contd*)

T/K	Substance and formula weight							
	NaCl	ThO$_2$	TiO$_2$ *rutile*	WC	UO$_2$	ZnO	ZnS	ZrO$_2$
	58.44	264.04	79.90	195.86	254.03	81.83	97.44	123.22
298 . . .	50.80	62.33	55.04	36.09	63.70	40.24	46.02	56.09
400 . . .	52.47	66.91	64.28	37.02	72.69	45.34	49.40	63.91
600 . . .	55.73	71.94	70.83	38.84	79.80	49.52	52.41	70.34
800 . . .	58.99	74.87	73.28	40.65	83.17	51.65	54.14	73.58
1000 . . .	62.26	77.66	74.54	42.47	85.45	53.19	55.50	75.91
1200 . . .	66.94	80.27	75.33	44.28	87.32	54.49	56.71	77.88
1500 . . .		84.05	76.13	47.01	89.76	56.25		74.48
2000 . . .				51.55				

E.F.G.H.

2.10.3 Molar heat capacities of gases at constant pressure at various temperatures

The values listed are the molar heat capacities, $C_p^{\ominus}$, in J mol^{-1} K^{-1} for the substances as ideal gases. Useful sources of information are *JANAF Thermochemical Tables*, 2nd edn, 1971, Project Directors D. R. Stull and H. Prophet, National Standard Reference Data Series, Nat. Bur. Stand. (US), and D. R. Stull, E. F. Westrum and G. C. Sinke, *The Chemical Thermodynamics of Organic Compounds*, 1969 (J. Wiley & Sons, Inc.).

T/K	Substance and formula weight							
	Acetone	Acetylene	Argon	Ammonia	Benzene	Bromine	Butane	Carbon dioxide
	58.0798	26.0378	39.948	17.0304	78.1134	159.808	58.123	44.0098
298 . . .	79.08	44.10	20.79	35.63	81.68	36.05	97.45	37.12
400 . . .	92.05	50.48	20.79	38.66	111.89	36.71	123.86	41.33
600 . . .	122.76	58.29	20.79	45.22	157.91	37.27	168.63	47.32
800 . . .	146.15	63.76	20.79	51.55	188.54	37.53	201.80	51.43
1000 . . .	163.80	68.27	20.79	56.35	209.89	37.70	226.87	54.31
1200 . . .		72.05	20.79	60.88		37.83		56.34

T/K	Substance and formula weight							
	Carbon disulphide	Carbon monoxide	Carbon tetrachloride	Chlorine	Chloroform	Deuterium	Diethyl ether	Ethane
	76.131	28.0104	153.823	70.906	119.3779	4.0282	74.1224	30.0694
298 . . .	45.49	29.14	83.40	33.94	65.38	29.20	112.51	52.63
400 . . .	49.44	29.34	91.70	35.30	74.25	29.24	138.11	65.61
600 . . .	54.27	30.44	99.68	36.57	85.25	29.62	183.76	89.33
800 . . .	57.04	31.90	103.09	37.15	91.51	30.05	218.66	108.07
1000 . . .	58.70	33.18	104.80	37.47	95.52	31.62	244.81	122.72
1200 . . .	59.75	34.17	105.77	37.70	98.30	32.73		

Molar heat capacities of gases at constant pressure at various temperatures (*contd*)

T/K	Substance and formula weight							
	Ethanol 46.0688	Ethylene 28.0536	Hydrochloric acid 36.4609	Hydrogen 2.0158	Hydrogen sulphide 34.0758	Methane 16.0426	Methanol 32.042	Nitric oxide 30.0061
298 . . .	65.44	42.89	29.14	28.84	34.19	35.64	43.89	29.84
400 . . .	81.00	53.05	29.17	29.18	35.58	40.50	51.42	29.94
600 . . .	107.49	70.66	29.58	29.33	38.94	52.23	67.03	31.24
800 . . .	126.90	85.31	30.50	29.65	42.52	62.93	79.66	32.77
1000 . . .	141.54	93.90	31.63	30.20	45.79	71.80	89.45	33.99
1200 . . .		101.63	32.71	30.92	48.47	78.83		34.88

T/K	Substance and formula weight							
	Nitrogen 28.0134	Nitrogen peroxide 46.0055	Nitrous oxide 44.0128	Oxygen 31.9988	Propane 44.0962	Sulphur dioxide 64.058	Sulphur trioxide 80.0582	Water 18.0152
298 . . .	29.12	36.97	38.62	29.37	73.51	39.87	50.66	33.58
400 . . .	29.25	40.17	42.62	30.11	94.31	43.49	57.67	34.25
600 . . .	30.11	45.84	48.39	32.09	129.20	49.05	67.26	36.30
800 . . .	31.43	49.71	52.24	33.74	155.14	52.43	72.76	38.69
1000 . . .	32.70	52.17	54.86	34.88	175.02	54.48	75.97	41.22
1200 . . .	33.73	53.75	56.67	35.68		55.79	77.94	43.70

E.F.G.H.

2.10.4 Cryoscopic and ebullioscopic constants, heats of fusion and vaporization for some common solvents

The cryoscopic constant of a solvent, k_f, would be the depression of freezing point in kelvin if one mole of any substance which does not dissociate or associate were dissolved in one kilogram of solvent and provided the laws of dilute solutions held at such a concentration. The analogous constant for boiling-point elevation is k_b. These constants k_f and k_b are connected with the heats of fusion, l_f, and vaporization l_v expressed in J g^{-1} by the equations

$$k_f = \boldsymbol{R} T_f^2 / 1000 \, l_f \text{ and } k_b = \boldsymbol{R} T_b^2 / 1000 \, l_v$$

where T_f and T_b are the freezing points and boiling points of the pure solvents and $\boldsymbol{R}$ is the gas constant $(8.314 \text{ J mol}^{-1} \text{ K}^{-1})$.

Cryoscopic and ebullioscopic constants, heats of fusion and vaporization for some common solvents (*contd*)

Solvent	$k_f/(\text{K kg mol}^{-1})$	$k_b/(\text{K kg mol}^{-1})$	$l_f/(\text{J g}^{-1})$	$l_v/(\text{J g}^{-1})$
Acetic acid	3.90	3.07	195	405
Acetone.	2.40	1.71	98	524
Aniline	5.87	3.22	88	460
Benzene.	5.12	2.53	126	394
Camphor	40	—	45	—
Carbon disulphide . . .	3.8	2.37	58	351
Carbon tetrachloride . .	30	4.95	16	195
Chloroform.	4.90	3.66	80	246
Cyclohexane	20.1	2.79	32	356
Diethyl ether	1.79	1.82	97	358
Naphthalene	6.94	5.8	152	316
Nitrobenzene	6.90	5.26	94	331
Phenol	7.27	3.04	122	485
Pyridine	4.75	—	94	460
Water	1.86	0.51	333	2257

Heats of vaporization at various temperatures

Approximate values of the heat of vaporization can be computed from vapour pressure–temperature tables (see p. 173) by the use of the Clapeyron equation. For conditions where the molar volume of the liquid is small compared with the molar volume of the vapour and where the vapour can be treated as an ideal gas the molar heat of vaporization H_v in J mol^{-1} can be calculated from the expression

$$H_v = \mathbf{R}T^2 p^{-1}(\mathrm{d}T/\mathrm{d}p)^{-1}$$

Values of $\mathrm{d}T/\mathrm{d}p$ for $p = 101.325$ kN m^{-2} for many substances are given on pp. 174–181.

The following tables give values for the heats of vaporization (J g^{-1}) for several substances at various temperatures (°C).

Substance	$\theta/°\text{C}$						
	0	20	40	60	80	120	140
Acetone	585	563	541	519	498	458	438
Carbon disulphide	375	367	357	345	331	300	286
Carbon tetrachloride	—	214	208	206	191	171	163
Chloroform	280	273	265	257	248	229	219
Diethyl ether	393	385	374	361	344	342	277
Sulphur dioxide	384	352	—	—	—	—	—

Some values for the heats of vaporization (J g^{-1}) at the normal boiling points are given in the preceding table.

Heats of vaporization for water at various temperatures

The subjoined highly accurate data for *water* which refer to ITS-68 are recalculated from those due to Osborne, Simpson and Ginnings, *J. Res. Bur. Stand.*, 1939, **23**, 197, 261.

$\theta/°C$	Heat vaporization/$(J\ g^{-1})$	$\theta/°C$	Heat vaporization/$(J\ g^{-1})$	$\theta/°C$	Heat vaporization/$(J\ g^{-1})$
0	2500.58	30	2429.95	90	2282.72
5	2488.87	40	2406.14	100	2256.67
10	2477.14	50	2382.10	130	2174.17
15	2465.39	60	2357.82	150	2114.39
20	2453.59	70	2333.20	170	2049.60
25	2441.78	80	2308.19	200	1940.56

E.F.G.H.

2.10.5 Standard free energies, heats of formation, entropies and molar heat capacities

The chemical species for which data are included in the following table are either of common occurrence or they have properties of particular interest. Most of the numerical values quoted are derived from the very comprehensive collection of data in *Selected Values of Chemical Thermodynamic Properties*, National Bureau of Standards Circular 500, Technical Notes 270–3, 270–4 and 270–5.

Under 'State' s, sI and sII refer to different crystalline forms; allomorphic forms are also indicated: s. solid; l, liquid; g, gas; a, amorphous; and aq., aqueous solution. The entry hyp. m = 1 in this table means a hypothetical concentration of 1 mole in 1 kg water.

$\Delta G_f^\ominus$ is the standard free energy of formation and $\Delta H_f^\ominus$ is the standard heat of formation of the compound. These are incremental values associated with the formation of the compound from its elements when the product and reactants are taken in their appropriate standard states at 298.15 K. The standard state for a liquid or solid is that of the component under 101.325 kN m^{-2} (1 standard atmosphere) pressure and for a gas that of the ideal gas at a pressure of 101.325 kN m^{-2}. By definition the free energy of formation and heat of formation of an element in the standard state is zero, and by convention the corresponding quantities for the hydrogen ion in the standard state of unit activity in aqueous solution (i.e. hyp. m = 1) are taken as equal to zero.

For entities other than ions, $S^\ominus$ is the entropy in the standard state as above defined taking as zero the entropy at 0 K of the crystalline form in which all the molecules are orientated regularly. Thus nuclear contributions and those arising from isotopic mixing are excluded. For ions the entropy (as well as $\Delta G_f^\ominus$ and $\Delta H_f^\ominus$) is taken by convention with respect to the hydrogen ion in the standard state as zero.

$C_p^\ominus$ is the molar heat capacity of the substance in the standard state at constant pressure.

Inorganic compounds

Substance and formula	State	$\Delta G_f^\ominus$ kJ mol^{-1}	$\Delta H_f^\ominus$ kJ mol^{-1}	$S^\ominus$ JK^{-1} mol^{-1}	$C_p^\ominus$ JK^{-1} mol^{-1}
Aluminium					
Al	s	0	0	28.3	24.4
Al$_2$O$_3$	s(α, corundum)	−1582.4	−1675.7	50.9	79.0
Al$_2$O$_3$	s, γ		−1652.7		
AlF$_3$	s	−1425.1	−1504.2	66.4	75.2
AlCl$_3$	s	−628.9	−704.2	110.7	91.8
AlBr$_3$	s		−527.2		101.7
Al(NO$_3$)$_3$.9H$_2$O . . .	s		−3757		
Al$_2$(SO$_4$)$_3$	s	−3100.1	−3440.8	239.3	259.4
NH$_4$Al(SO$_4$)$_2$.12H$_2$O . .	s	−4938.0	−5942.4	697.1	683.2
Al^{3+}	hyp. m = 1	−485.3	−531.4	−321.7	

Inorganic compounds (*contd*)

Substance and formula	State	$\Delta G_f^{\ominus}$ kJ mol^{-1}	$\Delta H_f^{\ominus}$ kJ mol^{-1}	$S^{\ominus}$ JK^{-1} mol^{-1}	$C_p^{\ominus}$ JK^{-1} mol^{-1}
Antimony					
Sb.	sIII	0	0	45.7	25.2
Sb_2O_4	s	−795.8	−907.5	127.2	114.6
Sb_2O_5	s	−829.3	−971.9	125.1	
SbF_3	s		−915.5		
$SbCl_3$	s	−323.7	−382.2	184.1	108.0
$SbCl_5$	l	−350.2	−440.2	301.3	
Sb_2S_3	s, black	−173.6	−174.8	182.0	119.9
Argon					
Ar.	g	0	0	154.7	20.8
Arsenic					
As.	s, α, grey metallic	0	0	35.1	24.6
As_2O_5	s	−782.4	−924.9	105.4	116.5
As_4O_6	s, octahedral	−1152.5	−1313.9	214.2	191.3
AsH_3	g	68.9	66.4	222.7	38.1
AsF_3	l	−909.1	−956.3	181.2	126.6
$AsCl_3$	l	−256.7	−305.0	207.5	
As_2S_3	s	−168.6	−169.0	163.6	116.3
H_3AsO_3	hyp. m = 1	−639.9	−742.2	195.0	
H_3AsO_4	hyp. m = 1	−766.1	−902.5	184.1	
$H_2AsO_3^-$	hyp. m = 1	−587.2	−714.8	110.5	
$H_2AsO_4^-$	hyp. m = 1	−753.3	−909.6	117.2	
$HAsO_4^{2-}$	hyp. m = 1	−714.7	−906.3	−1.7	
Barium					
Ba.	sII	0	0	66.9	26.4
BaO	s	−528.4	−558.1	70.3	47.4
BaF_2	s	−1148.5	−1200.4	96.2	71.2
$BaCl_2.2H_2O$. . . .	s	−1295.8	−1461.7	202.9	155.2
$Ba(OH)_2.8H_2O$. . .	s		−3345.1		
$BaSO_4$	s	−1353.1	−1465.2	132.2	101.8
$BaCO_3$	sII, witherite	−1138.9	−1218.8	112.1	85.4
Ba^{2+}	hyp. m = 1	−560.7	−538.4	12.6	
Beryllium					
Be.	s	0	0	9.5	17.8
BeO	s	−581.6	−610.9	14.1	25.4
Be^{2+}	aq. acid soln		−389.1		
Bismuth					
Bi.	s	0	0	56.7	25.5
Bi_2O_3.	s	−493.7	−573.9	151.5	113.5
$BiCl_3$.	s	−315.1	−379.1	177.0	104.6
BiOCl	s	−322.2	−366.9	120.5	
Bi_2S_3.	s	−140.6	−143.1	200.4	122.2
Boron					
B	s, β	0	0	5.9	11.1
B_2O_3	s	−1193.7	−1272.8	54.0	62.9
BF_3	g	−1120.3	−1137.0	254.0	50.5

Inorganic compounds (*contd*)

Substance and formula	State	$\Delta G_f^{\ominus}$ kJ mol^{-1}	$\Delta H_f^{\ominus}$ kJ mol^{-1}	$S^{\ominus}$ JK^{-1} mol^{-1}	$C_p^{\ominus}$ JK^{-1} mol^{-1}
BCl$_3$	l	−387.4	−427.2	206.3	106.7
B$_2$H$_6$	g	86.6	35.6	232.0	56.9
BN	s	−228.4	−254.4	14.8	19.7
H$_3$BO$_3$	s	−969.0	−1094.3	88.8	81.4
BF$_4^-$	hyp. m = 1	−1487.0	−1574.9	179.9	
Bromine					
Br$_2$	l	0	0	152.2	75.7
Br$_2$	g	3.1	30.9	245.4	36.0
HBr	g	−53.4	−36.4	198.6	29.1
Br$^-$	hyp. m = 1	−104.0	121.5	82.4	−141.8
BrO$_3^-$	hyp. m = 1	1.7	−83.7	163.2	
Cadmium					
Cd	s, γ	0	0	51.8	26.0
CdO	s	−228.4	−258.2	54.8	43.4
Cd(OH)$_2$	s, precipitated	−473.6	−560.7	96.2	
CdCl$_2$.2½H$_2$O	s	−944.1	−1131.9	227.2	
CdBr$_2$.4H$_2$O	s	−1248.0	−1492.6	316.3	
CdSO$_4$	s	−822.8	−933.3	123.0	99.6
CdS	s	−156.5	−161.9	64.9	
Cd^{2+}	hyp. m = 1	−77.6	−75.9	−73.2	
Caesium					
Cs	s	0	0	82.8	31.0
CsCl	sII		−433.0		
Cs$^+$	hyp. m = 1	−282.0	−247.7	133.1	
Calcium					
Ca	sII	0	0	41.6	26.3
CaO	s	−604.2	−635.5	39.7	42.8
Ca(OH)$_2$	s	−896.8	−986.6	76.1	84.5
CaH$_2$	s	−149.8	−188.7	41.8	
CaC$_2$	s	−67.8	−62.8	70.3	62.3
CaCO$_3$	s, calcite	−1128.8	−1206.9	92.9	81.9
CaCO$_3$	s, aragonite	−1127.7	−1207.0	88.7	81.3
CaF$_2$	s	−1161.9	−1214.6	68.9	67.0
CaCl$_2$	s	−750.2	−795.0	113.8	72.6
Ca$_3$(PO$_4$)$_2$	s, α	−3889.9	−4126.3	241.0	231.6
Ca$_3$(PO$_4$)$_2$	s, β	−3899.5	−4137.6	236.0	227.8
CaSO$_4$	s, anhydrite	−1320.3	−1432.7	106.7	99.6
CaSO$_4$	s, soluble α	−1311.8	−1423.7	108.4	100.0
CaSO$_4$	s, soluble β	−1307.3	−1419.3	108.4	99.6
CaSO$_4$.½H$_2$O	s, α	−1435.2	−1575.2	130.5	119.7
CaSO$_4$.½H$_2$O	s, β	−1434.2	−1573.1	134.3	123.8
CaSO$_4$.2H$_2$O	s	−1795.7	−2021.1	194.0	186.2
Ca(NH$_2$)$_2$	s		−383.2		
Ca(NO$_3$)$_2$.4H$_2$O . . .	sII	−1700.8	−2131.2	338.9	
Ca^{2+}	hyp. m = 1	−553.0	−543.0	−55.2	

Inorganic compounds (*contd*)

Substance and formula	State	$\Delta G_f^\ominus$ kJ mol^{-1}	$\Delta H_f^\ominus$ kJ mol^{-1}	$S^\ominus$ JK^{-1} mol^{-1}	$C_p^\ominus$ JK^{-1} mol^{-1}
Carbon					
C	s, graphite	0	0	5.7	8.5
C	s, diamond	2.9	1.9	2.4	6.1
CO	g	−137.2	−110.5	197.6	29.1
CO$_2$	g	−394.4	−393.5	213.6	37.1
CO$_2$	hyp. m = 1	−386.0	−413.8	117.6	
CF$_4$	g	−879.6	−924.7	261.5	61.1
CCl$_4$	g	−65.3	−135.4	216.4	131.8
COCl$_2$	g	−204.6	−218.8	283.4	57.7
CS$_2$	l	65.3	89.7	151.3	75.7
CNI	s	185.0	166.1	96.2	
HCN	g	124.7	135.1	201.7	35.9
HCN	l	124.9	108.9	112.8	70.6
C$_2$N$_2$	g	297.4	308.9	241.8	56.8
CO$_3^{2-}$	hyp. m = 1	−527.9	−677.1	−56.9	
HCO$_3^-$	hyp. m = 1	−586.8	−692.0	91.2	
CN$^-$	hyp. m = 1	172.4	150.6	94.1	
CNS$^-$	aq.	92.7	76.4	144.3	−40.2
H.COO$^-$ . . . . . formate	hyp. m = 1	−351.0	−426.6	92.0	−87.9
C$_2$H$_3$O$_2^-$ . . . . . acetate	hyp. m = 1	−369.4	−486.0	86.6	−6.3
C$_2$O$_4^{2-}$ . . . . . oxalate	hyp. m = 1	−674.0	−825.1	45.6	
HC$_2$O$_4^-$ . . . . . bioxalate	hyp. m = 1	−698.4	−818.4	149.4	
Cerium					
Ce	sIII	0	0	57.7	25.9
Ce^{3+}	hyp. m = 1	−713.4	−727.2	−184.1	
Chlorine					
Cl$_2$	g	0	0	223.0	33.9
Cl$_2$O	g	97.9	80.3	266.1	45.4
ClO$_2$	g	120.5	102.5	256.7	42.0
Cl$_2$O$_7$	g		272.0		
HCl	g	−95.3	−92.3	186.8	29.1
Cl$^-$	hyp. m = 1	−131.3	−167.2	56.5	−136.4
ClO$_3^-$	hyp. m = 1	−3.3	−99.2	162.3	
ClO$_4^-$	hyp. m = 1	−8.6	−129.3	182.0	
Chromium					
Cr	s	0	0	23.8	23.3
Cr$_2$O$_3$	s	−1058.1	−1139.7	81.2	118.7
CrO$_3$	s		−589.5		
CrCl$_2$	s	−356.1	−395.4	115.3	71.2
CrCl$_3$	s	−486.2	−556.5	123.0	91.8
CrO$_2$Cl$_2$	l	−510.9	−579.5	221.8	
Cr^{2+}	aq.		−143.5		
[Cr.6H$_2$O]$^{3+}$ . . .	aq. violet		−1999.1		
HCrO$_4^-$	hyp. m = 1	−764.8	−878.2	184.1	
CrO$_4^{2-}$	hyp. m = 1	−727.8	−881.2	50.2	
Cr$_2$O$_7^{2-}$	hyp. m = 1	−1300.2	−1490.3	261.9	

Inorganic compounds (*contd*)

Substance and formula	State	$\Delta G_f^{\ominus}$ kJ mol^{-1}	$\Delta H_f^{\ominus}$ kJ mol^{-1}	$S^{\ominus}$ JK^{-1} mol^{-1}	$C_p^{\ominus}$ JK^{-1} mol^{-1}
Cobalt					
Co	sIII	0	0	30.0	24.8
CoCl$_2$	s	-269.9	-312.5	109.2	78.5
CoCl$_2$.6H$_2$O	s	-1725.5	-2115.4	343.1	
CoSO$_4$.7H$_2$O	s	-2473.8	-2979.9	406.1	390.5
Co^{2+}	hyp. m = 1	-54.4	-58.2	-113.0	
Co^{3+}	hyp. m = 1	133.9	92.0	-305.4	
Copper					
Cu	s	0	0	33.1	24.4
Cu$_2$O	s	-146.0	-168.6	93.1	63.6
CuCl	s	-119.9	-137.2	86.2	48.5
CuCl$_2$	s	-175.7	-220.1	108.1	57.8
CuCl$_2$.2H$_2$O	s	-656.1	-821.3	167.4	
Cu(NO$_3$)$_2$.3H$_2$O . . .	s		-1217.1		
CuSO$_4$.5H$_2$O	sII	-1880.1	-2279.7	300.4	280.3
CuI	s	-69.5	-67.8	96.7	54.1
Cu$_2$S	s, α	-86.2	-79.5	120.9	76.3
CuS	s	-53.6	-53.1	66.5	47.8
Cu$^+$	hyp. m = 1	50.0	71.7	40.6	
Cu^{2+}	hyp. m = 1	65.5	64.8	-99.6	
Deuterium					
D$_2$	g	0	0	144.9	29.2
HD	g	-1.5	0.3	143.7	29.2
D$_2$O	g	-234.6	-249.2	198.2	34.3
D$_2$O	l	-243.5	-294.6	75.9	84.3
HDO	g	-233.1	-245.3	199.4	33.8
Erbium					
Er	s	0	0		28.2
Er^{3+}	hyp. m = 1	-659.0	-679.1		
Europium					
Eu	s	0	0		26.7
Eu^{3+}	hyp. m = 1	-690.8	-708.3		
Fluorine					
F$_2$	g	0	0	202.7	31.3
HF	g	-273.2	-271.1	173.7	29.1
F$^-$	hyp. m = 1	-278.8	-332.6	-13.8	-106.7
HF$_2^-$	aq.	-578.1	-649.9	92.5	
Gadolinium					
Gd	s	0	0	58.6	36.5
Gd^{3+}	hyp. m = 1	-688.7	-706.3	-197.1	
Gallium					
Ga	sI	0	0	40.9	25.9
Ga^{3+}	hyp. m = 1	-159.0	-211.7	-330.5	
GaO$_3^{3-}$	hyp. m = 1		-619.2		

Inorganic compounds (*contd*)

Substance and formula	State	ΔG_f° kJ mol^{-1}	ΔH_f° kJ mol^{-1}	S° JK^{-1} mol^{-1}	C_p° JK^{-1} mol^{-1}
Germanium					
Ge	s	0	0	31.1	23.3
GeO$_2$	s, hexagonal	−497.1	−551.0	55.3	52.1
GeH$_4$	g	113.4	90.8	217.0	45.0
Gold					
Au	s	0	0	47.4	25.4
Au(OH)$_3$	s, precipitated	−317.0	−424.7	189.5	
AuCl	s		−34.7		
AuCl$_3$	s		−117.6		
AuCl$_4^-$	hyp. m = 1	−235.2	−322.2	266.9	
Au(CN)$_2^-$	hyp. m = 1	285.8	242.3	171.5	
Hafnium					
Hf	s, α hexagonal	0	0	43.6	25.7
HfO$_2$	s	−1027.1	−1144.7	59.3	60.3
Helium					
He	g	0	0	126.0	20.8
Holmium					
Ho	s	0	0		27.3
Ho^{3+}	hyp. m = 1	−666.1	−684.9		
Hydrogen					
H$_2$	g	0	0	130.6	28.8
H	g	203.3	218.0	114.6	20.8
H$_2$O	g	−228.6	−241.8	188.7	33.6
H$_2$O	l	−237.2	−285.8	69.9	75.3
H$_2$O$_2$	g	−105.6	−136.3	232.6	43.1
H$_2$O$_2$	l	−120.4	−187.8	109.6	89.1
OH	g	34.2	39.0	183.6	29.0
H$^+$	hyp. m = 1	0	0	0	0
OH$^-$	hyp. m = 1	−157.3	−230.0	−10.8	−148.5
Indium					
In	s	0	0	57.8	26.7
InCl$_3$	s		−537.2		
In$_2$(SO$_4$)$_3$	s	−2439.3	−2786.5	272.0	280.3
In^{3+}	hyp. m = 1	−97.9	−104.6	−150.6	
Iodine					
I$_2$	s	0	0	116.1	54.4
I$_2$	g	19.4	62.4	260.6	36.9
I$_2$O$_5$	s		−158.1		
ICl	g	−5.4	17.8	247.4	35.6
ICl$_3$	s	−22.3	−89.5	167.4	
HI	g	1.7	26.5	206.5	29.2
HIO$_3$	s		−230.1		
I$^-$	hyp. m = 1	−51.6	−55.2	111.3	−142.3
I$_3^-$	hyp. m = 1	−51.5	−51.5	239.3	
IO$_3^-$	hyp. m = 1	−128.0	−221.3	118.4	

Inorganic compounds (*contd*)

Substance and formula	State	$\Delta G_f^{\ominus}$ kJ mol^{-1}	$\Delta H_f^{\ominus}$ kJ mol^{-1}	$S^{\ominus}$ JK^{-1} mol^{-1}	$C_p^{\ominus}$ JK^{-1} mol^{-1}
Iridium					
Ir	s	0	0	35.5	25.1
IrCl	s		−81.6		
IrCl$_3$	s		−245.6		
IrCl$_6^{2-}$	aq.		−619.7		
IrCl$_6^{3-}$	aq.		−751.0		
Iron					
Fe	s, α	0	0	27.3	25.1
Fe$_{0.947}$O	wustite	−245.1	−266.3	57.5	48.1
Fe$_2$O$_3$	hematite	−742.2	−824.2	87.4	103.8
Fe$_3$O$_4$	magnetite	−1015.5	−1118.4	146.4	143.4
FeCl$_2$	s	−302.3	−341.8	117.9	76.7
FeCl$_3$	s	−334.1	−399.5	142.3	96.7
FeCl$_3$.6H$_2$O	s		−2223.8		
FeSO$_4$.7H$_2$O	s	−2510.3	−3004.6	409.2	294.5
Fe$_{1.000}$S	Iron-rich pyrrhotite, α	−100.4	−100.0	60.3	50.5
FeS$_2$	s, pyrites	−166.9	−178.2	52.9	62.2
FeS$_2$	s, magkasite		−154.8		
Fe(NO$_3$)$_3$.9H$_2$O . . .	s		−3285.3		
Fe(CO)$_5$	l	−705.4	−774.0	338.1	240.6
Fe$_3$C	s, cementite	20.1	25.1	104.6	105.9
FeCO$_3$	s, siderite	−666.7	−740.6	92.9	82.1
Fe^{2+}	hyp. m = 1	−78.9	−89.1	−137.7	
Fe^{3+}	hyp. m = 1	−4.6	−48.5	−315.9	
Krypton					
Kr	g	0	0	164.0	20.8
KrF$_2$	g		58.6		
Lanthanum					
La	sIII	0	0	57.3	27.6
La$_2$(SO$_4$)$_3$.9H$_2$O . . .	s	−5870.6			
La^{3+}	hyp. m = 1	−723.4	−737.2	−184.1	
Lead					
Pb	s	0	0	64.8	26.4
PbO	s, red	−188.9	−219.0	66.5	45.8
PbO	s, yellow	−187.9	−217.3	68.7	46.8
PbO$_2$	s	−217.4	−277.4	68.6	64.6
Pb$_3$O$_4$	s	−601.2	−718.4	211.3	146.9
PbF$_2$	s	−617.1	−664.0	110.5	
PbCl$_2$	s	−314.1	−359.4	136.0	
PbI$_2$	s	−173.6	−175.5	174.8	77.4
Pb(NO$_3$)$_2$	s		−451.9		
PbSO$_4$	s	−813.2	−919.9	148.6	103.2
PbS	s	−98.7	−100.4	91.2	49.5
Pb(N$_3$)$_2$	s, monoclinic	624.7	478.2	148.1	
Pb(C$_2$H$_5$)$_4$	l		52.7		
Pb(CH$_3$COO)$_2$.3H$_2$O . .	s		−1851.5		
Pb^{2+}	hyp. m = 1	−24.4	−1.7	10.5	

Inorganic compounds (*contd*)

Substance and formula	State	$\Delta G_f^{\ominus}$ kJ mol^{-1}	$\Delta H_f^{\ominus}$ kJ mol^{-1}	$S^{\ominus}$ JK^{-1} mol^{-1}	$C_p^{\ominus}$ JK^{-1} mol^{-1}
Lithium					
Li	s	0	0	28.0	23.6
LiCl	s		−408.8		
LiAlH$_4$	s		−101.3		
LiBH$_4$	s		−186.6		
Li$^+$	hyp. m = 1	−293.8	−278.5	14.2	
Lutecium					
Lu	s	0	0		26.9
Lu^{3+}	hyp. m = 1	−648.5	−669.9		
Magnesium					
Mg	s	0	0	32.5	23.9
MgO	s	−569.6	−601.8	26.8	37.4
MgO	s, finely divided	−566.1	−598.1	27.9	37.8
MgCl$_2$.6H$_2$O	s	−2115.6	−2499.6	366.1	315.7
MgF$_2$	s	−1049.3	−1102.5	57.2	61.6
Mg(NO$_3$)$_2$.6H$_2$O . . .	s		−2612.3		
MgSO$_4$.7H$_2$O	s		−3383.6		
Mg(ClO$_4$)$_2$.2H$_2$O . . .	s		−1216.3		
Mg(ClO$_4$)$_2$	s		−588.3		
MgCO$_3$	s	−1029.3	−1112.9	65.7	75.5
Mg^{2+}	hyp. m = 1	−456.0	−462.0	−118.0	
Manganese					
Mn	s, α	0	0	32.0	26.3
Mn	s, γ	1.4	1.5	32.4	27.6
MnO	s	−362.9	−385.2	59.7	45.4
MnO$_2$	s	−465.2	−520.0	53.1	54.1
MnCl$_2$	s	−440.5	−481.3	118.2	72.9
MnSO$_4$.4H$_2$O	s		−2258.1		
Mn^{2+}	hyp. m = 1	−228.0	−220.7	−73.6	50.2
MnO$_4^-$	aq.	−447.3	−541.4	191.2	
Mercury					
Hg	l	0	0	76.0	28.0
Hg	g	31.9	61.3	174.8	20.8
HgO	s, red, orthorhombic	−58.6	−90.8	70.3	44.1
HgO	s, yellow	−58.4	−90.5	71.1	
Hg$_2$Cl$_2$	s	−210.8	−265.2	192.5	
HgCl$_2$	s	−178.7	−224.3	146.0	
Hg$_2$I$_2$	s	−111.0	−121.3	233.5	
HgI$_2$	s, red	−101.7	−105.4	179.9	
HgI$_2$	s, yellow		−102.9		
HgS	s, red	−50.6	−58.1	82.4	48.4
HgS	s, black	−47.7	−53.6	88.3	
Hg(ONC)$_2$	s, fulminate		267.8		
Hg$_2^{2+}$	hyp. m = 1	153.6	172.4	84.5	
Hg^{2+}	hyp. m = 1	164.4	171.1	−32.2	
Molybdenum					
Mo	s	0	0	28.7	24.1
H$_2$MoO$_4$.H$_2$O	s, yellow		−1359.8		
MoO$_3$	s	−668.0	−745.1	77.7	75.0

Inorganic compounds (*contd*)

Substance and formula	State	$\Delta G_f^\ominus$ kJ mol^{-1}	$\Delta H_f^\ominus$ kJ mol^{-1}	$S^\ominus$ JK^{-1} mol^{-1}	$C_p^\ominus$ JK^{-1} mol^{-1}
Neodymium					
Nd	sIII	0	0		30.1
Nd^{3+}	hyp. m = 1	−701.2	−716.3		
Neon					
Ne	g	0	0	146.2	20.8
Nickel					
Ni	s	0	0	29.9	26.1
NiO	s	−211.7	−239.7	38.0	44.3
NiCl$_2$.6H$_2$O	s	−1713.5	−2103.2	344.3	
Ni(NO$_3$)$_2$.6H$_2$O . . .	s		−2211.7		464.4
NiSO$_4$.6H$_2$O	s, α	−2225.0	−2682.8	334.5	327.9
Ni(CO)$_4$	tetragonal, green l	−588.3	−633.0	313.4	204.6
Ni^{2+}	hyp. m = 1	−45.6	−54.0	−128.9	
Ni(CN)$_4^{2-}$	hyp. m = 1	472.0	367.8	217.6	
Niobium					
Nb	s	0	0	36.4	24.6
Nb$_2$O$_5$	s, high temp. form	−1766.1	−1899.5	−137.2	132.1
Nitrogen					
N$_2$	g	0	0	191.5	29.1
NO	g	86.6	90.2	210.7	29.8
NO$_2$	g	51.3	33.2	240.0	37.2
N$_2$O$_4$	g	97.8	9.2	304.2	77.3
N$_2$O	g	104.2	82.0	219.7	38.5
N$_2$O$_5$	g	115.1	11.3	355.6	84.5
NH$_3$	g	−16.5	−46.1	192.3	35.1
NH$_3$	hyp. m = 1	−26.6	−80.3	111.3	
N$_2$H$_4$	l	149.2	50.6	121.2	98.9
N$_2$H$_4$.HCl	s		−196.7		
NH$_2$OH.HCl	s		−317.6		
HNO$_3$	l	−80.8	−174.1	155.6	109.9
HNO$_3$	hyp. m = 1	−111.3	−207.4	146.4	−86.6
NH$_4$NO$_3$	s	−184.0	−365.6	151.1	139.3
NH$_4$NO$_2$	s		−256.5		
NOCl	g	66.1	51.7	261.6	44.7
NH$_4$Cl	s	−203.0	−314.4	94.6	84.1
NH$_4$HSO$_4$	s		−1027.0		
(NH$_4$)$_2$SO$_4$	s	−901.9	−1180.9	220.1	187.5
HNO$_2$	hyp. m = 1	−55.6	−119.2	152.7	
NH$_4$HS	s	−50.6	−156.9	97.5	
(NH$_4$)$_2$S	hyp. m = 1	−72.8	−231.8	212.1	
(NH$_4$)$_2$S$_2$	hyp. m = 1	−79.1	−234.7	255.2	
NH$_4^+$	hyp. m = 1	−79.4	−132.5	113.4	79.9
NO$_3^-$	hyp. m = 1	−111.3	−207.4	146.4	−86.6
NO$_2^-$	hyp. m = 1	−37.2	−104.6	140.2	−97.5
Osmium					
Os	s	0	0	32.6	24.7
OsO$_4$	s, white	−303.8	−385.8	167.8	
OsO$_4$	s, yellow	−305.0	−394.1	143.9	

Inorganic compounds (*contd*)

Substance and formula	State	$\Delta G_f^{\ominus}$ kJ mol^{-1}	$\Delta H_f^{\ominus}$ kJ mol^{-1}	$S^{\ominus}$ JK^{-1} mol^{-1}	$C_p^{\ominus}$ JK^{-1} mol^{-1}
Oxygen					
O_2	g	0	0	205.0	29.4
O_3	g	163.2	142.7	238.8	39.2
Palladium					
Pd.	s	0	0	37.6	26.0
$PdCl_2$	s	−125.1	−171.5	104.6	
$PdCl_4^{2-}$	m = 1 in 1N HCl	−416.7	−522.2	259.4	
$PdCl_6^{2-}$	m = 1 in 1N HCl	−430.1	−598.3	272.0	
Phosphorus					
P	s, α, white	0	0	41.1	23.8
P	s, red, triclinic	−12.1	−17.6	22.8	21.2
P	s, black		−39.3		
P_4O_{10}	s, hexagonal	−2697.8	−2984.0	228.9	211.7
PH_3	g	13.4	5.4	210.1	37.1
PCl_3	g	−267.8	−287.0	311.7	71.8
PCl_3	l	−272.4	−319.7	217.1	
PCl_5	g		−443.5		
PCl_5	s	−305.0	−374.9	364.5	112.8
PBr_3	g	−162.8	−139.3	348.0	76.0
PBr_3	l	−175.7	−184.5	240.2	
$POCl_3$	g	−513.0	−558.5	325.3	84.9
$POCl_3$	l	−520.9	−597.1	222.5	138.8
PBr_5	s		−269.9		
H_3PO_4	s	−1119.22	−1279.0	110.5	106.1
$(NH_4)_2HPO_4$	s		−1566.9		
PO_4^{3-}	hyp. m = 1	−1018.8	−1277.4	−221.8	
HPO_4^{2-}	hyp. m = 1	−1089.3	−1292.1	−33.5	
$H_2PO_4^{-}$	hyp. m = 1	−1130.4	−1296.3	90.4	
Platinum					
Pt	s	0	0	41.63	25.85
$PtCl_2$	s		−109.4		
$PtCl_4^{2-}$	hyp. m = 1	−368.6	−503.3	167.4	
$PtCl_6^{2-}$	hyp. m = 1	−489.5	−673.6	220.1	
Potassium					
K	s	0	0	63.6	29.2
K_2O	s		−361.5		
KOH	s		−425.8		
KF	s	−533.1	−562.6	66.6	49.1
KCl	s	−408.3	−435.9	82.7	51.5
KBr	s	−379.2	−392.2	96.4	53.6
KI	s	−322.3	−327.6	104.3	55.1
KHF_2	s	−852.4	−920.4	104.3	76.9
KNO_3	s	−393.1	−492.7	132.9	96.3
KNO_2	s		−370.3		
K_2SO_4	s	−1316.4	−1433.7	175.7	130.1
$KClO_3$	s	−289.9	−391.2	143.0	100.2
$KClO_4$	s	−304.2	−433.5	151.0	110.2
$KBrO_3$	s	−243.5	−332.2	149.2	104.9

Inorganic compounds (*contd*)

Substance and formula	State	$\Delta G_f^{\ominus}$ kJ mol^{-1}	$\Delta H_f^{\ominus}$ kJ mol^{-1}	$S^{\ominus}$ JK^{-1} mol^{-1}	$C_p^{\ominus}$ JK^{-1} mol^{-1}
KIO$_3$	s	−423.4	−508.4	151.5	106.4
K$_2$CO$_3$	s		−1146.1		
KHCO$_3$	s		−959.4		
KCN	s		−112.5		
KCNO	s		−412.1		
KCNS	s		−203.4		
K$_3$Fe(CN)$_6$	s		−173.2		
K$_4$Fe(CN)$_6$	s		−523.4		
KMnO$_4$	s	−713.8	−813.4	171.7	119.2
K$_2$CrO$_4$	s		−1382.8		
K$_2$Cr$_2$O$_7$	s		−2033.0		
KH$_2$PO$_4$	s		−1568.6		
K$_2$S$_2$O$_8$	s		−1917.5		
K$^+$	hyp. m = 1	−282.3	−251.2	102.5	
Praseodymium					
Pr	s	0	0		28.5
Pr^{3+}	hyp. m = 1	−707.5	−722.6		
Radium					
Ra	s	0	0	71.1	27.0
Ra^{2+}	hyp. m = 1	−562.7	−527.2	54.4	
Radon					
Rn	g	0	0	176.1	20.8
Rhenium					
Re	s	0	0	36.9	25.5
ReO$_4^-$	aq.	−694.5	−787.4	201.3	−13.4
Rhodium					
Rh	s	0	0	31.5	25.0
RhCl$_3$	s		−299.2		
Rubidium					
Rb	s	0	0	69.5	30.4
RbCl	s		−430.6		
Rb$^+$	hyp. m = 1	−282.2	−246.4	124.3	
Ruthenium					
Ru	s	0	0	28.5	24.1
RuCl$_3$	s, black		−205.0		
Samarium					
Sm	s	0	0		27.1
SmCl$_3$	s		−1045.2		
Scandium					
Sc	s	0	0	34.6	25.5
Sc^{3+}	hyp. m = 1	−586.6	−614.2	−255.2	

Inorganic compounds (*contd*)

Substance and formula	State	$\Delta G_f^{\ominus}$ kJ mol^{-1}	$\Delta H_f^{\ominus}$ kJ mol^{-1}	$S^{\ominus}$ JK^{-1} mol^{-1}	$C_p^{\ominus}$ JK^{-1} mol^{-1}
Selenium					
Se	s, black hexagonal	0	0	42.4	25.4
SeO$_2$	s		− 225.4		
H$_2$Se	g	15.9	29.7	218.9	34.7
SeO$_3^{2-}$	hyp. m = 1	− 369.9	− 509.2	12.6	
SeO$_4^{2-}$	hyp. m = 1	− 441.4	− 599.1	54.0	
Silicon					
Si	s	0	0	18.8	20.0
SiO$_2$	s, α, quartz	− 856.7	− 910.9	41.8	44.4
SiO$_2$	s, α, cryostobalite	− 855.9	− 909.5	42.7	44.2
SiO$_2$	s, α, tridymite	− 855.3	− 909.1	43.5	44.6
SiO$_2$	a	− 850.7	− 903.5	48.9	44.4
SiH$_4$	g	56.9	34.3	204.5	42.8
SiF$_4$	g	− 1572.7	− 1614.9	282.4	73.6
SiCl$_4$	g	− 617.0	− 657.0	330.6	90.2
SiCl$_4$	l	− 619.9	− 687.0	239.7	145.3
SiC	s, β, cubic	− 62.8	− 65.3	16.6	26.9
SiF$_6^{2-}$	aq.	− 2199.5	− 2389.1	122.2	
Silver					
Ag	s	0	0	42.6	25.4
Ag$_2$O	s	− 11.2	− 31.0	121.3	65.9
AgCl	s	− 109.8	− 127.1	96.2	50.8
AgBr	s	− 96.9	− 100.4	107.1	52.4
AgI	s	− 66.2	− 61.8	115.5	56.8
AgNO$_3$	s	− 33.5	− 124.4	140.9	93.1
Ag$_2$S	s, α, orthorhombic	− 40.7	− 32.6	144.0	76.5
Ag$_2$S	s, β	− 39.5	− 29.4	150.6	
Ag$_2$SO$_4$	s	− 618.5	− 715.9	200.4	131.4
Ag$^+$	hyp. m = 1	77.1	105.6	72.7	21.8
Ag(CN$_2$)$^-$	hyp. m = 1	305.4	270.3	192.5	
Sodium					
Na	s	0	0	51.0	28.4
Na$_2$O	s	− 376.6	− 415.9	72.8	68.2
NaF	s	− 541.0	− 569.0	58.6	46.0
NaCl	s	− 384.0	− 411.0	72.4	49.7
NaBr	s		− 359.9		52.3
NaI	s		− 288.0		
NaOH	s		− 426.7		
NaNO$_3$	s	− 365.9	− 466.7	116.3	93.1
Na$_2$SO$_4$	s	− 1266.8	− 1384.5	149.5	127.6
Na$_2$CO$_3$	s	− 1047.7	− 1130.9	136.0	110.5
Na$_2$CO$_3$.10H$_2$O . . .	s		− 4081.9		
NaHCO$_3$	s	− 851.9	− 947.7	102.1	87.6
Na$_2$O$_2$	s		− 504.6		
NaClO$_3$	s		− 358.7		
NaClO$_4$	s		− 385.7		
Na$_2$SO$_3$.7H$_2$O . . .	s		− 3152.2		
Na$_2$SO$_4$.10H$_2$O. . . .	s	− 3644.0	− 4324.1	592.9	587.4
Na$_2$S$_2$O$_3$.5H$_2$O. . . .	sI		− 2602.0		

Inorganic compounds (*contd*)

Substance and formula	State	$\Delta G_f^{\ominus}$ kJ mol^{-1}	$\Delta H_f^{\ominus}$ kJ mol^{-1}	$S^{\ominus}$ JK^{-1} mol^{-1}	$C_p^{\ominus}$ JK^{-1} mol^{-1}
Na$_2$S$_2$O$_3$.5H$_2$O	sII		−2596.6		
NaNO$_2$	s		−359.4		
Na$_2$HPO$_4$.12H$_2$O . . .	s		−5298.6		
Na$_3$PO$_4$.12H$_2$O . . .	s		−5476.9		
NaC$_2$H$_3$O$_2$	s, acetate		−710.4		
NaCN	sIII		−89.8		
NaCNS	s		−174.6		
Na$_2$SiO$_3$	s	−1426.7	−1518.8	113.8	111.8
Na$_2$CrO$_4$	s		−1328.8		
Na$_2$B$_4$O$_7$.10H$_2$O . . .	s		−6264.3		
NaBH$_4$	sI	−119.5	−183.3	104.7	86.6
Na$^+$	hyp. m = 1	−261.9	−239.7	60.2	
Strontium					
Sr	s	0	0	54.4	25.1
Sr(NO$_3$)$_2$.4H$_2$O . . .	s		−2152.7		
Sr(OH)$_2$.8H$_2$O	s		−3352.2		
Sr^{2+}	hyp. m = 1	−557.3	−547.3	39.3	
Sulphur					
S	s, rhombic	0	0	31.8	22.6
S	s, monoclinic		0.3		
S	g	238.3	278.8	167.7	23.7
SO$_2$	g	−300.2	−296.8	248.1	39.9
SO$_3$	g	−371.1	−395.7	256.6	50.7
H$_2$S	g	−33.6	−20.6	205.7	34.2
H$_2$SO$_4$	l	−690.1	−814.0	156.9	138.9
S$_2$Cl$_2$	g	−31.8	−18.4	331.4	73.6
SO$_2$Cl$_2$	l		−394.1		133.9
S^{2-}	hyp. m = 1	85.8	33.1	−14.6	
HS$^-$	hyp. m = 1	12.0	−17.6	62.8	
SO$_3^{2-}$	hyp. m = 1	−486.6	−635.5	−29.3	
SO$_4^{2-}$	hyp. m = 1	−744.6	−909.3	20.1	−292.9
Tantalum					
Ta	s	0	0	41.5	25.4
Ta$_2$O$_5$	s, β	−1911.3	−2046.0	143.1	135.1
Tellurium					
Te	s	0	0	49.7	25.7
TeO$_2$	s	−270.3	−322.6	79.5	
H$_6$TeO$_6$	s		−1298.7		
H$_2$Te	g		99.6		
TeCl$_4$	s		−326.4		138.5
Thallium					
Tl	s	0	0	64.2	26.3
Tl$^+$	hyp. m = 1	−32.4	5.4	125.5	
Tl^{3+}	hyp. m = 1	214.6	196.6	−192.5	

Inorganic compounds (*contd*)

Substance and formula	State	$\Delta G_f^{\ominus}$ kJ mol^{-1}	$\Delta H_f^{\ominus}$ kJ mol^{-1}	$S^{\ominus}$ JK^{-1} mol^{-1}	$C_p^{\ominus}$ JK^{-1} mol^{-1}
Thorium					
Th	s	0	0	56.9	27.4
ThO$_2$	s		−1221.7		
Th(SO$_4$)$_2$.8H$_2$O . . .	s		−4889.4		
Tin					
Sn	sI, white	0	0	51.5	27.0
Sn	sII, grey	0.1	−2.1	44.1	25.8
SnO	s	−256.9	−285.8	56.5	44.3
SnO$_2$	s	−519.7	−580.7	52.3	52.6
SnCl$_4$	l	−440.2	−511.3	258.6	165.3
SnCl$_2$.2H$_2$O	s		−921.3		
Sn^{2+}	in aq. HCl	−27.2	−8.8	−16.7	
Sn^{4+}	in aq. HCl	2.5	30.5	−117.2	
Titanium					
Ti	sII	0	0	30.6	25.0
TiO$_2$	s, rutile	−889.5	−944.7	50.3	55.0
TiCl$_3$	s	−653.5	−720.9	139.7	97.2
TiCl$_4$	l	−737.2	−804.2	252.3	145.2
Tungsten					
W	s	0	0	32.6	24.3
WO$_3$	s	−764.1	−842.9	75.9	73.8
Uranium					
U	sIII	0	0	50.3	27.5
UO$_2$	s	−1075.3	−1129.7	77.8	
UO$_3$	s	−1184.1	−1263.6	98.6	
UF$_4$	s	−1761.5	−1853.5	151.0	117.7
UF$_6$	g	−2029.2	−2112.9	379.7	
UO$_2$(NO$_3$)$_2$.6H$_2$O . . .	s	−2615.0	−3197.8	505.6	
UO$_2$SO$_4$.3H$_2$O	s	−2451.8	−2789.9	263.6	
U^{3+}	hyp. m = 1	−520.5	−514.6	−125.5	
U^{4+}	hyp. m = 1	−579.1	−613.8	−326.4	
UO$_2^{2+}$	hyp. m = 1	−989.1	−1047.7	−71.1	
Vanadium					
V	s	0	0	28.9	24.9
V$_2$O$_5$	s	−1419.6	−1550.6	131.0	127.7
Xenon					
Xe	g	0	0	169.6	20.8
XeO$_3$	s		401.7		
XeO$_3$	g		527.2		
XeO$_4$	g		642.2		
XeF$_2$	g	−74.0	−107.6	259.5	54.1
XeF$_4$	g	−132.8	−206.5	327.8	90.4
XeF$_6$	g		−279.0		
Yttrium					
Y	s	0	0	44.4	26.5
Y$_2$O$_3$	s	−1816.7	−1905.3	99.1	102.5

Inorganic compounds (*contd*)

Substance and formula	State	$\Delta G_f^{\ominus}$ kJ mol^{-1}	$\Delta H_f^{\ominus}$ kJ mol^{-1}	$S^{\ominus}$ JK^{-1} mol^{-1}	$C_p^{\ominus}$ JK^{-1} mol^{-1}
Zinc					
Zn	s	0	0	41.6	25.4
ZnO	s	−318.3	−348.3	43.6	40.3
ZnCl$_2$.	s	−369.4	−415.1	111.5	71.3
ZnBr$_2$.	s	−312.1	−328.7	138.5	
ZnI$_2$	s	−208.9	−208.0	161.1	
ZnSO$_4$.7H$_2$O	s	−2563.1	−3077.8	388.7	383.4
ZnS	s, sphalerite	−201.3	−206.0	57.7	46.0
ZnS	s, wurtzite		−192.6		
Zn(NO$_3$)$_2$.6H$_2$O . . .	s	−1773.1	−2306.6	456.9	323.0
Zn^{2+}	hyp. m = 1	−147.0	−153.9	−112.1	46.0
Zirconium					
Zr	s, α, hexagonal	0	0	39.0	25.4
ZrOCl$_2$.8H$_2$O	s		−3471.9		
Zr(SO$_4$)$_2$.H$_2$O	s		−2553.9		

Organic compounds

Some additional sources of data are D. R. Stull, E. F. Westrum and G. C. Sinke, *The Chemical Thermodynamics of Organic Compounds*, 1969 (Wiley & Sons, Inc., New York) and J. D. Cox and G. Pilcher, *Thermochemistry of Organic and Organometallic Compounds*, 1970 (Academic Press, New York). See p. 238 for data on other carbon compounds.

Substance	State	$\Delta G_f^{\ominus}$ kJ mol^{-1}	$\Delta H_f^{\ominus}$ kJ mol^{-1}	$S^{\ominus}$ JK^{-1} mol^{-1}	$C_p^{\ominus}$ JK^{-1} mol^{-1}
Acetaldehyde	g	−128.9	166.2	250.2	57.3
Acetamide	s		−317.6		
Acetic acid	l	−389.9	−484.5	159.8	124.3
Acetone	l	−155.4	−248.1	200.4	124.7
Acetonitrile	l	99.2	53.6	149.6	91.5
Acetyl chloride	l	−208.1	−273.8	200.8	117.2
Acetylene	g	209.2	226.7	200.8	43.9
Benzene	g	129.7	82.9	269.2	81.7
Benzene	l	124.3	49.0	173.3	136.1
Benzoic acid	s	−245.3	−385.1	167.6	146.8
Bromoethane	l	−27.8	−92.0	198.7	100.8
Bromoethane	g	−25.9	−35.2	246.3	42.4
Butane	g	−17.2	−126.1	310.1	97.4
Buta-1,3-diene	g	150.7	110.2	278.7	79.5
Chloroacetic acid	sI		−513.0		
Chloroethane	g	−60.5	−112.2	275.9	62.8
Chloroform	l	−73.7	−134.5	201.7	113.8

Organic compounds (*contd*)

Substance	State	$\Delta G_f^{\ominus}$ kJ mol^{-1}	$\Delta H_f^{\ominus}$ kJ mol^{-1}	$S^{\ominus}$ JK^{-1} mol^{-1}	$C_p^{\ominus}$ JK^{-1} mol^{-1}
Chloromethane	g	−57.4	−80.8	234.5	40.8
Cyclohexane	l	6.4	−37.3	204.3	156.5
Diazomethane	g	217.8	192.5	242.8	52.5
1,2-Dichloroethane.	l	−79.6	−165.2	208.5	129.3
1,1-Dichloroethylene . . .	l	24.5	−24.3	201.5	111.3
Dichloromethane	l	−67.3	−121.5	177.8	100.0
Diethyl ether	g	−122.3	−252.2	342.7	112.5
Dimethylamine.	g	68.4	−18.4	273.0	70.7
Diphenyl ether	s	144.2	−31.7	233.9	216.6
Ethane	g	−32.9	−84.7	229.5	52.6
Ethanediol	l	−323.2	−454.8	166.9	149.8
Ethanol	l	−174.1	−277.0	160.7	113.4
Ethyl acetate	l	−332.7	−479.0	259.4	170.1
Ethylbenzene	g	130.6	29.8	360.5	128.4
Ethylene	g	68.1	52.3	219.5	43.6
Ethylene oxide	g	−13.1	−52.6	242.4	47.9
Ethyl hydrogen peroxide . .	l		−245.2		
Ethyl nitrate	l	−43.1	−190.4	247.2	170.3
Ethyl nitrite	g		−103.8		
Formaldehyde	g	−113.0	−117.2	218.7	35.4
Formamide	l		−257.7		107.6
Formic acid	l	−361.4	−424.7	129.0	99.0
Glycine	s	−373.4	−532.9	103.5	99.2
Heptane.	l	1.0	−224.4	328.6	224.3
Iodoform	g	177.9	210.9	355.6	74.9
Iodomethane	g	14.6	13.0	254.0	44.1
Ketene	g	−61.9	−61.1	247.5	51.8
Methane	g	−50.8	−74.8	186.2	35.3
Methanethiol	g	−9.3	−22.3	255.1	50.2
Methanol	l	−166.4	−238.7	126.8	81.6
Methylamine	g	32.1	−23.0	243.3	53.1
Methyl formate.	g	−297.2	−347.5	301.2	66.5
Nitroethane.	l		−140.2		134.2
Nitromethane	l	−14.5	−113.1	171.8	106.0
Oxalic acid	s	−701.2	−830.0	120.1	117.2
Pentane.	g	−8.4	−146.4	348.9	120.2
Phenol	s	−50.4	−165.0	144.0	127.4
Propane.	g	−23.5	−103.8	269.9	73.5
Propene.	g	62.7	20.4	266.9	63.9
Pyridine.	g	190.2	140.2	282.8	78.1
Styrene	g	213.8	147.4	345.1	122.1
Thiophen	g	126.8	115.7	278.9	72.9
Thiourea	s	21.7	−90.0	115.8	
Toluene.	g	122.0	50.0	320.7	103.6
Trichloroethylene	g	18.0	−7.8	324.7	80.3
Trimethylamine	g	98.9	−23.8	288.8	91.8
Urea	s	−197.2	−333.2	104.6	93.1
Vinyl chloride	g	51.9	35.6	263.9	53.7
o-Xylene	g	122.1	19.0	352.8	133.3
m-Xylene	g	118.9	17.2	357.7	127.6
p-Xylene	g	121.1	17.9	352.4	126.9

E.F.G.H.

2.11 Miscellaneous chemical data

2.11.1 Properties of polymers

As the physical properties of polymers vary from sample to sample and are dependent on composition and pretreatment of the specimen, only ranges of values are recorded. There are now available a large number of co-polymers and filled polymers concerning which information can be found in J. Brandrup and E. H. Immergut (eds), *Polymer Handbook*, 1966 (Interscience, New York) and in *Modern Plastics, 1970–71 Encyclopedia* (McGraw-Hill, Inc., New York).

Polymer	ρ $\overline{\text{kg m}^{-3}}$	Tensile strength $\overline{\text{MN m}^{-2}}$	10^5 Coeff. linear expansion $\overline{°C}$	Heat capacity $\overline{\text{J g}^{-1} \text{ K}^{-1}}$	Thermal conductivity $\overline{\text{W m}^{-1} \text{ K}^{-1}}$	n_D	Resistivity $\overline{\Omega \text{ cm}}$	Dielectric constant	Stability
Acetals	1420	65	8	1.46	0.23	1.48	10^{15}	—	Attacked by some acids and alkalies, good resistance to organic solvents.
Cellulose (cotton, wood pulp) and regenerated	1480–1530	80–240	—	1.3–1.5	0.23	1.53–1.56	10^7–10^{14}	6.7–7.5	Decomposed by oxid. agents and strong acids or by heating to 270 °C. Absorbs water.
Cellulose acetate: Moulded	1220–1340	12–58	8–18	1.26–1.8	0.17–0.33	1.46–1.50	10^{10}–10^{14}	3.5–7.5	Softens at ~130 °C, decomposed by strong acids and alkalies, soluble in ketones and esters, softened or dissolved by chlorinated or aromatic hydrocarbons.
Sheet	1280–1320	30–52	10–15	1.26–2.1	0.17–0.33	1.49–1.50	10^{11}–10^{15}	3.5–7.5	
Cellulose nitrate (Celluloid)	1350–1400	50	8–12	1.3–1.7	0.12–0.21	1.49–1.51	10^{10}	7.0–7.5	Softens at ~60 °C and ignites ~160 °C. Decomposed by strong acids and alkalies. Soluble in alcohol, ethers and esters.
Chlorinated polyether	1400	39	8	—	0.13	—	10^{15}	3.1	Unattacked by acids or alkalies except by oxid. acids. Resists most solvents.
Epoxy cast resins	1110–1400	26–85	4.5–6.5	1.0	0.17–0.21	1.55–1.61	10^{12}–10^{17}	3.5–5.0	Unattacked by weak acids and alkalies, some attack by strong acids and alkalies. Generally resistant to organic solvents.
Phenolic cast resins	1240–1320	33–59	6–8	1.25–1.67	0.15	1.58–1.66	10^{12}–10^{13}	6.5–17.5	Not affected by acids, attacked by strong alkalies, little affected by organic solvents.
Phenol–formaldehyde moulding compounds	1250–1300	45–52	2.5–6.0	1.59–1.76	0.13–0.25	1.5–1.7	10^{11}–10^{12}	5.0–6.5	Slightly attacked by acids and alkalies. Depending on composition soluble or insoluble in organic solvents.

Properties of polymers (contd)

Polymer	ρ / kg m⁻³	Tensile strength / MN m⁻²	10^5 Coeff. linear expansion / °C	Heat capacity / J g⁻¹ K⁻¹	Thermal conductivity / W m⁻¹ K⁻¹	n_D	Resistivity / Ω cm	Dielectric constant	Stability
Polyacrylonitrile	1160–1180	200 (fibres)	7	—	—	1.52	10^{14}	6.5	Fair resistance to weak alkalies but soluble in strong nitric and sulphuric acids.
Polycarbonates	1200	52–62	6.6	1.17–1.25	0.19	1.59	10^{16}	2.97–3.17	Unattacked by weak acids, some attack by strong acids and alkalies. Resistant to paraffinic solvents but soluble in aromatic and chlorinated hydrocarbons.
Poly-2-chloro-1:3 butadiene (Neoprene)	1230–1250	—	20	2.18	0.19	1.56	10^{10}–10^{13}	—	Soluble in many organic solvents but resistance to heat and oil greater than that of natural rubber.
Polychlorotrifluoro-ethylene	2100–2200	29–39	4.5–7.0	0.9	0.20–0.22	1.43	10^{18}	2.24–2.8	Not affected by acids or alkalies but halogenated organic solvents cause slight swelling.
Polyethylene Low density	910–925	4–15	10–20	—	0.33	1.51	10^{15}–10^{18}	2.25–2.35	All grades melt ~110 °C, insoluble in common solvents below 60 °C, resistant to acids and alkalies but attacked by oxid. acids.
Medium density	926–940	8–22	10–20	—	0.33–0.42	1.52	10^{15}–10^{18}	2.25–2.35	
High density	940–965	20–36	10–20	—	0.45–0.52	1.54	10^{15}–10^{18}	2.30–2.35	
Polyethylene terephthalate	1370–1380	66	6	1.25	0.29	1.54	10^{16}	3.25	Softens ~250 °C. Resistant to dilute acids and alkalies but attacked by oxid. acids. Insoluble in many common solvents but soluble in hot aromatic and chlorinated hydrocarbons.
Polyimide aromatics	1430	68	4–5	1.12	—	opaque	>10^{16}	3.4	Thermally very stable and does not burn. Resistant to acids but attacked by alkalies. Insoluble in organic solvents.
Polyisoprene Natural rubber	906–913	—	22	1.88	0.13	1.52	10^{6}	—	Swells in many organic solvents. More resistive to attack by chemicals than natural rubber.
Hard rubber	1130–1180	39	6	1.38	0.16	1.6	10^{16}	—	
Polymethylmethacrylate	1170–1200	45–72	5–9	1.5	0.17–0.25	1.48–1.50	>10^{14}	3.3–4.5	Softens at ~180 °C. Attacked by high concentrations of oxid. acids. Soluble in ketones, esters, aromatic and chlorinated hydrocarbons.

Properties of polymers (*contd*)

Polymer	ρ / kg m⁻³	Tensile strength / MN m⁻²	10^5 Coeff. linear expansion / °C	Heat capacity / J g⁻¹ K⁻¹	Thermal conductivity / W m⁻¹ K⁻¹	n_D	Resistivity / Ω cm	Dielectric constant	Stability
Polypropylene	902–906	28–36	6–10	1.92	0.12	1.49	$>10^{16}$	2.2–2.6	Softens at ~160 °C. Resistant to acids and alkalies but attacked by oxid. acids. Insoluble in common organic solvents below 80 °C.
Polystyrene	1040–1090	30–100	3.4–21	1.3–1.5	0.04–0.14	1.59–1.60	$>10^{16}$	2.45–2.65	Softens at ~98 °C and burns. Unaffected by acids or alkalies, attacked by strong oxid. acids. Soluble in aromatic and chlorinated hydrocarbons.
Polytetrafluoroethylene moulding compound	2140–2200	13–33	10	1.0	0.25	1.35	$>10^{18}$	—	Not affected by acids, alkalies or organic solvents.
Polyurethane Cast liquid	1100–1500	1–65	10–20	1.8	0.21	1.50–1.60	10^{11}–10^{15}	4.0–7.5	Slightly attacked by weak acids and alkalies. Strongly attacked by strong acids and alkalies. Only slightly affected by organic solvents.
Elastomer	1110–1250	29–55	10–20	1.8	0.07–0.31	—	10^{11}–10^{13}	5.4–7.6	Attacked by acids and alkalies and may dissolve. Resistant to most organic solvents.
Polyvinylchloride (PVC)	1300–1400	~50	7–8	0.84–1.17	0.12–0.17	1.53–1.55	10^{16}	3.2–3.6	Softens at 80 °C. Soluble in organic solvents.
Polyvinylidene chloride moulding compounds	1650–1720	20–33	19	1.34	0.13	1.60–1.63	10^{14}–10^{16}	4.5–6.0	Resistant to acids and alkalies, little or only slightly affected by organic solvents.
Nylon-6 (Poly-ϵ-caprolactam)	1120–1170	45–90	28	1.6	0.25	—	10^{12}–10^{15}	3.7–5.5	Stable at 100 °C. Softens at ~230 °C. Attacked by strong acids but resistant to alkalies. Resistant to common organic solvents but dissolves in phenol and formic acid.
Nylon-66 (Polyhexamethylene-adipamide)	1130–1150	60–80	8	1.7	0.25	1.53	10^{14}–10^{15}	4.0–4.6	Stable at 100 °C. Softens at ~230 °C. Attacked by strong acids but resistant to alkalies. Resistant to common organic solvents but dissolves in phenol and formic acid.
Silicone cast resin	1300	—	8–30	—	0.15–0.32	1.43	10^{14}–10^{15}	2.7–4.2	Excellent stability to heat and oxidation, water repellent, insoluble in many organic solvents.

Permeability constants for polymers

The permeability constant P is defined as

$$P = (\text{amount of permeant})/(\text{area})(\text{time})(\text{driving force across the film})$$

The unit employed here for P is the amount of vapour expressed in cm^3 measured at 0 °C and at 1 standard atmosphere (101.325 kN m^{-2}) passed in 1 second for an area of 1 cm^2 of film when the pressure across the film of 1 cm thickness is 101.325 kN m^{-2}.

Additional data can be found in an article by H. Yasuda, *Polymer Handbook* (eds J. Brandrup and E. H. Immergut), 1966 (Interscience, New York).

Polymer	$10^{10}P$									10^6P	
	H$_2$		N$_2$		O$_2$		CO$_2$			H$_2$O	
	$\theta/°C$		$\theta/°C$		$\theta/°C$		$\theta/°C$			$\theta/°C$	
Cellulose	25	0.494	25	0.243	25	0.160	25	0.357		—	—
(Cellophane)											
Cellulose acetate	20	266	30	21.3†	30	59.3†	30	181†		25	41.8
Cellulose nitrate	20	152	25	9.1	25	148	25	161		25	47.8
Polychlorotrifluoroethylene . .	20	71	25	0.380	25	3.0	40	16.0		25	0.0022
(Amorphous)											
Polyethylene:											
Low density	30	912	30	103	30	300	30	1269		25	0.684
High density	20	228	30	13.7	30	38.8	30	160		25	0.091
Polyethylene terephthalate . .	20	281	30	0.836	30	3.42	30	1.14		25	0.988
Polyisoprene	25	3952	25	623	25	1809	25	10260		—	—
(Natural rubber)											
Polypropylene	20	312	30	33.4	30	175	30	700		25	0.388
Polystyrene	—	—	30	22.0	30	83.6	30	669		25	9.12
Polytetrafluoroethylene . . .	20	152	—	—	—	—	20	357		—	—
Polyvinylchloride	20	182	30	8.36	30	22.8	30	114		25	1.19
Polyvinylidenechloride . . .	—	—	30	0.071	30	0.403	30	2.28		25	0.0038
Nylon-6	—	—	30	0.722	30	2.89	20	6.69		—	—

† Plasticized polymer was used for this measurement.

E.F.G.H.

2.11.2 Some characteristics of common glasses

Glass	Approximate composition	Annealing temperature/°C	$10^6 \times$ linear coefficient of expansion
Soda (S95) . . .	Sodium calcium silicate	515	9.2†
Lead (L92) . . .	Lead silicate	435	9.0†
Pyrex	Sodium aluminium silicate	530–600	3.2
Silica	Silica	—	0.45
GEC B37	Silica, boron, sodium	560	3.7‡
Monax	Silica (small amounts of sodium potassium, zinc and aluminium)	560	4.4§

† Suitable for sealing to platinum.
‡ Suitable for sealing to tungsten.
§ Suitable for sealing to platinum, tungsten and molybdenum, also impermeable to liquid helium.

For further information see W. E. S. Turner, *The Elements of Glass Technology for Scientific Glass Blowers (Lampworkers)*, 1954, The Glass Delegacy of the University of Sheffield, Sheffield, and J. H. Partridge, *Glass-to-Metal Seals*, 1949, The Society of Glass Technology, Sheffield. For information on the properties of American glasses see W. E. Barr and V. J. Anhorn, *Scientific and Industrial Glass Blowing and Laboratory Techniques*, 1959 (Instruments Publishing Company, Pittsburgh).

E.F.G.H.

2.11.3 Miscellaneous

Cooling agents

Composition	$\theta/°C$	Composition	$\theta/°C$
Crushed ice 2 parts, sodium chloride 1 part	-18	Liquid oxygen	-183
Crushed ice 3 parts, cryst. calcium chloride 4 parts	-48	Liquid nitrogen	-196
Solid CO_2 added to equal parts CCl_4 and $CHCl_3$	-80	Liquid helium	-269

Mohs's scale of mineral hardness

The numbers are not quantitative but merely indicate the sequence of hardness.

Mineral	Hardness	Mineral	Hardness	Mineral	Hardness
Talc	1	Apatite	5	Corundum . . .	9
Rock salt	2	Felspar	6	Diamond . . .	10
Calcspar	3	Quartz	7		
Fluorspar	4	Topaz	8	Finger-nail . . .	~ 2.5
				Penknife . . .	~ 6.5

E.F.G.H.

Atomic and
Nuclear Physics

3.1 Free electrons and ions in gases

3.1.1 Ionic mobility

The **ionic mobility** (k) is defined as the velocity attained by an ion moving through a gas under unit electric field. A suitable unit is therefore $m^2\ s^{-1}\ volt^{-1}$. In general the mobility is a function of E/p, where E is the field strength and p the gas pressure, but is constant for low values of E/p, the departure from constancy occurring when the ions attain drift velocities comparable with the agitation velocity of the gas molecules. The mobility is critically dependent on gas purity, as polar impurity molecules tend to cluster the ion and so reduce its mobility. Early measurements were frequently rendered quite valueless by inattention to this matter. The mobility depends little on the charge of the ion, a doubly charged ion having practically the same mobility as a singly charged ion (of either sign). By theory, $k \propto (\text{density})^{-1}$ so at constant temperature $k \propto (\text{pressure})^{-1}$. This is observed to hold from 10 to $6 \times 10^6\ N\ m^{-2}$. The reduced mobility K_0 $(= 273\ kp/1.01 \times 10^5\ T)$ is the mobility at 273 K and $1.01 \times 10^5\ N\ m^{-2}$ pressure in units of $m^2\ s^{-1}\ volt^{-1}$. K_0 is thus the value corresponding to a gas number density of $2.69 \times 10^{25}\ m^{-3}$. All values in the table below are for K_0. A dash indicates that either the particular ion does not exist or has not been examined. Where the mobility is not independent of E/p it is usual to take the mobility corresponding to the limit $E = 0$.

It should be noted that considerable confusion existed in the early measurements due to the production of molecular ions in the noble gases. Modern techniques have shown that the number of ion species produced is much greater and more complex than was thought earlier. (See McDaniel, *Collision Phenomena in Ionized Gases*, 1964, Wiley, London.)

Ionic mobilities for ions in their parent gas

$K_0/(10^{-4}\ m^2\ s^{-1}\ volt^{-1})$
at 273 K and $1.01 \times 10^5\ N\ m^{-2}$

Ionic species	H_2	He	N_2	O_2	Ne	Ar	Kr	Xe	CO	CH_4
Atomic positive ion . . .	—	10.8	3.3	—	4.2	1.5	0.9	0.6	—	—
Molecular positive ion . .	$12.3(H_3^+)$	20.3	1.8	2.2	6.5	2.3	1.2	0.8	1.6	3.5
Molecular negative ion . .	—	—	—	3.3	—	—	—	—	—	—

Ionic mobilities for ions in various gases

$k/(10^{-4}\ m^2\ s^{-1}\ volt^{-1})$

Ion	Gas									
	He	Ne	Ar	Kr	Xe	H_2	N_2	CO	CO_2	H_2O Vap.
Li^+	25.6	12	4.99	4.0	3.04	13.3	4.21	2.63		0.725
Na^+	23.4	8.70	3.23	2.34	1.80	15.0	3.40	2.44		0.715
K^+	22.7	8.0	2.81	1.98	1.44	14	2.7	2.32	1.46	0.705
Rb^+	21.2	7.18	2.40	1.59	1.10	13.4	2.39	2.08		0.700
Cs^+	19.1	6.50	2.23	1.42	0.99	13.4	2.25	1.98		0.695

These values, from Tyndall, *Mobility of Positive Ions in Gases*, 1938 (Cambridge University Press), are for gases at 293 K and $1.01 \times 10^5\ N\ m^{-2}$.

J.W.L.

3.1.2 Electron mobility

As an electron loses little energy on colliding with gas atoms or molecules, it rapidly attains energies in excess of the thermal agitation energy of the gas atoms and so a significant mobility value cannot be given (see ionic mobility). Instead the drift velocity as a function of E/p is normally found in the literature. For low E/p, the drift velocity is given by $v = \text{const } (E/p)^{1/2}$. At larger E/p values this relation is no longer valid, v increasing more rapidly than given by this relation. The breakdown is usually associated with the onset of inelastic collisions. Impurities, especially with the noble gases, are again of importance (see D. H. Wilkinson, *Ionization Chambers and Counters*, 1950, p. 31 (Cambridge University Press)).

The table below is compiled mainly from Pack, Voshall and Phelps, *Phys. Rev.*, 1962, **127**, 2084–2089, and McDaniel, *Collision Phenomena in Ionized Gases*, 1964 (Wiley, London).

Drift velocity/(10^3 m s^{-1})

Gas	Field strength per unit pressure $(E/p)/(\text{V mN}^{-1})$											
	0.075	0.15	0.3	0.45	0.6	0.75	1.5	3	4.5	6	7.5	15
H_2	3.1	5	6.8	8.0	9.2	10	15	23	30	36	42	70
He	2.7	4	6	7	8.5	9.6	15	36				
N_2	3.1	3.9	5	6	7	8	14	22	30	36	43	80
Ne	4	5	7	10	12	16	26	52				
Ar	2.3	2.7	3.2	3.5	3.9	4.1	6.2	13	30			
Kr	1.5	1.8	2.1	2.4	2.5	3.1						
Xe	1.0	1.2	1.4	1.6	1.7	1.8						
CO_2	0.56	1.1	2.3	3.4	4.6	5.7	11	33	66	94	109	137
CO	4.8	6.5	9.4	11	13	14	17	20	25	28	29	
BF_3	0.13	0.25	0.5	0.75	1.0	1.3	2.5	5.0	7.5	10	12.5	25
Ar/CH_4 . . .	34	60	62	58	51	47	40					
Ar/N_2	4	6	9	13								
Air	3.5	5	8	9.3	11	12	17		26	34	43	88
CH_4	8	24	60	80	95	100	100					

Ar/CH_4 and Ar/N_2 are 90% Argon, 10% CH_4 (or N_2) by volume.   J.W.L.

3.1.3 Ionic recombination

The neutralization of electrical carriers of opposite sign in a gas is termed recombination. The positive carrier is a positive ion. The negative carrier may be a negative ion or an electron depending on the nature of the gas. This table deals only with the former type of recombination, and only with the situation in which the ions are randomly distributed throughout the gas (see Loeb, *Basic Processes of Gaseous Electronics*, 1955, p. 477 *et seq.* California University Press). Then $dn_+/dt = dn_-/dt = -\alpha n_- . n_+$, where n_+ and n_- are the ion densities and α is the recombination coefficient. α is commonly taken as a constant. Experimentally, however, measurements show that α changes with ion age, presumably due to a change in the nature of the ions owing to the presence of small quantities of impurity. Further there is a small dependence on ion density, α decreasing with increasing ion density. α is dependent on pressure and temperature and from 10^5 N m^{-2} to a few hundred N m^{-2} is adequately described by the theory of Thomson. For pressures above 10^6 N m^{-2} the Langevin theory is employed. Three-body and mutual neutralization processes are by far the most important for ionic recombination. See the summary by Massey, *Advances in Physics*, 1952, Vol. I, p. 395. For general theoretical discussion, see Loeb (above); Jaffe, *Phys. Rev.*, 1940, **58**, 968; 1941, **59**, 652.

All values given below are at 1.01×10^5 N m^{-2} pressure unless otherwise stated.

Gas	$\alpha/(10^{-12}$ m^3 s$^{-1})$	Gas	$\alpha/(10^{-12}$ m^3 s$^{-1})$
Ar (impure) . .	1.06 (22 °C)	N$_2$	1.06 (22 °C)
O$_2$	1.32 (22 °C) 2.08 (25 °C)	H$_2$	0.28 (22 °C)
		Air	1.65 (18 °C)

Dependence on ion density, pressure and temperature

$n =$ No. of ions/m^3; $p =$ pressure; $T =$ temperature.

Air . . 15 °C .	$n/(10^{12}$ m$^{-3})$ $\alpha/(10^{-12}$ m^3 s$^{-1})$	1.5 2.65	2.7 2.5	3.1 2.5	4.1 2.3	4.4 2.3

Air . . 18 °C .	$p/(10^3$ N m$^{-2})$ $\alpha/(10^{-12}$ m^3 s$^{-1})$	5 0.12	10 0.21	15 0.29	20 0.37	40 0.7	60 1.05	80 1.4	90 1.55	100 1.72

Air . . . . 18 °C . . .	$p/(10^5$ N m$^{-2})$	25.7	23.6	21.1	17.8	15.1	12.5	9.7	8.0	6.4	4.6
	$\alpha/(10^{-13}$ m^3 s$^{-1})$	2.37	2.59	2.90	3.44	4.07	4.80	5.94	6.98	8.44	10.58
CO$_2$ . . . . 18 °C . . .	$p/(10^5$ N m$^{-2})$	22.0	17.4	12.5	7.9	4.6					
	$\alpha/(10^{-13}$ m^3 s$^{-1})$	1.24	1.56	2.24	3.58	5.76					

O$_2$ . . 20 °C .	$p/(10^3$ N m$^{-2})$	10	20	30	40	50	60	70	80	90	100	110
	$\alpha/(10^{-12}$ m^3 s$^{-1})$	0.7	1.3	1.6	1.75	1.88	1.96	2.01	2.05	2.08	2.08	2.08
O$_2$ . . 20 °C .	$T/$K	200	250	300	350	400						
	$\alpha/(10^{-12}$ m^3 s$^{-1})$	3.7	2.7	1.9	1.5	1.2						

3.1.4 Ionic diffusion

The coefficient of diffusion D is given by the equation $dn/dt = D(\nabla^2 n)$, where n is the ion density and ∇^2 is the Laplacian operator. Actual measurements of D are seldom undertaken as the coefficient can be calculated from the equation $k/D = Ne/p$, where k is the mobility, N is the number of molecules per m^3, p is the pressure in Newtons per square metre and e is the electronic charge. Thus, numerically, $D = 0.0234k$ m^2 s^{-1} at s.t.p. Values of k are given in the section on Ionic Mobility.

3.1.5 Electron diffusion

The coefficient of electron diffusion D_e may be computed from the equation $k_e/D_e = Ne/\eta P$, where k_e is the electron mobility, N is the number of gas molecules per m^3 at s.t.p., e is the electronic charge, P is the pressure in $N\ m^{-2}$ and η is the ratio of the mean agitation energy of an electron to that of a gas molecule. η is also known as the Townsend energy factor. Experimental values of η are given below as a function of E/p.

Field strength per unit pressure $(E/p)/(V\ mN^{-1})$

Gas	0.5	1	2	3	4	5	10	20	30	40
He . . .	37	74	126	152	176					
Ar. . . .	220	320	316	312	310	314	324			
H_2 . . .	6	10	18	23	28	32	55	100	130	150
N_2 . . .	16	25	34	39	42	44	52	68	89	114
O_2 . . .	11	18	32	41	47	52	68	96	133	
Air . . .	7	15	26	34	39	42	50	68	90	106
Cl_2 . . .							53	71	77	82
Br_2 . . .				40	52	59	74	82	89	91
I_2							27	76	91	99
CO . . .	5	8	14	19	23	27	38	48	68	
NO . . .	7	8	10	11	12	14	38			
CO_2 . . .	1.3	1.6	2	4	9	17	58	89	117	

Most of the values in the above table were obtained in the 1920s and are from Healey and Reed, *The Behaviour of Slow Electrons in Gases*, 1941 (Amalgamated Wireless, Sydney). Recent determinations have shown only small changes for H_2, He, N_2 and Air but the values for O_2 over a limited range are higher than those in the above table (see Huxley *et al.*, *Aus. J. Phys.*, 1959, **12**, 303).

3.1.6 Electron collisions

Observed total collision cross-section of electrons with atoms and molecules

Q is the total collision cross-section in units of πa_0^2, where a_0 is the radius of the first Bohr orbit of the hydrogen atom; $a_0 = 0.53 \times 10^{-10}$ m. Hence the unit of Q in the tables below is 0.88×10^{-20} m^2. The values are obtained by interpolation from graphs. General references: *Landolt-Bornstein Tables*, 1950, Vol. I (Springer-Verlag); Massey and Burhop, *Electronic and Ionic Impact Phenomena*, 1952 (Oxford University Press). For molecules, the data given relate to the Ramsauer method.

Q in units of πa_0^2

Gas	$(Energy)^{1/2}/V^{1/2}$											
	0.5	1	1.5	2	3	4	5	6	7	8	9	10
He . .	6.3	6.3	6.2	5.6	4.6	3.5	2.6	2.0	1.7	1.5	1.3	1.2
Ne . .	1.0	1.9	2.5	2.8	3.5	3.7	3.7	3.6	3.5	3.4	3.0	2.9
Ar . .	0.6	0.6	4.0	8.5	21	22	15	11	8	7.5	7	6.5
Kr . .	2.0	1.0	6.0	15	30	29	20	14	11.5	9	8	7.5
Xe . .	7.5	2.0	15	33	43	32.5	22.5	16	—	—	—	—
Na . .	—	280	380	180	130	110	85	65	50	48	46	35
K . .	—	450	420	270	220	160	125	80	60	50	50	40

Q in units of πa_0^2 (contd)

Gas	$(Energy)^{1/2}/V^{1/2}$											
	0.5	1	1.5	2	3	4	5	6	7	8	9	10
Rb . .	—	374	390	329	220	170	142	85	68	51	45	43
Cs . .	—	470	675	350	300	230	160	135	100	80	67	46
Cd . .	—	—	—	—	54	43	40	41.5	39	34	31	29
Zn . .	—	—	—	80	45	35	27	24	23	23	23	21.5
Hg . .	—	80	70	55	23	18	18	20	20	18.5	18	17.5
Tl . .	—	6	10	16	12.5	11	9.5	9.0	8.8	8.0	7.0	6.2
H . .	—	—	—	—	—	6	7.6	8.6	9.0	—	—	—
N$_2$ . .	17	10	29	13	12	12.5	14	13	11.5	—	—	—
O$_2$ . .	6	7	7.5	8	11	12	12	12	—	—	—	—
CO . .	15	14	38	18	14	14	14	—	—	—	—	—
H$_2$ . .	—	15	17	16	10.5	7	5.5	4	3.5	—	—	—

Cross-section (Q) for ionization by electron impact

Q is expressed in units of πa_0^2 (i.e. 0.88×10^{-20} m^2) for removal of the first electron.

Gas	$Energy/V$					
	50	100	200	300	400	500
He	0.23	0.4	0.34	0.28	0.25	0.23
Ne	0.46	0.85	0.89	0.77	0.68	0.61
Ar	3.6	3.3	2.7	2.4	1.9	1.7
H$_2$	1.1	1.1	0.8	0.66	0.56	0.48
Hg	6.6	5.7	4.0	3.15	2.6	—
N$_2$	2.5	3.3	2.9	2.4	2.1	1.8
O$_2$	2.5	3.3	3.0	2.5	2.1	1.8
CO	2.9	3.5	3.0	2.4	2.0	1.8
NO	2.9	3.7	3.2	2.7	2.5	2.0

J.W.L.

3.2 Work function

Energy required to extract an electron from a solid

The work function ϕ is the energy required to remove an electron from the highest filled level in the Fermi distribution of a solid to a point a great distance from the solid, at absolute zero. An estimate of the work function can be obtained *thermionically* from Richardson's equation

$$I = A T^2 \exp\left(-\phi/kT\right)$$

where I is the thermionic current, T the absolute temperature, k is Boltzmann's constant and A is a constant having the theoretical value 120 amp cm^{-2} deg^{-2}. Since A contains the reflection coefficient, which varies with temperature, observed A values usually differ from the theoretical value. ϕ too varies very slightly with temperature, but for most metals the difference between ϕ at absolute zero and at room temperature is less than experimental errors.

ϕ may also be estimated *photoelectrically* for metals. Einstein's expression for the photoelectric effect is $hv = e\phi + E$, where E is the kinetic energy of the ejected photoelectron. The photoelectric current J released when light of energy hv falls on the surface of a metal, for which the threshold frequency is given by $hv_0 = e\phi$ (for then $E = 0$), is given by the Fowler equation

$$J = B(kT^2) . f\{hv - hv_0)/kT\}$$

where f is a universal function of $(hv - hv_0)/kT$ and B is constant provided that hv is near to hv_0.

The third common method of measuring ϕ is by the *contact potential difference* (c.p.d.) V_{AB} that exists between the surfaces of two solids A and B of work functions ϕ_A and ϕ_B, when connected electrically, since

$$\phi_B - \phi_A = eV_{AB}$$

for the two solids at the same temperature. The method involves a prior knowledge of the work function of one of the solids if that of the other is to be measured absolutely.

Work functions

Metal	Work function $\phi\dagger$/eV		Metal	Work function $\phi\dagger$/eV		
	Photoelectric	*C.P.D.*		*Thermionic*	*Photoelectric*	*C.P.D.*
Li	—	2.32	W	4.55	4.55	4.55
K	2.22	—	Ta	4.33	4.30	4.22
Cs	2.14	—	Re	4.94	—	—
			Nb	—	—	4.37
Be	—	3.91	Mo	4.33	4.41	4.21
Mg	—	3.61				
Ba	—	2.35	Ni	5.24	5.15	5.25
			Fe	—	4.60	4.16
Zn	4.33	4.11	Cr	4.60	4.44	—
Cd	—	4.22	Co	—	4.97	—
			Mn	—	4.08	—
Al	—	4.19				
Ga	4.45	—	Ti	4.10	3.57	—
In	4.08	—	Zr	4.00	—	—
			Hf	3.65	—	—
Sn	4.28	4.43				
Pb	4.01	3.83	Ru	—	—	4.73
			Rh	4.72	—	—
Cu	4.40	4.51	Ir	4.57	—	—
Ag	4.44	4.29	Pd	—	5.40	—
Au	—	5.28	Pt	5.36	5.63	—
As	4.79	—	Th	—	—	3.71
Sb	4.56	—	U	3.47	3.47	3.63
Bi	4.25	—				

† ϕ values selected from the article by J. C. Rivière in *Solid State Surface Science* (ed. Mino Green), 1969, Vol. 1 (Marcel Dekker, New York).

J.C.R.

3.3 Electrons in atoms

3.3.1 Arrangement of electrons in atoms

In the following table (p. 266) the electrons in an atom are shown arranged in shells and sub-shells. The charge of the nucleus is $+Ze$ (Z = atomic number), and it is surrounded by Z electrons (charge $-e$) in a neutral atom.

Shells

The electrons with the same principal quantum number n are said to form a shell. Proceeding from the nucleus outwards the shells are called K, L, etc. Thus if $n = 1, 2, 3, 4$, the shell is the K, L, M or N shell respectively. This nomenclature had its origin in X-ray spectroscopy, in which a K-series line is due to an electron transition from an outer to the K shell.

Sub-shells. The electrons in each shell are arranged in sub-shells (often also called shells), specified by the value of l ($l = 0, 1, \ldots, n-1$). In Bohr's theory, n and l determine the size and shape of the orbit. The nl sub-shell is complete or closed when it contains $2(2l+1)$ electrons.

Quantum numbers for one electron

These are a development of those introduced by Bohr and Sommerfeld.

Principal quantum number n, in Bohr's theory of the H atom, specifies the major axis of the orbit and determines the energy $E = -Rhc/n^2$. With other atoms, the orbits with any given value of l are still numbered serially by n, starting with $n = l+1$ (circular orbit), but the energy now depends on l too. In wave mechanics, $n - l - 1$ is the number of nodes (zeros) in the radial factor in the wave function.

Azimuthal quantum number l specifies the angular momentum $lh/2\pi$ of the orbital motion, and is connected with the kind of symmetry possessed by the wave function ($l = 0$ implies spherical symmetry). Code letters for the value of l:

$l = 0$	1	2	3	4	5	6	7
Code letter s	p	d	f	g	h	i	k

and so on, omitting j (not used) and s, p (used out of turn). The value of the principal quantum number n is put in front of the code letter, e.g. 3d meaning $n = 3$, $l = 2$.

Magnetic quantum number m_l specifies one component of the orbital angular momentum: $m_l h/2\pi$ is the moment of momentum about some given line through the nucleus, e.g. drawn parallel to an external magnetic field. The $2l+1$ possible values of m_l are $l, (l-1), \ldots, -l$.

Spin quantum number m_s specifies the component of the intrinsic angular momentum (spin) of the electron; $m_s = \pm\frac{1}{2}$. The third and fourth quantum numbers are often of less direct significance owing to coupling effects. **Pauli's Exclusion Principle** states that no two electrons can have all their four quantum numbers the same.

Several electrons: spectral terms

In atoms with many electrons, it is usually a good approximation to assign values of (n, l) to each electron, corresponding to motion in an effective central field. Thus $1s^2 2s^2 2p^5$ means a **configuration** with two electrons in the K-shell, two with $n = 2$ and $l = 0$, and five with $n = 2$ and $l = 1$.

LS coupling. With the lighter atoms and with outer electrons, the energy of the various possible terms in any configuration depends on the extent to which the electrons keep apart (in other words on the probability of two electrons being found close together); and this in turn depends on the vector sums $L=\sum l$ and $S=\sum s$. The magnitude of L is specified by a capital letter (S, P, D...), and the resultant spin S is specified by writing the number $2S+1$ as a superscript: the relative orientation of L and S merely affects the magnetic energy and splits the terms into $2S+1$ (or fewer) levels. Thus a term with $L=2$ and $S=1$ is written ^3D (read 'triplet D'). Note that S is used in two senses, both as a code letter and as the resultant of spins.

Inner quantum number J specifies the total angular momentum $J=L+S$, and is written as a subscript ($^3D_3$, $^3D_2$, $^3D_1$).

X-ray spectra and jj coupling. With inner electrons or with heavy atoms, the coupling of the spin and orbital magnetic moments commonly makes it necessary to divide the nl 'orbits' into two groups according to the value of $j=|l+s|=l\pm\frac{1}{2}$, with a maximum number $2j+1$ of electrons in each (total $(2l+2)+2l=2(2l+1)$ as before), and L and S cease to be useful.

Electron configurations of the elements
Ground states of free neutral atoms

Element	K	L		M			N				O					Lowest state	
	1s	2s	2p	3s	3p	3d	4s	4p	4d	4f	5s	5p	5d	5f	5g		
1. H	1															$^2S_{1/2}$	
2. He	2																$^1S_0$
3. Li	2	1														$^2S_{1/2}$	
4. Be	2	2															$^1S_0$
5. B	2	2	1													$^2P_{1/2}$	
6. C	2	2	2														$^3P_0$
7. N	2	2	3													$^4S_{3/2}$	
8. O	2	2	4														$^3P_2$
9. F	2	2	5													$^2P_{3/2}$	
10. Ne	2	2	6														$^1S_0$
11. Na	2	2	6	1												$^2S_{1/2}$	
12. Mg	2	2	6	2													$^1S_0$
13. Al	2	2	6	2	1											$^2P_{1/2}$	
14. Si	2	2	6	2	2												$^3P_0$
15. P	2	2	6	2	3											$^4S_{3/2}$	
16. S	2	2	6	2	4												$^3P_2$
17. Cl	2	2	6	2	5											$^2P_{3/2}$	
18. Ar	2	2	6	2	6												$^1S_0$
19. K	2	2	6	2	6		1									$^2S_{1/2}$	
20. Ca	2	2	6	2	6		2										$^1S_0$
21. Sc	2	2	6	2	6	1	2									$^2D_{3/2}$	
22. Ti	2	2	6	2	6	2	2										$^3F_2$
23. V	2	2	6	2	6	3	2									$^4F_{3/2}$	
24. Cr	2	2	6	2	6	5	1										$^7S_3$
25. Mn	2	2	6	2	6	5	2									$^6S_{5/2}$	
26. Fe	2	2	6	2	6	6	2										$^5D_4$
27. Co	2	2	6	2	6	7	2									$^4F_{9/2}$	
28. Ni	2	2	6	2	6	8	2										$^3F_4$

Electron configurations of the elements (*contd*)

Element	K	L		M			N				O					Lowest state	
	1s	2s	2p	3s	3p	3d	4s	4p	4d	4f	5s	5p	5d	5f	5g		
29. Cu	2	2	6	2	6	10	1									$^2S_{1/2}$	
30. Zn	2	2	6	2	6	10	2										$^1S_0$
31. Ga	2	2	6	2	6	10	2	1								$^2P_{1/2}$	
32. Ge	2	2	6	2	6	10	2	2									$^3P_0$
33. As	2	2	6	2	6	10	2	3								$^4S_{3/2}$	
34. Se	2	2	6	2	6	10	2	4									$^3P_2$
35. Br	2	2	6	2	6	10	2	5								$^2P_{3/2}$	
36. Kr	2	2	6	2	6	10	2	6									$^1S_0$
37. Rb	2	2	6	2	6	10	2	6			1					$^2S_{1/2}$	
38. Sr	2	2	6	2	6	10	2	6			2						$^1S_0$
39. Y	2	2	6	2	6	10	2	6	1		2					$^2D_{3/2}$	
40. Zr	2	2	6	2	6	10	2	6	2		2						$^3F_2$
41. Nb	2	2	6	2	6	10	2	6	4		1					$^6D_{1/2}$	
42. Mo	2	2	6	2	6	10	2	6	5		1						$^7S_3$
43. Tc	2	2	6	2	6	10	2	6	5		2					$^6S_{5/2}$	
44. Ru	2	2	6	2	6	10	2	6	7		1						$^5F_5$
45. Rh	2	2	6	2	6	10	2	6	8		1					$^4F_{9/2}$	
46. Pd	2	2	6	2	6	10	2	6	10								$^1S_0$
47. Ag	2	2	6	2	6	10	2	6	10		1					$^2S_{1/2}$	
48. Cd	2	2	6	2	6	10	2	6	10		2						$^1S_0$
49. In	2	2	6	2	6	10	2	6	10		2	1				$^2P_{1/2}$	
50. Sn	2	2	6	2	6	10	2	6	10		2	2					$^3P_0$
51. Sb	2	2	6	2	6	10	2	6	10		2	3				$^4S_{3/2}$	
52. Te	2	2	6	2	6	10	2	6	10		2	4					$^3P_2$
53. I	2	2	6	2	6	10	2	6	10		2	5				$^2P_{3/2}$	
54. Xe	2	2	6	2	6	10	2	6	10		2	6					$^1S_0$

      2   8   18

Element	K	L	M	N				O					P						Q
				4s	4p	4d	4f	5s	5p	5d	5f	5g	6s	6p	6d	6f	6g	6h	7s
55. Cs . .	2	8	18	2	6	10		2	6				1						
56. Ba . .	2	8	18	2	6	10		2	6				2						
57. La . .	2	8	18	2	6	10		2	6	1			2						
58. Ce . .	2	8	18	2	6	10	2	2	6				2						
59. Pr . .	2	8	18	2	6	10	3	2	6				2						
60. Nd . .	2	8	18	2	6	10	4	2	6				2						
61. Pm . .	2	8	18	2	6	10	5	2	6				2						
62. Sm . .	2	8	18	2	6	10	6	2	6				2						
63. Eu . .	2	8	18	2	6	10	7	2	6				2						
64. Gd . .	2	8	18	2	6	10	7	2	6	1			2						
65. Tb . .	2	8	18	2	6	10	9	2	6				2						
66. Dy . .	2	8	18	2	6	10	10	2	6				2						
67. Ho . .	2	8	18	2	6	10	11	2	6				2						

Electron configurations of the elements (*contd*)

Element	K	L	M	4s	4p	4d	4f	5s	5p	5d	5f	5g	6s	6p	6d	6f	6g	6h	7s
						N				O					P				Q
68. Er	2	8	18	2	6	10	12	2	6				2						
69. Tm	2	8	18	2	6	10	13	2	6				2						
70. Yb	2	8	18	2	6	10	14	2	6				2						
71. Lu	2	8	18	2	6	10	14	2	6	1			2						
72. Hf	2	8	18	2	6	10	14	2	6	2			2						
73. Ta	2	8	18	2	6	10	14	2	6	3			2						
74. W	2	8	18	2	6	10	14	2	6	4			2						
75. Re	2	8	18	2	6	10	14	2	6	5			2						
76. Os	2	8	18	2	6	10	14	2	6	6			2						
77. Ir	2	8	18	2	6	10	14	2	6	7			2						
78. Pt	2	8	18	2	6	10	14	2	6	9			1						
79. Au	2	8	18	2	6	10	14	2	6	10			1						
80. Hg	2	8	18	2	6	10	14	2	6	10			2						
81. Tl	2	8	18	2	6	10	14	2	6	10			2	1					
82. Pb	2	8	18	2	6	10	14	2	6	10			2	2					
83. Bi	2	8	18	2	6	10	14	2	6	10			2	3					
84. Po	2	8	18	2	6	10	14	2	6	10			2	4					
85. At	2	8	18	2	6	10	14	2	6	10			2	5					
86. Em	2	8	18	2	6	10	14	2	6	10			2	6					
87. Fr	2	8	18	2	6	10	14	2	6	10			2	6					1
88. Ra	2	8	18	2	6	10	14	2	6	10			2	6					2
89. Ac	2	8	18	2	6	10	14	2	6	10			2	6	1				2
90. Th	2	8	18	2	6	10	14	2	6	10			2	6	2				2
91. Pa	2	8	18	2	6	10	14	2	6	10	2		2	6	1				2
92. U	2	8	18	2	6	10	14	2	6	10	3		2	6	1				2
93. Np	2	8	18	2	6	10	14	2	6	10	4		2	6	1				2
94. Pu	2	8	18	2	6	10	14	2	6	10	6		2	6					2
95. Am	2	8	18	2	6	10	14	2	6	10	7		2	6					2
96. Cm	2	8	18	2	6	10	14	2	6	10	7		2	6	1				2
97. Bk	2	8	18	2	6	10	14	2	6	10	9		2	6					2
98. Cf	2	8	18	2	6	10	14	2	6	10	10		2	6					2
99. Es	2	8	18	2	6	10	14	2	6	10	11		2	6					2
100. Fm	2	8	18	2	6	10	14	2	6	10	12		2	6					2
101. Md	2	8	18	2	6	10	14	2	6	10	13		2	6					2
102. No	2	8	18	2	6	10	14	2	6	10	14		2	6					2
103. Lr	2	8	18	2	6	10	14	2	6	10	14		2	6	1				2
104. Ku	2	8	18	2	6	10	14	2	6	10	14		2	6	2				2
105. Ha	2	8	18	2	6	10	14	2	6	10	14		2	6	3				2

3.3.2 Ionization potentials

The ionization potential is the least energy, in electron volts, which is necessary to remove an electron from a free unexcited neutral atom (or additional electron from an ionized atom). It can be determined either directly from the speed of the slowest electrons which will ionize the atom by collision, or more accurately from the limit of a spectroscopic series.

H	13.598	Ne	21.56	K	4.34	Ni^+	18.15	Tc	7.28	Ba^+	10.00
He	24.587	Ne^+	41.08	K^+	31.71	Ni^{2+}	35.20	Ru	7.36	La	5.61
He^+	54.42	Ne^{2+}	63.45	Ca	6.11	Cu	7.72	Rh	7.46	Ta	7.88
Li	5.39	Na	5.14	Ca^+	11.87	Zn	9.39	Pd	8.33	W	7.98
Li^+	75.60	Na^+	47.30	Sc	6.54	Ga	6.00	Ag	7.57	Re	7.87
Be	9.32	Mg	7.65	Ti	6.82	Ge	7.88	Cd	8.99	Os	8.70
B	8.30	Mg^+	15.04	V	6.74	As	9.81	In	5.79	Pt	9.00
C	11.26	Al	5.99	Cr	6.77	Se	9.75	Sn	7.34	Au	9.22
C^+	24.38	Al^+	18.83	Mn	7.43	Br	10.56	Sb	8.64	Hg	10.43
C^{2+}	47.88	Si	8.15	Fe	7.87	Kr	14.00	Te	9.01	Tl	6.11
N	14.53	P	10.49	Fe^+	16.18	Rb	4.18	I	10.45	Pb	7.42
N^+	29.60	S	10.36	Fe^{2+}	30.65	Sr	5.69	Xe	12.13	Ra	5.28
N^{2+}	47.44	S^+	23.41	Co	7.86	Y	6.25	Xe^+	23.39	Th	7.50
O	13.62	Cl	12.97	Co^+	17.06	Zr	6.95	Cs	3.89	U	6.19
O^+	35.12	A	15.75	Co^{2+}	33.50	Nb	6.77	Cs^+	25.10	Np	6.16
O^{2+}	54.90	A^+	27.63	Ni	7.64	Mo	7.06	Ba	5.21	Pu	5.71
F	17.42										

References

(1) W. Lotz, *J. Opt. Soc. Am.*, 1967, **57**, 873.
(2) *Handbook of Physics*, 2nd edn, 1967.
(3) *J. Chem. Phys.*, 1969, **51**, 3105.

3.3.3 Auger spectroscopy

Impact of electrons in the energy range 1000–3000 eV with the surface of a solid causes ionization of some atomic energy levels of atoms in the surface. An ionized atom can relax by the ejection either of an X-ray photon or of an Auger electron; the latter process is much more probable at low primary energies. In the Auger process the ionized level, of energy E_A, is filled by an electron from an outer level, of energy E_B, and the excess energy $(E_A - E_B)$ can be given to another electron either in the same level or in a still more shallow one of energy E_C. If the ionized atom is of atomic number Z, and if all energies are measured relative to zero at the vacuum level, then a good approximation to the energy E_{ABC} of the Auger transition ABC, that is, the kinetic energy of the ejected electron, is

$$E_{ABC}(Z) = E_A(Z) - E_B(Z) - E_C(Z+1) \qquad \ldots (1)$$

The energy of the third level is taken as that of the same level for the next element up the Periodic Table in order to allow empirically for the adjustment in Coulombic screening of the second level. Since the equation can be used in conjunction with tables of atomic energy levels to calculate the expected energies of Auger transitions, analysis of the observed spectrum of Auger energies from a solid surface enables its atomic composition to be established. In general, the Auger energies are less than 600 eV, so that the Auger electrons must originate either at, or very close to, the true surface if they are to escape and be observed externally.

Auger electrons of the same energies can also be produced by excitation with X-rays or with energetic ions.

The table sets out the experimentally observed energies of the most intense Auger transitions of the more common elements in solid form. Energy levels from the tables published by Bearden and Burr[1] have been used in the allocation of Auger transitions on the basis of equation (1).

For more comprehensive reading see the article by Gallon and Matthew[2].

References

(1) J. A. Bearden and A. F. Burr, *Rev. Mod. Phys.*, 1967, **39**, 125.
(2) T. E. Gallon and J. A. D. Matthew, *Rev. Phys. Tech.*, 1972, **3**, 31.

Energies of the major Auger lines for the more common elements

Element	Z	E/eV	Transition†	Element	Z	E/eV	Transition†
C . . .	6	270	KVV	Zn . . .	30	58	M_2VV
N . . .	7	380	KVV			830	$L_3M_{2,3}M_{2,3}$
O . . .	8	510	KVV			910	$L_3M_{2,3}V$
Na . . .	11	30	$L_{2,3}VV$			990	L_3VV
		990	$KL_{2,3}L_{2,3}$	Ge . . .	32	24	$M_{4,5}VV$
Mg . . .	12	47	$L_{2,3}VV$			44	$M_3M_{4,5}M_{4,5}$
		1180	$KL_{2,3}L_{2,3}$			48	$M_2M_{4,5}M_{4,5}$
Al . . .	13	66	$L_{2,3}VV$			84	$M_{2,3}M_{4,5}V$
		1380	$KL_{2,3}L_{2,3}$			1140	$L_3M_{4,5}M_{4,5}$
Si . . .	14	91	$L_{2,3}VV$			1173	$L_2M_{4,5}M_{4,5}$
		1495	KL_1L_1	As . . .	33	36	$M_{4,5}VV$
P . . .	15	116	$L_{2,3}VV$			42	$M_3M_{4,5}M_{4,5}$
S . . .	16	150	$L_{2,3}VV$			47	$M_2M_{4,5}M_{4,5}$
Cl . . .	17	180	$L_{2,3}VV$			91	$M_{2,3}M_{4,5}V$
K . . .	19	250	$L_{2,3}VV$	Zr . . .	40	20	$N_{2,3}VV$
Ca . . .	20	290	$L_{2,3}M_{2,3}M_{2,3}$			90	$M_{4,5}N_1N_{2,3}$
Cr . . .	24	35	$M_{2,3}VV$			115	$M_{4,5}N_{2,3}N_{2,3}$
		486	$L_3M_{2,3}M_{2,3}$			145	$M_{4,5}N_{2,3}V$
		527	$L_3M_{2,3}V$	Nb . . .	41	22	$N_{2,3}VV$
Mn . . .	25	40	$M_{2,3}VV$			102	$M_{4,5}N_1N_{2,3}$
		540	$L_3M_{2,3}M_{2,3}$			164	$M_{4,5}N_{2,3}N_{2,3}$
		588	$L_3M_{2,3}V$			194	$M_{4,5}N_{2,3}V$
		637	L_3VV	Mo . . .	42	27	$N_{2,3}VV$
Fe . . .	26	44	$M_{2,3}VV$			120	$M_{4,5}N_1N_{2,3}$
		594	$L_3M_{2,3}M_{2,3}$			185	$M_{4,5}N_{2,3}N_{2,3}$
		647	$L_3M_{2,3}V$			220	$M_{4,5}N_{2,3}V$
		700	L_3VV	Ag . . .	47	48	N_3VV
Co . . .	27	52	$M_{2,3}VV$			56	N_2VV
		651	$L_3M_{2,3}M_{2,3}$			270	$M_{4,5}N_1V$
		711	$L_3M_{2,3}V$			308	$M_{4,5}N_{2,3}V$
		771	L_3VV			362	$M_{4,5}V$
Ni . . .	28	60	$M_{2,3}VV$	Sn . . .	50	22	$N_{4,5}VV$
		716	$L_3M_{2,3}M_{2,3}$			64	$N_{2,3}N_{4,5}V$
		781	$L_3M_{2,3}V$			310	$M_5N_1N_{4,5}$
		847	L_3VV			360	$M_5N_{2,3}N_{4,5}$
Cu . . .	29	60	M_2VV			423	$M_5N_{4,5}N_{4,5}$
		772	$L_3M_{2,3}M_{2,3}$	Ta . . .	73	26	$N_{6,7}VV$
		844	$L_3M_{2,3}V$			163	$N_5N_{6,7}N_{6,7}$
		917	L_3VV			173	$N_4N_{6,7}N_{6,7}$
						199	$N_5N_{6,7}V$
						209	$N_4N_{6,7}V$
						341	$N_3N_{6,7}N_{6,7}$

Energies of the major Auger lines for the more common elements (*contd*)

Element	Z	E/eV	Transition†	Element	Z	E/eV	Transition†
W . . .	74	20	$N_{6,7}VV$	Pb . . .	82	46	$O_3O_{4,5}O_{4,5}$
		164	$N_5N_{6,7}N_{6,7}$			57	$O_2O_{4,5}O_{4,5}$
		176	$N_4N_{6,7}N_{6,7}$			92	$N_7O_{4,5}O_{4,5}$
		205	$N_5N_{6,7}V$			116	$N_7O_{4,5}V$
		216	$N_4N_{6,7}V$			246	$N_5N_7O_{4,5}$
		347	$N_3N_{6,7}N_{6,7}$			265	$N_4N_7O_{4,5}$
Pt . . .	78	13	N_7O_3V	Bi . . .	83	60	$O_2O_{4,5}O_{4,5}$
		45	O_3VV			102	$N_7O_{4,5}O_{4,5}$
		68	N_7VV			128	$N_7O_{4,5}V$
		160	$N_5N_{6,7}N_{6,7}$			250	$N_5N_7O_{4,5}$
		171	$N_4N_{6,7}N_{6,7}$			270	$N_4N_7O_{4,5}$
		238	$N_5N_{6,7}V$				
		254	$N_4N_{6,7}V$				
Au . . .	79	42	O_3VV				
		71	N_7VV				
		151	$N_5N_{6,7}N_{6,7}$				
		165	$N_4N_{6,7}N_{6,7}$				
		244	$N_5N_{6,7}V$				
		261	$N_4N_{6,7}V$				

† The symbols K, L, M, N, and O are in the usual X-ray notation for atomic energy levels; the symbol V refers to the valence band of the solid.

J.C.R.

3.4 X-rays

3.4.1 X-ray absorption edges, characteristic X-ray lines and fluorescence yields

Characteristic X-ray lines are generated by the movement of an electron from an 'initial' level or edge, to a 'terminal' level in which a vacancy exists as a result of excitation of some sort (e.g. X-ray, electron, etc.); the energy of the line is equal to the difference in energy of the terminal and initial levels. Depending on atomic number, the X-ray spectra from the elements can include lines from the K, L, M, N and O series corresponding to excitation of the K, L, M, N or O levels; the table lists the energies—in KeV—of the principal lines of the common K, L and M series along with the corresponding terminal edge, or excitation, energy. Lines are identified both by the common labels— e.g. $K\alpha_1$, $K\alpha_2$, etc.—and the term labels giving, first, the terminal and, secondly, the initial edges— e.g. KL_{III}, KL_{II}, etc. With the exception of the elements 92–103 the table has been prepared by calculation of line energies from the compilation of smoothed edge energies given by Dewey, Mapes and Reynolds, *Progress in Nuclear Energy*, 1969, Series IX, Vol. 9, *Analytical Chemistry*, eds Elion and Stewart (Pergamon Press), which in turn drew extensively on the compilation by Bearden, *Rev. Mod. Phys.*, 1967, **39** (1), 78; the edge data for elements 92–103 is taken directly from Bearden. The energies of the softer radiations may be affected by the chemical state of the elements concerned; generally the shifts do not exceed a few electron volts. The wavelength λ, in pm, can be derived from the tabulated energy E, in keV, by the relationship $\lambda = 1239.81/E$.

Approximate line intensities are given, where available, relative to the strongest line within a given series and follow the pattern given by Clark (*Encyclopaedia of X-rays and Gamma Rays*, 1963 (Reinhold). These intensities are only intended as a guide since the exact values vary considerably with atomic number, and available data are sparse and inconsistent; further details can be found in the book by Sagel (*Tabellen zur Röntgenspektralanalyse*, 1959 (Springer-Verlag) and the article by Sandström (*Handbuch der Physik*, Vol. XXX, 'X-Rays', 1957 (Springer-Verlag)).

In addition to the lines given in the table, satellite, or non-diagram, lines also occur; these are generally only of significance in the case of the K-satellites of the lighter elements Al, Mg, Si, etc. where their intensities may be a few per cent of that of the Kα line (see Clark or Sandström).

The transition of an electron to fill, for example, a vacancy in the K-shell may be accompanied by either the emission of an X-ray photon or the transfer of energy to another electron which is then

Atomic number and element	K *edge*	KN_{III} / $K\beta_2$	KM_{III} / $K\beta_1$	KM_{II} / $K\beta_3$	KL_{III} / $K\alpha_1$	KL_{II} / $K\alpha_2$	L_I *edge*	$L_I N_{III}$ / $L\gamma_3$	$L_I M_{III}$ / $L\beta_3$	L_{II} *edge*	$L_{II}N_{IV}$ / $L\gamma_1$	$L_{II}M_{IV}$ / $L\beta_1$
Intensity	—	~5	~30	~12	100	~50	—	~2	~6	—	~10	~50
4 Be	0.115				0.109							
5 B	0.188				0.183							
6 C	0.282				0.277							
7 N	0.397				0.393							
8 O	0.533				0.525							
9 F	0.692				0.677							
10 Ne	0.874		0.858		0.848							
11 Na	1.080		1.071		1.041							
12 Mg	1.309		1.302		1.253		0.062					
13 Al	1.562		1.557		1.487	1.486	0.087			0.076		
14 Si	1.840		1.836		1.740	1.739	0.118			0.101		
15 P	2.143		2.139		2.014	2.013	0.153			0.130		
16 S	2.471		2.464		2.308	2.307	0.193			0.164		
17 Cl	2.824		2.816		2.622	2.620	0.237			0.204		
18 Ar	3.203		3.190		2.958	2.956	0.286			0.247		
19 K	3.607		3.590		3.314	3.311	0.340			0.296		
20 Ca	4.034		4.013		3.692	3.688	0.403			0.346		
21 Sc	4.486		4.461		4.090	4.086	0.462			0.400		0.400
22 Ti	4.965		4.932		4.511	4.505	0.529			0.460		0.458
23 V	5.463		5.427		4.952	4.944	0.626		0.585[3]	0.519		0.519
24 Cr	5.987		5.947		5.415	5.405	0.694		0.654[3]	0.582		0.583
25 Mn	6.537		6.490		5.899	5.888	0.768		0.721[3]	0.649		0.649
26 Fe	7.112		7.058		6.404	6.391	0.846		0.792[3]	0.721		0.719
27 Co	7.712		7.649		6.930	6.915	0.929		0.870[3]	0.797		0.791
28 Ni	8.339		8.265		7.478	7.461	1.016		0.941[3]	0.878		0.869
29 Cu	8.993		8.905	8.903	8.048	8.028	1.109		1.023[3]	0.965		0.950
30 Zn	9.673	9.658[1]	9.572	9.567	8.639	8.616	1.208		1.107[3]	1.057		1.035
31 Ga	10.386	10.366[1]	10.271	10.261	9.252	9.231	1.316		1.197[3]	1.155		1.125
32 Ge	11.115	11.101[1]	10.983	10.978	9.887	9.856	1.426		1.294	1.259		1.218
33 As	11.877	11.864[1]	11.727	11.721	10.544	10.509	1.536		1.386	1.368		1.316
34 Se	12.666	12.652[1]	12.496	12.489	11.222	11.181	1.662		1.492	1.485		1.419
35 Br	13.483	13.470[1]	13.292	13.285	11.924	11.878	1.791		1.600	1.605		1.523
36 Kr	14.330	14.315[1]	14.113	14.105	12.650	12.598	1.923		1.706	1.732		1.637
37 Rb	15.202	15.185[1]	14.962	14.952	13.396	13.336	2.067	2.051[2]	1.827	1.866		1.752
38 Sr	16.106	16.085[1]	15.836	15.826	14.166	14.098	2.217	2.197[2]	1.947	2.008		1.872
39 Y	17.037	17.015[1]	16.737	16.725	14.958	14.882	2.372	2.347[2]	2.072	2.155		1.996
40 Zr	17.997	17.963[1]	17.662	17.649	15.770	15.692	2.535	2.503[2]	2.200	2.305	2.292	2.118
41 Nb	18.985	18.947[1]	18.623	18.606	16.615	16.521	2.698	2.660[2]	2.336	2.464	2.449	2.257
42 Mo	20.002	19.960	19.608	19.590	17.479	17.374	2.867	2.825[2]	2.473	2.628	2.611	2.396
43 Tc	21.048	21.002	20.619	20.599	18.367	18.251	3.047	3.001[2]	2.618	2.797	2.778	2.537
44 Ru	22.123	22.072	21.656	21.637	19.279	19.150	3.230	3.179[2]	2.763	2.973	2.952	2.683
45 Rh	23.229	23.173	22.723	22.698	20.216	20.073	3.421	3.365[2]	2.915	3.156	3.132	2.835

emitted (an Auger electron); the probability that a vacancy in a given shell will result in emission of an X-ray is the fluorescence yield of that shell. The table lists fluorescence yields (ω_K and ω_L) for the K and L shells based on the tabulation by Birks, *Spectrochemical Analysis*, 1969 (Wiley), which was derived from the extensive compilation by Fink *et al.*, *Rev. Mod. Phys.*, 1966, **38** (3), 513, and extended to include the data for the very light elements given by Dick and Lucas, *Phys. Rev. A*, 1970, **2** (3), 2.

Z	L-series (contd)			M-series							Average fluorescence yields	
	L_{III}	$L_{III}N_V$	$L_{III}M_V$	M_{III}	$M_{III}N_V$	M_{IV}	$M_{IV}N_{VI}$	M_V	M_VN_{VII}	M_VN_{VI}		
	edge	$L\beta_2$	$L\alpha_1$	edge	$M\gamma$	edge	$M\beta$	edge	$M\alpha_1$	$M\alpha_2$		
Int.	—	~ 20	100	—		—		—			ω_K	ω_L
4	0.006										0.0003	
5	0.005										0.0007	
6	0.005										0.0013	
7	0.004										0.0015	
8	0.008										0.002	
9	0.015										0.01	
10	0.026										0.015	
11	0.039										0.02	
12	0.056										0.03	
13	0.075										0.04	
14	0.100										0.055	
15	0.129										0.07	
16	0.163										0.09	
17	0.202										0.105	
18	0.245										0.125	
19	0.293										0.14	
20	0.342										0.165	
21	0.396		0.395[4]								0.19	
22	0.454		0.452[4]								0.22	
23	0.511		0.511[4]								0.24	
24	0.572		0.573[4]								0.26	
25	0.638		0.637[4]								0.285	
26	0.708		0.705[4]								0.32	
27	0.782		0.776[4]								0.345	
28	0.861		0.852[4]								0.375	
29	0.945		0.930[4]			0.015					0.41	
30	1.034		1.012[4]			0.022					0.435	
31	1.134		1.098[4]	0.115		0.030					0.47	
32	1.228		1.188[4]	0.132		0.041					0.50	
33	1.333		1.282[4]	0.150		0.052					0.53	
34	1.444		1.379[4]	0.170		0.066					0.565	
35	1.559		1.480[4]	0.191		0.082					0.60	
36	1.680		1.586[4]	0.217		0.095					0.635	
37	1.806		1.694	0.240		0.114		0.112			0.665	
38	1.940		1.806	0.270		0.136		0.134			0.685	
39	2.079		1.923	0.300		0.159		0.156			0.71	
40	2.227	2.215	2.043	0.335	0.323	0.187		0.184			0.73	0.055
41	2.370	2.357	2.166	0.362	0.349	0.207		0.204			0.755	0.06
42	2.523	2.508	2.295	0.394	0.379	0.232		0.228			0.77	0.065
43	2.681	2.664	2.424	0.429	0.412	0.260		0.257			0.785	0.075
44	2.844	2.825	2.556	0.467	0.448	0.290		0.288			0.80	0.08
45	3.013	2.992	2.698	0.506	0.485	0.321		0.315			0.81	0.085

X-ray absorption edges, characteristic X-ray lines and fluorescence yields (contd)

Atomic number and element	K-series						L-series					
	K edge	KN$_{III}$	KM$_{III}$	KM$_{II}$	KL$_{III}$	KL$_{II}$	L$_I$ edge	L$_I$N$_{III}$	L$_I$M$_{III}$	L$_{II}$ edge	L$_{II}$N$_{IV}$	L$_{II}$M$_{IV}$
		Kβ$_2$	Kβ$_1$	Kβ$_3$	Kα$_1$	Kα$_2$		Lγ$_3$	Lβ$_3$		Lγ$_1$	Lβ$_1$
Intensity	—	~5	~30	~12	100	~50	—	~2	~6	—	~10	~50
46 Pd	24.365	24.303	23.819	23.792	21.178	21.021	3.619	3.557	3.073	3.344	3.318	2.990
47 Ag	25.531	25.463	24.943	24.912	22.163	21.991	3.822	3.754	3.234	3.540	3.511	3.151
48 Cd	26.727	26.653	26.095	26.061	23.173	22.985	4.034	3.960	3.402	3.742	3.710	3.319
49 In	27.953	27.872	27.275	27.237	24.209	24.002	4.250	4.169	3.572	3.951	3.915	3.487
50 Sn	29.211	29.122	28.491	28.439	25.272	25.044	4.475	4.377	3.750	4.167	4.127	3.661
51 Sb	30.499	30.402	29.725	29.677	26.359	26.110	4.706	4.609	3.932	4.389	4.345	3.843
52 Te	31.817	31.712	30.995	30.944	27.472	27.201	4.942	4.837	4.120	4.616	4.568	4.030
53 I	33.168	33.054	32.295	32.239	28.612	28.317	5.186	5.072	4.313	4.851	4.799	4.221
54 Xe	34.551	34.428	33.625	33.562	29.779	29.459	5.442	5.319	4.516	5.092	5.035	4.415
55 Cs	35.966	35.833	34.985	34.918	30.973	30.625	5.700	5.567	4.719	5.341	5.278	4.619
56 Ba	37.414	37.270	36.378	36.303	32.194	31.817	5.964	5.820	4.928	5.597	5.529	4.827
57 La	38.894	38.739	37.802	37.721	33.442	33.034	6.235	6.080	5.143	5.860	5.786	5.037
58 Ce	40.410	40.243	39.258	39.170	34.720	34.279	6.516	6.349	5.364	6.131	6.051	5.261
59 Pr	41.958	41.778	40.748	40.653	36.026	35.550	6.802	6.622	5.592	6.408	6.321	5.485
60 Nd	43.538	43.345	42.272	42.166	37.361	36.847	7.095	6.902	5.829	6.691	6.597	5.722
61 Pm	45.152	44.947	43.825	43.713	38.725	38.171	7.398	7.193	6.071	6.981	6.880	5.962
62 Sm	46.801	46.584	45.413	45.289	40.118	39.523	7.707	7.490	6.319	7.278	7.169	6.205
63 Eu	48.486	48.256	47.036	46.902	41.542	40.902	8.024	7.794	6.574	7.584	7.467	6.455
64 Gd	50.207	49.964	48.696	48.554	42.996	42.309	8.343	8.100	6.832	7.898	7.772	6.713
65 Tb	51.965	51.709	50.382	50.228	44.481	43.744	8.679	8.423	7.096	8.221	8.086	6.976
66 Dy	53.761	53.491	52.119	51.956	45.999	45.208	9.013	8.743	7.371	8.553	8.409	7.249
67 Ho	55.593	55.308	53.878	53.707	47.547	46.699	9.365	9.080	7.650	8.894	8.740	7.529
68 Er	57.464	57.164	55.681	55.491	49.128	48.221	9.725	9.425	7.942	9.243	9.078	7.813
69 Tm	59.374	59.059	57.513	57.303	50.742	49.773	10.097	9.782	8.236	9.601	9.426	8.103
70 Yb	61.322	60.991	59.374	59.157	52.389	51.354	10.479	10.148	8.531	9.968	9.781	8.402
71 Lu	63.311	62.960	61.286	61.049	54.070	52.965	10.869	10.518	8.844	10.346	10.144	8.709
72 Hf	65.345	64.973	63.236	62.979	55.790	54.611	11.262	10.890	9.153	10.734	10.517	9.016
73 Ta	67.405	67.011	65.221	64.946	57.533	56.277	11.672	11.278	9.488	11.128	10.894	9.345
74 W	69.517	69.100	67.244	66.951	59.318	57.982	12.092	11.675	9.819	11.535	11.284	9.671
75 Re	71.670	71.230	69.309	68.994	61.140	59.718	12.522	12.082	10.161	11.952	11.682	10.006
76 Os	73.869	73.404	71.416	71.077	63.001	61.487	12.968	12.503	10.515	12.382	12.092	10.349
77 Ir	76.111	75.620	73.560	73.203	64.896	63.287	13.416	12.925	10.865	12.824	12.514	10.705
78 Pt	78.400	77.883	75.751	75.364	66.832	65.123	13.880	13.363	11.231	13.277	12.944	11.073
79 Au	80.729	80.182	77.985	77.580	68.804	66.990	14.353	13.806	11.609	13.739	13.383	11.432
80 Hg	83.109	82.532	80.261	79.822	70.819	68.894	14.835	14.258	11.987	14.215	13.834	11.823
81 Tl	85.532	84.924	82.575	82.384	72.872	70.832	15.344	14.736	12.387	14.700	14.293	12.217
82 Pb	88.008	87.367	84.936	84.450	74.969	72.804	15.863	15.222	12.791	15.204	14.769	12.618
83 Bi	90.540	89.866	87.354	86.831	77.118	74.815	16.391	15.717	13.205	15.725	15.261	13.031
84 Po	93.113	92.403	89.801	89.250	79.301	76.863	16.940	16.230	13.628	16.250	15.756	13.452
85 At	95.730	94.983	92.302	91.722	81.523	78.943	17.495	16.748	14.067	16.787	16.262	13.882
86 Rn	98.402	97.617	94.866	94.246	83.793	81.065	18.047	17.262	14.511	17.337	16.777	14.323
87 Fr	101.131	100.306	97.477	96.807	86.114	83.231	18.630	17.805	14.976	17.900	17.307	14.775
88 Ra	103.909	103.039	100.130	99.432	88.476	85.434	19.222	18.352	15.443	18.475	17.848	15.238
89 Ac	106.738	105.837	102.846	102.101	90.884	87.675	19.823	18.922	15.931	19.063	18.402	15.711
90 Th	109.641	108.690	105.611	104.831	93.358	89.952	20.449	19.498	16.419	19.689	18.993	16.215

Z	L-series (contd)			M-series							Average fluorescence yields	
	L_{III}	$L_{III}N_V$	$L_{III}M_V$	M_{III}	$M_{III}N_V$	M_{IV}	$M_{IV}N_{VI}$	M_V	$M_V N_{VII}$	$M_V N_{VI}$		
	edge	$L\beta_2$	$L\alpha_1$	edge	$M\gamma$	edge	$M\beta$	edge	$M\alpha_1$	$M\alpha_2$	ω_K	ω_L
Int.	—	~20	100	—		—		—			ω_K	ω_L
46	3.187	3.163	2.838	0.546	0.522	0.354		0.349			0.82	0.09
47	3.368	3.342	2.985	0.588	0.562	0.389		0.383			0.83	0.095
48	3.554	3.525	3.134	0.632	0.603	0.423		0.420			0.84	0.105
49	3.744	3.712	3.288	0.678	0.646	0.464		0.456			0.85	0.11
50	3.939	3.903	3.442	0.720	0.684	0.506		0.497			0.86	0.115
51	4.140	4.101	3.604	0.774	0.735	0.546		0.536			0.87	0.12
52	4.345	4.302	3.770	0.822	0.779	0.586		0.575			0.875	0.125
53	4.556	4.509	3.938	0.873	0.826	0.630		0.618			0.88	0.13
54	4.772	4.720	4.110	0.926	0.874	0.677		0.662			0.89	0.14
55	4.993	4.936	4.289	0.981	0.924	0.722		0.704			0.895	0.145
56	5.220	5.158	4.470	1.036	0.974	0.770		0.750			0.90	0.15
57	5.452	5.385	4.651	1.092	1.025	0.823	0.854	0.801		0.833	0.905	0.155
58	5.690	5.617	4.839	1.152	1.079	0.870	0.902	0.851		0.883	0.91	0.165
59	5.932	5.853	5.034	1.210	1.131	0.923	0.950	0.898		0.929	0.915	0.17
60	6.177	6.091	5.231	1.266	1.180	0.969	0.997	0.946		0.978	0.915	0.175
61	6.427	6.334	5.433	1.327	1.234	1.019		0.994			0.92	0.18
62	6.683	6.582	5.635	1.388	1.287	1.073	1.100	1.048		1.081	0.925	0.185
63	6.944	6.835	5.843	1.450	1.341	1.129	1.153	1.101		1.131	0.93	0.19
64	7.211	7.034	6.058	1.511	1.334	1.185	1.209	1.153		1.185	0.93	0.195
65	7.484	7.358	6.273	1.583	1.457	1.245	1.266	1.211		1.240	0.935	0.205
66	7.762	7.627	6.496	1.642	1.507	1.304	1.325	1.266		1.293	0.935	0.21
67	8.046	7.901	6.719	1.715	1.570	1.365	1.383	1.327		1.348	0.94	0.215
68	8.336	8.180	6.951	1.783	1.627	1.430	1.443	1.385		1.406	0.94	0.22
69	8.632	8.465	7.181	1.861	1.694	1.498	1.503	1.451		1.462	0.94	0.225
70	8.933	8.755	7.415	1.948	1.770	1.566	1.568	1.518	1.521	1.507	0.945	0.235
71	9.241	9.049	7.655	2.025	1.833	1.637	1.623	1.586		1.572	0.945	0.25
72	9.555	9.348	7.891	2.109	1.902	1.718	1.700	1.664		1.646	0.95	0.27
73	9.872	9.649	8.147	2.184	1.961	1.783	1.760	1.725		1.702	0.95	0.285
74	10.199	9.959	8.396	2.273	2.033	1.864	1.835	1.803	1.776	1.774	0.95	0.30
75	10.530	10.273	8.651	2.361	2.104	1.946	1.910	1.879	1.845	1.843	0.955	0.315
76	10.868	10.592	8.905	2.453	2.177	2.033	1.988	1.963	1.921	1.918	0.955	0.33
77	11.215	10.919	9.175	2.551	2.255	2.119	2.062	2.040	1.988	1.983	0.96	0.34
78	11.568	11.251	9.439	2.649	2.332	2.204	2.134	2.129	2.065	2.059	0.96	0.355
79	11.925	11.585	9.705	2.744	2.404	2.307	2.220	2.220	2.142	2.133	0.96	0.365
80	12.290	11.927	9.999	2.848	2.485	2.392	2.285	2.291	2.195	2.184	0.96	0.375
81	12.660	12.272	10.271	2.957	2.569	2.483	2.360	2.389	2.270	2.266	0.96	0.38
82	13.039	12.625	10.555	3.072	2.658	2.586	2.442	2.484	2.345	2.340	0.965	0.39
83	13.422	12.981	10.836	3.186	2.745	2.694	2.534	2.586	2.422	2.426	0.965	0.40
84	13.812	13.342	11.131	3.312	2.842	2.798	2.620	2.681	2.501	2.503	0.965	0.405
85	14.207	13.708	11.427	3.428	2.929	2.905	2.707	2.780	2.581	2.582	0.965	0.41
86	14.609	14.079	11.727	3.536	3.006	3.014	2.794	2.882	2.663	2.662	0.965	0.42
87	15.017	14.456	12.031	3.654	3.093	3.125	2.881	2.986	2.746	2.742	0.965	0.425
88	15.433	14.839	12.340	3.779	3.185	3.237	2.967	3.093	2.829	2.823	0.970	0.43
89	15.854	15.227	12.652	3.892	3.265	3.352	3.054	3.202	2.913	2.904	0.970	0.43
90	16.283	15.622	12.970	4.030	3.369	3.474	3.145	3.313	2.996	2.984	0.970	0.435

X-ray absorption edges, characteristic X-ray lines and fluorescence yields (*contd*)

Atomic number and element	K-series						L-series					
	K *edge*	KN_{III}	KM_{III}	KM_{II}	KL_{III}	KL_{II}	L_I *edge*	L_IN_{III}	L_IM_{III}	L_{II} *edge*	$L_{II}N_{IV}$	$L_{II}M_{IV}$
		$K\beta_2$	$K\beta_1$	$K\beta_3$	$K\alpha_1$	$K\alpha_2$		$L\gamma_3$	$L\beta_3$		$L\gamma_1$	$L\beta_1$
Intensity	—	~5	~30	~12	100	~50	—	~2	~6	—	~10	~50
91 Pa	112.599	111.606	108.435	107.606	95.883	92.287	21.088	20.095	16.924	20.312	19.581	16.715
92 U	115.606	114.561	111.303	110.424	98.440	94.659	21.757	20.712	17.454	20.947	20.167	17.219
93 Np	118.678	117.591	114.243	113.312	101.068	97.077	22.427	21.340	17.992	21.601	20.785	17.751
94 Pu	121.818	120.703	117.261	116.277	103.761	99.552	23.097	21.982	18.540	22.266	21.417	18.293
95 Am	125.027	123.891	120.360	119.317	106.523	102.083	23.773	22.637	19.106	22.944	22.065	18.852
96 Cm	128.220	127.066	123.423	122.325	109.290	104.441	24.460	23.306	19.663	23.779		19.552
97 Bk	131.590	130.355	126.663	125.443	112.138	107.205	25.275	24.040	20.348	24.385		20.019
98 Cf	135.960	134.681	130.851	129.601	116.030	110.710	26.110	24.831	21.001	25.250		20.763
99 Es	139.490	138.169	134.238	132.916	119.080	113.470	26.900	25.579	21.648	26.020		21.390
100 Fm	143.090	141.724	137.693	136.347	122.190	116.280	27.700	26.334	22.303	26.810		22.044
101 Md	146.780	145.370	141.234	139.761	125.390	119.170	28.530	27.120	22.984	27.610		22.707
102 No	150.540	149.092	144.852	143.295	128.660	122.100	29.380	27.932	23.692	28.440		23.403
103 Lw	154.380	152.900	148.670	146.920	132.020	125.100	30.240	28.760	24.530	29.280		24.130

Unresolved lines:

1—K $N_{II, III}$ ($K\beta_2$); 2—L_I $N_{II, III}$ ($L\gamma_{2, 3}$); 3—L_I $M_{II, III}$ ($L\beta_{3,4}$); 4—L_{III} $M_{IV, V}$ ($L\alpha_{1, 2}$).

Depending on the resolving power of the dispersing system used (e.g. crystal spectrometer, solid state energy dispersive detector) line pairs shown separately in the table may not be resolved and the effective energy of the doublet will be close to the mean value weighted by the relative intensity of the components.

D.M.P.

3.4.2 Attenuation of photons

The intensity $I(t)$ of a photon beam after passing through t kg m^{-2} of matter is given by

$$I(t) = I(o) \exp(-\mu t/\rho)$$

where μ/ρ is the **mass absorption coefficient**. The values of μ/ρ given in the accompanying figure and table refer to the attenuation of a collimated beam of photons striking the attenuating material with an emergent beam consisting only of those photons which are undeflected and have suffered no energy loss. Inclusion of secondary products in the transmitted beam reduces this attenuation. The present values are taken from Storm and Israel, *Nuclear Data Tables*, 1970, A7, 565.

The four main interaction processes of photons with matter are photoelectric absorption and coherent (Rayleigh) scattering which dominate at low energies, incoherent (Compton) scattering important between 0.2 and 5 MeV, and pair production which is dominant at high energies. Storm and Israel also list these components by element and by energy.

Mass absorption coefficients for composite materials may be calculated by taking the average value of the absorption coefficients for the constituents weighted in proportion to their abundance by weight.

Up to 0.2 MeV the error of the coefficients may be as large as 10%. Between 0.2 and 5 MeV the accuracy is within 5%. Above 5 MeV the errors again may be as large as 10%.

Z	L-series (contd)			M-series							Average fluorescence yields	
	L_{III}	$L_{III}N_V$	$L_{III}M_V$	M_{III}	$M_{IV}N_V$	M_{IV}	$M_{IV}N_{VI}$	M_V	M_VN_{VII}	M_VN_{VI}		
	edge	$L\beta_2$	$L\alpha_1$	edge	$M\gamma$	edge	$M\beta$	edge	$M\alpha_1$	$M\alpha_2$	ω_K	ω_L
Int.	—	~20	100	—		—		—			ω_K	ω_L
91	16.716	16.022	13.300	4.164	3.470	3.597	3.251	3.416	3.083	3.070	0.970	0.44
92	17.166	16.429	13.614	4.303	3.566	3.728	3.337	3.552	3.171	3.161	0.970	0.445
93	17.610	16.840	13.944	4.435	3.665	3.850	3.435	3.666	3.262	3.251	0.970	0.445
94	18.057	17.256	14.279	4.557	3.756	3.973	3.527	3.778	3.346	3.332	0.970	0.45
95	18.504	17.676	14.617	4.667	3.839	4.092	0.0	3.887	0.0	0.0		
96	18.930		14.959	4.797		4.227		3.971				
97	19.452		15.320	4.927		4.366		4.132				
98	19.930		15.677	5.109		4.487		4.253				
99	20.410		16.036	5.252		4.630		4.374				
100	20.900		16.402	5.397		4.766		4.498				
101	21.390		16.768	5.546		4.903		4.622				
102	21.880		17.139	5.688		5.037		4.741				
103	22.360		17.500	5.710		5.150		4.860				

Mass absorption coefficient as a function of element and photon energy in units of $m^2\ kg^{-1} \times 10^{-2}$.

MeV	Element											
	H	Be	C	Al	Fe	Ge	Ag	I	W	Pb	U	Concrete†
0.2	2.425	1.089	1.228	1.223	1.445	1.659	2.948	3.626	7.797	9.942	12.88	1.24
0.3	2.109	0.942	1.068	1.042	1.100	1.120	1.545	1.751	3.200	3.983	5.111	1.07
0.4	1.894	0.849	0.953	0.926	0.937	0.929	1.122	1.206	1.903	2.291	2.884	0.95
0.5	1.726	0.775	0.867	0.844	0.839	0.819	0.927	0.968	1.362	1.590	1.943	0.87
0.6	1.595	0.715	0.802	0.779	0.768	0.743	0.809	0.826	1.081	1.233	1.465	0.80
0.8	1.404	0.628	0.707	0.683	0.668	0.641	0.676	0.674	0.799	0.875	0.999	0.71
1.0	1.260	0.565	0.637	0.614	0.598	0.571	0.592	0.579	0.655	0.704	0.779	0.64
1.5	1.027	0.459	0.516	0.500	0.487	0.465	0.474	0.463	0.498	0.517	0.549	0.52
2.0	0.872	0.393	0.443	0.431	0.425	0.408	0.419	0.410	0.436	0.454	0.476	0.45
3.0	0.693	0.313	0.355	0.355	0.361	0.352	0.374	0.369	0.403	0.416	0.438	0.36
4.0	0.580	0.267	0.305	0.310	0.331	0.328	0.360	0.359	0.400	0.416	0.435	0.32
5.0	0.564	0.235	0.271	0.283	0.315	0.316	0.357	0.359	0.406	0.424	0.445	0.29
6.0	0.449	0.213	0.247	0.266	0.305	0.312	0.358	0.364	0.416	0.436	0.453	0.27
8.0	0.374	0.182	0.215	0.243	0.298	0.310	0.367	0.378	0.439	0.459	0.481	0.24
10.0	0.325	0.163	0.196	0.232	0.296	0.314	0.384	0.395	0.465	0.486	0.506	0.23
15.0	0.253	0.136	0.169	0.219	0.307	0.331	0.422	0.438	0.524	0.547	0.574	0.21
20.0	0.215	0.123	0.157	0.216	0.321	0.350	0.456	0.474	0.577	0.608	0.635	0.21
30.0	0.174	0.110	0.147	0.219	0.345	0.383	0.509	0.536	0.659	0.695	0.734	0.21
40.0	0.153	0.104	0.143	0.225	0.367	0.411	0.549	0.579	0.717	0.759	0.800	0.21
50.0	0.141	0.102	0.142	0.230	0.382	0.431	0.581	0.612	0.760	0.802	0.848	0.22
60.0	0.132	0.100	0.142	0.234	0.395	0.446	0.603	0.636	0.796	0.840	0.891	0.22
80.0	0.123	0.099	0.143	0.243	0.417	0.470	0.642	0.674	0.842	0.895	0.946	0.23
100.0	0.118	0.099	0.145	0.252	0.427	0.489	0.664	0.702	0.881	0.933	0.984	0.24

† Approximate composition of concrete: oxygen 50%, silicon 31%, calcium 8%, aluminium 5%, other materials 6%.

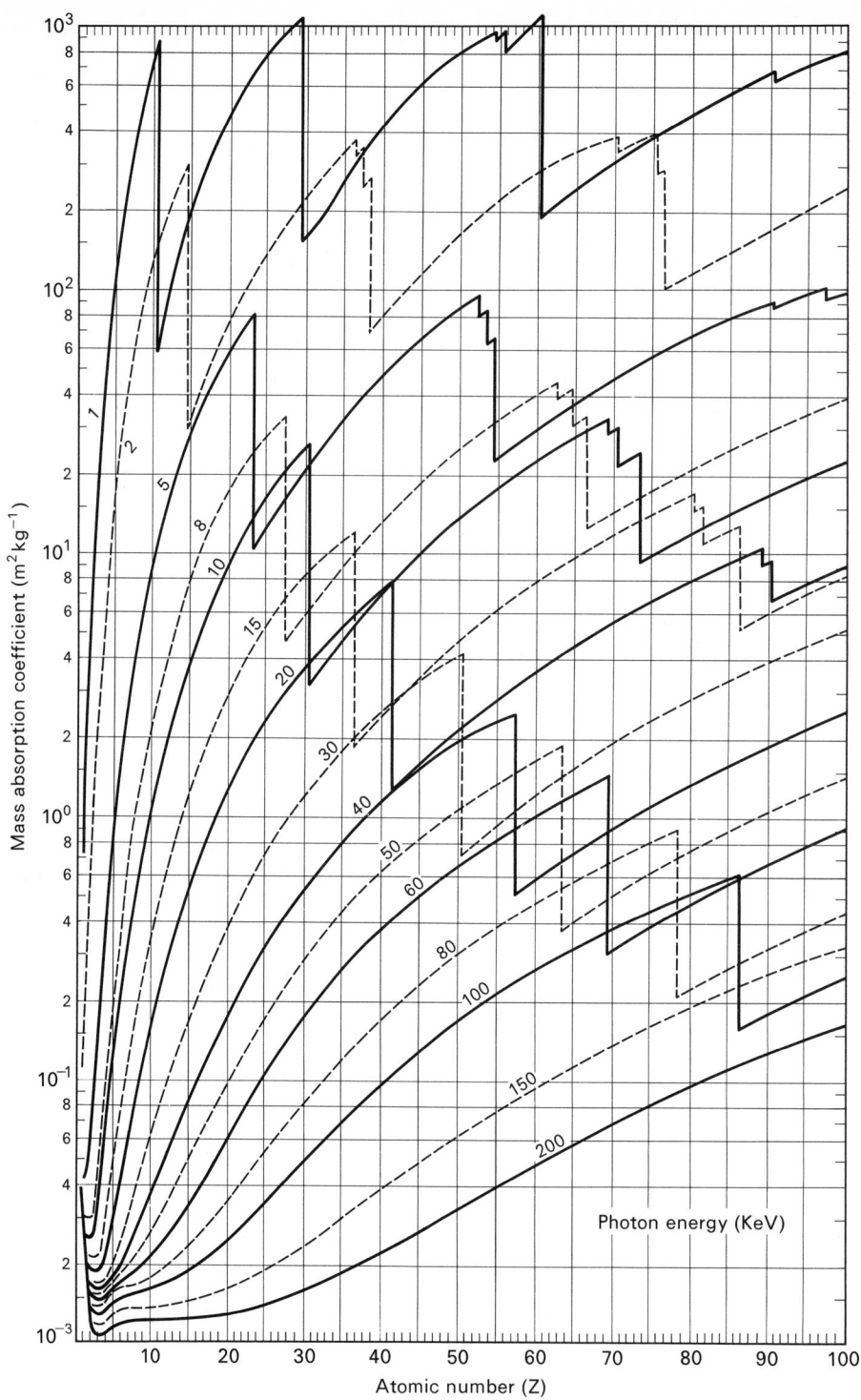

Mass absorption coefficient (μ/ρ) as a function of element and photon energy.

O.N.J.   C.W.

3.5 Absorption of particles and dosimetry

3.5.1 Range and stopping power of ions in various materials

The **stopping power** of a material for an ion of a given energy is defined as the energy loss per unit thickness (in terms of the mass per unit area) of the stopping material as measured in an ideally thin sample. The **range** of an ion is defined as the distance (again in terms of mass per unit area) which the particle traverses in the material before it comes to rest. Fluctuations in the processes of energy loss result in variations in range and stopping power. For light ions the range variation is of the order of 10% for low energies and a few per cent for high energies. The energy loss variation for high energies and thin absorbers can rise to 40% and is never less than a few per cent. Methods of estimating these straggling effects are given in *Studies in Penetration of Charged Particles in Matter* (ed. U. Fano), Publication 1133, NAS–NRC, 1964. In the case of heavy ions the fluctuating processes give rise to asymmetric range distributions and the phenomena are not well understood.

In the following tables the mean values of range and stopping power are given and interpolation will not give rise to errors in excess of the uncertainties in the data.

Range and stopping power of light particles

Energy MeV	Range/(kg m^{-2}) for protons in:							
	H	C	Al	Cu	Pb	U	Water	
2.0	2.89	8.41	11.5	16.2	30.8	33.7	7.13	$\times 10^{-2}$
6.0	2.09	5.46	6.97	9.21	15.8	17.0	4.71	$\times 10^{-1}$
10.0	5.31	13.7	17.0	21.8	35.6	38.1	11.8	$\times 10^{-1}$
22.0	2.27	5.65	6.79	8.38	13.1	14.1	4.93	$\times 10^{0}$
50.0	10.3	24.9	29.2	35.0	52.3	55.7	21.8	$\times 10^{0}$
100.0	3.62	8.62	9.98	11.8	16.8	17.8	7.57	$\times 10^{1}$
200.0	12.3	29.0	33.2	38.7	53.5	56.2	25.5	$\times 10^{1}$
500.0	5.64	13.1	14.8	17.1	22.9	23.9	11.5	$\times 10^{2}$

Energy MeV	Range/(kg m^{-2}) for alphas in:						
	H	C	Al	Cu	Pb	U	
2.0	2.84	11.6	18.5	31.1	61.5	66.2	$\times 10^{-3}$
6.0	1.70	5.40	7.53	11.6	22.8	24.7	$\times 10^{-2}$
10.0	4.18	12.2	16.5	23.9	45.4	49.6	$\times 10^{-2}$
22.0	1.74	4.83	6.07	8.18	14.2	15.3	$\times 10^{-1}$
50.0	7.85	20.7	25.1	32.1	51.1	54.5	$\times 10^{-1}$
100.0	2.80	7.16	8.50	10.6	15.8	16.8	$\times 10^{0}$
200.0	9.98	24.9	29.0	35.2	50.6	53.3	$\times 10^{0}$
500.0	5.21	12.7	14.6	17.3	23.9	25.1	$\times 10^{1}$

Range and stopping power of light particles (*contd*)

Energy MeV	Range/(kg m^{-2}) for pi-mesons in:							
	H	C	Al	Cu	Pb	U	Water	
2.0	1.36	3.46	4.25	5.35	8.58	9.18	3.00	$\times 10^{-1}$
6.0	10.3	25.1	29.6	35.8	54.1	57.8	22.0	$\times 10^{-1}$
10.0	2.63	6.31	7.35	8.75	12.8	13.6	5.53	$\times 10^{0}$
22.0	10.8	25.6	29.4	34.4	48.2	50.8	22.5	$\times 10^{0}$
50.0	4.42	10.3	11.7	13.6	18.4	19.3	9.07	$\times 10^{1}$
100.0	13.2	30.6	34.5	39.6	52.8	55.1	26.9	$\times 10^{1}$
200.0	3.52	8.13	9.09	10.4	13.6	14.2	7.13	$\times 10^{2}$
500.0	10.8	24.9	27.5	31.1	39.9	41.5	21.8	$\times 10^{2}$

Energy MeV	Range/(kg m^{-2}) for mu-mesons in:							
	H	C	Al	Cu	Pb	U	Water	
2.0	1.72	4.33	5.25	6.54	10.3	11.1	3.76	$\times 10^{-1}$
6.0	13.0	31.4	36.8	44.1	65.5	69.8	27.5	$\times 10^{-1}$
10.0	3.29	7.85	9.11	10.8	15.5	16.4	6.90	$\times 10^{0}$
22.0	13.4	31.4	35.9	41.9	57.9	60.9	27.6	$\times 10^{0}$
50.0	5.26	12.2	13.9	16.0	21.5	22.5	10.8	$\times 10^{1}$
100.0	15.1	34.7	39.1	44.8	59.2	61.8	30.6	$\times 10^{1}$
200.0	3.82	8.81	9.82	11.2	14.5	15.2	7.72	$\times 10^{2}$
500.0	11.1	25.6	28.2	31.9	40.5	42.2	22.4	$\times 10^{2}$
1000.0	22.5	52.7	57.6	64.7	80.8	84.2	46.2	$\times 10^{2}$
2000.0	4.38	10.4	11.3	12.6	15.5	16.1	9.15	$\times 10^{3}$
5000.0	10.2	24.8	26.7	29.6	35.7	37.2	21.9	$\times 10^{3}$

Energy MeV	Stopping power/(MeV kg^{-1} m^2) for protons in:						
	H	He	C	Al	Cu	Air	
2.0	38.6	16.9	14.6	11.2	8.32	14.2	$\times 10^{0}$
6.0	15.8	7.01	6.17	4.96	3.88	8.42	$\times 10^{0}$
10.0	10.2	4.53	4.07	3.38	2.73	3.99	$\times 10^{0}$
22.0	5.20	2.37	2.16	1.83	1.52	2.12	$\times 10^{0}$
50.0	26.6	12.1	11.4	9.62	8.15	11.0	$\times 10^{-1}$
200.0	9.35	4.32	4.02	3.54	3.06	3.97	$\times 10^{-1}$
500.0	5.64	2.62	2.45	2.17	1.91	2.42	$\times 10^{-1}$
5000.0	4.25	2.01	1.81	1.67	1.49	1.90	$\times 10^{-1}$

Energy MeV	Stopping power/(MeV kg^{-1} m^2) for alphas in:					
	H	He	C	Al	Cu	
2.0	47.6	18.0	14.4	10.0	6.21	$\times 10^{1}$
6.0	20.3	8.19	6.99	5.38	3.79	$\times 10^{1}$
10.0	13.4	5.51	4.79	3.80	2.81	$\times 10^{1}$
22.0	6.92	2.93	2.60	2.13	1.66	$\times 10^{1}$
50.0	34.7	15.0	13.5	11.4	9.17	$\times 10^{0}$
200.0	11.0	4.88	4.49	3.89	3.26	$\times 10^{0}$
500.0	5.39	2.42	2.25	1.97	1.68	$\times 10^{0}$
5000.0						

Range and stopping power of light particles (*contd*)

Energy	Stopping power/(MeV kg^{-1} m^2) for pi-mesons in:						
MeV	H	He	C	Al	Cu	Air	
2.0	7.96	3.55	3.21	2.69	2.19	3.15	× 10^0
6.0	31.6	14.4	13.2	11.4	9.59	13.0	× 10^{-1}
10.0	21.0	9.57	8.83	7.66	6.53	8.70	× 10^{-1}
22.0	11.5	5.30	4.92	4.32	3.88	4.86	× 10^{-1}
50.0	6.84	3.17	2.96	2.62	2.28	2.93	× 10^{-1}
200.0	4.28	2.01	1.86	1.68	1.48	1.87	× 10^{-1}
1000.0	4.36	2.07	1.85	1.71	1.53	1.96	× 10^{-1}
5000.0	5.22	2.50	2.10	1.97	1.78	2.39	× 10^{-1}

Energy	Stopping power/(MeV kg^{-1} m^2) for mu-mesons in:						
MeV	H	He	C	Al	Cu	Air	
2.0	6.29	2.85	2.56	2.17	1.78	2.52	× 10^0
6.0	25.2	11.5	10.6	9.16	7.77	10.4	× 10^{-1}
10.0	16.8	7.71	7.13	6.21	5.32	7.03	× 10^{-1}
22.0	9.50	4.39	4.08	3.59	3.11	4.03	× 10^{-1}
50.0	5.95	2.77	2.58	2.29	2.00	2.56	× 10^{-1}
200.0	4.16	1.96	1.80	1.64	1.44	1.83	× 10^{-1}
1000.0	4.49	2.14	1.89	1.75	1.57	2.02	× 10^{-1}
5000.0	5.39	2.59	2.14	2.01	1.83	2.47	× 10^{-1}

For protons, pi-mesons and mu-mesons the data in the tables were derived from W. H. Barkas and M. J. Berger, Nuclear Science Series, Report Number 39, National Research Council, 1964. The data for alpha-particles were derived from C. F. Williamson, J-P. Boujot and J. Picard, Report CEA–R3042, 1966. These data agree to a few per cent with available experimental measurements.

In that the rate of energy loss of a particle of given charge in a given medium depends only on its velocity, the data in the above tables can be extended to other isotopes of hydrogen and helium and to other mesons and hyperons.

For a particle of charge ze, mass M and energy E, the stopping power for a given material may be calculated from

$$[S(E)]_{M, ze} = z^2 \left[S\left(\frac{M_p}{M} \cdot E \right) \right]_{proton}$$

where M_p is the proton mass.

Similarly its range in a given material is given by

$$[R(E)]_{M, ze} = \frac{1}{z^2} \cdot \frac{M}{M_p} \left[R\left(\frac{M_p}{M} \cdot E \right) \right]_{proton}$$

These formulae are only accurate up to $z = 2$.

Range and stopping power of heavy particles

Energy per mass unit	Ranges of heavy ions						
MeV/a.m.u.	Range/(kg m^{-2} × 10^{-2}) for $^4_2$He ions in:						
	C	Al	Ti	Ge	Ag	Au	U
0.0125	0.099	0.142	0.219	0.307	0.371	0.746	0.874
0.0500	0.235	0.327	0.498	0.684	0.812	1.620	1.895
0.2000	0.559	0.802	1.175	1.579	1.854	3.539	4.110
0.8000	2.103	3.244	4.337	5.560	6.544	11.12	12.71
3.2000	18.66	24.86	29.86	35.85	42.95	63.34	70.59
12.0000	189.5	233.3	261.7	299.8	356.9	472.5	515.3
MeV/a.m.u.	Range/(kg m^{-2} × 10^{-2}) for $^{12}_6$C ions in:						
	C	Al	Ti	Ge	Ag	Au	U
0.0125	0.085	0.119	0.181	0.255	0.313	0.635	0.758
0.0500	0.221	0.304	0.457	0.627	0.749	1.495	1.764
0.2000	0.515	0.733	1.070	1.438	1.692	3.232	3.771
0.8000	1.462	2.234	3.020	3.898	4.589	7.938	9.109
3.2000	7.795	10.54	12.85	15.57	18.62	28.12	31.49
12.0000	65.54	80.99	91.20	104.8	124.7	166.4	181.8
MeV/a.m.u.	Range/(kg m^{-2} × 10^{-2}) for $^{27}_{13}$Al ions in:						
	C	Al	Ti	Ge	Ag	Au	U
0.0125	0.091	0.124	0.183	0.254	0.313	0.618	0.745
0.0500	0.266	0.361	0.531	0.720	0.857	1.672	1.980
0.2000	0.630	0.885	1.278	1:706	2.002	3.772	4.406
0.8000	1.501	2.268	3.078	3.978	4.675	8.121	9.341
3.2000	5.927	8.086	9.974	12.18	14.53	22.32	25.09
12.0000	38.31	47.62	53.96	62.28	74.14	100.1	109.7
MeV/a.m.u.	Range/(kg m^{-2} × 10^{-2}) for $^{40}_{18}$Ar ions in:						
	C	Al	Ti	Ge	Ag	Au	U
0.0125	0.097	0.130	0.187	0.256	0.313	0.603	0.727
0.0500	0.300	0.402	0.583	0.783	0.926	1.771	2.095
0.2000	0.724	1.016	1.457	1.935	2.262	4.207	4.908
0.8000	1.694	2.553	3.458	4.464	5.235	9.043	10.40
3.2000	5.993	8.214	10.17	12.45	14.83	22.89	25.76
12.0000	34.58	43.12	49.03	56.71	67.51	91.65	100.5
MeV/a.m.u.	Range/(kg m^{-2} × 10^{-2}) for $^{58}_{28}$Ni ions in:						
	C	Al	Ti	Ge	Ag	Au	U
0.0125	0.095	0.125	0.174	0.234	0.285	0.526	0.632
0.0500	0.313	0.414	0.589	0.782	0.919	1.706	2.016
0.2000	0.780	1.087	1.542	2.034	2.366	4.319	5.031
0.8000	1.752	2.627	3.548	4.569	5.340	9.160	10.53
3.2000	5.507	7.582	9.431	11.57	13.75	21.31	24.02
12.0000	27.17	34.05	38.91	45.16	53.73	73.54	80.82

Range and stopping power of heavy particles (*contd*)

Energy per mass unit	Ranges of heavy ions						
MeV/a.m.u.	$Range/(\text{kg m}^{-2} \times 10^{-2})$ *for* $^{74}_{32}\text{Ge}$ *ions in:*						
	C	Al	Ti	Ge	Ag	Au	U
0.0125	0.108	0.139	0.190	0.251	0.302	0.541	0.647
0.0500	0.363	0.476	0.669	0.881	1.027	1.865	2.196
0.2000	0.918	1.273	1.795	2.358	2.730	4.921	5.719
0.8000	2.034	3.041	4.095	5.263	6.142	10.47	12.02
3.2000	6.188	8.526	10.61	13.02	15.46	23.93	26.96
12.0000	29.16	36.58	41.86	48.63	57.85	79.32	87.21

MeV/a.m.u.	$Range/(\text{kg m}^{-2} \times 10^{-2})$ *for* $^{90}_{40}\text{Zr}$ *ions in:*						
	C	Al	Ti	Ge	Ag	Au	U
0.0125	0.108	0.138	0.185	0.241	0.287	0.498	0.591
0.0500	0.377	0.490	0.679	0.885	1.026	1.821	2.139
0.2000	0.975	1.346	1.883	2.460	2.837	5.041	5.847
0.8000	2.120	3.155	4.239	5.437	6.331	10.71	12.28
3.2000	6.120	8.441	10.52	12.92	15.31	23.70	26.70
12.0000	26.72	33.62	38.56	44.88	53.37	73.44	80.81

MeV/a.m.u.	$Range/(\text{kg m}^{-2} \times 10^{-2})$ *for* $^{107}_{47}\text{Ag}$ *ions in:*						
	C	Al	Ti	Ge	Ag	Au	U
0.0125	0.112	0.141	0.186	0.239	0.282	0.473	0.557
0.0500	0.398	0.512	0.701	0.906	1.044	1.816	2.127
0.2000	1.046	1.437	1.998	2.599	2.986	5.238	6.063
0.8000	2.253	3.345	4.482	5.738	6.661	11.21	12.83
3.2000	6.276	8.665	10.81	13.27	15.70	24.29	27.36
12.0000	25.96	32.71	37.60	43.81	52.07	71.84	79.09

MeV/a.m.u.	$Range/(\text{kg m}^{-2} \times 10^{-2})$ *for* $^{153}_{63}\text{Eu}$ *ions in:*						
	C	Al	Ti	Ge	Ag	Au	U
0.0125	0.125	0.152	0.194	0.244	0.279	0.439	0.505
0.0500	0.451	0.570	0.762	0.968	1.100	1.828	2.125
0.2000	1.220	1.661	2.279	2.936	3.345	5.715	6.581
0.8000	2.620	3.870	5.152	6.564	7.594	12.58	14.37
3.2000	6.862	9.486	11.83	14.53	17.15	26.42	29.74
12.0000	25.66	32.46	37.44	43.72	51.92	71.90	79.22

MeV/a.m.u.	$Range/(\text{kg m}^{-2} \times 10^{-2})$ *for* $^{181}_{73}\text{Ta}$ *ions in:*						
	C	Al	Ti	Ge	Ag	Au	U
0.0125	0.130	0.157	0.198	0.245	0.276	0.419	0.476
0.0500	0.475	0.595	0.785	0.988	1.115	1.813	2.099
0.2000	1.303	1.765	2.404	3.082	3.497	5.891	6.766
0.8000	2.804	4.130	5.479	6.963	8.027	13.21	15.07
3.2000	7.114	9.842	12.28	15.07	17.75	27.30	30.72
12.0000	25.20	31.95	36.92	43.17	51.22	71.11	78.38

Range and stopping power of heavy particles (*contd*)

Energy per mass unit	Ranges of heavy ions						
MeV/a.m.u.	$Range/(\mathrm{kg\ m}^{-2} \times 10^{-2})$ *for* $^{197}_{79}$Au *ions in:*						
	C	Al	Ti	Ge	Ag	Au	U
0.0125	0.133	0.159	0.199	0.244	0.273	0.407	0.458
0.0500	0.486	0.606	0.794	0.995	1.119	1.798	2.076
0.2000	1.344	1.815	2.462	3.148	3.563	5.957	6.832
0.8000	2.896	4.259	5.639	7.149	8.239	13.51	15.39
3.2000	7.225	9.998	12.47	15.29	18.02	27.68	31.13
12.0000	24.83	31.53	36.47	42.66	50.61	70.34	77.55
MeV/a.m.u.	$Range/(\mathrm{kg\ m}^{-2} \times 10^{-2})$ *for* $^{238}_{92}$U *ions in:*						
	C	Al	Ti	Ge	Ag	Au	U
0.0125	0.141	0.167	0.205	0.248	0.272	0.391	0.433
0.0500	0.520	0.642	0.830	1.029	1.148	1.801	2.068
0.2000	1.456	1.955	2.631	3.345	3.767	6.200	7.087
0.8000	3.154	4.623	6.096	7.712	8.865	14.41	16.39
3.2000	7.650	10.59	13.20	16.19	19.03	29.15	32.76
12.0000	24.81	31.57	36.60	42.87	50.81	70.75	78.03

The data in the table are derived from L. C. Northcliffe and R. F. Schilling, *Nuclear Data Tables*, 1970, Vol. 7, section A, Nos 3 and 4, who list the calculated ranges and stopping powers for 103 species of ion in 24 materials. At the lower energies, where some comparison with experimental determinations is possible, deviations from measurements are no larger than disagreements between different measurements of the same parameter, which may reach 10%. The higher energy regions are completely devoid of experimental data.

C.W.

3.5.2 Range of electrons and beta rays

The slowing down and stopping of electrons (except perhaps in H and He) is so much influenced by multiple scattering that the process almost resembles diffusion. The practical maximum range tabulated below is roughly that thickness of aluminium through which very few particles will penetrate. It is nearly the same for beta rays of maximum energy E and for homogeneous electrons of energy E, and depends little on counter geometry and on the absorbing material (except for H and heavy elements). For detailed definitions and methods, see Katz and Penfold, *Rev. Mod. Phys.*, 1952, **24**, 28; Glendenin, *Nucleonics*, 1948, **2**, 12; and Feather, *Proc. Camb. Phil. Soc.*, 1938, **34**, 599.

Energy E	Range in Al	Energy E	Range in Al	Energy E	Range in Al	Energy E	Range in Al	Energy E	Range in Al
keV	$\mathrm{mg\ cm}^{-2}$	keV	$\mathrm{mg\ cm}^{-2}$	MeV	$\mathrm{mg\ cm}^{-2}$	MeV	$\mathrm{g\ cm}^{-2}$	MeV	$\mathrm{g\ cm}^{-2}$
		30	1.5	0.15	26.5	0.8	0.31	4	2.02
		40	2.6	0.20	42	1.0	0.42	5	2.54
10	0.16	50	4.0	0.25	59	1.2	0.52	6.5	3.3
12	0.24	65	6.4	0.30	78	1.5	0.68	8	4.1
15	0.38	80	9.2	0.40	120	2.0	0.95	10	5.2
20	0.68	100	13.5	0.50	165	2.5	1.21	12	6.2
25	1.06	120	18.5	0.65	235	3.0	1.48	15	7.8

J.G.C.

3.5.3 Magnetic deflection of charged particles

When particles of charge ze and momentum p enter a uniform magnetic field B perpendicular to their direction of motion, they are deflected into a circular path whose radius ρ is given by

$$B\rho = \frac{10^{-2}}{2.9979z}\,p$$

for p in MeV c^{-1} B in Wb m^{-2} and ρ in metres. The relationship between the momentum and the total energy of a particle of rest mass m_0c^2 is

$$pc = \beta\gamma\,m_0c^2$$

and the kinetic energy is equal to $(\gamma-1)m_0c^2$, where β is the ratio of the velocity of the particle to that of light and $\gamma = (1-\beta^2)^{-1/2}$.

To facilitate conversion between kinetic energy, momentum and the product $B\rho$ a table of $\beta\gamma$ versus $(\gamma-1)$ is provided. For kinetic energies less than $0.001m_0c^2$, $\beta\gamma=\beta$ to better than 0.1% and for kinetic energies greater than $10.0m_0c^2$, $\beta\gamma=\gamma$ to better than 0.5%.

Values of m_0c^2 for the stable particles are given in the table of the same name (p. 344); values for atomic nuclei may be derived from mass defects given in the table of nuclides (p. 286).

$\beta\gamma$	$\gamma-1$	$\beta\gamma$	$\gamma-1$	$\beta\gamma$	$\gamma-1$
0.04	0.000 800	0.30	0.044 031	2.00	1.236 068
0.05	0.001 249	0.40	0.077 033	3.00	2.162 278
0.06	0.001 798	0.50	0.118 034	4.00	3.123 106
0.08	0.003 195	0.60	0.166 190	5.00	4.099 020
0.10	0.004 988	0.80	0.280 625	6.00	5.082 763
0.15	0.011 187	1.00	0.414 214	8.00	7.062 258
0.20	0.019 804	1.50	0.802 776	10.00	9.049 876

O.N.J.

3.5.4 Radiation units and quantities

The quantity of particulate radiation may be defined in terms of **fluence**, which is the number of particles which enter a sphere of unit cross-sectional area. For a unidirectional beam the fluence so defined is equal to the number of particles which cross a unit area placed perpendicular to the beam. When the radiation is distributed isotropically the number of particles which cross a unit area of any plane is one half the fluence. The **fluence rate** is known as the **flux density** or **flux**.

The quantity of X- or gamma-radiation to which an object is exposed can be specified in terms of **exposure**, which is the ionization that the radiation would produce in air. Exposure is defined as the quotient of ΔQ by Δm where ΔQ is the sum of the electrical charges on all ions of one sign produced in air when all the electrons (negatrons and positrons) liberated by photons in a volume of air whose mass is Δm are completely stopped in air. The ionization arising from the absorption of bremsstrahlung or annihilation radiation emitted by the electrons is not to be included in ΔQ. Owing to the difficulty of measurement, exposure is not normally used when the quantum energy exceeds 3 MeV. The unit of exposure is the **roentgen** (R) which is equal to 2.58×10^{-4} coulomb per kilogram of air (i.e. 1 electrostatic unit of charge per 0.001 293 gram).

The energy imparted to matter by ionizing radiation is known as the **absorbed dose** of which the unit is the **rad** which is equivalent to 0.01 joule per kilogram (100 erg per gram), or 6.242×10^{10} MeV per kilogram. An exposure of one roentgen produces an absorbed dose in air of 0.869 rad. This exposure also produces 2.082×10^{15} ion pairs per metre^3 of air at s.t.p., or 1.610×10^{15} ion pairs per kilogram.

The sum of the kinetic energies of all charged particles liberated in a medium by indirectly ionizing particles per unit mass of the medium is known as **kerma**. The kinetic energy includes the energy that the secondary particles may lose subsequently as bremsstrahlung or annihilation radiation and also the kinetic energy of charged particles produced by other secondary processes such as Auger electrons. Since the charged particles do not dissipate their energy at the point at which they are produced, kerma is not identical with absorbed dose unless complete secondary equilibrium exists in the medium. Exposure for X- and gamma-radiation is the ionization equivalent of kerma in air, except at very high energies when a difference arises due to the bremsstrahlung and annihilation radiation effects.

The **linear energy transfer** or **restricted linear collision stopping power** (L_Δ) of charged particles is the differential mean energy loss per unit distance traversed due to collisions with energy transfers less than some specified value Δ in electron volts. L_{100}, for example, designates the linear energy transfer when $\Delta = 100$ electron volts. The symbol L_∞ is used when all possible energy transfers are included, although the subscript does not mean that infinitely large energy transfers are possible. L_∞ is sometimes referred to as the **unrestricted linear collision stopping power**. Linear energy transfer is usually expressed in units of kiloelectron volts per micrometre. In a medium with a density of 1000 kilograms per metre^3 (1 gram per cm^3) a L_∞ of 1 keV per µm is equivalent to a specific energy loss (or stopping power, (p. 279)) of 1 MeV m^2 per kilogram or 10 MeV cm^2 per gram.

For protection purposes the quantity **dose equivalent** is employed to express on a common scale the risk to exposed persons from all ionizing radiations. The unit of dose equivalent is the **rem** which is equal to the absorbed dose to tissue in rads multiplied by appropriate modifying factors. The modifying factors depend on the quality of the radiation and the distribution in the body of absorbed dose. The most important modifying factor is the **quality factor** which is related to the unrestricted linear collision stopping power, L_∞, in water of the charged particles responsible for the absorbed dose in tissue.

The number of nuclear transformations in a quantity of material in a unit time is known as its **activity**. The unit of activity is the **curie** (Ci) which is defined as 3.7×10^{10} transformations per second. In its original definition the curie was that quantity of radon which is in radioactive equilibrium with 1 gram of radium. As the total number of transformations in 1 gram of radium and hence in the radon is about 3.7×10^{10} per second, the unit was unofficially applied to the quantity of any material in which this rate of transformation occurred and led in 1962 to the present definition.

References

(1) *Radiation Quantities and Units*, ICRU Report 11, 1968.
(2) *Report of Committee IV on Protection Against Electromagnetic Radiation above 3 MeV and Electrons, Neutrons and Protons*, ICRP Publication 4, 1964.

J.A.D.

3.6 Radioactive elements

3.6.1 Table of the nuclides

The nuclei contained in this table have well-established atomic number and mass with the exception of some of the nuclei in the transuranic region. It should be noted that the natural isotopic abundance of elements in the Earth's crust can vary according to the source, particularly in the case of the lighter elements.

The first column specifies the nuclide (isotope) giving the atomic number Z, the chemical symbol and the mass number A. Long-lived isomeric states are indicated by the symbol m following the mass number. In a few cases energy measurements are insufficiently precise to say which of the two entries of a given mass number should be considered as the ground state. The second column gives the half-life or natural abundance in atoms percent; in certain cases both quantities are relevant.

The third column gives the major radiations emitted by a radioactive nuclide in decaying to the ground state of its daughter. Percentages refer to percentages of the total decay events. For β^- and β^+ decay the highest end point energy is specified in MeV, followed by more detailed information if the highest energy branch is weak. K capture is always energetically possible if β^+ decay can occur, with an energy release—largely carried away by the neutrino—up to 1.02 MeV higher than the corresponding maximum β^+ energy. K capture is therefore only particularly specified when β^+ decay is weak or energetically impossible. There are two cases where L capture predominates and these are indicated by L. Details of γ decay are given in cases where the intensity as a percentage of total decay events is known and the three most intense γ-rays are quoted, provided their intensity exceeds 1%; annihilation radiation, which invariably results from positron capture, is not specifically mentioned. Gamma-rays are invariably accompanied to a greater or lesser extent by conversion electrons, X-rays and Auger electrons but these have not been included, with rare exceptions in the case of some important conversion electron lines (e^-). For α decay the energies and percentages of the three most intense groups are given. The strength of any spontaneous fission branch is also given.

The fourth column gives the mass excess $(M - A)$, expressed in MeV, i.e. the difference between the actual atomic mass M and A atomic mass units on the ^{12}C scale (a.m.u. $= 931.478$ MeV). The fifth column gives the neutron cross-section in barns for various processes for a neutron velocity of 2200 m s^{-1}, or an energy $0.025\,30$ eV (see section on Neutron Cross Sections, p. 332). The different processes are indicated as follows: γ for (n, γ), (γm) for (n, γ) leading to an isomeric state, (α) for (n, α), (p) for (n, p), (f) for fission (n, f) and (t) for the sum of all such absorption processes. The sixth column gives the spin (I) and parity (π).

It should be re-emphasized that this list refers only to well-established species. The total of nuclides grows steadily as new species are identified both in the region of proton-rich and neutron-rich isotopes of the elements listed and in the region of heavy elements.

References

(1) C. H. Lederer, J. M. Hollander and I. Perlman, *Table of Isotopes* (6th ed), 1967.
(2) *Fourth United Nations Conference on the Peaceful Uses of Transuranic Elements*, 1971, p. 841 (Data on the transuranium elements).
(3) G. T. Seaborg, *Ann. Rev. Nucl. Sci.*, 1968, **18**, 53.

Table of the nuclides

Nuclide		Natural abundance or half-life	Major radiations	Mass excess MeV	σ/b	I$^\pi$
Z	A					
0 n	1	12 min	β^- 0.78	8.0714		$\frac{1}{2}^+$
1 H	1	99.985%		7.2890	0.332 (γ)	$\frac{1}{2}^+$
H	2	0.015%		13.136	0.0005 (γ)	1$^+$
H	3	12.3 yr	β^- 0.0186	14.950		$\frac{1}{2}^+$
2 He	3	0.000 13%		14.9313	5330 (p)	$\frac{1}{2}^+$
He	4	100%		2.4248		0$^+$
He	6	0.8 s	β^- 3.508	17.598		0$^+$
He	8	0.122 s	β^- 9.7; 88% γ 0.98	31.7		0$^+$
3 Li	6	7.42%		14.088	953 (α)	1$^+$
Li	7	92.58%		14.907	0.037 (γ)	$\frac{3}{2}^-$
Li	8	0.84 s	β^- 13	20.946		2$^+$
Li	9	0.17 s	β^- 13.61; n; α	24.97		
4 Be	7	53 d	K 0.86; 10% γ 0.477	15.769	54 000 (p)	
Be	8	2×10^{-16} s	α			
Be	9	100%		11.351	0.009 (γ)	$\frac{3}{2}^-$
Be	10	2.5×10^6 yr	β^- 0.555	12.607		0$^+$

Table of the nuclides (*contd*)

Nuclide		Natural abundance or half-life	Major radiations	Mass excess MeV	σ/b	Iπ
Z	A					
Be	11	13.6 s	β⁻ 11.5; 32% γ 2.14; 4.4% γ 6.79; 2.4% γ 5.85	20.18	0.005 (γ)	
Be	12	0.011 s		25		0⁺
5 B	8	0.77 s	β⁺ 14; α	22.923		2⁺
B	9	8 × 10⁻¹⁹ s	p			
B	10	19.6%		12.052		3⁺
B	11	80.4%		8.6677		3/2⁻
B	12	0.02 s	β⁻ 13.37; α; 1.3% γ 4.43	13.370		1⁺
B	13	0.02 s	β⁻ 13.44; 7% γ 3.68	16.562		
6 C	9	0.127 s	p; α	29.0		
C	10	19.4 s	β⁺ 1.87; 100% γ 0.717; 1.7% γ 1.023	15.66		0⁺
C	11	20.3 min	β⁺ 0.97	10.648		3/2⁻
C	12	98.892%		0.0	0.0034 (γ)	0⁺
C	13	1.108%		3.125	0.0009 (γ)	1/2⁻
C	14	5730 yr	β⁻ 0.156	3.0198		0⁺
C	15	2.4 s	β⁻ 9.82; 68% β 4.51; 68% γ 5.299	9.873		1/2⁺
C	16	0.74 s		13.69		0⁺
7 N	12	0.011 s	β⁺ 16.4	17.36		1⁺
N	13	10.0 min	β⁺ 1.20	5.345		1/2⁻
N	14	99.635%		2.8637	1.81 (p)	1⁺
N	15	0.365%		0.100	2.4 × 10⁻⁵ (γ)	1/2⁻
N	16	7.2 s	β⁻ 10.40; 69% γ 6.13; 5% γ 7.11	5.685		2⁻
N	17	4.16 s	β⁻ 8.68; 95% β⁻ 4.1, n; 3% γ 0.87	7.87		
N	18	0.63 s	β⁻ 9.4; 59% γ 1.65, 0.82; 100% γ 1.98	13.1		
8 O	13	0.009 s	p	23.1		
O	14	71.0 s	β⁺ 4.12; 99% β⁺ 1.811; 99% γ 2.312	8.0080		0⁺
O	15	124 s	β⁺ 1.74	2.860		1/2⁻
O	16	99.759%		−4.7366	0.000 18 (γ)	0⁺
O	17	0.037%		−0.808	0.24 (α)	5/2⁺
O	18	0.204%		−0.7824	0.000 21 (γ)	0⁺
O	19	29 s	β⁻ 4.60; 97% γ 0.197; 59% γ 1.37	3.333		5/2⁺
O	20	14 s	β⁻; 100% γ 1.06	3.80		0⁺
9 F	17	66 s	β⁺ 1.74	1.952		5/2⁺
F	18	109.7 min	β⁺ 0.635	0.872		1⁺
F	19	100%		−1.486	0.010 (γ)	1/2⁺
F	20	11.4 s	β⁻ 5.41; 100% γ 1.63	−0.012		2⁺
F	21	4.4 s	β⁻ 5.4	−0.05		5/2⁺
F	22	4.0 s	β⁻ 11; 100% γ 1.28; 67% γ 2.06	4		
10 Ne	17	0.10 s	p	33.9		
Ne	18	1.5 s	β⁺ 3.42; 7% γ 1.04	5.319		0⁺
Ne	19	17.5 s	β⁺ 2.22	1.752		1/2⁺
Ne	20	90.92%		−7.042		0⁺
Ne	21	0.257%		−5.730		3/2⁺

Table of the nuclides (*contd*)

Nuclide Z	A	Natural abundance or half-life	Major radiations	Mass excess MeV	σ/b	I^π
Ne	22	8.82%		−8.025	0.04 (γ)	0^+
Ne	23	37.6 s	β^- 4.38; 33% γ 0.439	−5.148		
Ne	24	3.38 min	β^- 1.99; 100% γ 0.472; 8% γ 0.88	−5.95		0^+
11 Na	20	0.39 s	β^+ 11.4; α, γ	7.0		
Na	21	23 s	β^+ 2.52; 2.3% γ 0.35	−2.19		$\frac{3}{2}^+$
Na	22	2.60 yr	β^+ 1.820; 90% β^+ 0.545; 100% γ 1.275	−5.182		3^+
Na	23	100%		−9.528	0.53 (γ)	$\frac{3}{2}^+$
Na	24	15.0 h	100% γ 1.369; 100% γ 2.754	−8.418		4^+
Na	24m	0.02 s	β^- 6	−7.945		1^+
Na	25	60 s	β^- 3.83; 15% γ 0.98; 14% γ 0.58, 0.39	−9.36		$\frac{5}{2}^+$
Na	26	1.0 s	β^- 6.7; 100% γ 1.82	−7.7		$2, 3^+$
12 Mg	20	0.6 s		16		0^+
Mg	21	0.12 s	p	10.9		
Mg	23	12.1 s	β^+ 3.03; 9% γ 0.44	−5.472		$\frac{3}{2}^+$
Mg	24	78.60%		−13.933	0.03 (t)	0^+
Mg	25	10.11%		−13.191	0.3 (t)	$\frac{5}{2}^+$
Mg	26	11.29%		−16.214	0.027 (γ)	0^+
Mg	27	9.5 min	β^- 1.75; 70% γ 0.84; 30% γ 1.013	−14.583		$\frac{1}{2}^+$
Mg	28	21 h	β^- 0.46; 96% γ 0.031; 70% γ 1.35; 30% γ 0.95, 0.40	−15.02		0^+
13 Al	24	2.1 s	β^+ 8.5; γ	−0.1		4^+
Al	25	7.2 s	β^+ 3.24; γ	−8.93		$\frac{5}{2}^+$
Al	26	7.4×10^5 yr	β^+ 1.17; 100% γ 1.81; 4% γ 1.12	−12.211		5^+
Al	26m	6.4 s	β^+ 3.21	−11.982		0^+
Al	27	100%		−17.196	0.235 (γ)	$\frac{5}{2}^+$
Al	28	2.31 min	β^- 2.85; 100% γ 1.780	−16.855		3^+
Al	29	6.6 min	β^- 2.40; 94% γ 1.28; 6% γ 2.43	−18.22		$\frac{5}{2}^+$
Al	30	3.3 s	β^- 5.0; 61% γ 2.23; 39% γ 3.51	−17.2		$2, 3^+$
Al	30m	72 s	γ			
14 Si	25	0.23 s	p	4.0		
Si	26	2.1 s	β^+ 3.83	−7.13		0^+
Si	27	4.2 s	β^+ 3.85	−12.386		$\frac{5}{2}^+$
Si	28	92.18%		−21.490	0.08 (t)	0^+
Si	29	4.71%		−21.894	0.3 (t)	$\frac{1}{2}^+$
Si	30	3.12%		−24.439	0.11 (γ)	0^+
Si	31	2.62 h	β^- 1.48; γ	−22.96		$\frac{3}{2}^+$
Si	32	650 yr	β^- 0.21	−24.08		0^+
15 P	28	0.28 s	β^+ 11.0; 75% γ 1.780; 10% γ 4.44, 2.6	−7.7		
P	29	4.4 s	β^+ 3.95; γ	−16.95		$\frac{1}{2}^+$
P	30	2.50 min	β^+ 3.24; γ	−20.20		1^+
P	31	100%		−24.438	0.19 (γ)	$\frac{1}{2}^+$
P	32	14.3 d	β^- 1.710	−24.303		1^+
P	33	25 d	β^- 0.248	−26.335		

Table of the nuclides (*contd*)

Nuclide		Natural abundance or half-life	Major radiations	Mass excess MeV	σ/b	I^π
Z	A					
P	34	12.4 s	β^- 5.1; 25% γ 2.13	−24.8		1$^+$
16 S	29	0.19 s	p	−2.9		
S	30	1.4 s	β^+ 5.09; 80% β^+ 4.42; 80% γ 0.687	−14.09		0$^+$
S	31	2.7 s	β^+ 4.42; 1.1% γ 1.27	−18.99		$\frac{1}{2}^+$
S	32	95.0%		−26.013		0$^+$
S	33	0.760%		−26.583		$\frac{3}{2}^+$
S	34	4.22%		−29.934	0.27 (γ)	0$^+$
S	35	88 d	β^- 0.167	−28.847		$\frac{3}{2}^+$
S	36	0.014%		−30.66	0.14 (γ)	0$^+$
S	37	5.06 min	β^- 4.7; 90% β^- 1.6; 90% γ 3.09	−27.0		
S	38	2.87 h	β^- 3.0; 95% β^- 1.1; 95% γ 1.88	−26.8		0$^+$
17 Cl	32	0.31 s	β^+ 9.9; 70% γ 2.24; 14% γ 4.77; 7% γ 4.29	−12.8		
Cl	33	2.5 s	β^+ 4.55; γ	−21.01		$\frac{3}{2}^+$
Cl	34	1.56 s	β^+ 4.46	−24.45		0$^+$
Cl	34m	32.0 min	β^+ 2.48; 45% γ 0.145; 38% γ 2.12; 12% γ 1.17	−24.31		3$^+$
Cl	35	75.79%		−29.015	44 (γ)	$\frac{3}{2}^+$
Cl	36	3.1×10^5 yr	β^- 0.714; 2% K, γ	−29.520	100 (γ)	2$^+$
Cl	37	24.20%		−31.765	0.4 (γ)	$\frac{3}{2}^+$
Cl	38	37.3 min	β^- 4.91; 47% γ 2.170; 38% γ 1.60	−29.80		2$^-$
Cl	38m	0.74 s	100% γ 0.66	−29.13		5$^-$
Cl	39	55.5 min	β^- 3.45; 50% γ 1.27; 44% γ 0.246; 42% γ 1.52	−29.80		
Cl	40	1.4 min	β^- 7.5, γ	−27.5		
18 Ar	33	0.18 s	p 3.16	−9.5		
Ar	35	1.83 s	β^+ 4.94; 5% γ 1.22; 2% γ 1.76	−23.05		$\frac{3}{5}^+$
Ar	36	0.337%		−30.232	6 (γ)	0$^+$
Ar	37	35 d	K 0.814	−30.951		$\frac{3}{2}^+$
Ar	38	0.063%		−34.718	0.8 (γ)	0$^+$
Ar	39	269 yr	β^- 0.565	−33.24		
Ar	40	99.600%		−35.038	0.61 (γ)	0$^+$
Ar	41	1.83 h	β^- 2.49; 99% β^- 1.20; 99% γ 1.293	−33.061	0.5 (γ)	$\frac{7}{2}^-$
Ar	42	33 yr	β^-	−34.42		0$^+$
19 K	37	1.23 s	β^+ 5.14; 2% γ 2.79	−24.79		$\frac{3}{2}^+$
K	38	7.71 min	β^+ 2.68; 100% γ 2.170	−28.79		3$^+$
K	38m	0.95 s	β^+ 5.0	−28.66		0$^+$
K	39	93.22%		−33.803	2.0 (γ)	$\frac{3}{2}^+$
K	40	1.26×10^9 yr 0.118%	89% β^- 1.314; 11% K; 0.001% β^+ 0.483; 11% γ 1.46	−33.533	70 (γ)	4
K	41	6.77%		−35.552	1.2 (γ)	$\frac{3}{2}^+$
K	42	12.4 h	β^- 3.52; 18% γ 1.524	−35.02		2$^-$
K	43	22.4 h	β^- 1.82; 85% γ 0.373; 81% γ 0.619. Others	−36.58		$\frac{3}{2}^+$

Table of the nuclides (*contd*)

Z	Nuclide A	Natural abundance or half-life	Major radiations	Mass excess MeV	σ/b	Iπ
	K 44	22.0 min	β⁻ 5.2; 61% γ 1.156; 37% γ 2.1 complex	−36.3		
	K 45	16 min	β⁻ 4.0, γ	−36.6		
	K 47	17.5 s	β⁻ 6.1; 84% γ 2.0; 15% γ 2.6	−36.3		
20	Ca 37	0.173 s	p 3.1	−13.3		
	Ca 38	0.66 s	β⁺	−22		0⁺
	Ca 39	0.87 s	β⁺ 5.49	−27.30		3/2 ⁺
	Ca 40	96.97%		−34.848	0.23 (t)	0⁺
	Ca 41	8 × 10⁴ yr	K	−35.125		7/2 −
	Ca 42	0.64%		−38.540	42 (t)	0⁺
	Ca 43	0.145%		−38.396		7/2 −
	Ca 44	2.06%		−41.460	0.7 (γ)	0⁺
	Ca 45	165 d	β⁻ 0.252	−40.809		
	Ca 46	0.0033%		−43.14	0.3 (γ)	0⁺
	Ca 47	4.53 d	β⁻ 1.98; 82% β⁻ 0.67; 74% γ 1.308; 5% γ 0.815, 0.49	−42.35		7/2 −
	Ca 48	0.185%		−44.22	1.1 (γ)	0⁺
	Ca 49	8.8 min	β⁻ 1.95; 89% γ 3.10; 10% γ 4.1	−41.29		
	Ca 50	9 s	β⁻, γ	−41		0⁺
21	Sc 40	0.179 s	β⁺ 9.1; 100% γ 3.75	−20.3		
	Sc 41	0.60 s	β⁺ 5.47	−28.63		7/2 −
	Sc 42	0.68 s	β⁺	−32.109		0⁺
	Sc 42m	61 s	β⁺ 2.82; 100% γ 1.52, 1.22, 0.438	−31.58		7, 6⁺
	Sc 43	3.92 h	β⁺ 1.20	−36.17		7/2 −
	Sc 44	3.92 h	β⁺ 1.47; 100% γ 1.159	−37.81		2⁺
	Sc 44m	2.44 d	K; 98% γ 0.271	−37.54		6
	Sc 45	100%		−41.061	13 (γ); 10 (γm)	7/2 −
	Sc 46	83.9 d	β⁻ 1.48; 100% β⁻ 0.357; 100% γ 1.120, 0.889	−41.756		4⁺
	Sc 46m	20 s	γ 0.142	−41.614		7
	Sc 47	3.43 d	β⁻ 0.600; 73% γ 0.160	−44.326		7/2 −
	Sc 48	1.83 d	β⁻ 0.65; 100% γ 1.314, 1.040, 0.983	−44.51		6⁺
	Sc 49	57.5 min	β⁻ 2.01	−46.55		
	Sc 50	1.72 min	β⁻ 3.6; 100% γ 1.55, 1.12, 0.520	−45.0		
	Sc 50m	0.35 s	γ 0.258	−44.7		
22	Ti 41	0.09 s.	p	−15.9		
	Ti 43	0.56 s	β⁺ 5.8	−29.3		7/2 −
	Ti 44	48 yr	K, 98% γ 0.078; 90% γ 0.068	−37.66		0⁺
	Ti 45	3.09 h	β⁺ 1.04	−39.002		7/2 −
	Ti 46	7.99%		−44.123	0.6 (t)	0⁺
	Ti 47	7.32%		−44.927	1.7 (t)	5/2 −
	Ti 48	73.99%		−48.483	8.0 (t)	0⁺
	Ti 49	5.46%		−48.558	1.9 (t)	7/2 −
	Ti 50	5.25%		−51.431	0.14 (γ)	0⁺
	Ti 51	5.8 min	β⁻ 2.14; 95% γ 0.320; 5% γ 0.928; 1.5% γ 0.605	−49.74		3/2 −
23	V 46	0.426 s	β⁺ 6.03	−37.069		

Table of the nuclides (*contd*)

Nuclide		Natural abundance or half-life	Major radiations	Mass excess MeV	σ/b	I^π
Z	A					
V	47	33 min	β^+ 1.89, γ	−42.01		
V	48	16.0 d	β^+ 0.696; 100% γ 0.983; 97% γ 1.312; 10% γ 0.945	−44.470		4^+
V	49	330 d	K	−47.950		$\frac{7}{2}^-$
V	50	6×10^{15} yr 0.25%	K; β^-	−49.216	130 (γ)	6^+
V	51	99.75%		−52.199	4.9 (γ)	$\frac{7}{2}^-$
V	52	3.76 min	β^- 2.47; 100% γ 1.434	−51.44		3^+
V	53	2.0 min	β^- 2.50; 100% γ 1.00	−51.8		
V	54	55 s	β^- 3.3; 100% γ 0.99, 0.84, 2.21	−50		
24 Cr	48	23 h	K; 99% γ 0.31, 98% γ 0.116	−43.1		0^+
Cr	49	41.9 min	β^+ 1.54; 28% γ 0.091; 14% γ 0.063; 13% γ 0.153	−45.39		
Cr	50	4.31%		−50.249	17 (γ)	0^+
Cr	51	27.8 d	K; 9% γ 0.320	−51.447		
Cr	52	83.76%		−55.411	0.8 (γ)	0^+
Cr	53	9.55%		−55.281	18 (γ)	$\frac{3}{2}^-$
Cr	54	2.38%		−56.931	0.38 (γ)	0^+
Cr	55	3.5 min	β^- 2.59	−55.11		$\frac{3}{2}^-$
Cr	56	5.9 min	β^- 1.5	−55.3		0^+
25 Mn	50m	0.286 s	β^+ 6.61			
Mn	51	45 min	β^+ 2.17	−48.26		
Mn	52	5.7 d	β^+ 0.575; 100% γ 1.434; 84% γ 0.935; 82% γ 0.744	−50.70		6^+
Mn	52m	21 min	β^+ 1.63; 100% γ 1.434; 2% γ 0.383	−50.32		2^+
Mn	53	2×10^6 yr	K	−54.683	170 (γ)	$\frac{7}{2}^+$
Mn	54	303 d	K; 100% 0.835	−55.55		3^+
Mn	55	100%		−57.705	13.3 (γ)	$\frac{5}{2}^-$
Mn	56	2.576 h	β^- 2.85; 99% γ 0.847; 29% γ 1.811; 15% γ 2.110	−56.904		3^+
Mn	57	1.7 min	β^- 2.55; γ.	−57.5		
Mn	58	1.1 min	β^-	−56		
26 Fe	52	8.2 h	β^+ 0.8; 100% γ 0.165	−48.33		0^+
Fe	53	8.5 min	β^+ 3.0; 32% γ 0.38	−50.70		
Fe	54	5.84%		−56.246	2.9 (γ)	0^+
Fe	55	2.6 yr	K	−57.474		$\frac{3}{2}^-$
Fe	56	91.68%		−60.605	2.7 (γ)	0^+
Fe	57	2.17%		−60.176	2.5 (γ)	$\frac{1}{2}^-$
Fe	58	0.31%		−62.147	1.1 (γ)	0^+
Fe	59	45 d	β^- 1.57; 56% γ 1.095, 44% γ 1.292; 2.8% γ 0.192	−60.660		$\frac{3}{2}^-$
Fe	60	3×10^5 yr	β^-	−61.51		0^+
Fe	61	6.0 min	β^- 2.8; γ	−59		
27 Co	54	0.194 s	β^+	−47.99		
Co	55	18.2 h	β^+ 1.5; 80% γ 0.930; 13% γ 1.41; 12% γ 0.480	−54.01		$\frac{7}{2}^-$
Co	56	77 d	β^+ 1.49; 80% K; 100% γ 0.847; 66% γ 1.24; 17% γ 2.60	−56.03		4^+

Table of the nuclides (*contd*)

Nuclide Z	A	Natural abundance or half-life	Major radiations	Mass excess MeV	σ/b	I^π
Co	57	270 d	K; 87% γ 0.122; 11% γ 0.136; 9% γ 0.004	−59.339		$\frac{7}{2}^-$
Co	58	71.3 d	β⁺ 0.474; 85% K; 99% γ 0.810; 1.4% γ 0.865	−59.84	2500 (γ)	2^+
Co	58m	9.0 h	e⁻ 0.017, 0.024	−59.81	1.4 × 10⁵ (γ)	5^+
Co	59	100%		−62.233	19 (γ); 18 (γm)	$\frac{7}{2}^-$
Co	60	5.26 yr	β⁻ 1.48; 99.9% β⁻ 0.314; 100% γ 1.332; 99.9% γ 1.173	−61.651	6 (γ)	5^+
Co	60m	10.5 min	β⁻ 1.55	−61.593	100 (γ)	2^+
Co	61	99 min	β⁻ 1.22; 89% γ 0.067	−62.93		
Co	62	1.6 min	β⁻ 2.88; 180% γ 1.17 complex; 20% γ 1.47	−61.53		
28 Ni	56	6.10 d	K; 99% γ 0.163; 85% γ 0.812, 48% γ 0.748	−53.92		0^+
	57	36.0 h	β⁺ 0.85; 86% γ 1.37; 14% γ 1.89, 0.127	−56.10		$\frac{3}{2}^-$
	58	67.76%		−60.23	4.4 (γ)	0^+
Ni	59	8 × 10⁴ yr	K	−61.159		$\frac{3}{2}^-$
Ni	60	26.16%		−64.471	2.6 (γ)	0^+
Ni	61	1.25%		−64.22	2 (γ)	$\frac{3}{2}^-$
Ni	62	3.66%		−66.75	15 (γ)	0^+
Ni	63	92 yr	β⁻ 0.067	−65.52		
Ni	64	1.16%		−67.11	1.5 (γ)	0^+
Ni	65	2.56 h	β⁻ 2.13; 25% γ 1.481; 16% γ 1.115; 4.5% γ 0.368	−65.14		$\frac{5}{2}^-$
Ni	66	55 h	β⁻ 0.20	−66.06		0^+
Ni	67	50 s	β⁻ 4.1; 51% γ 0.9; 15% γ 1.26	−63.2		
29 Cu	58	3.20 s	β⁺ 8.2	−51.66		
Cu	59	81 s	β⁺ 3.7; 11% γ 1.305; 9% γ 0.872; 5% γ 0.463, 0.343	−56.36		
Cu	60	24 min	β⁺ 3.92; 80% γ 1.332; 52% γ 1.76; 15% γ 0.85	−58.35		2^+
Cu	61	3.32 h	β⁺ 1.22; 12% γ 0.284; 4% γ 0.067; 3% γ 0.38	−61.98		$\frac{3}{2}^-$
Cu	62	9.8 min	β⁺ 2.91	−62.81		1^+
Cu	63	69.1%		−65.583	4.5 (γ)	$\frac{3}{2}^-$
Cu	64	12.8 h	38% β⁻ 0.573; 19% β⁺ 0.656; K; γ	−65.428		1^+
Cu	65	30.9%		−67.27	2.3 (γ)	$\frac{3}{2}^-$
Cu	66	5.1 min	β⁻ 2.63; 9% γ 1.039	−66.26	130 (γ)	1^+
Cu	67	59 h	β⁻ 0.57; 40% γ 0.184; 23% γ 0.092 doublet	−67.29		
Cu	68	30 s	β⁻ 3.5; 95% γ 1.078; 17% γ 0.80; 5% γ 1.88	−65.4		
30 Zn	60	2.1 min	β⁺			0^+
Zn	61	1.48 min	β⁺ 4.4; 11% γ 0.48; 6% γ 1.64; 3% γ 0.98	−56.6		
Zn	62	9.3 h	β⁺ 0.66; 82% K	−61.12		0^+
Zn	63	38.4 min	β⁺ 2.34; 8% γ 0.669; 6% γ 0.962	−62.22		$\frac{3}{2}^-$

Table of the nuclides (*contd*)

Nuclide			Natural abundance or half-life	Major radiations	Mass excess MeV	σ/b	Iπ
Z		A					
	Zn	64	48.89%		−66.000	0.46 (γ)	0+
	Zn	65	245 d	β+ 0.327; 98% K; 49% γ 1.115	−65.92		5/2−
	Zn	66	27.81%		−68.88		0+
	Zn	67	4.11%		−67.86		5/2−
	Zn	68	18.56%		−69.99	1.1 (γ)	0+
	Zn	69	57 min	β− 0.90	−68.43		1/2−
	Zn	69m	13.8 h	95% γ 0.439; 5% e− 0.429	−67.99		9/2+
	Zn	70	0.62%		−69.55	0.11 (γ)	0+
	Zn	71	2.4 min	β− 2.61; 13% γ 0.510; 3% γ 0.92	−67.5		
	Zn	71m	4.0 h	β− 1.46; 94% γ 0.385; 75% γ 0.495; 65% 0.609	−67.2		
	Zn	72	46.5 h	β− 0.30; 90% γ 0.145; 10% γ 0.192; 8% γ 0.015	−68.14		0+
31	Ga	63	33 s	β+	−57		
	Ga	64	2.6 min	β+ 6.05; 43% γ 0.992; 18% γ 3.32; 15% γ 0.80	−58.93		
	Ga	65	15.2 min	β+ 2.24; 55% γ 0.115; 12% γ 0.061; 10% γ 0.75	−62.66		
	Ga	66	9.4 h	β+ 4.153; 37% γ 1.039; 25% γ 2.748; 5% 0.828	−63.71		0
	Ga	67	78 h	K; 40% γ 0.093; 24% γ 0.184; 22% γ 0.296	−66.87		3/2−
	Ga	68	68.3 min	β+ 1.90; 3.5% γ 1.078	−67.07		1+
	Ga	69	60.5%		−69.326	1.9 (γ)	3/2−
	Ga	70	21.1 min	β− 1.65	−68.90		1+
	Ga	71	39.5%		−70.135	5.0 (γ)	3/2−
	Ga	72	14.1 h	β− 3.15; 96% γ 0.835; 27% γ 0.630; 26% γ 2.201	−68.58		3−
	Ga	73	4.9 h	β− 1.19; 94% γ 0.295; 9% γ 0.054; 6% γ 0.74	−69.74		
	Ga	74	7.9 min	β− 2.5; 100% γ 0.6 doublet; 45% γ 2.35	−67.8		
	Ga	75	2 min	β− 3.3; 3% γ 0.58	−68.5		
	Ga	76	32 s	β− 6; γ			
32	Ge	65	1.5 min	β+ 3.7; 3% γ 0.67; 2% γ 1.72	−56		
	Ge	66	2.4 h	β+ 2.0; 48% γ 0.38; 37% γ 0.046; 23% γ 0.185	−60.7		0+
	Ge	67	19 min	β+ 3.1; 105% γ 0.170 doublet; 7% γ 0.92	−62.5		
	Ge	68	275 d	K	−67		0+
	Ge	69	38 h	β+ 1.22; 28% γ 1.107; 13% γ 0.573; 10% γ 0.872	−67.101		
	Ge	70	20.55%		−70.558	3.2 (γ)	0+
	Ge	71	11.4 d	K	−69.90		1/2−
	Ge	72	27.37%		−72.579	1.0 (γ)	0+
	Ge	73	7.67%		−71.293	14 (γ)	9/2+
	Ge	73m	0.53 s	9% γ 0.054	−71.226		1/2−
	Ge	74	36.74%		−73.419	0.3 (γ); 0.2 (γm)	0+

Table of the nuclides (*contd*)

Nuclide		Natural abundance or half-life	Major radiations	Mass excess MeV	σ/b	Iπ
Z	A					
Ge	75	82 min	β⁻ 1.19; 11% γ 0.265; 1.4% γ 0.199	−71.83		
Ge	75m	48 s	34% γ 0.139	−71.69		
Ge	76	7.67%		−73.209	0.1 (γ); 0.1 (γm)	0⁺
Ge	77	1.3 h	β⁻ 2.2; 61% γ 0.21 doublet; 45% γ 0.263	−71.2		
Ge	77m	54 s	β⁻ 2.9; 21% γ 0.215; 12% γ 0.159	−71.0		
Ge	78	1.47 h	β⁻ 0.71; 94% γ 0.277	−71.8		0⁺
33 As	69	15 min	β⁺ 2.9; γ	−63.2		
As	70	52 min	β⁺ 2.89; 78% γ 1.040; 25% γ 0.67; 23% γ 0.60	−64.32		
As	71	62 h	β⁺ 0.81; 90% γ 0.175	−67.89		
As	72	26 h	β⁺ 3.34; 78% γ 0.835; 8% γ 0.630	−68.22		2⁻
As	73	80.3 d	K; 9% γ 0.054	−70.92		3/2⁻
As	74	17.9 d	32% β⁻ 1.36; 29% β⁺ 1.54; 39% K; 61% γ 0.596; 14% γ 0.635	−70.855		2⁻
As	74m	8.0 s	γ 0.283	−70.572		
As	75	100%		−73.031	4.5 (γ)	3/2⁻
As	76	26.5 h	β⁻ 2.97; 43% γ 0.559; 6% γ 0.657	−72.29		2⁻
As	77	38.7 h	β⁻ 0.68; 2.5% γ 0.239	−73.92		3/2
As	78	91 min	β⁻ 4.1	−72.8		
As	79	9.0 min	β⁻ 2.15; 2% γ 0.36, 0.43	−73.7		
As	80	15.3 s	β⁻ 6.0; 42% γ 0.666; 4% γ 1.22	−71.8		1
As	81	33 s	β⁻ 3.8	−72.6		
34 Se	71	5 min	β⁺ 3.4; γ	−63.5		
Se	72	8.4 d	K; 59% γ 0.046	−68		0⁺
Se	73	7.1 h	β⁺ 1.66; 99% γ 0.359; 65% γ 0.066	−68.17		
Se	73	42 min	β⁺ 1.7;			
Se	74	0.87%		−72.212	30 (γ)	0⁺
Se	75	120.4 d	K; 60% γ 0.265; 57% γ 0.136; 25% γ 0.280	−72.166		5/2⁺
Se	76	9.02%		−75.26	63 (γ); 22 (γm)	0⁺
Se	77	7.58%		−74.60	42 (γ)	1/2⁻
Se	77m	17.5 s	50% γ 0.161	−74.44		7/2⁺
Se	78	23.52%		−77.021	0.05 (γ); 0.36 (γm)	0⁺
Se	79	6.5 × 10⁴ yr	β⁻ 0.16	−75.921		7/2⁺
Se	79m	3.9 min	9% γ 0.096	−75.825		1/2
Se	80	49.82%		−77.753	0.5 (γ); 0.1 (γm)	0⁺
Se	81	18.6 min	β⁻ 1.58	−76.40		1/2⁻
Se	81m	57 min	8% γ 0.103	−76.29		
Se	82	9.19%		−77.59	0.004 (γ); 0.05 (γm)	0⁺
Se	83	25 min	β⁻ 1.8; 69% γ 0.36; 59% γ 0.52; 44% γ 0.22	−75.4		
Se	83m	70 s	β⁻ 3.8; γ	−75.2		
Se	84	3.3 min	β⁻			0⁺

Table of the nuclides (*contd*)

Nuclide			Natural abundance or half-life	Major radiations	Mass excess MeV	σ/b	Iπ
Z		A					
	Se	85	39 s	β⁻			
	Se	87	16 s	β⁻			
35	Br	74	36 min	β⁺ 4.7; γ	−65		
	Br	75	1.7 h	β⁺ 1.70; γ	−69.44		
	Br	76	16.1 h	β⁺ 3.6; 63% γ 0.559; 13% γ 1.21; 11% γ 1.86	−70.6		1
	Br	77	57 h	β⁺ 0.34, 24% γ 0.52; 30% γ 0.24 complex	−73.24		$\frac{3}{2}-$
	Br	77m	4.2 min	γ 0.108	−73.13		
	Br	78	6.5 min	β⁺ 2.55; 14% γ 0.614	−73.45		1⁺
	Br	79	50.52%		−76.075	8.5 (γ); 2.9 (γm)	$\frac{3}{2}-$
	Br	79m	4.8 s	γ 0.21	−75.87		
	Br	80	17.6 min	92% β⁻ 2.0; 2.6% β⁺ 0.87; 5.7% K; 7% γ 0.618	−75.882		1⁺
	Br	80m	4.38 h	36% γ 0.037	−75.796		$5-$
	Br	81	49.48%		−77.97	3 (γ)	$\frac{3}{2}-$
	Br	82	35.34 h	β⁻ 0.444; 83% γ 0.777; 66% γ 0.554; 41% γ 0.619	−77.50		5⁻
	Br	82m	6.1 min	γ	−77.45		2⁻
	Br	83	2.41 h	β⁻ 0.93; 1.4% γ 0.530	−79.02		
	Br	84	31.8 min	β⁻ 4.68; 51% γ 0.88; 18% γ 1.90; 13% γ 3.93	−77.7		
	Br	84	6.0 min	β⁻ 1.9; 75% γ 1.46, 0.88; 68% γ 0.44			
	Br	85	3.0 min	β⁻ 2.5	−78.7		
	Br	86	54 s	β⁻ 7.1, γ	−76		
	Br	87	55 s	β⁻ 8.0; n, γ	−74.6		
	Br	88	16 s	β⁻; γ			
	Br	89	4.5 s	n			
36	Kr	74	20 min	β⁺ 3.1	−62		0⁺
	Kr	76	14.8 h	K; γ	−69		0⁺
	Kr	77	1.19 h	β⁺ 1.86; γ	−70.4		
	Kr	78	0.354%		−74.14	2 (γ)	0⁺
	Kr	79	34.9 h	β⁺ 0.60; 92% K; 10% γ 0.398, 0.606; 9% γ 0.261	−74.46		$\frac{1}{2}-$
	Kr	79m	55 s	γ 0.127	−74.33		
	Kr	80	2.27%		−77.89	15 (γ)	0⁺
	Kr	81	2.1 × 10⁵ yr	K	−77.7		
	Kr	81m	13 s	65% γ 0.190	−77.5		
	Kr	82	11.56%		−80.589	42 (γ); 3 (γm)	0⁺
	Kr	83	11.55%		−79.985	180 (γ)	$\frac{9}{2}+$
	Kr	83m	1.86 h	9% γ 0.009	−79.943		$\frac{1}{2}-$
	Kr	84	56.90%		−82.433	0.04 (γ); 0.10 (γm)	0⁺
	Kr	85	10.76 yr	β⁻ 0.67; γ	−81.48		$\frac{9}{2}+$
	Kr	85m	4.4 h	β⁻ 0.82; 74% γ 0.150; 13% γ 0.305	−81.18		$\frac{1}{2}-$
	Kr	86	17.37%		−83.259	0.06 (γ)	0⁺
	Kr	87	76 min	β⁻ 3.8; 84% γ 0.403; 35% γ 2.57; 16% γ 0.85	−80.70		
	Kr	88	2.80 h	β⁻ 2.8; 35% γ 2.40, 0.191; 23% γ 0.85	−79.9		0⁺

Table of the nuclides (*contd*)

Nuclide		Natural abundance or half-life	Major radiations	Mass excess MeV	σ/b	Iπ
Z	A					
Kr	89	3.2 min	β⁻ 4.0; γ	−78		
Kr	90	33 s	β⁻ 2.80; 65% γ 0.120; 48% γ 1.11, 0.536	−74.8		0⁺
Kr	91	10 s	β⁻ 3.6			
Kr	92	3.0 s	β⁻			0⁺
Kr	93	2.0 s	β⁻			
Kr	94	1.4 s	β⁻			0⁺
37 Rb	79	24 min	β⁺; 73% γ 0.15; 29% γ 0.19			1⁺
Rb	80	34 s	β⁺ 4.1; 39% γ 0.618	−73		1⁺
Rb	81	4.7 h	β⁺ 1.03; 87% K; γ	−75.4		3/2−
Rb	81m	31 min	β⁺ 1.4	−75.3		9/2+
Rb	82	1.3 min	β⁺ 3.15; 9% γ 0.777	−76.42		1⁺
Rb	82m	6.3 h	β⁺ 0.78; 94% K; 83% γ 0.777; 66% γ 0.554	−76.14		5−
Rb	83	83 d	K; 93% γ 0.53 complex	−79		5/2−
Rb	84	33.0 d	β⁺ 1.66; 74% γ 0.88	−79.753		2−
Rb	84m	20 min	65% γ 0.250; 37% γ 0.216; 32% γ 0.464	−79.289		6⁺
Rb	85	72.15%		−82.16	0.9 (γ); 0.1 (γm)	5/2−
Rb	86	18.66 d	β⁻ 1.78; 8.8% γ 1.078	−82.72		2−
Rb	86m	1.04 min	γ 0.56	−82.16		
Rb	87	5 × 10¹⁰ yr 27.85%	β⁻ 0.274	−84.591	0.12 (γ)	3/2−
Rb	88	12.8 min	β⁻ 5.3; 21% γ 1.863; 13% γ 0.898; 2.3% γ 2.68	−82.7	1.0 (γ)	2−
Rb	89	15.4 min	β⁻ 3.92; 75% γ 1.05; 54% γ 1.26; 17% γ 0.66	−82.3		
Rb	90	2.9 min	β⁻ 6.6; 61% γ 0.83 doublet; 18% γ 4.34	−79.3		
Rb	91	1.2 min	β⁻ 4.6	−78		
Rb	92	5.3 s	β⁻	−75		
Rb	93	5.6 s	β⁻			
Rb	94	2.9 s	β⁻			
38 Sr	80	1.7 h	K; γ			0⁺
Sr	81	29 min	β⁺			
Sr	82	25 d	K	−76		0⁺
Sr	83	32.4 h	β⁺ 1.15; 84% K; 40% γ 0.76; 35% γ 0.38; 24% γ 0.04	−77		
Sr	84	0.56%		−80.638	0.8 (γ); 0.65 (γm)	0⁺
Sr	85	64 d	K; 100% γ 0.514	−81.05		
Sr	85m	70 min	K; 85% γ 0.231; 14% γ 0.150	−80.81		
Sr	86	9.86%		−84.499	1.3 (γm)	0⁺
Sr	87	7.02%		−84.865		9/2+
Sr	87m	2.83 h	80% γ 0.388	−84.477		1/2−
Sr	88	82.56%		−87.89	0.006 (γ)	0⁺
Sr	89	52 d	β⁻ 1.463; γ	−86.22	0.4 (γ)	5/2+
Sr	90	28.1 yr	β⁻ 0.546	−85.95	1 (γ)	0⁺
Sr	91	9.67 h	β⁻ 2.67; 30% γ 1.025; 27% γ 0.748; 15% γ 0.645	−83.68		5/2+
Sr	92	2.71 h	β⁻ 1.5; 90% γ 1.37; 4% γ 0.44; 3% γ 0.23	−82.9		0⁺

Table of the nuclides (*contd*)

Z	Nuclide	A	Natural abundance or half-life	Major radiations	Mass excess MeV	σ/b	Iπ
	Sr	93	8 min	β⁻ 4.8; γ	−79.4		
	Sr	94	1.3 min	β⁻ 2.1; 100% γ 1.42	−78.8		0⁺
	Sr	95	0.8 min	β⁻			
39	Y	82	12 min	β⁺			
	Y	83	7.4 min	β⁺			
	Y	84	41 min	β⁺ 3.5; 100% γ 0.982, 0.795; 50% γ 1.041	−74.3		
	Y	85	5.0 h	β⁺ 2.24; 13% γ 0.231; 9% γ 2.16; 8% γ 0.77	−77.79		
	Y	85m	2.7 h	β⁺ 1.54; 9% γ 0.92	−77.75		
	Y	86	14.6 h	β⁺ 3.15, 82% γ 1.077; 37% γ 0.63 doublet	−79.23		4⁻
	Y	86m	48 min	94% γ 0.208	−79.01		8⁺
	Y	87	80 h	β⁺ 0.7; 99% K; γ	−83.2		
	Y	87m	14 h	β⁺ 1.60; 74% γ 0.381	−82.8		
	Y	88	108 d	β⁺ 0.76; 99% K; 100% γ 1.836; 91% 0.898	−84.27		
	Y	89	100%		−87.678	1.3 (γ)	½⁻
	Y	89m	16 s	99% γ 0.91	−86.77		9/2⁺
	Y	90	64 h	β⁻ 2.27	−86.50		2⁻
	Y	90m	3.1 h	97% γ 0.202; 91% γ 0.482	−85.81		
	Y	91	58.8 d	β⁻ 1.545; γ	−86.35	1.4 (γ)	½⁻
	Y	91m	50 min	95% γ 0.551	−85.80		9/2⁺
	Y	92	3.53 h	β⁻ 3.63; 14% γ 0.934; 4.7% γ 1.40; 2.6% γ 0.56	−84.83		2⁻
	Y	93	10.2 h	β⁻ 2.89; 6% γ 0.267; 2.3% γ 0.94; 1.8% γ 1.90	−84.22		
	Y	94	20.3 min	β⁻ 5.0; 43% γ 0.92; 6% γ 0.56	−82.3		
	Y	95	10.9 min	β⁻	−81		
	Y	96	2.3 min	β⁻ 3.5; γ	−79		
40	Zr	82	10 min	β⁺			0⁺
	Zr	84	16 min	β⁺			0⁺
	Zr	85	1.4 h	β⁺			
	Zr	86	16.5 h	K; 96% γ 0.243; 20% γ 0.028; 5% γ 0.612	−78		0⁺
	Zr	87	1.6 h	β⁺ 2.10; γ	−79.7		
	Zr	88	85 d	K; 97% γ 0.394	−84		0⁺
	Zr	89	78.4 h	β⁺ 0.90; 78% K; 99% γ 0.91	−84.85		
	Zr	89m	4.18 min	β⁺ 2.40; 87% γ 0.588; 6% γ 1.51	−84.26		
	Zr	90	51.46%		−88.770	0.1 (γ)	0⁺
	Zr	90m	0.81 s	86% γ 2.32; 14% γ 2.18; 4% γ 0.133	−86.45		5⁻
	Zr	91	11.23%		−87.893	1 (γ)	5/2⁺
	Zr	92	17.11%		−88.462	0.2 (γ)	0⁺
	Zr	93	1.5 × 10⁶ yr	β⁻ 0.06	−87.11		5/2⁺
	Zr	94	17.40%		−87.267	0.08 (γ)	0⁺
	Zr	95	65 d	β⁻ 0.89; 98% β⁻ 0.396; 49% γ 0.756, 0.724	−85.663		5/2⁺
	Zr	96	2.80%		−85.430	0.05 (γ)	0⁺
	Zr	97	17.0 h	β⁻ 1.91; 92% γ 0.747	−82.93		

Table of the nuclides (*contd*)

Nuclide		Natural abundance or half-life	Major radiations	Mass excess MeV	σ/b	I^π
Z	A					
41 Nb	88	14 min	β^+ 3.2; γ	-77		
Nb	89	1.9 h	β^+ 2.9; γ	-81.0		
Nb	89m	42 min	β^+ 3.1; 93% γ 0.588	-80.2		
Nb	90	14.6 h	β^+ 1.50; 97% γ 1.14; 82% γ 2.32; 75% γ 0.142	-82.66		
Nb	90m	24 s	71% γ 0.122	-82.54		
Nb	91	Long		-86.8		
Nb	91m	62 d	K, 3% γ 1.21	-86.6		
Nb	92	1.2×10^8 yr	K; γ			
Nb	92m	10.16 d	K, 99% γ 0.934	-86.32		2^+
Nb	93	100%		-87.204	0.1 (γ); 1.0 (γm)	$\frac{9}{2}^+$
Nb	93m	13.6 yr	γ	-87.173		$\frac{1}{2}^-$
Nb	94	2.0×10^4 yr	β^- 0.49; 100% γ 0.871, 0.702	-86.35	15 (γ)	6^+
Nb	94m	6.3 min	e^-; γ	-86.31		3^+
Nb	95	35 d	β^- 0.160; 100% γ 0.765	-86.784	7 (γ)	
Nb	95m	90 h	e^-	-86.549		
Nb	96	23.4 h	β^- 0.7; 97% γ 0.778; 59% γ 0.569; 49% γ 1.092	-85.64		
Nb	97	72 min	β^- 1.27; 98% γ 0.665	-85.61		
Nb	97m	1.0 min	98% γ 0.747	-84.86		
Nb	98	51 min	β^- 3.1; 100% γ 0.787; 75% γ 0.720; 30% γ 1.16	-83.5		
Nb	99	2.4 min	β^- 3.2; γ	-83		
Nb	100	11 min	β^-; γ	-80		
Nb	101	1.0 min	β^-; γ			
42 Mo	88	27 min	β^+ 2.5; γ			0^+
Mo	89	7 min	β^+ 4.9			
Mo	90	5.7 h	β^+ 1.2; 85% γ 0.257; 71% γ 0.122; 10% γ 0.945	-80.17		0^+
Mo	91	15.49 min	β^+ 3.44	-82.3		
Mo	91m	66 s	β^+ 3.99; 54% γ 0.658; 22% γ 1.21; 15% γ 1.53	-81.6		
Mo	92	15.86%		-86.804		0^+
Mo	93	> 100 yr	K	-86.79		$\frac{5}{2}^+$
Mo	93m	6.9 h	100% γ 1.479, 0.685; 58% γ 0.264	-84.36		
Mo	94	9.12%		-88.407		0^+
Mo	95	15.70%		-87.709	14 (γ)	$\frac{5}{2}^+$
Mo	96	16.50%		-88.794	1 (γ)	0^+
Mo	97	9.45%		-87.539	2 (γ)	$\frac{5}{2}^+$
Mo	98	23.75%		-88.110	0.51 (γ)	0^+
Mo	99	67 h	β^- 1.23; 12% γ 0.740; 7% γ 0.181; 4% γ 0.780	-85.96		$\frac{1}{2}^+$
Mo	100	9.62%		-86.185	0.2 (γ)	0^+
Mo	101	14.6 min	β^- 2.23; 25% γ 1.02, 0.191; 21% γ 0.59	-83.50		
Mo	102	11 min	β^- 1.2	-84		0^+
Mo	103	62 s	β^-			
Mo	104	1.3 min	β^- 4.8; γ			0^+
Mo	105	40 s	β^-			

Table of the nuclides (*contd*)

Nuclide		Natural abundance or half-life	Major radiations	Mass excess MeV	σ/b	Iπ
Z	A					
43 Tc	92	4.4 min	β⁺ 4.1; 100% γ 1.54; 95% γ 0.79; 90% γ 0.33; 67% γ 0.14	−78.8		
Tc	93	2.7 h	β⁺ 0.8; 65% γ 1.35; 33% γ 1.49	−83.60		
Tc	93m	43 min	63% γ 0.390; K	−83.21		
Tc	94	293 min	β⁺ 0.816; 100% γ 0.871, 0.849, 0.702	−84.15		6, 7⁺
Tc	94m	53 min	β⁺ 2.47; 91% γ 0.871; 10% γ 1.53; 9% γ 1.87	−84.04		2⁺
Tc	95	20.0 h	K; 82% γ 0.768; 11% γ 0.84; 4% γ 1.06	−86.05		
Tc	95m	61 d	β⁺ 0.68; 70% γ 0.204; 36% γ 0.584; 27% γ 0.838	−86.01		
Tc	96	4.3 d	K; 100% γ 0.851, 0.778; 84% γ 0.81	−85.9		
Tc	96m	52 min	γ	−85.8		
Tc	97	2.6 × 10⁶ yr	K	−87		
Tc	97m	90 d	γ	−87		
Tc	98	1.5 × 10⁶ yr	β⁻ 0.30 100% γ 0.76, 0.66	−86.5	3 (γ)	
Tc	99	2.12 × 10⁵ yr	β⁻ 0.292	−87.33	22 (γ)	9/2⁺
Tc	99m	6.0 h	90% γ 0.140	−87.18		1/2⁻
Tc	100	17 s	β⁻ 3.38; γ	−85.9		1⁺
Tc	101	14.0 min	β⁻ 1.32; γ	−86.32		
Tc	102	5 s	β⁻ 2; γ	−85		
Tc	103	50 s	β⁻ 2.2; γ	−84.9		
Tc	104	18 min	β⁻; γ	−82.2		
Tc	105	8 min	β⁻ 3.4; γ	−82.6		
Tc	106	37 s	β⁻			
Tc	107	29 s	β⁻			
44 Ru	95	1.7 h	β⁺ 1.33; 85% K; 70% γ 0.340; 21% γ 1.09; 13% γ 0.625	−84.02		
Ru	96	5.46%		−86.07	0.2 (γ)	0⁺
Ru	97	2.9 d	K, 91% γ 0.215; 8% γ 0.324	−86		
Ru	98	1.868%		−88.222		0⁺
Ru	99	12.63%		−87.619	11 (γ)	5/2⁺
Ru	100	12.53%		−89.219	10 (γ)	0⁺
Ru	101	17.02%		−87.953	3 (γ)	5/2⁺
Ru	102	31.6%		−89.098	1.4 (γ)	0⁺
Ru	103	39.6 d	β⁻ 0.70; 88% γ 0.497; 6% γ 0.610	−87.27		
Ru	104	18.87%		−88.090	0.48 (γ)	0⁺
Ru	105	4.44 h	β⁻ 1.87; 48% γ 0.726; 20% γ 0.475 doublet	−86.00	0.2 (γ)	
Ru	106	367 d	β⁻ 0.039	−86.33	0.15 (γ)	0⁺
Ru	107	22 min	β⁻ 3.2; 14% γ 0.195; 7% γ 0.86; 4% γ 1.29	−83.7		
Ru	108	4.5 min	β⁻ 1.3; 28% γ 0.165	−84		0⁺
45 Rh	97	32 min	β⁺ 2.47; γ	−83		
Rh	98	8.7 min	β⁺ 2.5; 100% γ 0.65	−84.0		

Table of the nuclides (*contd*)

Nuclide		Natural abundance or half-life	Major radiations	Mass excess MeV	σ/b	I^π
Z	A					
Rh	99	16 d	β^+ 1.03 γ	−85.57		
Rh	99	4.7 h	β^+ 0.74; 90% K; 70% γ 0.34; 20% γ 0.62	−85.52		
Rh	100	20 h	β^+ 2.62; 93% K; 88% γ 0.540; 39% γ 2.37 complex	−85.58		1, 2$^-$
Rh	101	3 yr	K; 88% γ 0.127; 75% γ 0.198; 11% γ 0.325	−87.39		
Rh	101m	4.5 d	K; 83% γ 0.307; 6% γ 0.545	−87.24		
Rh	102	206 d	β^- 1.15; β^+ 1.29; 57% γ γ 0.475; 4% γ 0.628	−86.77		0, 1, 2$^-$
Rh	102	2.9 yr	K; 95% γ 0.475; 54% γ 0.632 doublet			
Rh	103	100%		−88.014	144 (γ); 11 (γm)	$\frac{1}{2}^-$
Rh	103m	57 min	γ	−87.974		$\frac{7}{2}^+$
Rh	104	43 s	β^- 2.44; 2% γ 0.56	−86.95	40 (γ)	1$^+$
Rh	104m	4.41 min	47% γ 0.051; 2.6% γ 0.097; 2.5% γ 0.078	−86.82	800 (γ)	5$^+$
Rh	105	35.9 h	β^- 0.568; 19% γ 0.319; 5% γ 0.306	−87.87	21 000 (γ)	
Rh	105m	45 s	γ	−87.74		
Rh	106	30 s	β^- 3.54; 21% γ 0.512; 11% γ 0.622 doublet	−86.37		1$^+$
Rh	106	130 min	β^- 1.62; 88% γ 0.512; 35% γ 0.82, 0.451	−86.3		
Rh	107	22 min	β^- 1.20; 73% γ 0.305; 11% γ 0.390; 3% γ 0.68	−86.86		
Rh	108	17 s	β^- 4.5; 43% γ 0.434; 22% γ 0.62	−85		1$^+$
Rh	110	5 s	β^- 5.5; γ	−83		1$^+$
46 Pd	98	17 min	K; γ			0$^+$
Pd	99	22 min	β^+ 2.0; γ	−81.7		
Pd	100	4.0 d	K; 49% γ 0.084; 34% γ 0.074; 16% γ 0.126	−85		0$^+$
Pd	101	8.4 h	β^+ 0.78; 30% γ 0.296; 24% γ 0.590; 8% γ 0.270	−85.40		
Pd	102	0.96%		−87.92	4.8 (γ)	0$^+$
Pd	103	17 d	K	−87.46		
Pd	104	10.97%		−89.41		0$^+$
Pd	105	22.2%		−88.43		$\frac{5}{2}^+$
Pd	106	27.3%		−89.91	0.29 (γ)	0$^+$
Pd	107	7×10^6 yr	β^- 0.04	−88.368		$\frac{11}{2}^-$
Pd	107m	22 s	γ	−88.16		
Pd	108	26.7%		−89.52	12 (γ); 0.2 (γm)	0$^+$
Pd	109	13.47 h	β^- 1.028; γ	−87.60		$\frac{5}{2}^+$
Pd	109m	4.7 min	58% γ 0.188	−87.41		
Pd	110	11.8%		−88.34	0.2 (γ); 0.04 (γm)	0$^+$
Pd	111	22 min	β^- 2.2; γ	−86.0		
Pd	111m	5.5 h	β^- 2.0; γ	−85.8		
Pd	112	21 h	β^- 0.28; 20% γ 0.019	−86.27		0$^+$
Pd	113	1.4 min	β^-			
Pd	114	2.4 min	β^-			0$^+$

Table of the nuclides (*contd*)

Nuclide			Major radiations	Mass excess MeV	σ/b	I^π
Z		A	Natural abundance or half-life			
	Pd	115	45 s	β^-			
47	Ag	102	15 min	β^+	-83		
	Ag	103	66 min	β^+ 1.6; 70% K; γ	-84.9		$\frac{7}{2}^+$
	Ag	103m	5.7 s	γ	-84.7		
	Ag	104	67 min	β^+ 0.99; 84% γ 0.556; 48% γ 0.764; 30% γ 0.845	-85.14		5^+
	Ag	104m	30 min	β^+ 2.70; 100% γ 0.556	-85.12		2^+
	Ag	105	40 d	K; 42% γ 0.344 complex; 32% γ 0.280	-87		$\frac{1}{2}^-$
	Ag	106	24.0 min	β^+ 1.96	-86.94		1^+
	Ag	106m	8.4 d	K; 86% γ 0.512; 31% γ 0.717 complex	-86.6		6^+
	Ag	107	51.35%		-88.403	35 (γ)	$\frac{1}{2}^-$
	Ag	107m	44.3 s	5% γ 0.094	-88.310		$\frac{7}{2}^+$
	Ag	108	2.42 min	97.5% β^- 1.64; 0.28% β^+ 0.90; 2.2% K; γ	-87.61		1^+
	Ag	108m	>5 yr	K; 90% γ 0.722, 0.614; 89% γ 0.434	-87.50		
	Ag	109	48.65%		-88.717	89 (γ); 3 (γm)	$\frac{1}{2}^-$
	Ag	109m	40 s	5% γ 0.088	-88.630		$\frac{7}{2}^+$
	Ag	110	24.4 s	β^- 2.87; 4.5% γ 0.658	-87.47		1^+
	Ag	110m	253 d	β^- 1.5; 96% γ 0.658; 71% γ 0.885; 32% γ 0.937	-87.35	80 (γ)	6^+
	Ag	111	7.5 d	β^- 1.05; 6% γ 0.342	-88.20		$\frac{1}{2}^-$
	Ag	111m	74 s	γ	-88.13		
	Ag	112	3.2 h	β^- 3.94; 41% γ 0.617; 5% γ 1.40; 3% γ 2.11, 1.63	-86.57		2^-
	Ag	113	1.2 min	β^-; γ			
	Ag	113	5.3 h	β^- 2.0; γ	-87.04		$\frac{1}{2}^-$
	Ag	114	4.5 s	β^- 4.6; γ	-85.4		
	Ag	115	20 min	β^- 3.2; 49% γ 0.22 complex; 13% γ 2.12	-84.8		
	Ag	115	20 s	β^-			
	Ag	117	1.1 min	β^-			
48	Cd	103	10 min	β^+; γ			
	Cd	104	57 min	K; γ	-84		0^+
	Cd	105	55 min	β^+ 1.69; γ	-84		
	Cd	106	1.22%		-87.128	1 (γ)	0^+
	Cd	107	6.5 h	β^+ 0.302; 99% K; γ	-86.99		$\frac{5}{2}^+$
	Cd	108	0.88%		-89.248	3 (γ)	0^+
	Cd	109	453 d	K; γ	-88.55		$\frac{5}{2}^+$
	Cd	110	12.39%		-90.342	0.1 (γm)	0^+
	Cd	111	12.75%		-89.246		$\frac{1}{2}^+$
	Cd	111m	48.6 min	94% γ 0.247; 30% γ 0.150	-88.850		$\frac{11}{2}^-$
	Cd	112	24.07%		-90.575	0.03 (γm)	0^+
	Cd	113	>10^{15} yr 12.26%		-89.041	20 000 (γ)	$\frac{1}{2}^+$
	Cd	113m	14 yr	β^- 0.58; γ	-88.77		$\frac{11}{2}^-$
	Cd	114	28.86%		-90.018	1.1 (γ); 0.14 (γm)	0^+
	Cd	115	53.5 h	β^- 1.11; 26% γ 0.53; 10% γ 0.49	-88.09		$\frac{1}{2}^+$

Table of the nuclides (*contd*)

Nuclide		Natural abundance or half-life	Major radiations	Mass excess MeV	σ/b	Iᵖ
Cd	115m	43 d	β^- 1.62; 1.9% γ 0.935	−87.91		$\frac{11}{2}-$
Cd	116	7.58%		−88.712	1.4 (γ); 0.7 (γm)	0^+
Cd	117	2.4 h	β^- 2.23; 31% γ 0.273; complex γ	−86.41		$\frac{1}{2}+$
Cd	117m	3.4 h	β^- 1.91; complex γ	−86.27		$\frac{11}{2}-$
Cd	118	49 min	β^-	−87		0^+
Cd	119	10 min	β^- 3.5	−84.1		
Cd	119	2.7 min	β^- 3.5	−84.1		
Cd	121	13 s	β^-			
49 In	106	5.3 min	β^+ 4.9; γ	−80.6		
In	107	32 min	β^+ 2.2; 46% γ 0.22	−83.5		
In	108	56 min	β^+ 1.29	−84.14		
In	108	39 min	β^+ 3.5; γ	−84.10		$2, 3^+$
In	109	4.3 h	β^+ 0.79; γ	−86.53		$\frac{9}{2}+$
In	109m1	1.3 min	γ	−85.87		
In	109m2	0.20 s	100% γ 0.68; 77% γ 1.43; 20% γ 0.40	−84.42		
In	110	66 min	β^+ 2.25; 95% γ 0.658	−86.41		
In	110	4.9 h	K; γ	−86.41		
In	111	2.81 d	K; 94% γ 0.247; 89% γ 0.173	−88.2		$\frac{9}{2}+$
In	112	14 min	44% β^- 0.66; 22% β^+ 1.56; 34% K; 6% γ 0.617	−87.98		1^+
In	112m	20.7 min	9% γ 0.156	−87.83		4^+
In	113	4.23%		−89.34	4 (γ); 8 (γm)	$\frac{9}{2}+$
In	113m	100 min	64% γ 0.393	−88.95		$\frac{1}{2}-$
In	114	72 s	98% β^- 1.988; β^+; γ	−88.58		1^+
In	114m	50.0 d	17% γ 0.192; 3.5% γ 0.724, 0.558	−88.39		5^+
In	115	95.77%		−89.54	45 (γ); 145 (γm1); 4 (γm2)	$\frac{9}{2}+$
In	115m	4.5 h	5% β^- 0.83; 50% γ 0.335	−89.21		$\frac{1}{2}-$
In	116	14 s	β^- 3.3	−88.20		1^+
In	116m1	54 min	β^- 1.0; 80% γ 1.293; 53% γ 1.09; 36% γ 0.417	−88.14		5^+
In	116m2	2.16 s	γ	−87.98		
In	117	44 min	β^- 0.74; 87% γ 0.158; 100% γ 0.565	−88.93		$\frac{9}{2}+$
In	117m	1.93 h	γ; β^- 1.78; 31% γ 0.314; 14% γ 0.158	−88.61		$\frac{1}{2}-$
In	118	5 s	β^- 4.2; 15% γ 1.230	−87.5		1^+
In	118	4.4 min	β^- 2.0; 97% γ 1.230; 80% γ 1.05; 41% γ 0.69	−87.4		
In	119	2.1 min	β^- 1.6; 95% γ 0.82	−87.6		
In	119m	18 min	β^- 2.7; γ	−87.3		
In	120	46 s	β^- 3.1; 100% γ 1.171; 61% γ 1.02; 34% γ 0.86	−86		
In	120	3.2 s	β^- 5.6; 15% γ 1.171	−86		
In	121	3.1 min	β^- 3.7	−86		
In	121	30 s	β^-; γ	−86		
In	122	8 s		−83		
In	124	4 s	β^- 5; γ	−81		

Table of the nuclides (*contd*)

Z	Nuclide A	Natural abundance or half-life	Major radiations	Mass excess MeV	σ/b	Iπ
50 Sn	108	9 min	K; γ			0^+
Sn	109	18.1 min	β^+ 1.6; γ			
Sn	110	4.0 h	K; 95% γ 0.283			0^+
Sn	111	35 min	β^+ 1.51; γ	−85.6		
Sn	112	0.95%		−88.64	0.9 (γ); 0.4 (γm)	0^+
Sn	113	115 d	K; 1.8% γ 0.255	−88.32		
Sn	113m	20 min	K; γ	−88.24		
Sn	114	0.65%		−90.57		0^+
Sn	115	0.34%		−90.03		$\frac{1}{2}^+$
Sn	116	14.24%		−91.523	0.006 (γm)	0^+
Sn	117	7.57%		−90.392		$\frac{1}{2}^+$
Sn	117m	14.0 d	87% γ 0.158	−90.075		$\frac{11}{2}^-$
Sn	118	24.01%		−91.652	0.01 (γm)	0^+
Sn	119	8.58%		−90.062		$\frac{1}{2}^+$
Sn	119m	250 d	16% γ 0.024	−89.973		$\frac{11}{2}^-$
Sn	120	32.97%		−91.100	0.14 (γ)	0^+
Sn	121	27 h	β^- 0.383	−89.21		
Sn	121m	76 yr	β^- 0.42; γ	−89.14		
Sn	122	4.71%		−89.943	0.001 (γ); 0.2 (γm)	0^+
Sn	123	125 d	β^- 1.42	−87.80		$\frac{11}{2}^-$
Sn	123m	40 min	β^- 1.26; 84% γ 0.160	−87.78		
Sn	124	5.98%		−88.237	0.004 (γ); 0.1 (γm)	0^+
Sn	125	9.4 d	β^- 2.34; 4% γ 1.068	−85.93		
Sn	125m	9.7 min	β^- 2.04; 97% γ 0.325	−85.91		
Sn	126	10^5 yr	β^-; γ	−86		0^+
Sn	127	2.1 h	β^- 1.45; γ	−84		
Sn	127	4 min	β^- 2.7; 100% γ 0.49	−83.5		
Sn	128	59 min	β^- 0.80; 61% γ 0.50; 22% γ 0.57; 19% γ 0.072	−83.4		0^+
Sn	129	9 min	β^-; γ			
Sn	129	1.0 h	β^-			
Sn	130	2.6 min	β^-			0^+
Sn	132	2.2 min	β^-			0^+
51 Sb	112	0.9 min	β^+; γ			
Sb	113	6.7 min	β^+ 2.42; γ	−83.85		
Sb	114	3.3 min	β^+ 2.7; γ	−84.3		0^+
Sb	115	31 min	β^+ 1.51; 100% γ 0.499; 5% γ 1.24, 0.98	−87.00		
Sb	116	15 min	β^+ 2.3; 85% γ 1.293; 26% γ 0.93; 14% γ 2.23	−87.0		
Sb	116m	60 min	β^+ 1.16; 100% γ 1.293; 75% γ 0.96; 68% γ 0.545	−86.5		
Sb	117	2.8 h	β^+ 0.57; 97% K; 87% γ 0.158	−88.57		
Sb	118	3.5 min	β^+ 2.67	−87.96		1^+
Sb	118m	5.1 h	K; 0.16% β^+; 100% γ 1.230, 1.049; 93% γ 0.254	−87.77		
Sb	119	38.0 h	K; 16% γ 0.024	−89.48		
Sb	120	15.9 min	β^+ 1.70; 1.3% γ 1.171	−88.42		1^+
Sb	120	5.8 d	K; 100% γ 1.171; 99% γ 1.03; 88% γ 0.200	−88.42		
Sb	121	57.25%		−89.593	6.06 (γ)	$\frac{5}{2}^+$

Table of the nuclides (*contd*)

Nuclide		Natural abundance or half-life	Major radiations	Mass excess MeV	σ/b	I^π
Z	A					
Sb	122	2.8 d	97% β⁻ 1.97; β⁺ 0.56; 66% γ 0.564; 3.4% γ 0.686	−88.32		2⁻
Sb	122m	4.2 min	50% γ 0.061; 17% γ 0.075	−88.16		
Sb	123	42.75%		−89.224	3.3 (γ); 0.03 (γm1)	7/2⁺
Sb	124	60 d	β⁻ 2.31; 50% γ 1.692; 97% γ 0.603; 14% γ 0.72	−87.58	2000 (γ)	3⁻
Sb	124m1	93 s	β⁻ 1.19; 20% γ 0.644, 0.603, 0.505	−87.57		
Sb	124m2	24 min	γ	−87.55		
Sb	125	2.7 yr	β⁻ 0.61; 31% γ 0.427; 24% γ 0.599 doublet	−88.28		7/2⁺
Sb	126	12.5 d	β⁻ 1.9; γ	−86.3		
Sb	126	19.0 min	β⁻ 1.9; γ	−86.3		
Sb	127	93 h	β⁻ 1.5; γ	−86.70		
Sb	128	11 min	β⁻ 2.6; 83% γ 0.320; 200% γ 0.75 doublet	−84.7		
Sb	128	9 h	β⁻; γ			
Sb	129	4.3 h	β⁻ 1.87; γ	−85		
Sb	130	33 min	β⁻; γ	−82		
Sb	130	7 min	β⁻; γ			
Sb	131	25 min	β⁻; 48% γ 0.95; 37% γ 0.64			
Sb	132	2.1 min	β⁻			
Sb	133	4.2 min	β⁻			
Sb	135	2 s	β⁻			
52 Te	107	2.2 s	α 3.28			
Te	108	5.3 s	α 3.08; p			0⁺
Te	115	6.0 min	β⁺ 2.8; 34% γ 0.72; 32% γ 1.38, 1.28	−82.5		
Te	116	2.50 h	β⁺; γ	−85.4		0⁺
Te	117	61 min	β⁺ 1.81; 65% γ 0.72; 9% γ 1.78; 6% γ 0.93	−85.1		1/2⁺
Te	118	6.00 d	K	−88		0⁺
Te	119	15.9 h	β⁺ 0.627; 85% γ 0.645; 11% γ 0.70; 3.6% γ 1.76	−87.19		1/2⁺
Te	119m	4.7 d	K; 67% γ 1.221; 62% γ 0.153; 36% γ 1.0 complex	−86.9		11/2⁻
Te	120	0.089%		−89.40	0.3 (γ); 2.0 (γ)	0⁺
Te	121	17 d	K; 80% γ 0.573; 18% γ 0.508	−88.31		1/2⁺
Te	121m	154 d	82% γ 0.212; 3% γ 1.10	−88.01		11/2⁻
Te	122	2.46%		−90.29	2 (γ); 1 (γm)	0⁺
Te	123	1.2 × 10¹³ yr 0.87%		−89.16	400 (γ)	1/2⁺
Te	123m	117 d	84% γ 0.159	−88.92		11/2⁻
Te	124	4.61%		−90.50	2 (γ); 5 (γm)	0⁺
Te	125	6.99%		−89.03	1.5 (γ)	1/2⁺
Te	125m	58 d	7% γ 0.035	−88.89		11/2⁻
Te	126	18.71%		−90.05	0.9 (γ); 0.1 (γm)	0⁺
Te	127	9.4 h	β⁻ 0.70; γ	−88.30		
Te	127m	109 d	γ	−88.21		
Te	128	31.79%		−88.98	0.157 (γ)	0⁺

Table of the nuclides (contd)

Nuclide		Natural abundance or half-life	Major radiations	Mass excess MeV	σ/b	Iπ
	Te 129	69 min	β⁻ 1.45; 19% γ 0.027; 15% γ 0.455	−87.02		$\frac{3}{2}-$
	Te 129m	34 d	β⁻ 1.60; 6% γ 0.69	−86.92		$\frac{11}{2}-$
	Te 130	34.49%		−87.34	0.24 (γ)	0^+
	Te 131	25 min	β⁻ 2.14; 68% γ 0.150; 16% γ 0.453	−85.16		$\frac{3}{2}+$
	Te 131m	30 h	β⁻ 2.46; 31% γ 0.85 complex	−84.98		$\frac{11}{2}-$
	Te 132	78 h	β⁻ 0.22; 90% γ 0.230; 17% γ 0.053	−85.21		0^+
	Te 133	12.5 min	β⁻; γ			
	Te 133m	50 min	β⁻ 2.4; 85% γ 0.754; 57% γ 0.91; 50% γ 0.432			
	Te 134	42 min	β⁻; 21% γ 0.204; 19% γ 0.262; 16% γ 0.17			0^+
53	I 117	7 min	β⁺; γ			
	I 118	14 min	β⁺; γ	−81		
	I 119	19 min	β⁺; γ			
	I 120	1.3 h	β⁺ 4.0; γ	−83.8		
	I 121	2.1 h	β⁺ 1.2; 90% γ 0.212; 6% γ 0.32; 3% γ 0.27	−86.0		
	I 122	3.5 min	β⁺ 3.1; γ	−86.15		1^+
	I 123	13.3 h	K; 83% γ 0.159	−88		$\frac{5}{2}+$
	I 124	4.2 d	β⁺ 2.14; 67% γ 0.605; 14% γ 1.69, 0.73	−87.33		2^-
	I 125	60 d	K; 7% γ 0.035	−88.88	900 (γ)	$\frac{5}{2}+$
	I 126	13 d	44% β⁻ 1.25; 1.3% β⁺ 1.13; 55% K; 34% γ 0.386; 33% γ 0.667	−87.90		2^-
	I 127	100%		−88.984	6.4 (γ)	
	I 128	25.0 min	β⁻ 2.12; 14% γ 0.441; 1.4% γ 0.528	−87.71		1^+
	I 129	1.7 × 10⁷ yr	β⁻ 0.150; 9% γ 0.040	−88.50	28 (γ)	$\frac{7}{2}+$
	I 130	12.4 h	β⁻ 1.7; 100% γ 0.669; 99% γ 0.538; 87% γ 0.743	−86.89	18 (γ)	5
	I 131	8.05 d	β⁻ 0.806; 82% γ 0.364; 6.8% γ 0.637; 5.4% γ 0.284	−87.441	0.7 (γ)	$\frac{7}{2}+$
	I 132	2.3 h	β⁻ 2.12; 89% γ 0.773; 144% γ 0.67 complex	−85.71		4^+
	I 133	21 h	β⁻ 1.27; 90% γ 0.53	−85.9		$\frac{7}{2}+$
	I 134	52.8 min	β⁻ 2.43; 95% γ 0.85; 65% γ 0.89; 18% γ 0.61	−84.0		
	I 135	6.7 h	β⁻ 1.4; 37% γ 1.14; 34% γ 1.28; 19% γ 1.72	−84		$\frac{7}{2}+$
	I 136	83 s	β⁻; 95% γ 1.32 complex; 19% γ 2.3 complex	−79.4		
	I 137	23 s	β⁻; n			
	I 138	5.9 s	β⁻			
	I 139	2 s	β⁻			
54	Xe 118	6 min	β⁺; γ			0^+
	Xe 119	6 min	β⁺; γ			

Table of the nuclides (*contd*)

Nuclide		Natural abundance or half-life	Major radiations	Mass excess MeV	σ/b	Iπ
Z	A					
Xe	120	40 min	K; γ			0+
Xe	121	39 min	β+ 2.8; γ	−82.2		
Xe	122	20 h	K; γ			0+
Xe	123	2.1 h	β+ 1.51; γ	85		
Xe	124	0.096%		−87.5	110 (γ)	0+
Xe	125	17 h	K; γ	−87		
Xe	125m	55 s	γ			
Xe	126	0.090%		−89.15	2 (γ)	0+
Xe	127	36.4 d	K; 65% γ 0.203; 22% γ 0.172; 20% γ 0.375	−88.54		
Xe	127m	75 s	γ			
Xe	128	1.919%		−89.85	<5 (γ)	0+
Xe	129	26.44%		−88.692	25 (γ)	1⁄2+
Xe	129m	8.0 d	9% γ 0.040; 6% γ 0.197	−88.456		11⁄2−
Xe	130	4.08%		−89.88	<5 (γ)	0+
Xe	131	21.18%		−88.411	85 (γ)	3⁄2+
Xe	131m	11.8 d	2% γ 0.164	−88.247		11⁄2−
Xe	132	26.89%		−89.272		0+
Xe	133	5.27 d	β− 0.346; 37% γ 0.081	−87.73	190 (γ)	
Xe	133m	2.26 d	14% γ 0.233	−87.50		
Xe	134	10.4%		−88.121		0+
Xe	135	9.2 h	β− 0.92; 91% γ 0.250; 3% γ 0.61	−86.6	2.7 × 10⁶ (γ)	
Xe	135m	15.6 m	80% γ 0.527	−86.1		
Xe	136	8.87%		−86.42	0.15 (γ)	0+
Xe	137	3.9 min	β− 4.1; 33% γ 0.455	−82.8		
Xe	138	17 min	β− 2.4; γ	−80.9		0+
Xe	139	43 s	β−; γ	−76.5		
Xe	140	16 s	β−; γ			0+
Xe	141	2 s	β−			
Xe	142	1.5 s	β−			0+
Xe	143	1.0 s	β−			
Xe	144	1 s	β−			0+
55 Cs	123	8 min	β+			
Cs	125	45 min	β+ 2.05	−84		
Cs	126	1.6 min	β+ 3.8; 38% γ 0.386	−84.4		1+
Cs	127	6.2 h	β+ 1.08; 72% γ 0.406; 10% γ 0.125	−86.4		1⁄2+
Cs	128	3.8 min	β+ 2.9; 27% γ 0.441	−85.92		1+
Cs	129	32 h	K; 48% γ 0.375; 25% γ 0.416; 5% γ 0.550	−88		1⁄2+
Cs	130	30 min	β+ 1.97	−86.89		1+
Cs	131	9.70 d	K	−88.06		5⁄2+
Cs	132	6.5 d	0.6% β+ 0.4; 2% β− 0.7; 97% K; 99% γ 0.668; 4% γ 0.48	−87.19		2
Cs	133	100%		−88.16	28 (γ); 2.6 (γm)	7⁄2+
Cs	134	2.05 yr	β− 0.662; 99% γ 0.796; 98% γ 0.605; 3.4% γ 1.365	−86.79	136 (γ)	4+
Cs	134m	2.90 h	1% β− 0.55; 14% γ 0.128	−86.65		8−
Cs	135	3 × 10⁶ yr	β− 0.21	−87.8	8.7 (γ)	7⁄2+
Cs	135m	53 min	100% γ 0.781; 96% γ 0.840	−86.2		

Table of the nuclides (contd)

Nuclide Z	A	Natural abundance or half-life	Major radiations	Mass excess MeV	σ/b	Iπ
Cs	136	13 d	β⁻ 0.657; 100% γ 0.818; 53% γ 0.340; 82% γ 1.05	−86.6		
Cs	137	30.0 yr	β⁻ 1.176; 93.5% β⁻ 0.514; 85% γ 0.662	−86.9	0.11 (γ)	$\frac{7}{2}^+$
Cs	138	32.2 min	β⁻ 3.4; 73% γ 1.426; 25% γ 1.01; 23% γ 0.463	−83.7		
Cs	139	9.5 min	β⁻; γ	−81.1		
Cs	140	66 s	β⁻; γ	−77		
Cs	141	24 s	β⁻			
Cs	142	2.3 s	β⁻			
Cs	143	2.0 s	β⁻			
56 Ba	123	2 min	β⁺			
Ba	125	6 min	β⁺			
Ba	126	97 min	K; γ			0⁺
Ba	127	10 min	β⁺	−83		
Ba	128	2.4 d	K; γ	−85		0⁺
Ba	129	2 h	β⁺ 1.42; γ	−85		
Ba	130	0.101%		−87.33	8.8 (γ)	0⁺
Ba	131	12 d	K; 48% γ 0.496 complex; 28% γ 0.124	−86.89		
Ba	131m	15 min	40% γ 0.107	−86.71		
Ba	132	0.097%		−88.4	7 (γ)	0⁺
Ba	133	7.2 yr	K; 69% γ 0.356; 36% γ 0.080 complex	−87.67		$\frac{1}{2}^+$
Ba	133m	38.9 h	17% γ 0.276	−87.39		$\frac{11}{2}^-$
Ba	134	2.42%		−88.85	>0.16 (γ)	0⁺
Ba	135	6.59%		−88.0	5 (γ)	$\frac{3}{2}^+$
Ba	135m	28.7 h	16% γ 0.268	−87.7		$\frac{11}{2}^-$
Ba	136	7.81%		−89.1		0⁺
Ba	136m	0.32 s	100% γ 1.05, 0.818; 40% γ 0.164	−87.1		7⁻
Ba	137	11.32%		−88.0	4 (γ)	$\frac{3}{2}^+$
Ba	137m	2.55 min	89% γ 0.662	−87.4		$\frac{11}{2}^-$
Ba	138	71.66%		−88.5	0.4 (γ)	0⁺
Ba	139	82.9 min	β⁻ 2.3; 23% γ 0.166	−85.1	4 (γ)	
Ba	140	12.8 d	β⁻ 1.02; 34% γ 0.537; 11% γ 0.030; 6% γ 0.305; 0.163	−83.31	<20 (γ)	
Ba	141	18 min	β⁻ 3.0; γ	−80.1		
Ba	142	11 min	β⁻ 1.7; γ	−77.9		
Ba	143	12 s	β⁻			
57 La	126	1.0 min	β⁺; γ			
La	127	3.5 min	β⁺			
La	128	4.4 min	β⁺; γ			
La	129	10 min	β⁺	−81		
La	130	8.7 min	β⁺; γ	−82		
La	131	59 min	β⁺ 1.94; 20% γ 0.417, 0.364; 23% γ 0.115	−83.9		
La	132	4.5 h	β⁺ 3.8; γ	−83.1		
La	133	4.0 h	β⁺ 1.2; γ	−85.5		
La	134	67 min	β⁺ 2.7; 6% γ 0.605	−85.1		1⁺
La	135	19.5 h	K; 1.9% γ 0.481	−87.0		

Table of the nuclides (*contd*)

Nuclide		Natural abundance or half-life	Major radiations	Mass excess MeV	σ/b	Iπ
Z	A					
La 136		9.5 min	β⁺ 1.9; 2.5% γ 0.818	−86.3		1⁺
La 137		6 × 10⁴ yr	K	−88		7/2⁺
La 138		1.12 × 10¹¹ yr 0.089%	β⁻ 0.21; 70% γ 1.426; 30% γ 0.81	−86.7		5⁻
La 139		99.911%		−87.43	8.9 (γ)	7/2⁺
La 140		40.22 h	β⁻ 2.175; 96% γ 1.596; 40% γ 0.487; 19% γ 0.815	−84.36		3⁻
La 141		3.9 h	β⁻ 2.43; 2% γ 1.37	−83.06		
La 142		92 min	β⁻ 4.51; 48% γ 0.65; 15% γ 2.41; 11% γ 2.55	−80.1		
La 143		14.0 min	β⁻ 3.3; γ	−78.4		
58 Ce 130		30 min	β⁺; γ			0⁺
Ce 132		4.2 h	K; γ	−82		0⁺
Ce 133		6.3 h	β⁺ 1.3; γ	−83		
Ce 134		72 h	K	−84.9		0⁺
Ce 135		17.2 h	β⁺ 0.81; γ	−85		
Ce 136		0.193%		−86.6	6.0 (γ); 0.6 (γm)	0⁺
Ce 137		9.0 h	K; 2.3% γ 0.446 complex	−86		3/2⁺
Ce 137m		34.4 h	11% γ 0.255	−87		11/2⁻
Ce 138		0.250%		−87.7	1.0 (γ); 0.4 (γm)	0⁺
Ce 139		140 d	K; 80% γ 0.165	−87.16		3/2⁺
Ce 139m		55 s	93% γ 0.746	−86.41		
Ce 140		88.48%		−88.13	0.6 (γ)	0⁺
Ce 141		33 d	β⁻ 0.581; 48% γ 0.145	−85.49	30 (γ)	7/2⁻
Ce 142		11.07%		−84.63	1 (γ)	0⁺
Ce 143		33 h	β⁻ 1.39; 46% γ 0.293; 11% γ 0.057; 8% γ 0.725	−81.67	6 (γ)	3/2⁻
Ce 144		284 d	β⁻ 0.31; 11% γ 0.134; 2% γ 0.080	−80.49	1.0 (γ)	0⁺
Ce 145		3.0 min	β⁻ 2.0	−77		
Ce 146		14 min	β⁻ 0.7; γ	−75.8		0⁺
Ce 147		65 s	β⁻			
Ce 148		43 s	β⁻			
59 Pr 134		17 min	β⁺; γ			
Pr 135		22 min	β⁺ 2.5; γ			
Pr 136		1.1 h	β⁺ 2.0; γ			
Pr 137		1.5 h	β⁺ 1.7	−84		
Pr 138		2.1 h	β⁺ 1.65; 100% γ 1.04, 0.79; 77% γ 0.298	−82.9		
Pr 139		4.5 h	β⁺ 1.09; 89% K; γ	−85.0		
Pr 140		3.39 min	β⁺ 2.32; γ	−84.78		
Pr 141		100%		−86.07	12 (γ)	5/2⁺
Pr 142		19.2 h	β⁻ 2.16; 3.7% γ 1.57	−83.85	20 (γ)	2⁻
Pr 143		13.6 d	β⁻ 0.933	−83.11	89 (γ)	7/2⁺
Pr 144		17.3 min	β⁻ 2.99; 1.5% γ 0.695	−80.81		0⁻
Pr 145		5.98 h	β⁻ 1.80; γ	−79.66		
Pr 146		24.0 min	β⁻ 3.7; 77% γ 0.455; 27% γ 1.51; 16% γ 0.74	−76.8		
Pr 147		12.0 min	β⁻ 2.1; 47% γ 0.32 complex; 39% γ 0.56	−75.5		
Pr 148		2.0 min	β⁻ 4.2; γ	−72.9		

Table of the nuclides (*contd*)

Nuclide			Natural abundance or half-life	Major radiations	Mass excess MeV	σ/n	I^π
Z		A					
60	Nd	137	55 min	β^+ 3; γ			
	Nd	138	22 min	β^+			0^+
	Nd	139	30 min	β^+; γ			
	Nd	139m	5.5 h	β^+ 3.1; γ	-82		
	Nd	140	3.3 d	K	-84		0^+
	Nd	141	2.5 h	β^+ 0.79; γ	-84.27		$\frac{3}{2}^+$
	Nd	141m	63 s	γ 0.755	-83.52		
	Nd	142	27.13%		-86.01	17 (γ)	0^+
	Nd	143	12.20%		-84.04	330 (γ)	$\frac{7}{2}^-$
	Nd	144	2.4×10^{15} yr 23.87%	α 1.83	-83.80	5 (γ)	0^+
	Nd	145	8.29%		-81.47	50 (γ)	$\frac{7}{2}^-$
	Nd	146	17.18%		-80.96	2 (γ)	0^+
	Nd	147	11.1 d	β^- 0.81; 28% γ 0.091; 13% γ 0.533	-78.18		$\frac{5}{2}^-$
	Nd	148	5.72%		-77.44	4 (γ)	0^+
	Nd	149	1.8 h	β^- 1.5; 26% γ 0.27; 27% γ 0.210; 18% γ 0.114	-74.41		$\frac{5}{2}^-$
	Nd	150	5.60%		-73.67	1.5 (γ)	0^+
	Nd	151	12 min	β^- 2.0; 40% γ 0.118; 11% γ 0.256; 10% γ 0.174	-71.0		
61	Pm	141	22 min	β^+ 2.6; 13% γ 0.195	-80.7		
	Pm	142	40 s	β^+ 3.78	-81.2		
	Pm	143	0.73 yr	K; 47% γ 0.742	-82.9		
	Pm	144	1.1 yr	K; 99% γ 0.695, 0.615; 45% γ 0.474	-82		0^-
	Pm	145	17.7 yr	K; 2.3% γ 0.072	-81.33		
	Pm	146	4 yr	β^- 0.78; 65% γ 0.75 doublet; 65% γ 0.453	-79.52		
	Pm	147	2.62 yr	β^- 0.224	-79.08	120 (γ); 110 (γm)	$\frac{7}{2}^+$
	Pm	148	5.4 d	β^- 2.48; 27% γ 0.551; 23% γ 1.465; 15% γ 0.914	-76.89	2000 (γ)	1^-
	Pm	148m	42 d	β^- 0.69; 95% γ 0.551; 87% γ 0.630; 36% γ 0.727	-76.75	30 000 (γ)	6^-
	Pm	149	53.1 h	β^- 1.07; 2% γ 0.286	-76.07		$\frac{7}{2}^+$
	Pm	150	2.7 h	β^- 3.05; 71% γ 0.334; 23% γ 1.165; 22% γ 1.33	-73.6		
	Pm	151	28 h	β^- 1.19; 21% γ 0.340; 18% γ 0.17 complex	-73.4		$\frac{5}{2}^+$
	Pm	152	6 min	β^- 2.2; γ	-71		
	Pm	154	2.5 min	β^- 2.5			
62	Sm	142	73 min	β^+; γ			0^+
	Sm	143	8.9 min	β^+	-79.6		
	Sm	143m	64 s	γ 0.748	-78.8		
	Sm	144	3.16%		-81.98	0.7 (γ)	0^+
	Sm	145	340 d	K; 13% γ 0.061	-80.67	100 (γ)	
	Sm	146	7×10^7 yr	α 2.46	-81.05		0^+
	Sm	147	1.05×10^{11} yr 15.07%	α 2.23	-79.30	90 (γ)	$\frac{7}{2}^-$
	Sm	148	11.27%		-79.37		0^+
	Sm	149	13.84%		-77.15	41 500 (γ)	$\frac{7}{2}^-$

Table of the nuclides (*contd*)

Nuclide		Natural abundance or half-life	Major radiations	Mass excess MeV	σ/b	Iπ
Z	A					
Sm	150	7·47%		−77.06	100 (γ)	0+
Sm	151	87 yr	β− 0.076; 4% γ 0.022	−74.59	15 000 (γ)	
Sm	152	26.63%		−74.75	210 (γ)	0+
Sm	153	47 h	β− 0.8; 28% γ 0.103; 5.4% γ 0.070	−72.56		3/2+
Sm	154	22.53%		−72.39	5 (γ)	0+
Sm	155	23 min	β− 1.53, 73% γ 0.104	−70.14		
Sm	156	9.4 h	β− 0.72; 30% γ 0.088; 20% γ 0.204; 10% γ 0.166	−69.33		0+
Sm	157	0.5 min	β−; γ			
63 Eu	144	10 s	β+ 5.2	−75.66		1+
Eu	145	5.9 d	β+; γ	−77.9		
Eu	146	4.6 d	β+ 2.11; 100% γ 0.749; 77% γ 0.634 doublet	−77.18		
Eu	147	22 d	99.5% K; β+; α 2.91; 24% γ 0.198; 20% γ 0.122; 11% γ 0.680	−77.5		
Eu	148	54 d	β+ 0.92; α 2.63; 90% γ 0.62 complex; 120% γ 0.551	−76.26		
Eu	149	106 d	K; γ	−76		
Eu	150	12.6 h	90% β− 1.01; 0.4% β+ 1.24; 4% γ 0.334	−74.81		
Eu	150	5 yr	K; 96% γ 0.334; 86% γ 0.439; 60% γ 0.584		5900 (γ); 2800 (γm1)	
Eu	151	47·77%		−74.67		5/2
Eu	152	12 yr	28% β− 1.48; β+ 0.71; 72% K; 37% γ 0.122; 27% γ 0.344	−72.89	5000 (γ)	3−
Eu	152m1	9.3 h	77% β− 1.88; β+ 0.89; 23% K; 12% γ 0.963; 8% γ 0.122	−72.84		
Eu	152m2	96 min	74% γ 0.090	−72.74		8−
Eu	153	52.23%		−73.36	320 (γ)	5/2+
Eu	154	16 yr	β− 1.85; 38% γ 0.123; 37% γ 1.278; 31% γ 1.0 doublet	−71.68	1400 (γ)	3−
Eu	155	1.81 yr	β− 0.25; 32% γ 0.087; 20% γ 0.105	−71.79	13 000 (γ)	
Eu	156	15 d	β− 2.45; 16% γ 1.24 complex; 14% γ 1.15 complex	−70.05		0+
Eu	157	15.2 h	β− 1.3; 27% γ 0.413, 0.064; 14% γ 0.37	−69.43		
Eu	158	46 min	β− 2.5; γ	−67.1		
Eu	159	18 min	β− 2.6; 42% γ 0.07; 21% γ 0.67; 18% γ 0.09	−66.02		
64 Gd	145	25 min	β+ 2.4; γ			
Gd	146	50 d	β+; 99% K; γ	−76		0+
Gd	147	35 h	K; γ	−75		
Gd	148	84 yr	α 3.18	−76.29		0+
Gd	149	9.4 d	99% K; α 3.01; 48% γ 0.150; 26% γ 0.299	−75.2		
Gd	150	2.1 × 10⁶ yr	α 2.73	−75.82		0+

Table of the nuclides (*contd*)

Nuclide			Natural abundance or half-life	Major radiations	Mass excess MeV	σ/b	Iπ
Z		A					
	Gd	151	120 d	K; 7% γ 0.244, 0.154; 3% γ 0.022, 0.175	−74		
	Gd	152	1.1×10^{14} yr 0.20%	α 2.1	−74.71	<180 (γ)	0^+
	Gd	153	242 d	K; 55% γ 0.099 complex	−73.12		$\frac{3}{2}^+$
	Gd	154	2.15%		−73.65		0^+
	Gd	155	14.7%		−72.04	58 000 (γ)	$\frac{3}{2}^-$
	Gd	156	20.47%		−72.49		0^+
	Gd	157	15.68%		−70.77	2.4×10^5 (γ)	$\frac{3}{2}^-$
	Gd	158	24.9%		−70.63	3.4 (γ)	0^+
	Gd	159	18.0 h	β⁻ 0.95; 9% γ 0.363; 3% γ 0.058	−68.59		$\frac{3}{2}^-$
	Gd	160	21.9%		−67.89	0.8 (γ)	0^+
	Gd	161	3.7 min	β⁻ 1.6; 66% γ 0.361; 25% γ 0.315; 11% γ 0.102	−65.5		
65	Tb	147	24 min	β⁺			
	Tb	148	70 min	β⁺ 4.6; γ	−70.7		
	Tb	149	4.1 h	84% K; 16% α 3.95; γ	−71.4		
	Tb	149m	4.3 min	β⁺; α 3.99			
	Tb	150	3.1 h	β⁺ 3.6; γ	−71.03		
	Tb	151	18 h	99% K; α 3.42; 35% γ 0.108, 0.252; 32% γ 0.288	−71.6		
	Tb	152	18 h	β⁺ 2.82; γ	−70.5		
	Tb	153	55 h	K; 30% γ 0.212; 12% γ 0.11 complex	−71		
	Tb	154	21 h	β⁺; γ	−70		
	Tb	154	8.5 h	β⁺; γ	−70		
	Tb	155	5.6 d	K; 37% γ 0.087; 25% γ 0.105%; 8% γ 0.180	−71		
	Tb	156	5.4 d	K; 70% γ 0.535; 40% γ 0.199; 29% γ 1.22	−70		
	Tb	156m	5.5 h	γ	−70		
	Tb	157	150 yr	K	−70.71		
	Tb	158	1.2×10^3 yr	β⁻ 0.85; 69% γ 0.95; 12% γ 0.080; 10% γ 0.782, 0.182	−69.43		
	Tb	158m	11 s	γ	−69.32		
	Tb	159	100%		−69.53	46 (γ)	$\frac{3}{2}^+$
	Tb	160	72.1 d	β⁻ 1.74; 31% γ 0.966 complex; 31% γ 0.879	−67.85	525 (γ)	3^-
	Tb	161	6.9 d	β⁻ 0.59; 21% γ 0.026; 19% γ 0.049	−67.47		$\frac{3}{2}^+$
	Tb	162	7.5 min	β⁻; γ	−65		
	Tb	163	6.5 h	β⁻ 1.65; γ	−64.7		
66	Dy	149	15 min	K			
	Dy	150	7.2 min	β⁺, 18% α 4.23; γ	−69		0^+
	Dy	151	18 min	β⁺; 6% α 4.06; γ	−69		
	Dy	152	2.4 h	β⁺; α 3.65; γ	−70.11		0^+
	Dy	153	6.4 h	K; 0.003% α 3.48; γ	−69.2		
	Dy	154	10^6 yr	α 2.85	−70.5		
	Dy	154m	13 h	α 3.37			

Table of the nuclides (*contd*)

Z	Nuclide	A	Natural abundance or half-life	Major radiations	Mass excess MeV	σ/b	Iπ
	Dy	155	10.2 h	β⁺ 1.08; 98% K; 68% γ 0.227; 8% γ 0.52	−69		
	Dy	156	0.0524%		−70.9	3 (γ)	0⁺
	Dy	157	8.1 h	K; 91% γ 0.326	−70		
	Dy	158	0.0902%		−70.37	100 (γ)	0⁺
	Dy	159	144 d	K; 4% γ 0.058	−69.15		$\frac{3}{2}-$
	Dy	160	2.294%		−69.67		0⁺
	Dy	161	18.88%		−68.05	600 (γ)	$\frac{5}{2}+$
	Dy	162	25.53%		−68.18	140 (γ)	0⁺
	Dy	163	24.97%		−66.36	130 (γ)	$\frac{5}{2}-$
	Dy	164	28.18%		−65.95	800 (γ), 2000 (γm)	0⁺
	Dy	165	139.2 min	β⁻ 1.29; 4% γ 0.095	−63.51	4700 (γ)	$\frac{7}{2}+$
	Dy	165m	1.26 min	β⁻ 1.04; 3% γ 0.108	−63.40		$\frac{1}{2}-$
	Dy	166	81.5 h	β⁻ 0.48; 12% γ 0.082	−62.59		0⁺
	Dy	167	4.4 min	β⁻			
67	Ho	151	42 s	β⁺; 20% α 4.51			
	Ho	152	2.4 min	β⁺; 30% α 4.38	−63.8		
	Ho	152	52 s	β⁺; 19% α 4.45			
	Ho	153	9 min	β⁺; 0.3% α 3.92	−65.0		
	Ho	154	7 min	β⁺; γ	−65		
	Ho	155	50 min	β⁺ 2.1; γ			
	Ho	156	55 min	β⁺ 2.9; γ			
	Ho	157	14 min	β⁺; γ			
	Ho	158	11 min	K; γ	−66.33		
	Ho	158m	29 min	β⁺ 1.32; γ	−66.26		
	Ho	159	33 min	K; γ	−67		
	Ho	159m	6.9 s	γ	−67		
	Ho	160	26 min	K, γ	−66.4		
	Ho	160m	5.0 h	β⁺ 1.9; 50% γ 0.729; 37% γ 0.965 complex	−66.3		
	Ho	161	2.5 h	K; 23% γ 0.026; 15% γ 0.075	−67		$\frac{7}{2}-$
	Ho	161m	6 s	53% γ 0.211	−67		$\frac{1}{2}+$
	Ho	162	15 min	β⁺ 1.10; 8% γ 0.081	−66.02		1⁺
	Ho	162m	68 min	K; 26% γ 0.185; 24% γ 1.224; 13% γ 0.940	−65.92		
	Ho	163	>10³ yr		−66.35		$\frac{7}{2}-$
	Ho	163m	1.1 s	γ	−66.05		$\frac{1}{2}+$
	Ho	164	37 min	β⁻ 0.99; K; γ	−64.84		
	Ho	165	100%		−64.81	64 (γ); 1 (γm)	$\frac{7}{2}-$
	Ho	166	26.9 h	β⁻ 1.84; 5.4% γ 0.081	−63.07		0⁻
	Ho	166m	1.2 × 10³ yr	β⁻; 90% γ 0.184; 60% γ 0.810; 58% γ 0.711	−63.06		
	Ho	167	3.1 h	β⁻ 0.96; γ	−62.3		
	Ho	168	3.3 min	β⁻ 2.2; γ	−59.7		
	Ho	169	4.8 min	β⁻ 1.95; γ	−58.8		
	Ho	170	45 s	β⁻ 3.1; γ	−55.8		
68	Er	152	11 s	β⁺; 90% α 4.80			0⁺
	Er	153	36 s	β⁺; α 4.67			
	Er	154	5 min	α 4.15	−63		
	Er	157	25 min	β⁺; γ			
	Er	158	2.3 h	β⁺ 0.8; γ			0⁺

Table of nuclides (*contd*)

Nuclide			Natural abundance or half-life	Major radiations	Mass excess MeV	σ/b	Iπ
Z		*A*					
	Er	159	36 min	β⁺; γ			
	Er	160	29 h	K			0⁺
	Er	161	3.1 h	β⁺ 1.2; 63% γ 0.826; 9% γ 0.211; 8% γ 0.592	−65		
	Er	162	0.136%		−66.4	2 (γ)	0⁺
	Er	163	75 min	β⁺ 0.19; 99% K; γ	−65.14		5/2−
	Er	164	1.56%		−65.87	1.7 (γ)	0⁺
	Er	165	10.3 h	K	−64.44		5/2−
	Er	166	33.41%		−64.92	12 (γ)	0⁺
	Er	167	22.94%		−63.29	700 (γ)	7/2+
	Er	167m	2.3 s	43% γ 0.208	−63.08		1/2−
	Er	168	27.07%		−62.98	2 (γ)	0⁺
	Er	169	9.4 d	β⁻ 0.34	−60.91		1/2−
	Er	170	14.88%		−60.0	9 (γ)	0⁺
	Er	171	7.52 h	β⁻ 1.49; 63% γ 0.308; 28% γ 0.296; 25% γ 0.112	−57.6		5/2−
	Er	172	49 h	β⁻ 0.89; 40% γ 0.610, 0.407	−56.5		0⁺
69	Tm	153	1.6 s	α 5.10			
	Tm	154	5 s	α 5.04			
	Tm	161	30 min	K; γ	−62		
	Tm	162	22 min	K; γ	−61.5		
	Tm	163	1.8 h	β⁺ 1.1; γ	−62.87		
	Tm	164	1.9 min	β⁺ 2.94; γ	−61.91		
	Tm	165	30.1 h	K; γ	−62.87		
	Tm	166	7.7 h	β⁺ 1.94; γ	−61.88		2⁺
	Tm	167	9.6 d	K; 43% γ 0.208; 4% γ 0.057; 2% γ 0.532	−62.13		1/2+
	Tm	168	85 d	K; 88% γ 0.82 complex; 77% γ 0.19 complex	−61.27		
	Tm	169	100%		−61.25	125 (γ)	1/2−
	Tm	170	130 d	β⁻ 0.97; 3.3% γ 0.084	−59.6	150 (γ)	1⁻
	Tm	171	1.92 yr	β⁻ 0.097	−59.1		1/2+
	Tm	172	63.6 h	β⁻ 1.88; 7% γ 1.39, 1.09; 6% γ 1.53	−57.4		2⁻
	Tm	173	8.2 h	β⁻ 1.3; 89% γ 0.399; 8% γ 0.465	−56.4		
	Tm	174	5.2 min	β⁻ 1.2; 93% γ 0.366; 89% γ 0.99; 85% γ 0.273	−54.6		
	Tm	175	20 min	β⁻ 2.0	−52.3		
70	Yb	154	0.39 s	α 5.33			
	Yb	155	1.6 s	α 5.21			
	Yb	162	24 min	K			
	Yb	164	76 min	K			0⁺
	Yb	165	10 min	β⁺	−60		
	Yb	166	57.5 h	K; 17% γ 0.082	−61.6		0⁺
	Yb	167	18 min	K, 90% γ 0.113 complex; 15% γ 0.176	−60.17		
	Yb	168	0.140%		−61.3	11 000 (γ)	0⁺
	Yb	169	32 d	K; 45% γ 0.063; 35% γ 0.198; 22% γ 0.177	−60		
	Yb	169m	46 s	γ	−60		

Table of the nuclides (*contd*)

Z	A	Natural abundance or half-life	Major radiations	Mass excess MeV	σ/b	Iπ
	Yb 170	3.03%		−60.5		0+
	Yb 171	14.31%		−59.2		$\frac{1}{2}$−
	Yb 172	21.82%		−59.3		0+
	Yb 173	16.13%		−57.7		$\frac{5}{2}$−
	Yb 174	31.84%		−57.1	55 (γ)	0+
	Yb 175	101 h	β− 0.466; 3.7% γ 0.283; 1.9% γ 0.114	−54.8		$\frac{7}{2}$−
	Yb 176	12.73%		−53.4	7 (γ)	0+
	Yb 176m	11 s	γ	−52.4		
	Yb 177	1.9 h	β− 1.40; 16% γ 0.151; 5% γ 1.080	−50.8		$\frac{9}{2}$+
	Yb 177m	6.5 s	65% γ 0.104; 13% γ 0.228	−50.5		$\frac{1}{2}$−
71	Lu 155	0.07 s	α 5.63			
	Lu 156	0.5 s	α 5.5			
	Lu 167	54 min	β+ 1.5; γ	−57.1		
	Lu 168	7.1 min	β+ 1.2; γ	−57		
	Lu 169	34 h	β+ 1.2; γ	−58		
	Lu 169m	2.7 min	γ	−58		
	Lu 170	2.0 d	β+ 2.4; γ	−57.1		0+
	Lu 170m	0.7 s	γ	−57.0		4−
	Lu 171	8.3 d	K; 68% γ 0.741; 14% γ 0.668; 20% γ 0.019	−58		
	Lu 171m	76 s	γ	−58		
	Lu 172	6.7 d	K; 60% γ 1.09; 26% γ 0.182; 21% γ 0.81	−57		
	Lu 172m	3.7 min	γ	−57		
	Lu 173	1.37 yr	K; 18% γ 0.272; 14% γ 0.079; 7% γ 0.101	−57.0		
	Lu 174	3.6 yr	K; 9% γ 1.24; 6% γ 0.076	−55.6		1−
	Lu 174m	140 d	K; γ	−55.4		6−
	Lu 175	97.40%		−55.3	5 (γ)	$\frac{7}{2}$+
	Lu 176	3 × 10¹⁰ yr 2.60%	β− 0.43; 95% γ 0.306; 85% γ 0.202; 15% γ 0.088	−53.4	21 000 (γ)	7
	Lu 176m	3.7 h	β− 1.31; 10% γ 0.088	−53.1		8
	Lu 177	6.7 d	β− 0.497; 6.1% γ 0.208; 2.8% γ 0.113	−52.2		$\frac{7}{2}$+
	Lu 177m	155 d	78% β− 0.165; 62% γ 0.208; 37% γ 0.228; 29% γ 0.378	−51.3		$\frac{23}{2}$−
	Lu 178	20 min	β− 1.50; γ	−50.0		
	Lu 179	4.6 h	β− 1.35; γ	−49.6		
72	Hf 157	0.12 s	α 5.68			
	Hf 158	3 s	α 5.27			0+
	Hf 168	22 min	β+; γ			0+
	Hf 169	1.5 h	β+ 1.3; γ			
	Hf 170	12.2 h	K; γ			0+
	Hf 171	11 h	K; γ			
	Hf 172	5 yr	K; 22% γ 0.024; 21% γ 0.125 complex			0+
	Hf 173	23.6 h	K; 96% γ 0.13 complex; 52% γ 0.30 complex			

Table of the nuclides (contd)

Nuclide		Natural abundance or half-life	Major radiations	Mass excess MeV	σ/n	I^π
Z	A					
Hf	174	2×10^{15} yr 0.163%		-55.6	400 (γ)	0^+
Hf	175	70 d	K; 85% γ 0.343; 3.4% γ 0.089; 1.4% γ 0.433	-54.7		
Hf	176	5.21%		-54.4	<30 (γ)	0^+
Hf	177	18.56%		-52.7	371.4 (γ)	$\frac{7}{2}^-$
Hf	177m	1.1 s	81% γ 0.208; 48% γ 0.228; 37% γ 0.378	-51.4		$\frac{23}{2}^+$
Hf	178	27.1%		-52.3	80 (γ)	0^+
Hf	178m	4.3 s	97% γ 0.427; 94% γ 0.326; 75% γ 0.214	-51.1		
Hf	179	13.75%		-50.3	65 (γ); 0.2 (γm)	$\frac{9}{2}^+$
Hf	179m	18.6 s	94% γ 0.217	-49.9		$\frac{1}{2}^-$
Hf	180	35.22%		-49.5	10 (γ)	0^+
Hf	180m	5.5 h	93% γ 0.333; 82% γ 0.215; 80% γ 0.444	-48.4		8^-
Hf	181	42.5 d	β^- 0.41; 81% γ 0.482; 48% γ 0.133 complex	-47.41	40 (γ)	
Hf	182	9×10^6 yr	β^- 0.5; 84% γ 0.271	-45.8		0^+
Hf	183	65 min	β^- 1.6; γ	-43.0		
73 Ta	172	44 min	β^+; γ			
Ta	173	3.7 h	K; γ			
Ta	174	1.2 h	β^+; γ			
Ta	175	10.5 h	K; γ			
Ta	176	8.0 h	K; γ	-51		
Ta	177	56.6 h	K; 6% γ 0.113; 1% γ 0.208	-51.6		
Ta	178	9.4 min	1% β^+ 0.89; 99% K; γ	-50.4		
Ta	178	2.3 h	K; 120% γ 0.328; 97% γ 0.427; 75% γ 0.214			
Ta	179	600 d	K	-50.2		
Ta	180	$>10^{13}$ yr 0.0123%		-48.86		
Ta	180m	8.1 h	87% K; 13% β^- 0.71; 4% γ 0.093	-48.65		
Ta	181	99.9877%		-48.43	21 (γ); 0.07 (γm)	$\frac{7}{2}^+$
Ta	182	115 d	β^- 1.17; 42% β^- 0.522; 34% γ 1.122	-46.35	8000 (γ)	
Ta	182m	16.5 min	40% γ 0.172, 0.147	-45.84		
Ta	183	5.1 d	β^- 0.62; 33% γ 0.246 complex; 17% γ 0.161 complex	-45.20		$\frac{7}{2}^+$
Ta	184	8.7 h	β^- 2.64; 93% β^- 1.19; 49% γ 0.9 complex; 42% γ 0.25	-42.9		
Ta	185	50 min	β^- 1.7; 60% γ 0.175; 6% γ 0.100; 5% γ 0.245	-41.3		
Ta	186	10 min	β^- 2.2; 74% γ 0.20; 48% γ 0.73; 33% γ 0.61	-38.7		
74 W	173	16 min	K			
W	174	31 min	K			0^+
W	175	34 min	K; γ			
W	176	2.3 h	99% K: β^+; γ	-50		0^+
W	177	135 min	K; γ	-50		

Table of the nuclides (*contd*)

Nuclide Z	Nuclide A	Natural abundance or half-life	Major radiations	Mass excess MeV	σ/b	I$^\pi$
	W 178	21.5 d	K	−50		0$^+$
	W 179	38 min	K ; 22% γ 0.031	−49		
	W 179m	5.2 min	γ	−49		
	W 180	0.135%		−49.37	< 20 (γ)	0$^+$
	W 181	140 d	K ; γ	−48.24		
	W 182	26.4%		−48.16	20 (γ) ; 0.5 (γm)	0$^+$
	W 183	14.4%		−46.27	11 (γ)	$\frac{1}{2}^-$
	W 183m	5.3 s	11% γ 0.053 ; 19% γ 0.108	−45.96		
	W 184	30.6%		−45.62	2.11 (γ)	0$^+$
	W 185	75 d	β$^-$ 0.429	−43.30		$\frac{3}{2}^-$
	W 185m	1.6 min	γ	−42.93		
	W 186	28.4%		−42.44	40 (γ)	0$^+$
	W 187	23.9 h	β$^-$ 1.31 ; 27% γ 0.686 ; 23% γ 0.479 ; 11% γ 0.072	−39.83	90 (γ)	$\frac{3}{2}^-$
	W 188	69 d	β$^-$ 0.349 ; γ	−38.44		0$^+$
	W 189	11.5 min	β$^-$ 2.5 ; γ	−35.3		
75	Re 177	17 min	β$^+$	−47		
	Re 178	15 min	β$^+$ 3.1			
	Re 179	20 min	K	−46		
	Re 180	20 h	β$^+$ 1.1 ; γ			
	Re 181	18 h	K ; γ	−47		
	Re 182	13 h	0.3% β$^+$ 1.74 ; K ; γ	−45.30		
	Re 182	64 h	K ; γ			
	Re 183	71 d	K ; γ	−45		$\frac{5}{2}^+$
	Re 184	38 d	K ; γ	−44		
	Re 184m	169 d	K ; γ	−44		
	Re 185	37.07%		−43.73	110 (γ)	$\frac{5}{2}^+$
	Re 186	90 h	β$^-$ 1.07 ; 5% K ; 9% γ 0.137	−41.9		1$^-$
	Re 187	4 × 10^{10} yr 62.93%	β$^-$ 0.003	−41.14	70 (γ) ; 1.3 (γm)	$\frac{5}{2}^+$
	Re 188	16.7 h	β$^-$ 2.12 ; 10% γ 0.155	−38.79	< 2 (γ)	1$^-$
	Re 188m	18.7 min	10% γ 0.106 ; 5% γ 0.092	−38.62		6$^-$
	Re 189	24 h	β$^-$ 1.00 ; 10% γ 0.218 doublet ; 4% γ 0.245	−37.8		
	Re 190	2.8 min	β$^-$ 1.6 ; γ	−35.4		
	Re 190m	2.8 h	β$^-$ 1.6 ; γ			
	Re 192	6 s	β$^-$ 2.5 ; γ			
76	Os 181	23 min	K ; γ	−44		
	Os 182	22 h	K ; γ	−44		0$^+$
	Os 183	12 h	K ; 90% γ 0.382 ; 27% γ 0.114 ; 10% γ 0.168	−43		
	Os 183m	9.9 h	K ; 48% γ 1.105 complex ; 6% γ 1.035	−43		
	Os 184	0.018%		−44.0	< 200 (γ)	0$^+$
	Os 185	94 d	K ; 80% γ 0.646 ; 14% γ 0.875	−42.74		$\frac{1}{2}^-$
	Os 186	1.59%		−43.0		0$^+$
	Os 187	1.64%		−41.14		$\frac{1}{2}^-$
	Os 188	13.3%		−40.91		0$^+$
	Os 189	16.1%		−38.8	0.008 (γm)	$\frac{3}{2}^-$
	Os 189m	5.7 h	γ	−38.8		$\frac{9}{2}^-$
	Os 190	26.4%		−38.5	3.9 (γ) ; 8.6 (γm)	0$^+$

Table of the nuclides (*contd*)

Nuclide		Natural abundance or half-life	Major radiations	Mass excess MeV	σ/b	I^π
Z	A					
Os	190m	9.9 min	99% γ 0.616; 98% γ 0.502; 94% γ 0.361	−38.5		
Os	191	15 d	β⁻ 0.143; 25% γ 0.129	−36.4		
Os	191m	13 h	γ	−36.3		
Os	192	41.0%		−35.9	1.6 (γ)	0⁺
Os	193	31 h	β⁻ 1.13; 3.9% γ 0.460; 3% γ 0.139	−33.32	200 (γ)	
Os	194	6.0 yr	β⁻ 0.053; 10% γ 0.043	−32.39		0⁺
Os	195	6.5 min	β⁻ 2	−30		
77 Ir	182	15 min	K; γ	−39		
Ir	183	0.9 hr	K; γ			
Ir	184	3.2 hr	β⁺; γ	−40		
Ir	185	14 h	K; γ	−40		
Ir	186	16 h	97% K; 3% β⁺ 1.94; 74% γ 0.297; 45% γ 0.137	−39.1		
Ir	186	2 h	β⁺ 2.6; γ			
Ir	187	10.5 h	K; γ	−40		
Ir	188	41 h	99% K; β⁺ 1.66; 34% γ 0.155; 29% γ 0.633 doublet	−38.08		
Ir	189	13.3 d	K; 18% γ 0.245	−38		
Ir	190	11 d	K; 72% γ 0.56 complex	−36.5		
Ir	190m1	1.2 h	γ	−36.5		
Ir	190m2	3.2 h	93% γ 0.616; 92% γ 0.502; 88% γ 0.361	−36.3		
Ir	191	38.5%		−36.7	750 (γ); 250 (γm1)	$\frac{3}{2}^+$
Ir	191m	4.9 s	25% γ 0.129	−36.5		$\frac{11}{2}^-$
Ir	192	74.2 d	K; 95% β⁻ 0.67; 81% γ 0.317; 49% γ 0.468	−34.7	700 (γm)	4
Ir	192m1	1.4 min	0.017% β⁻; γ	−34.7		1
Ir	192m2	>5 yr	γ	−34.6		9
Ir	193	61.5%		−34.45	110 (γ)	$\frac{3}{2}^+$
Ir	193m	12 d	γ	−34.37		$\frac{11}{2}^-$
Ir	194	17.4 h	β⁻ 2.24; 10% γ 0.328	−32.49		1⁻
Ir	195	4.2 h	β⁻ 1.0; γ	−31.8		
Ir	196	120 min	β⁻ 0.95; 100% γ 0.65; 99% γ 0.522; 95% γ 0.44	−29.23		
Ir	198	50 s	β⁻ 3.6; γ	−25.5		
78 Pt	174	0.7 s	α 6.03			0⁺
Pt	175	2.1 s	α 5.95			
Pt	176	6 s	α 5.74			
Pt	177	6.6 s	α 5.51			
Pt	178	21 s	α 5.44			
Pt	179	33 s	α 5.15			
Pt	180	50 s	α 5.14			
Pt	181	51 s	α 5.02			
Pt	182	3.0 min	α 4.84	−36		0⁺
Pt	183	7 min	α 4.73			
Pt	184	20 min	α 4.50			0⁺
Pt	185	1.2 h	K; γ			
Pt	186	3 h	K; α 4.23; γ			0⁺
Pt	187	2 h	K; γ			

Table of the nuclides (*contd*)

Z	Nuclide	A	Natural abundance or half-life	Major radiations	Mass excess MeV	σ/b	I^π
	Pt	188	10.2 d	K; 10^{-5}% α 3.93; γ	−37.6		0^+
	Pt	189	10.9 h	K; γ	−37		
	Pt	190	6×10^{11} yr 0.0127%	α 3.18	−37.3	150 (γ)	0^+
	Pt	191	3.0 d	K; 9% γ 0.539; 5% γ 0.36 complex	−36		
	Pt	192	0.78%		−36.2		0^+
	Pt	193	<500 yr	K	−34.41		
	Pt	193m	4.3 d	γ	−34.26		
	Pt	194	32.9%		−34.72	1.1 (γ); 0.09 (γm)	0^+
	Pt	195	33.8%		−32.78	27 (γ)	$\frac{1}{2}^-$
	Pt	195m	4.1 d	11% γ 0.099; 1% γ 0.129	−32.52		$\frac{13}{2}^+$
	Pt	196	25.2%		−32.63	0.9 (γ); 0.05 (γm)	0^+
	Pt	197	18 h	β⁻ 0.670; 20% γ 0.077; 6% γ 0.191	−30.42		
	Pt	197m	80 min	3% β⁻ 0.737; 13% γ 0.346; 2.6% γ 0.279	−30.02		
	Pt	198	7.19%		−29.91	4 (γ)	0^+
	Pt	199	30 min	β⁻ 1.69; 24% γ 0.540; 12% γ 0.475 doublet	−27.40	15 (γ)	
	Pt	199m	14.1 s	90% γ 0.393	−26.98		
	Pt	200	11.5 h	β⁻	−27		0^+
	Pt	201	2.5 min	β⁻ 2.66; γ	−23.5		
79	Au	177	1.4 s	α 6.1			
	Au	178	2.7 s	α 5.91			
	Au	179	7 s	α 5.84			
	Au	181	10 s	α 5.60, 5.47			
	Au	183	44 s	α 5.34			
	Au	185	7 min	K			
	Au	185	4.3 min	β⁺; α 5.07			
	Au	186	12 min	K; γ			
	Au	187	8 min	K			
	Au	188	8 min	β⁺; γ			
	Au	189	30 min	K			
	Au	190	39 min	98% K; 2% β⁺; γ	−33		1^-
	Au	191	3.2 h	K; γ	−34		$\frac{3}{2}^+$
	Au	192	4.1 h	K; γ	−33.0		1
	Au	193	16 h	K; 11% γ 0.18 complex	−33		$\frac{3}{2}^+$
	Au	193m	3.9 s	65% γ 0.258	−33		$\frac{11}{2}^-$
	Au	194	39.5 h	β⁺ 1.49; 68% γ 0.328; 12% γ 0.294; 8% γ 1.469	−32.21		1^-
	Au	195	183 d	K; 10% γ 0.099; 1% γ 0.129	−32.55		$\frac{3}{2}^+$
	Au	195m	31 s	77% γ 0.261	−32.23		$\frac{11}{2}^-$
	Au	196	6.18 d	93.8% K; 6% β⁻ 0.259; 94% γ 0.356; 25% γ 0.333	−31.15		2^-
	Au	196m	9.7 h	32% γ 0.188; 42% γ 0.148; 5% γ 0.316, 0.285	−30.56		12
	Au	197	100%		−31.17	98.8 (γ)	$\frac{3}{2}^+$
	Au	197m	7.2 s	75% γ 0.279; 8% γ 0.130	−30.76		$\frac{11}{2}^-$
	Au	198	2.698 d	β⁻ 0.962; 95% γ 0.412	−29.59	26 000 (γ)	2^-

Table of the nuclides (*contd*)

Nuclide Z	Nuclide A	Natural abundance or half-life	Major radiations	Mass excess MeV	σ/b	I^π
	Au 199	3.15 d	β^- 0.46; 37% γ 0.158; 8% γ 0.208	−29.09	30 (γ)	$\frac{3}{2}^+$
	Au 200	48.4 min	β^- 2.2; 24% γ 0.368; 23% γ 1.227; 1% γ 1.593	−27.3		
	Au 201	26 min	β^- 1.5; γ	−26.2		
	Au 203	55 s	β^- 1.9; γ	−23		
80	Hg 185	50 s	K			
	Hg 186	1.5 min	K; γ			0^+
	Hg 187	3 min	K; γ			
	Hg 188	3.7 min	K; γ			0^+
	Hg 189	10 min	K; γ			
	Hg 190	20 min	K; γ	−31		0^+
	Hg 191	55 min	K; γ			
	Hg 192	4.8 h	β^+; 99% K; γ	−32		0^+
	Hg 193	6 h	K; γ	−31		$\frac{3}{2}^-$
	Hg 193m	10 h	β^+; 84% K; γ	−31		$\frac{13}{2}^+$
	Hg 194	1.9 yr	K; γ	−32.2		0^+
	Hg 195	9.5 h	K; γ	−31		$\frac{1}{2}^-$
	Hg 195m	40 h	K; 35% γ 0.200; 20% γ 0.560, 0.261	−31		$\frac{13}{2}^+$
	Hg 196	0.146%		−31.84	880 (γ); 25 (γm)	0^+
	Hg 197	65 h	K; 18% γ 0.077; 2% γ 0.191	−30.75		$\frac{1}{2}^-$
	Hg 197m	24 h	6% K; 42% γ 0.134; 7% γ 0.279	−30.45		$\frac{13}{2}^+$
	Hg 198	10.02%		30.97	0.02 (γ)	0^+
	Hg 199	16.84%		−29.55	2000 (γ)	$\frac{1}{2}^-$
	Hg 199m	43 min	53% γ 0.158; 15% γ 0.375	−29.01		$\frac{13}{2}^+$
	Hg 200	23.13%		−29.50	<50 (γ)	0^+
	Hg 201	13.22%		−27.66	<50 (γ)	$\frac{3}{2}^-$
	Hg 202	29.80%		−27.35	4 (γ)	0^+
	Hg 203	46.9 d	β^- 0.214; 77% γ 0.279	−25.26		
	Hg 204	6.85%		−24.69	0.4 (γ)	0^+
	Hg 205	5.5 min	β^- 1.7; γ	−22.2		
	Hg 206	8 min	β^- 1.3; γ	−20.95		0^+
81	Tl 191	10 min	β^+			
	Tl 192	11 min	β^+			
	Tl 193	23 min	β^+; γ			
	Tl 193m	2.1 min	γ			
	Tl 194	33 min	K; γ	−26		
	Tl 194m	32.8 min	K; γ			
	Tl 195	1.2 h	β^+ 1.8	−28		$\frac{1}{2}^+$
	Tl 195m	3.5 s	95% γ 0.383	−28		
	Tl 196	1.84 h	K; γ	−27.2		
	Tl 196m	1.41 h	96% K	−26.8		
	Tl 197	2.8 h	K; γ	−28.5		$\frac{1}{2}^+$
	Tl 197m	0.54 s	90% γ 0.385; 40% γ 0.222	−27.9		
	Tl 198	5.3 h	β^+ 2.4; 99% K; 90% γ 0.412; 40% γ 0.65 complex	−27.5		2^-
	Tl 198m	1.87 h	K; 45% γ 0.412; 35% γ 0.635, 0.586	−27.0		7^+

Table of the nuclides (*contd*)

Nuclide		Natural abundance or half-life	Major radiations	Mass excess MeV	σ/b	Iᵗ
	Z	A				
	Tl	199	7.4 h	K; 12% γ 0.208; 14% γ 0.455; 9% γ 0.247	− 28.5		$\frac{1}{2}^+$
	Tl	200	26.1 h	K; 0.37% β⁺; 88% γ 0.368; 35% γ 1.21 complex	− 27.05		2^-
	Tl	201	73 h	K; 8% γ 0.167; 2% γ 0.135	− 27.3		$\frac{1}{2}^+$
	Tl	202	12.0 d	K; 95% γ 0.439	− 26.13		2^-
	Tl	203	29.50%		− 25.75	11 (γ)	$\frac{1}{2}^+$
	Tl	204	3.8 yr	98% β⁻ 0.766; 2% K	− 24.34		2^-
	Tl	205	70.50%		− 23.81	0.11 (γ)	$\frac{1}{2}^+$
(RaEᴵᴵ)	Tl	206	4.19 min	β⁻ 1.52	− 22.26		
(AcCᴵᴵ)	Tl	207	4.79 min	β⁻ 1.44; γ	− 21.01		
(ThCᴵᴵ)	Tl	208	3.10 min	β⁻ 1.80; 100% γ 2.614; 86% γ 0.583; 23% γ 0.511	− 16.76		
	Tl	209	2.2 min	β⁻ 1.99; 100% γ 1.56, 0.45; 50% γ 0.12	− 13.65		
(RaCᴵᴵ)	Tl	210	1.3 min	β⁻ 2.3; 100% γ 0.795; 80% γ 0.296; 19% γ 1.08	− 9.23		
82	Pb	194	11 min	K; γ			0^+
	Pb	195	17 min	K; γ			
	Pb	196	37 min	K; γ	− 24		0^+
	Pb	197m	42 min	K; γ	− 24		
	Pb	198	2.4 h	K; 40% γ 0.38 complex; 28% γ 0.173	− 26		0^+
	Pb	199	90 min	K; 80% γ 0.367; 17% γ 0.353; 10% γ 0.720	− 25		
	Pb	199m	12.2 min	20% γ 0.424	− 25		
	Pb	200	21.5 h	K; γ	− 26		0^+
	Pb	201	9.4 h	K; γ	− 25		
	Pb	201m	61 s	51% γ 0.629	− 25		
	Pb	202	3 × 10⁵ yr	L	− 26.08		0^+
	Pb	202m	3.62 h	K; 90% γ 0.961, 0.422; 45% γ 0.787	− 23.91		9^-
	Pb	203	52.1 h	K; 81% γ 0.279; 5% γ 0.401	− 24.94		
	Pb	203m	6.1 s	70% γ 0.825	− 24.11		
	Pb	204	1.40%		− 25.11	0.7 (γ)	0^+
	Pb	204m	66.9 min	189% γ 0.90 doublet; 93% γ 0.375	− 22.92		9^-
	Pb	205	3 × 10⁷ yr	L	− 23.77		
	Pb	206	25.1%		− 23.79	0.03 (γ)	0^+
	Pb	207	21.7%		− 22.45	0.72 (γ)	$\frac{1}{2}^-$
	Pb	207m	0.80 s	98% γ 0.570; 83% γ 1.064	− 20.81		$\frac{13}{2}^+$
	Pb	208	52.3%		− 21.75	0.0005 (γ)	0^+
	Pb	209	3.30 h	β⁻ 0.635	− 17.63		
(RaD)	Pb	210	21 yr	β⁻ 0.061; α 3.72; γ	− 14.73		0^+
(AcB)	Pb	211	36.1 min	β⁻ 1.36; 3.4% γ 0.405; 1.8% γ 0.427	− 10.46		
(ThB)	Pb	212	10.64 h	β⁻ 0.58; 47% γ 0.239; 3.2% γ 0.300	− 7.55		0^+
	Pb	213	10 min	β⁻	− 3		
(RaB)	Pb	214	26.8 min	β⁻ 1.03; 36% γ 0.352; 19% γ 0.295; 4% γ 0.242	− 0.15		0^+

Table of the nuclides (*contd*)

Nuclide			Natural abundance or half-life	Major radiations	Mass excess MeV	σ/b	I^π
Z		A					
83	Bi	199	24 min	K; α 5.53	− 20		$\frac{9}{2}$
	Bi	200	35 min	K	− 20		7
	Bi	201	1.8 h	K	− 21		$\frac{9}{2}-$
	Bi	201m	52 min	γ; α 5.28			
	Bi	202	95 min	K; γ	− 21		5
	Bi	203	11.8 h	β^+ 1.35; 78% γ 0.82; 35% γ 1.87 doublet	− 21.8		$\frac{9}{2}-$
	Bi	204	11.2 h	K; γ	− 21		6^+
	Bi	205	15.3 d	K; 0.06% β^+ 0.98; 28% γ 0.703; 27% γ 1.766	− 21.07		$\frac{9}{2}-$
	Bi	206	6.24 d	K; 99% γ 0.803; 72% γ 0.880; 46% γ 0.516	− 20.18		6^+
	Bi	207	30 yr	K; 98% γ 0.570; 77% γ 1.063; 9% γ 1.771	− 20.04		
	Bi	208	3.7×10^5 yr	K; 100% γ 2.614	− 18.88		
	Bi	209	100%		− 18.26	0.015 (γ)	$\frac{9}{2}$
(RaE)	Bi	210	5.01 d	99% β^- 1.160; α 4.69, 4.65	− 14.79		1^-
	Bi	210m	3×10^6 yr	0.4% β^-; 58% α 4.96; 36% α 4.92; 6% α 4.57; 25% γ 0.262	− 14.52		
(AcC)	Bi	211	2.15 min	0.27% β^-; 84% α 6.62; 16% α 6.28; 14% γ 0.351	− 11.84		
(ThC)	Bi	212	60.6 min	64% β^- 2.25; 10% α 6.09; 25% α 6.05; 7% γ 0.727	− 8.13		1
	Bi	213	47 min	98% β^- 1.39; 2% α 5.87	− 5.24		
(RaC)	Bi	214	19.7 min	99% β^- 3.26; α 5.51, 5.45; 47% γ 0.609	− 1.19		
	Bi	215	7 min	β^-	1.7		
84	Po	194	0.5 s	α 6.85			0^+
	Po	195	3 s	α 6.63			
	Po	195m	1.4 s	α 6.72			
	Po	196	6 s	α 6.53			0^+
	Po	197	54 s	α 6.30			
	Po	197m	25 s	α 6.39			
	Po	198	1.7 min	α 6.16			0^+
	Po	199	5.0 min	97% K; α 5.94			
	Po	199m	4.2 min	74% K; α 6.05			
	Po	200	11 min	88% K; α 5.86	− 16		0^+
	Po	201	15.1 min	99% K; α 5.68	− 16		$\frac{3}{2}$
	Po	201m	9 min	97% K; α 5.78			
	Po	202	45 min	98% K; α 5.58	− 18		0^+
	Po	203	42 min	99% K; α 5.49	− 17		$\frac{5}{2}$
	Po	204	3.6 h	99% K; α 5.38	− 18		0^+
	Po	205	1.8 h	99% K; α 5.25	− 18		$\frac{5}{2}-$
	Po	206	88 d	95% K; α 5.22; γ	− 18.33		0^+
	Po	207	5.7 h	99% K; 0.5% β^+ 1.14; α 5.11; γ	− 17.14		$\frac{5}{2}-$
	Po	207m	2.8 s	100% γ 0.82; 42% γ 0.26; 40% γ 0.31	− 15.75		
	Po	208	2.93 yr	α 5.11; γ	− 17.47		0^+
	Po	209	103 yr	99% α 4.88; K	− 16.37		$\frac{1}{2}$

Table of the nuclides (*contd*)

Nuclide			Natural abundance or half-life	Major radiations	Mass excess MeV	σ/b	I$^\pi$
	Z	*A*					
(RaF)	Po	210	138.4 d	α 5.305	−15.95	<0.03 (γ)	0$^+$
(AcC^I)	Po	211	0.52 s	α 7.45; γ	−12.43		
	Po	211m	25 s	91% α 7.28; 7% α 8.88; 92% γ 0.570; 77% γ 1.063	−11.00		
(ThC^I)	Po	212	3 × 10^{-7} s	α 8.78	−10.37		0$^+$
	Po	212m	45 s	97% α 11.65; 2.6% γ 2.61; 2% γ 0.57	−7.44		
	Po	213	4.2 × 10^{-6} s	α 8.38	−6.66		
(RaC^I)	Po	214	1.6 × 10^{-6} s	α 7.69; γ	−4.47		0$^+$
(AcA)	Po	215	1.78 × 10^{-3} s	α 7.38	−0.52		
(ThA)	Po	216	0.15 s	α 6.78	1.78		0$^+$
	Po	217	<10 s	α 6.55	6		
(RaA)	Po	218	3.05 min	99.9% α 6.00; β$^-$	8.38		0$^+$
85	At	200	0.9 min	α 6.47, 6.42			
	At	201	1.5 min	α 6.35; K			
	At	202	3.0 min	88% K; α 6.23, 6.12	−10		
	At	203	7.4 min	86% K; α 6.09	−11		
	At	204	9.3 min	95.5% K; α 5.95	−11		
	At	205	26 min	82% K; α 5.90	−13		
	At	206	32 min	88% α 5.70; K	−12		
	At	207	1.8 h	90% K; α 5.76	−13.41		
	At	208	1.6 h	99% K; α 5.65; γ	−12		
	At	209	5.5 h	95% K; 5% α 5.65; 94% γ 0.780; 62% γ 0.545; 23% γ 0.195	−12.89		
	At	210	8.3 h	99% K; α 5.52, 5.44, 5.36; 100% γ 1.180; 79% γ 0.245; 48% γ 1.483	−12.12		
	At	211	7.21 h	59% K; α 5.868	−11.64		$\frac{9}{2}$
	At	212	0.30 s	80% α 7.66; 20% α 7.60	−8.64		
	At	212m	0.12 s	80% α 7.82; 20% α 7.88	−8.42		
	At	214	10^{-6} s	99% α 8.78	−3.42		
	At	215	10^{-4} s	α 8.01	−1.25		
	At	216	3 × 10^{-4} s	97% α 7.80; β$^-$	2.25		
	At	217	0.032 s	α 7.07	4.38		
	At	218	2 s	94% α 6.70; 6% α 6.65	8.11		
	At	219	0.9 min	α 6.28, β$^-$	10.5		
86	Em	204	75 s	α 6.42	−7		0$^+$
	Em	205	1.8 min	α 6.26	−7		
	Em	206	6.5 min	35% K; α 6.26	−9		0$^+$
	Em	207	11 min	96% K; α 6.15	−9		
	Em	208	23 min	80% K; α 6.15	−10		0$^+$
	Em	209	30 min	83% K; α 6.04	−9		
	Em	210	2.42 h	4% K; 96% α 6.04	−9.74		0$^+$
	Em	211	15 h	74% K; 17% α 5.78; 9% α 5.85; 74% γ 0.680; 38% γ 1.37	−8.75		
	Em	212	25 min	α 6.27	−8.66		0$^+$
	Em	215	10^{-6} s	α 8.6	−1.2		
	Em	216	4.5 × 10^{-5} s	α 8.05	0.25		0$^+$
	Em	217	5.4 × 10^{-4} s	α 7.74	3.65		

Table of the nuclides (*contd*)

Z		A	Natural abundance or half-life	Major radiations	Mass excess MeV	σ/b	Iᵖ
	Em	218	0.035 s	α 7.14; γ	5.22		0⁺
(An)	Em	219	4.00 s	81% α 6.82; 11% α 6.55; 8% α 6.42; 9% γ 0.272	8.85		
(Tn)	Em	220	55 s	α 6.29; γ	10.61	<0.2 (γ)	0⁺
	Em	221	25 min	80% β⁻; α 6.0	14		
(Rn)	Em	222	3.823 d	α 5.49; γ	16.39	0.7 (γ)	0⁺
	Em	223	43 min	β⁻			
	Em	224	1.9 h	β⁻			
87	Fr	204	2.0 s	α 7.03			
	Fr	205	3.7 s	α 6.92			
	Fr	206	15.8 s	α 6.80	−0		
	Fr	207	19 s	α 6.78	−2		
	Fr	208	37 s	α 6.66	−2		
	Fr	209	55 s	α 6.66	−3		
	Fr	210	2.6 min	α 6.56	−3		
	Fr	211	3.1 min	α 6.56	−4.3		
	Fr	212	19 min	56% K; 17% α 6.39; 16% α 6.42; 11% α 6.35	−4		
	Fr	213	34 s	α 6.78	−3.55		
	Fr	218		α 7.85	7.00		
	Fr	219	0.02 s	α 7.31	8.61		
	Fr	220	27.5 s	85% α 6.68; 13% α 6.64	11.47		
	Fr	221	4.8 min	82% α 6.34; 15% α 6.12; 14% γ 0.218	13.27		
	Fr	222	14.8 min	β⁻	16.34		
(AcK)	Fr	223	22 min	β⁻ 1.15; 40% γ 0.050; 13% γ 0.080	18.40		
88	Ra	213	2.7 min	α 6.91	−0		
	Ra	219	10⁻³ s	α 8.0	9.4		
	Ra	220	0.023 s	99% α 7.46; 1% γ 0.465	10.27		0⁺
	Ra	221	30 s	34% α 6.61; 30% α 6.76; 20% α 6.67; 13% γ 0.151	12.96		
	Ra	222	38 s	96% α 6.56; 4% γ 0.325	14.32		0⁺
(AcX)	Ra	223	11.43 d	54% α 5.71; 26% α 5.61; 9% α 5.75, 5.54; 10% γ 0.270	17.26	130 (γ)	½⁺
(ThX)	Ra	224	3.64 d	94% α 5.68; 6% α 5.45; 3.7% γ 0.241	18.82	12 (γ)	0⁺
	Ra	225	14.8 d	β⁻ 0.36; 33% γ 0.040	22.01		
(Ra)	Ra	226	1602 yr	95% α 4.78; 6% α 4.60; 4% γ 0.186	23.69	20 (γ)	0⁺
	Ra	227	41.2 min	β⁻ 1.31; 4% γ 0.291	27.18		
(MsTh₁)	Ra	228	6.7 yr	β⁻ 0.05	28.96	36 (γ)	0⁺
89	Ac	222	5 s	93% α 7.00	16.55		
	Ac	223	2.2 min	42% α 6.65; 38% α 6.66; 13% α 6.57	17.82		
	Ac	224	2.9 h	90% K; 3% α 6.20, 6.14, 6.04; 28% γ 0.132	20.21		
	Ac	225	10.0 d	54% α 5.83; 28% α 5.79; 10% α 5.73 doublet; γ	21.62		
	Ac	226	29 h	β⁻ 1.2; 20% K; 47% γ 0.230; 32% γ 0.158; 11% γ 0.253	24.31		

Table of the nuclides (*contd*)

Nuclide			Natural abundance or half-life	Major radiations	Mass excess MeV	σ/b	I^π
Z		A					
(Ac)	Ac	227	21.6 yr	99% β^- 0.046; α 4.95; γ	25.87	830 (γ)	$\frac{3}{2}^+$
(MsTh$_2$)	Ac	228	6.13 h	β^- 2.11; 25% γ 0.908; 20% γ 0.96 complex	28.91		
	Ac	229	66 min	β^-	31		
	Ac	231	15 min	β^- 2.1; γ	35.9		
90	Th	223	0.9 s	α 7.56	19.5		
	Th	224	1.05 s	79% α 7.18; 19% α 6.91; 9% γ 0.177	20.00		0^+
	Th	225	8.0 min	39% α 6.48; 13% α 6.44; 12% α 6.50; 27% γ 0.322; 10% K	22.30		
	Th	226	30.9 min	79% α 6.34; 19% α 6.22; 3.4% γ 0.111	23.19		0^+
(RdAc)	Th	227	18.2 d	24% α 5.98; 23% α 6.04; 21% α 5.76; γ	25.82	1500 (f)	$\frac{3}{2}^+$
(RdTh)	Th	228	1.910 yr	71% α 5.43; 28% α 5.34; γ	26.77	123 (γ) <0.3 (f)	0^+
	Th	229	7340 yr	58% α 4.84; 11% α 4.90; 10% α 4.97; 10% γ 0.2	29.61	32 (f)	$\frac{5}{2}^+$
(Io)	Th	230	8×10^4 yr	76% α 4.68; 24% α 4.62; γ	30.87	23 (γ) <0.001 (f)	0^+
(UY)	Th	231	25.5 h	β^- 0.30; 10% γ 0.084; 2% γ 0.026	33.83		
(Th)	Th	232	1.4×10^{10} yr 100%	76% α 4.01; 24% α 3.95	35.47	7.4 (γ) <0.0002 (f)	0^+
	Th	233	22.2 min	β^- 1.23; 2.7% γ 0.087	38.76	1500 (γ) 15 (f)	
(UX$_1$)	Th	234	24.1 d	β^- 0.191; 4% γ 0.093; 3.5% γ 0.063	40.64	1.8 (γ) <0.01 (f)	0^+
91	Pa	226	1.8 min	26% K; 38% α 6.86; 34% α 6.82	25.96		
	Pa	227	38.3 min	15% K; 43% α 6.47; 23% α 6.42 complex	26.83		
	Pa	228	22 h	98% K; α 6.11; 93% γ 0.95; 32% γ 0.46	28.86		
	Pa	229	1.5 d	99% K; α 5.67	29.88		
	Pa	230	17.7 d	89.6% K; β^- 0.41; α 5.34 complex; 50% γ 0.954	32.17	1500 (f)	
(Pa)	Pa	231	3.25×10^4 yr	24% α 5.01; 23% α 5.02; 22% α 4.95	33.44	200 (γ) 0.010 (f)	$\frac{3}{2}^-$
	Pa	232	1.31 d	β^- 1.3; 70% β^- 0.32; 51% γ 0.87 complex	35.95	760 (γ) 700 (f)	
	Pa	233	27.0 d	β^- 0.568; 27% β^- 0.257; 44% γ 0.31 complex	37.51	43 (γ) <0.1 (f)	$\frac{3}{2}^-$
(UZ)	Pa	234	6.75 h	β^- 1.3; 13% β^- 1.13; 70% γ 0.90 complex	40.38	<5000 (f)	
(UX$_2$)	Pa	234m	1.17 min	β^- 2.29; γ	40.45	<500 (f)	
	Pa	235	23.7 min	β^- 1.4	42.3		
	Pa	237	39 min	β^- 2.3; γ	47.7		
92	U	227	1.3 min	α 6.8	29		
	U	228	9.1 min	α 6.69; K; γ	29.23		0^+

Table of the nuclides (contd)

Nuclide			Natural abundance or half-life	Major radiations	Mass excess MeV	σ/b	Iπ
	Z	A					
	U	229	58 min	80% K; 13% α 6.36; 4% α 6.33; 3% α 6.30	31.20		
	U	230	20.8 d	67% α 5.89; 32% α 5.82; γ	31.60	25 (f)	0⁺
	U	231	4.3 d	99% K; α 5.46; 7% γ 0.084	33.8	400 (f)	
	U	232	72 yr	68% α 5.32; 32% α 5.27; γ	34.60	78 (γ); 77 (f)	0⁺
	U	233	1.62 × 10⁵ yr	83% α 4.82; 15% α 4.78; γ	36.94	49 (γ); 524 (f)	5/2⁺
(U_II)	U	234	2.47 × 10⁵ yr 0.0057%	72% α 4.77; 28% α 4.72; γ	38.16	95 (γ)	0⁺
(AcU)	U	235	7.1 × 10⁸ yr 0.7196%	57% α 4.40; 18% α 4.37; 8% α 4.58 doublet	40.93	101 (γ); 577 (f)	7/2⁻
	U	235m	26.1 m	e⁻	40.93		1/2⁺
	U	236	2.39 × 10⁷ yr	76% α 4.49; 24% α 4.44	42.46	6 (γ)	0⁺
	U	237	6.75 d	β⁻ 0.248; 36% γ 0.060; 23% γ 0.208	45.41		1/2⁺
(U_I)	U	238	4.51 × 10⁹ yr 99.276%	75% α 4.20; 25% α 4.15	47.33	2.73 (γ); <0.005 (f)	0⁺
	U	239	23.5 min	β⁻ 1.29; 51% γ 0.075	50.60	22 (γ); 14 (f)	
	U	240	14.1 h	β⁻ 0.36	52.74		0⁺
93	Np	231	50 min	α 6.29	35.7		
	Np	233	35 min	99% K; α 5.54; γ	38		
	Np	234	4.40 d	K; 0.05% β⁺ 0.8; γ	40.0	900 (f)	
	Np	235	410 d	99% K; α 5.02	41.05		5/2⁺
	Np	236	22 h	51% K; 49% β⁻ 0.52	43.41		
	Np	236	>5 × 10³ yr			2500 (f)	
	Np	237	2.14 × 10⁶ yr	75% α 4.78 complex; γ	44.89	170 (γ); 0.019 (f)	5/2⁺
	Np	238	2.1 d	β⁻ 1.25; 42% γ 1.01 complex	47.47	1600 (f)	2
	Np	239	2.35 d	β⁻ 0.713; 23% γ 0.106; 12% γ 0.228	49.32	60 (γ); <1 (f)	5/2⁺
	Np	240	63 min	β⁻ 0.89; γ	52.2		
	Np	240m	7.3 min	β⁻ 2.16; 21% γ 0.56; 13% γ 0.60	52.3		1
	Np	241	16 min	β⁻ 1.4	54.3		
94	Pu	232	36 min	α 6.59	38.4		0⁺
	Pu	233	20 min	99% K; α 6.31	40.04		
	Pu	234	9.0 h	94% K; 4% α 6.20; 1.9% α 6.15	40.34		0⁺
	Pu	235	26 min	99% K; α 5.86	42.2		
	Pu	236	2.85 yr	69% α 5.77; 31% α 5.72; γ	42.90	170 (f)	0⁺
	Pu	237	45.6 d	99% K; α 5.66; 5% γ 0.060	45.12	2500 (f)	
	Pu	237m	0.18 s	2% γ 0.145	45.26		
	Pu	238	86 yr	72% α 5.50; 28% α 5.46	46.18	500 (γ); 16.8 (f)	0⁺
	Pu	239	2.44 × 10⁴ yr	88% α 5.16 doublet; 11% α 5.11; γ	48.60	274 (γ); 741 (f)	1/2⁺
	Pu	240	6580 yr	76% α 5.17; 24% α 5.12; γ	50.14	286 (γ); 0.08 (f)	0⁺
	Pu	241	13.2 yr	β⁻ 0.021; α 4.90; γ	52.98	425 (γ); 950 (f)	5/2⁺
	Pu	242	3.79 × 10⁵ yr	76% α 4.90; 24% α 4.86	54.74	19 (γ); <0.2 (f)	0⁺
	Pu	243	4.98 h	β⁻ 0.58; 21% γ 0.084	57.77	170 (γ)	
	Pu	244	8 × 10⁷ yr	α	59.83	1.8 (γ)	0⁺
	Pu	245	10 h	β⁻	63	260 (γ)	
	Pu	246	10.9 d	β⁻ 0.33; 30% γ 0.044; 25% γ 0.224	65.3		0⁺

Table of the nuclides (*contd*)

Nuclide		Natural abundance or half-life	Major radiations	Mass excess MeV	σ/b	Iπ
Z	A					
95 Am	237	1.3 h	99% K; α 6.02	47		
Am	238	1.9 h	K; 80% γ 0.98 doublet; 76% γ 1.35	48		
Am	239	12.1 h	99% K; α 5.78; 18% γ 0.228 doublet; 17% γ 0.278	49.41		
Am	240	51 h	K; 77% γ 1.00; 23% γ 0.90	51		
Am	241	458 yr	85% α 5.49; 13% α 5.44; 36% γ 0.060	52.96	800 (γ); 3.0 (f)	$\frac{5}{2}-$
Am	242	16.0 h	β⁻ 0.67	55.48	2900 (f)	1
Am	242m	152 yr	γ; α 5.21		2000 (γ); 6000 (f)	5
Am	243	7.95 × 10³ yr	87% α 5.28; 11.5% α 5.23; γ	57.18	74 (γ); <0.07 (f)	$\frac{5}{2}-$
Am	244	10.1 h	β⁻ 0.387; 66% γ 0.746; 25% γ 0.900; 19% γ 0.154	59.90	2300 (f)	
Am	244m	26 min	β⁻ 1.50; 0.4% K	60.02		
Am	245	2.1 h	β⁻ 0.91; γ	61.93		
Am	246	25 min	β⁻ 2.10; 65% γ 1.07; 29% γ 0.799	64.9		
96 Cm	238	2.5 h	α 6.51; K	49.39		0+
Cm	239	2.9 h	K; γ	51		
Cm	240	26.8 d	72% α 6.29; 28% α 6.25	51.72		0+
Cm	241	35 d	99% K; α 5.94; 95% γ 0.475	53.73		
Cm	242	163 d	74% α 6.12; 26% α 6.07; γ	54.82	20 (γ); <5 (f)	0+
Cm	243	32 yr	α 6.06; K; 14% γ 0.278; 12% γ 0.228; 4% γ 0.209	57.19	250 (γ); 660 (f)	$\frac{5}{2}+$
Cm	244	17.6 yr	77% α 5.81; 23% α 5.77; γ	58.47	15 (γ)	0+
Cm	245	9.3 × 10³ yr	80% α 5.36; 7% α 5.31; 14% γ 0.173; 5% γ 0.13	61.02	200 (γ); 1900 (f)	$\frac{7}{2}+$
Cm	246	5.5 × 10³ yr	81% α 5.39; 19% α 5.34	62.64	15 (γ)	0+
Cm	247	1.6 × 10⁷ yr	α	65.56	180 (γ)	
Cm	248	4.7 × 10⁵ yr	82% α 5.08; 18% α 5.04	67.43	6 (γ)	0+
Cm	249	64 min	β⁻ 0.9	70.8		
Cm	250	1.7 × 10⁴ yr	Spon fission	73		0+
97 Bk	243	4.6 h	99% K; α 6.76; γ	58.7		
Bk	244	4.4 h	99% K; α; γ	61		
Bk	245	4.98 d	99% K; α 6.36; γ	61.84		
Bk	246	1.8 d	K; 40% γ 0.800; 12% γ 1.07 complex	64		
Bk	247	1.4 × 10³ yr	58% α 5.52; 37% α 5.68; 30% γ 0.27	65.47		
Bk	248	16 h	K; β⁻			
Bk	249	314 d	β⁻ 0.125; α 5.42; γ	69.86	500 (γ)	$\frac{7}{2}+$
Bk	250	3.22 h	β⁻ 1.76; 47% γ 0.990; 39% γ 1.032	72.95		
Bk	251	57 min	β⁻			
98 Cf	244	25 min	α 7.18	61.43		0+
Cf	245	44 min	70% K; α 7.12	63.38		
Cf	246	36 h	78% α 6.76; 22% α 6.72	64.11		0+
Cf	247	2.5 h	K; 1% γ 0.295	66		
Cf	248	350 d	82% α 6.27; 18% α 6.22	67.26		0+
Cf	249	360 yr	84% α 5.81; 72% γ 0.388; 16% γ 0.333	69.74	270 (γ); 1735 (f)	$\frac{9}{2}-$

Table of the nuclides (*contd*)

Nuclide			Natural abundance or half-life	Major radiations	Mass excess MeV	σ/b	Iπ
	Z	A					
	Cf	250	13 yr	83% α 6.03; 17% α 5.99	71.19	1500 (γ)	0+
	Cf	251	800 yr	45% α 5.85; 55% α 5.67	74.15	3000 (γ); 3000 (f)	½+
	Cf	252	2.65 yr	82% α 6.12; 15% α 608; 3.1% spon fission	76.05	30 (γ)	0+
	Cf	253	17.6 d	β⁻ 0.27; α 5.98	79.3		
	Cf	254	60.5 d	99% spon fission; α 5.84	81		0+
99	Es	243	20 s	α 7.9			
	Es	245	1.3 min	83% K; α 7.70	66		
	Es	246	7.3 min	90% K; α 7.33	68		
	Es	247	5.0 min	93% K; α 7.33	68		
	Es	248	25 min	99% K; α 6.88	70		
	Es	249	2 h	99% K; α 6.77	71.15		
	Es	250	8 h	K	73		
	Es	251	1.5 d	99% K; α 6.49	74.5		
	Es	252	140 d	82% α 6.64; 13% α 6.58; γ	77.1		
	Es	253	20.47 d	90% α 6.64; γ	79.03	300 (γ)	7/2+
	Es	254	276 d	93% α 6.44; γ	82.00		
	Es	254m	39.3 h	β⁻ 1.13; γ	82.10		
	Es	255	38.3 d	91% β; α 6.31	84		
100	Fm	244	3.3 × 10⁻³ s	Spon fission			
	Fm	245	4.2 s	α 8.1			
	Fm	246	1.2 s	92% α 8.2; 8% spon fission; γ			
	Fm	247	35 s	α 7.93			
	Fm	248	38 s	80% α 7.87; 20% α 7.83; spon fission; γ	72		
	Fm	249	2.5 min	α 7.9	73.8		
	Fm	250	30 min	α 7.44	74.1		0+
	Fm	251	7 h	99% K; α 6.89	76		
	Fm	252	23 h	α 7.05	76.84		0+
	Fm	253	3 d	89% K; 9% α 6.96; 2% α 6.91	80		7/2+
	Fm	254	3.24 h	82% α 7.20; 17% α 7.16	80.93		
	Fm	255	20.1 h	93% α 7.03	83.82		
	Fm	256	2.7 h	97% spon fission; 3% α 6.86	85.44		
	Fm	257	80 d	94% α 6.53; spon fission; γ	88.6		
101	Md	248	6 s	α 8.32			
	Md	249	24 s	α 8.03			
	Md	250	53 s	α 7.73			
	Md	251	4 min	α 7.53			
	Md	252	8 min	K			
	Md	255	0.6 h	90% K; α 7.34	84.4		
	Md	256	1.5 h	97% K; α 7.18	86.9		
	Md	257	3 h	92% K; α 7.25	89		
	Md	258	53 d	α 6.7			
102	No	251	0.5 s	α 8.6			
	No	252	4.5 s	70% α 8.4; 30% spon fission			
	No	253	95 s	α 8.0			
	No	254	60 s	α 8.1			
	No	254m	0.2 s	γ			
	No	255	180 s	α 8.1			
	No	256	7 s	α 8.4			
	No	257	23 s	α 8.3			

Table of the nuclides (*contd*)

Nuclide			Natural abundance or half-life	Major radiations	Mass excess MeV	σ/b	I$^\pi$
Z		A					
103	Lr	255	22 s	50% α 8.37; 50% α 8.35			
	Lr	256	31 s	34% α 8.43; 23% α 8.39; 19% α 8.52			
	Lr	257	0.6 s	81% α 8.87; 19% α 8.81			
	Lr	258	4.2 s	47% α 8.62; 30% α 8.59; 16% α 8.65			
	Lr	259	5.4 s	100% α 8.45			
	Lr	260	180 s	100% α 8.03			
104	(Ku, Rf)	257	4.5 s	35% α 9.0; 30% α 8.95; 20% α 8.78			
		259	3 s	60% α 8.77; 40% α 8.86			
		261	65 s	α 8.4			
105	(Ha)	260	1.6 s	55% α 9.06; 25% α 9.10; 20% α 9.14			
		261	1.8 s	α 8.93			
		262	40 s	α 8.6			

B.W.H.

3.6.2 Radioactive sources

The sources listed are commercially available with the exception of Co^{56} which can be made using the Fe^{56}(p, n)Co^{56} reaction. The alpha source Pb^{212} is obtained by collecting radio thorium recoil products; it gives rise to 6090 and 6050 keV alphas from Bi^{212} and 8785 keV alphas from Po^{212}. The list contains only the major radiations and excludes annihilation radiation generated by β$^+$ sources; more detailed information can be obtained from the table of nuclides (p. 286).

Tables of γ-rays in order of increasing energy, half-life and atomic number can be found in references (1) and (2), which also catalogue the response of NaI and Ge(Li) counters to radioactive sources.

References

(1) C. E. Crouthamel, *Applied Gamma-Ray Spectrometry*, 1970.
(2) R. L. Heath, *Scintillation Spectrometry*, IDO–16880 (1964).

Radioactive sources

Sources with monoenergetic electrons			Beta-ray sources			
Source	Half-life	Energy/keV	Source	Half-life	Radiation	End point/keV
Cs^{137} . . .	30 yr	624	C^{14} . .	5730 yr	β$^-$	156
		656	Sr^{90}/Y^{90}	28 yr	β$^-$	2270
Bi^{207} . . .	30 yr	482	^{3}H . .	12.26 yr	β$^-$	18
		975	Na^{22} . .	2.6 yr	β$^+$	545
		1048	Co^{58} . .	71 d	β$^+$	485

Radioactive sources (*contd*)

Source	Half-life	Energy/keV	Relative intensity	Source	Half-life	Energy/keV	Relative intensity
Gamma-ray source							
Na^{22} . .	2.60 yr	1274.55				1175.26	1.86
Mn^{54} . .	314 d	834.81				1238.34	69.35
Co^{56} . .	77 d	733.79	0.1			1360.35	4.38
		787.92	0.4			1771.57	15.30
		846.76	100			1964.88	0.72
		977.47	1.52			2015.49	2.93
		1037.97	13.02			2035.03	7.33
		2113.00	0.37	Tl^{208} . .	(1.91 yr)	510.723	23
		2598.80	16.77	(ThC^{II}) .		583.139	86
		3009.99	0.84			860	12
		3202.25	3.15			2614.47	100
		3253.82	7.70				
		3273.38	1.55	*Alpha-particle source*			
		3452.18	0.88	Am^{241} . .	458 yr	5490	100
		3548.11	0.18			5440	15
Co^{60} . .	5.26 yr	1173.23	100	Cm^{242} . .	163 d	6115	100
		1332.49	100			6071	36.1
Zn^{65} . .	246 d	1115.4				5971	0.03
Y^{88} . .	107 d	898.04	91	Pb^{210} . .	22 yr	5304.5	
		1836.13	100	Pb^{212} . .	10 h	8785	
Cs^{137} . .	30 yr	661.635				6090	
Bi^{207} . .	30 yr	569.62	100			6050	
		1063.44	78				
		1769.71	9				

B.W.H.

3.7 Nuclear fission and neutron interactions

3.7.1 Fission fundamentals

When nuclei of certain of the heavy elements capture neutrons, fission takes place; that is to say, the nuclei divide usually into two parts, though in a small fraction ($<1\%$) of cases a third light particle is emitted. In this process about 200 MeV of energy is released, largely in the form of kinetic energy of the two fission fragments. Fission is also accompanied by the emission of two or three neutrons and about ten gamma-rays, emitted from the fission fragments during their recoil. An approximate formula for the fission neutron spectrum is given in the section on Neutron Cross-sections (p. 332) which also contains information on the total number of neutrons per fission for various fissile nuclei in three different neutron spectra (table, p. 335).

The mass partition between the two fragments is in general asymmetrical, particularly for fission induced by low energy neutrons. The special cases of the fission of ^{235}U and ^{239}Pu by thermal and fast neutrons are illustrated in the diagram and it will be observed that despite its characteristic

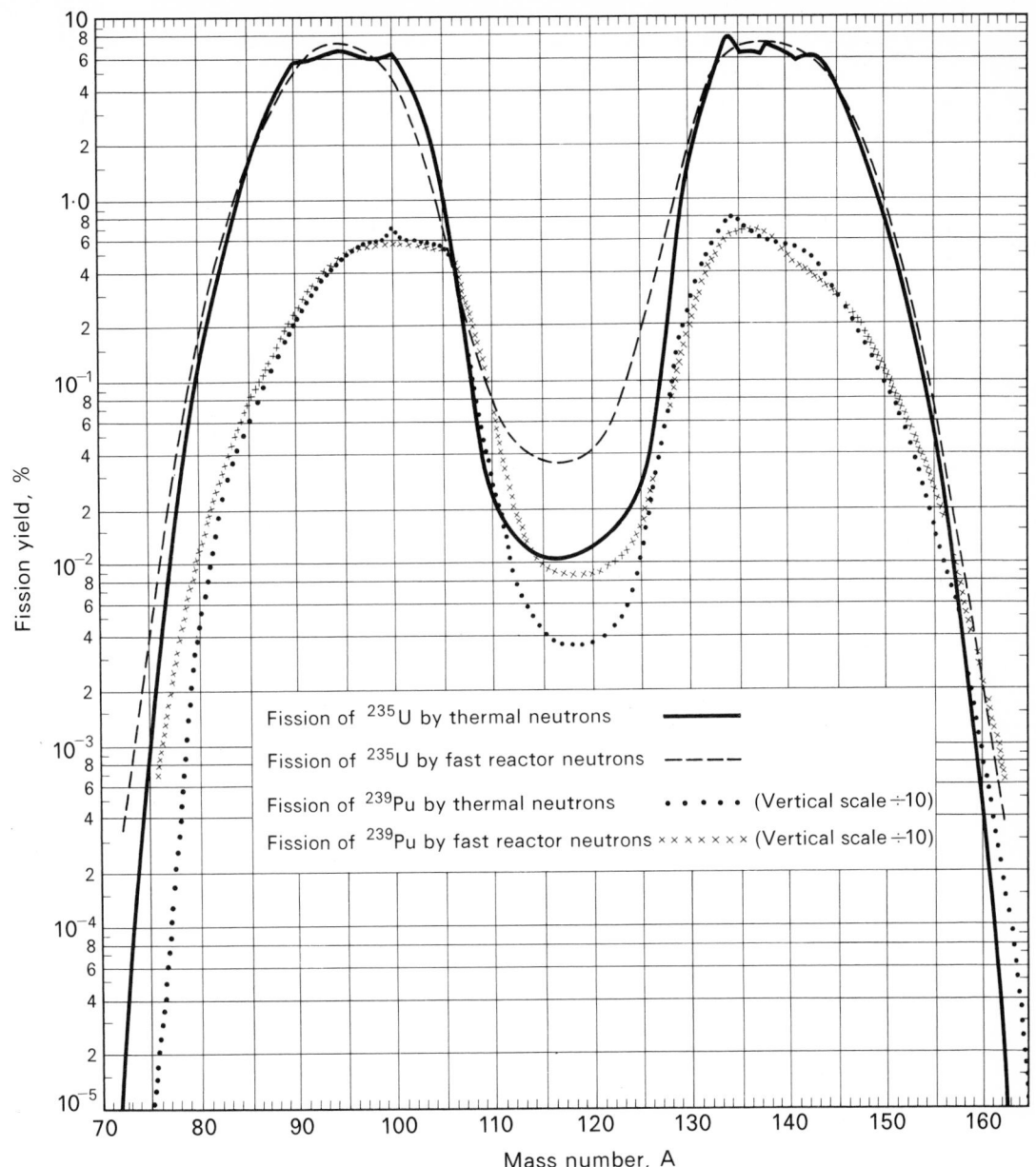

Fission yield for ^{235}U and ^{239}Pu fissioned by thermal and fast reactor neutrons.

shape the fission yield curve is by no means a smooth function of mass. As the neutron energy increases, the valley between the peaks fills in and the wings of the peaks rise, whilst at very high energies, and also for very heavy elements, the mass yield curve becomes symmetrical and single humped.

The products of fission are themselves generally radioactive and the Table of Nuclides (p. 286) should be consulted for their properties.

The average total energy released by slow-neutron induced fission in the fissile materials ^{235}U and ^{239}Pu is given below.

	^{235}U	^{239}Pu
Instantaneously released energy		
Kinetic energy of fission fragments	166 MeV	172 MeV
Kinetic energy of prompt neutrons	5	6
Energy carried by prompt γ-rays	8	8
Energy from decaying fission products		
Energy of β⁻-particles	8	8
Energy of anti-neutrinos	12	12
Energy of delayed γ-rays	7	7
	206	213

In an operating reactor, all of this energy except that carried by the anti-neutrinos is converted into heat. There is, however, an additional source of energy (~ 12 MeV for ^{235}U, ~ 14 MeV for ^{239}Pu) arising from the binding energy released when those prompt neutrons which do not produce fission are finally captured. Alternatively, after a fission in free space there would be a further release of ~ 2 MeV following the β⁻-decay of the neutrons.

References

Nuclear Fission : J. G. Cuninghame, *Encyclopaedia Britannica*, 1973.
Fission Yields: K. F. Flynn and L. E. Glendenin, USAEC Report ANL 7749, 1970.
                E. Sidebotham, UKAEA (Risley) Report 2143(R), 1971.
                M. E. Meek and B. F. Rider, USAEC Report NEDO 12154, 1972.
                E. A. Crouch UKAEA (Harwell) Report R7209, 1972.

J.G.C.

3.7.2 Neutron cross-sections

If $n(E)\,\mathrm{d}E$ is the number of neutrons per unit volume of a material having an energy between E and $E+\mathrm{d}E$, then the neutron flux, φ, is defined as $\varphi(E) = n(E)v$, where v is the neutron velocity. The number of neutrons giving a particular type of interaction per unit volume per unit time in a material containing N nuclei per unit volume is $\int N\phi(E)\sigma_x(E)\,\mathrm{d}E$, where $\sigma_x(R)$ is the nuclear cross-section for the particular type of interaction x. The product $N\sigma_x$ is called the macroscopic cross-section, Σ_x, and its reciprocal can be shown to be the mean free path.

After a neutron enters a nucleus the resulting compound nucleus may emit a charged particle (e.g. a proton or α particle), a γ-ray, a neutron or a pair of neutrons. In heavy materials fission may take place. These reactions are known respectively as (n, p) (n, α) (n, γ) (n, n) (n, 2n) (n, f) events. The (n, γ) reaction is referred to as radiative capture and the corresponding cross-section written σ_γ. The (n, n) interaction indicates scattering. If after the interaction the nucleus remains in the ground state the event is referred to as elastic scattering and the cross-section is written σ_n. If it is left in an excited state it is referred to as inelastic scattering and is denoted by (n, n′) and the cross-section is written $\sigma_{n'}$. The fission cross-section is written σ_f and the total cross-section for all events is written σ_T. The absorption cross-section, σ_A, is defined by $\sigma_A = \sigma_T - \sigma_n - \sigma_{n'}$.

The energy range of interest is from 0.001 eV to 10 MeV and values of $\sigma(E)$ are shown graphically in BNL 325 (available from Office of Technical Services, Department of Commerce, Washington, D.C.) for most elements and for important individual isotopes. Point values are held in a number of nuclear data files prepared at several centres, e.g. the UK Nuclear Data Library (Ref. AEEW–M824 available from HM Stationery Office, London) gives a point by point energy representation (up to 4000 points) covering the range 0.0001 eV to 15 MeV and including all the important cross-sections of 88 nuclides. In Figures A, B, C, D, E and F a selection of cross-sections is shown. Certain features may be distinguished:

At low energies absorption cross-sections tend to be proportional to $1/v$ (i.e. to $E^{-1/2}$) and for ^{10}B and ^{6}Li (n, α) reactions this law is good to 5% in the region up to about 100 keV (see Figure A).

For light nuclides scattering is dominated by potential scattering which can be considered as the deflection of a neutron by the nuclear field without the formation of a compound nucleus, and the cross-section tends to be independent of energy. Departures, however, arise at low energies in crystalline materials (e.g. C and Be) when the neutron wavelength is comparable to the lattice spacing, and in molecular compounds (H_2, H_2O) when the collision energy is comparable to the vibrational quanta of the molecular oscillations (see Figure B).

At intermediate energies the cross-sections show sharp maxima at neutron energies at which the reacting particles have the same energy (including the binding energy) as one of the resonance levels in the compound nucleus. Light nuclei have larger energy level spacings than heavy nuclei (cf. Figures C and D). It is thus more probable to find resonances at low energies in heavy nuclei and such resonances can be seen in Figures A and B for U and Pu.

For heavy nuclei having an odd number of neutrons (e.g. ^{235}U) fission is possible at all energies and such materials are called fissile (see Figures A, D and E).

For heavy nuclei having an even number of neutrons (e.g. ^{238}U) fission is only significant above a threshold energy (see Figure F). Since neutron capture in such materials produces a fissile material they are called fertile.

All (n, n$'$) and (n, 2n) reactions have threshold energies. In some light nuclides (n, p) and (n, α) interactions have appreciable cross-sections at all energies but in the heavier nuclides these cross-sections always show an effective threshold at an MeV or so, even when the reaction is theoretically possible with low energy neutrons (see Figure F).

Because of the strong dependence of cross-section on energy it is usually necessary in practical calculations to make use of group cross-sections which are values averaged over a chosen energy region and weighted according to the energy spectrum in the system, i.e.

$$\bar{\sigma} = \frac{\displaystyle\int_{E_1}^{E_2} \sigma(E)\varphi(E)\,\mathrm{d}E}{\displaystyle\int_{E_1}^{E_2} \varphi(E)\,\mathrm{d}E}$$

For a general comparison of the interaction of neutrons in various materials it is often appropriate to consider three energy regions.

At high energies the cross-sections can be averaged over a fission neutron spectrum which is conventionally expressed as $\varphi_f = (E/b^2)\exp(-E/b)$, where $b = 2.0$ MeV. In the tables below the average values have been taken over the range 1 keV to 20 MeV although most of the contribution comes from a much narrower range around 2 MeV.

At intermediate energies the spectrum in a well-moderated system is nearly proportional to $1/E$ and the average cross-section is equal to:

$$\frac{\displaystyle\int_{E_1}^{E_2} (\sigma/E)\,\mathrm{d}E}{\ln(E_2/E_1)}$$

The numerator is referred to as the resonance integral (R.I.) since it gives the cumulative absorption in the various resonances as the neutron slows down from energy E_2 to E_1. For absorption processes it is customary to quote the resonance integrals rather than average values, and the tables show values where $E_2 = 0.55$ eV and $E_1 = 2$ MeV although the values are usually insensitive to the upper limit.

At lower energies, when the neutrons are in thermal equilibrium with their surroundings, $\varphi_{th} = E \exp\left(-E/kT_n\right)/(kT_n)^2$, where T_n is the temperature of the surroundings. If the cross-section varies as $1/v$ then the average cross-section in a thermal flux can be shown to be equal to $\sigma_0(\pi T_0/4 T_n)^{1/2}$, where σ_0 is the cross-section corresponding to an energy kT_0. It is customary to refer cross-sections to the value σ_0 at a standard neutron velocity of 2200 m s^{-1}, which corresponds to a neutron energy kT_0 of 0.02530 eV and a temperature of 293.59 K. Values of σ_0 have been given in the Table of Nuclides (p. 287) and values averaged over a thermal spectrum with $T_n = T_0$ are given in the following tables. The departure of these average values from the value of $\sigma_0(\pi/4)^{1/2}$ is indicative of the departure of the particular cross-section from the $1/v$ characteristic. The integrations were taken over the range 0.0001 eV to 1.0 eV although most of the contribution comes from a narrower range around 0.025 eV.

The table gives properties of nuclides which have been grouped according to their function in nuclear reactors. Cross-sections are given in barns (10^{-28} m^2). For fissionable materials the number of neutrons produced per fission and denoted by ν are given as average values over the three spectral regions.

Neutrons can be detected by both active and passive devices. Active detectors include fission chambers, BF$_3$ ion chambers, proton recoil counters and their response to neutrons is determined by the appropriate cross-section for the particular interaction employed in the device. The table of neutron cross-sections for reactor materials includes materials used in these detectors. The passive detectors contain materials which become radioactive when irradiated in a neutron flux and on

(continued on p. 339)

Activation cross-section for neutron foil detectors

Nuclide and reaction	Half-life of principal activity	Principal resonance energy (R) or mean response energy in fission spectrum (T.D.)/eV	Activation cross-section/b		
			Average over thermal spectrum	Resonance integral	Average over fission spectrum
^{164}Dy(n, γ)^{165}Dy .	2.35 h	−1.89 (R)	2100†	300†	98 m†
^{176}Lu(n, γ)^{177}Lu .	6.74 d	0.142 ,,	3120†	900†	
^{103}Rh(n, γ)^{104m}Rh .	4.4 min	1.26 ,,	9.8†	83†	15 m†
^{115}In(n, γ)^{116m}In .	54 min	1.46 ,,	142†	2600†	120 m†
^{197}Au(n, γ)^{198}Au .	2.7 d	4.9 ,,	87.9†	1560†	97 m
^{152}Sm(n, γ)^{153}Sm .	47 h	8.0 ,,	190†	3000†	
^{107}Ag(n, γ)^{108}Ag .	2.3 min	16.5 ,,	27†	74†	96 m†
^{186}W(n, γ)^{187}W .	23.7 h	18.8 ,,	37†	500†	41 m†
^{59}Co(n, γ)^{60}Co .	5.27 yr	130 ,,	33.2	71	3.1 m
^{55}Mn(n, γ)^{56}Mn .	2.58 h	337 ,,	11.7	14.1†	3.8 m
^{63}Cu(n, γ)^{64}Cu .	12.8 h	580 ,,	4.0	3.2	12 m
^{23}Na(n, γ)^{24}Na .	15.4 h	2.8 k ,,	0.47	0.31	0.26 m
^{115}In(n, n′)^{115m}In .	4.5 h	2.8 M (T.D.)	0	70 m	173 m
^{31}P(n, p)^{31}S .	2.63 h	3.7 M ,,	0	0	34 m
^{32}S(n, p)^{32}P .	14.2 d	4.2 M ,,	0	0.087 m	65 m
^{58}Ni(n, p)^{58}Co .	71.3 d	4.3 M ,,	0	9.7 m	109 m
^{27}Al(n, p)^{27}Mg .	9.4 min	5.8 M ,,	0	0	3.8 m
^{56}Fe(n, p)^{56}Mn .	2.58 h	7.4 M ,,	0	0	0.81 m
^{27}Al(n, α)^{24}Na .	15 h	8.5 M ,,	0	0	0.62 m
^{63}Cu(n, 2n)^{62}Cu .	12.8 h	13.7 M ,,	0	0	0.05 m

(R) = Resonance Detector.   (T.D.) = Threshold Detector.   D.J.
m indicates millibarns.
k and M represent keV and MeV respectively.
All values from UK Nuclear Data Library except those marked †.

Neutron cross-sections

Nuclide or element	Average over thermal spectrum				Slowing-down region				Average over fission neutron spectrum				
	σ_n/b	σ_γ/b	σ_f/b	$\bar\nu$	σ_n/b	Capture R.I./b	Fission R.I./b	$\bar\nu$	σ_n/b	σ_γ/b	σ_f/b	$\sigma_{n'}$/b	$\bar\nu$
Fissile materials													
^{233}U	11.6†	42.2†	472.6†	2.487‡	10.3	144	740	2.487	4.44	0.043	1.89	1.08	2.673
^{235}U	17.0†	86.3†	502.2†	2.432‡	10.7	139	264	2.424	4.73	0.081	1.24	1.46	2.658
^{239}Pu	8.6†	271.9†	693.2†	2.880‡	11.5	171	280	2.882	4.41	0.049	1.78	1.79	3.169
^{241}Pu	12.0†	328.7†	936.1†	2.934‡	10.8	165	555	2.934	4.76	0.040	1.66	1.33	3.24
Fertile materials													
^{238}U	7.94	2.42			19.4	269	0.17	2.62	5.06	0.087	0.300†	2.38	2.81
^{240}Pu	2.12	256			57.1	8011	2.41	3.055	5.26	0.0907	1.24	1.40	3.30
^{232}Th	12.1	6.53			10.6	88†	0.035	2.28†	4.75	0.12	0.074	2.4	2.28†
^{234}U					11.4	665†	1.78	2.46	5.14	0.028	1.15	1.46	2.62
^{236}U			15		11.3	320†	0.51	2.60	5.09	0.17	0.57	1.94	2.77
^{238}Pu	13.8	323			11.1	150†	72	2.88	4.7	0.079	2.19	0.63	3.11
^{233}Pa	11.0	36.8			11.1	918	1.3	2.62	5.0	0.12	0.96	2.17	2.78
^{237}Np		170†			11.1	720†	2.1				1.35		
										σ_A/b			
Cladding and structural materials													
Al	1.4	0.20			2.3	0.18			2.9	4.8 m		0.29	
Si	2.1	0.16†			2.5	0.010†			3.0	9.3 m		0.15	
Cr	4.1	2.7			5.1	1.4			2.8	3.1 m		0.50	
Mn	1.7	11.7			41	14.1†			2.8	5.8 m		0.92	
Fe	11.3	2.26			10.3	1.28			2.5	7.9 m		0.62	
Co		33.2				71				3.0 m			
Ni	17.4	3.42			16	2.03			3.1	8.5 m		0.42	
Cu	7.0	3.3			8.4	4.1			2.9	28 m		0.74	
Zr	6.2	0.16			7.8	1.0			5.3	6.3 m		0.56	
Nb	5.7	1.0			6.9	11			4.5	30 m		1.3	
Mo	4.5	2.4			7.9	38			4.6	30 m		1.3	
Ta	5.1	18			15	720			4.1	100 m		2.5	
W	5.0	16			6.6	373†			4.6	51 m		2.0	
Pb	8.6	0.15			10	0.20			5.6	2.9 m		0.70	

	Nuclide or element	Average over thermal spectrum				Slowing-down region				Average over fission neutron spectrum				
		$\bar{\sigma}_n$/b	$\bar{\sigma}_A$/b	$\bar{\sigma}_f$/b	$\bar{\nu}$	σ_n/b	Absorption R.I./b	Fission R.I./b	$\bar{\nu}$	$\bar{\sigma}_n$/b	$\bar{\sigma}_\gamma$/b	$\bar{\sigma}_f$/b	$\bar{\sigma}_{n'}$/b	$\bar{\nu}$
Components of coolants	H in H$_2$O	45.3	0.294			17.2	0.1416			4.00	0.039 m		0	
	D in D$_2$O	5.10	0.46 m			3.27	0.24 m			2.56	0.007 m		0	
	He	0.73	6.1 m			1.12	3.0 m			3.66	0.001 m		0	
	C	4.69	3.0 m			4.42	1.5 m			2.35	0.62 m		0.011	
	N	11	1.69			7.7	0.90			1.9	0.13		5.8 m	
	O	3.35	0.28 m			3.74	0.13 m			2.8	7.1 m		3.1 m	
	Na	3.3	0.47			8.8	0.31			2.72	2.1 m		0.48	
	K	2.2	1.8			2.2	120			2.6	0.1		0.10	
Absorbers and poisons	^{6}Li	0.72	833			0.99	495							
	Li	1.04	63			1.2	30							
	^{10}B	2.13	3398			2.3	1640							
	B	4.4	672			4.2	325							
	Cd	10.1	2900			6.8	61							
	^{135}Xe	3.8 × 10^5	2.7 × 10^6			560	5890							
	Hf	8†	84†			16†	2000†							

m indicates that the cross-section is in mb (10^{-31} m^2).

From UK Nuclear Data Library except values marked †.

Values marked ‡ are recommended thermal cross-sections from the 1969 I.A.E.A. fit (G. C. Hanna et al., Atomic Energy Review, **7**, No. 4, p. 3).

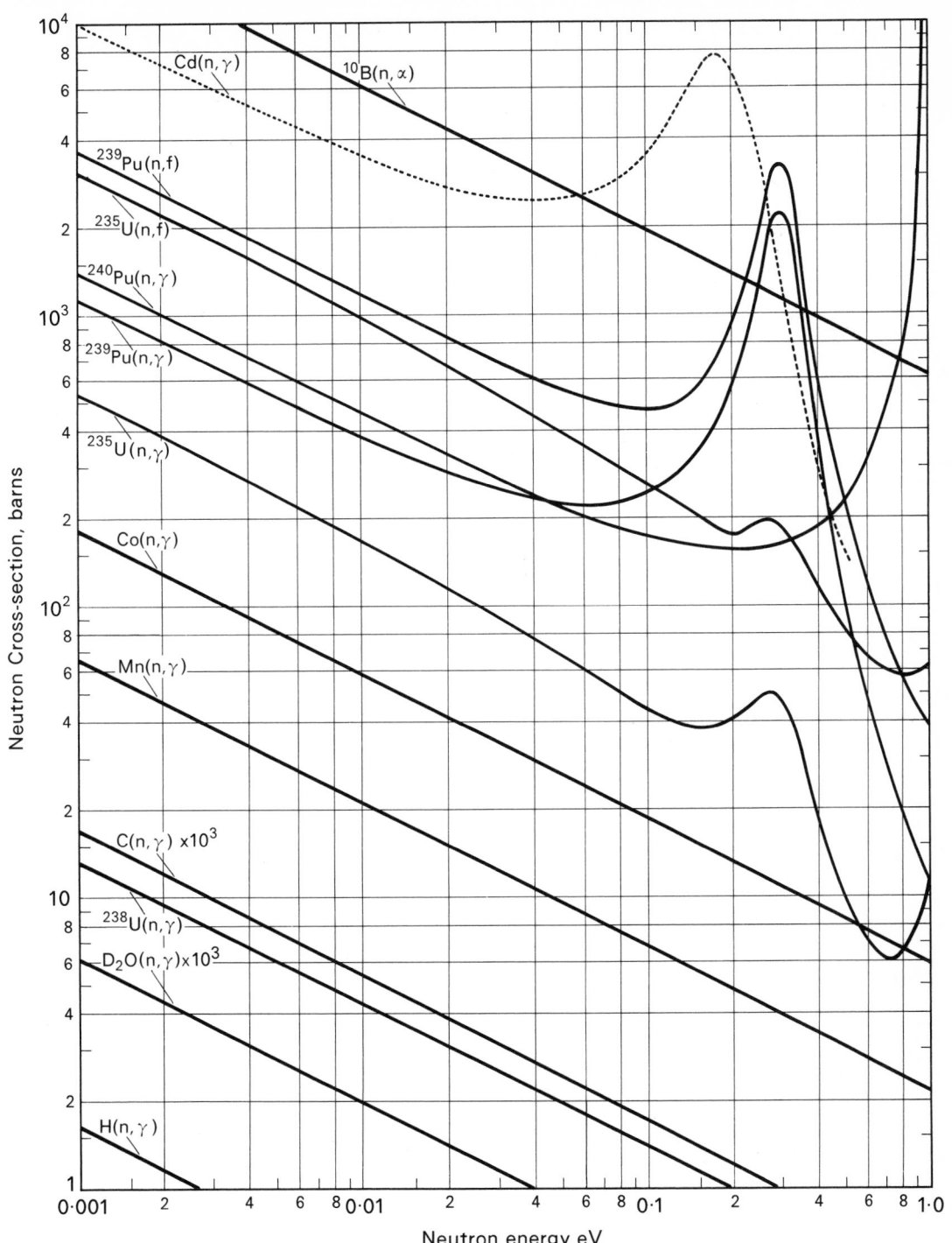

A   Absorption, capture and fission cross-sections in thermal region

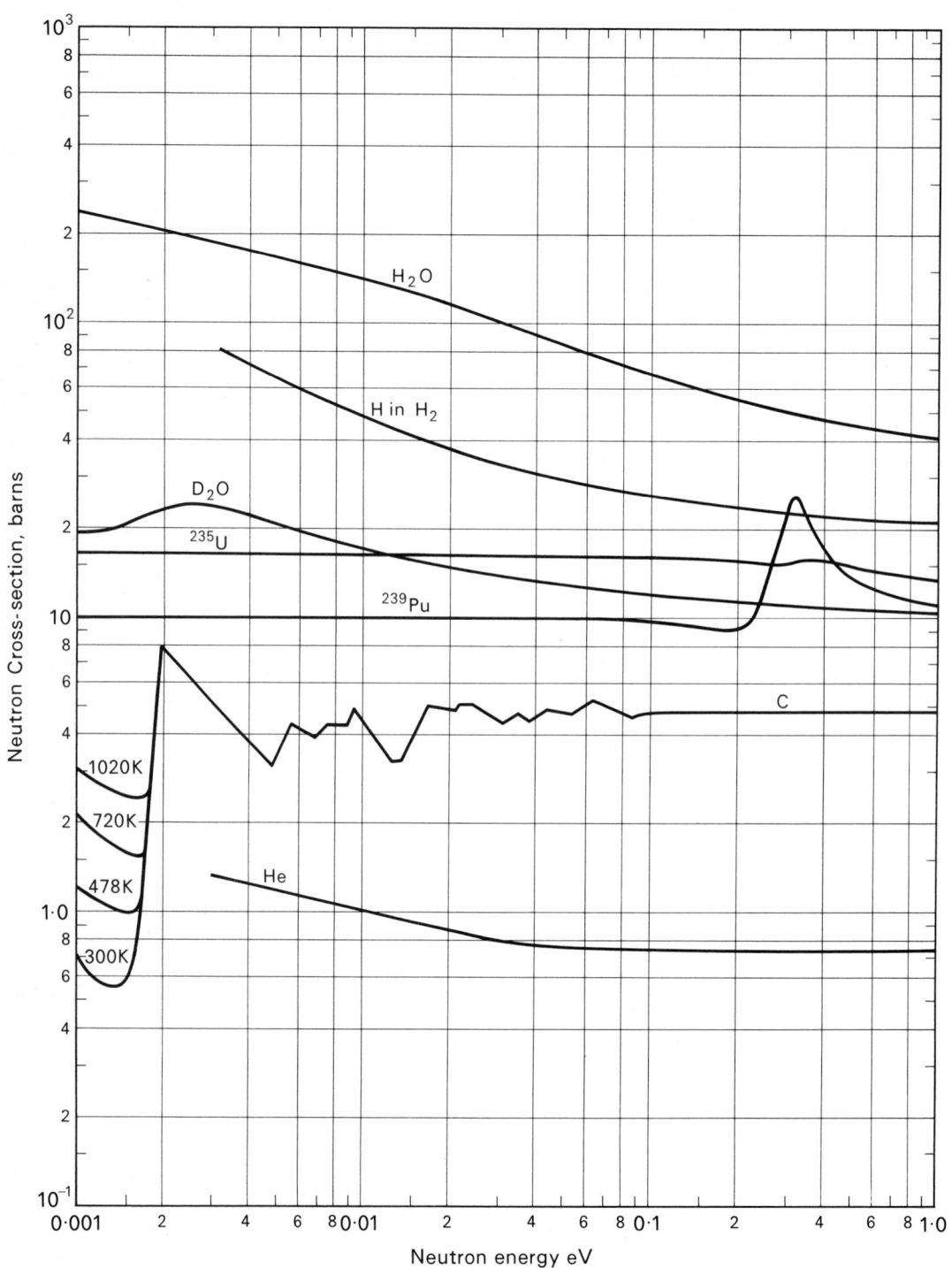

B   Elastic scattering cross-sections in thermal region

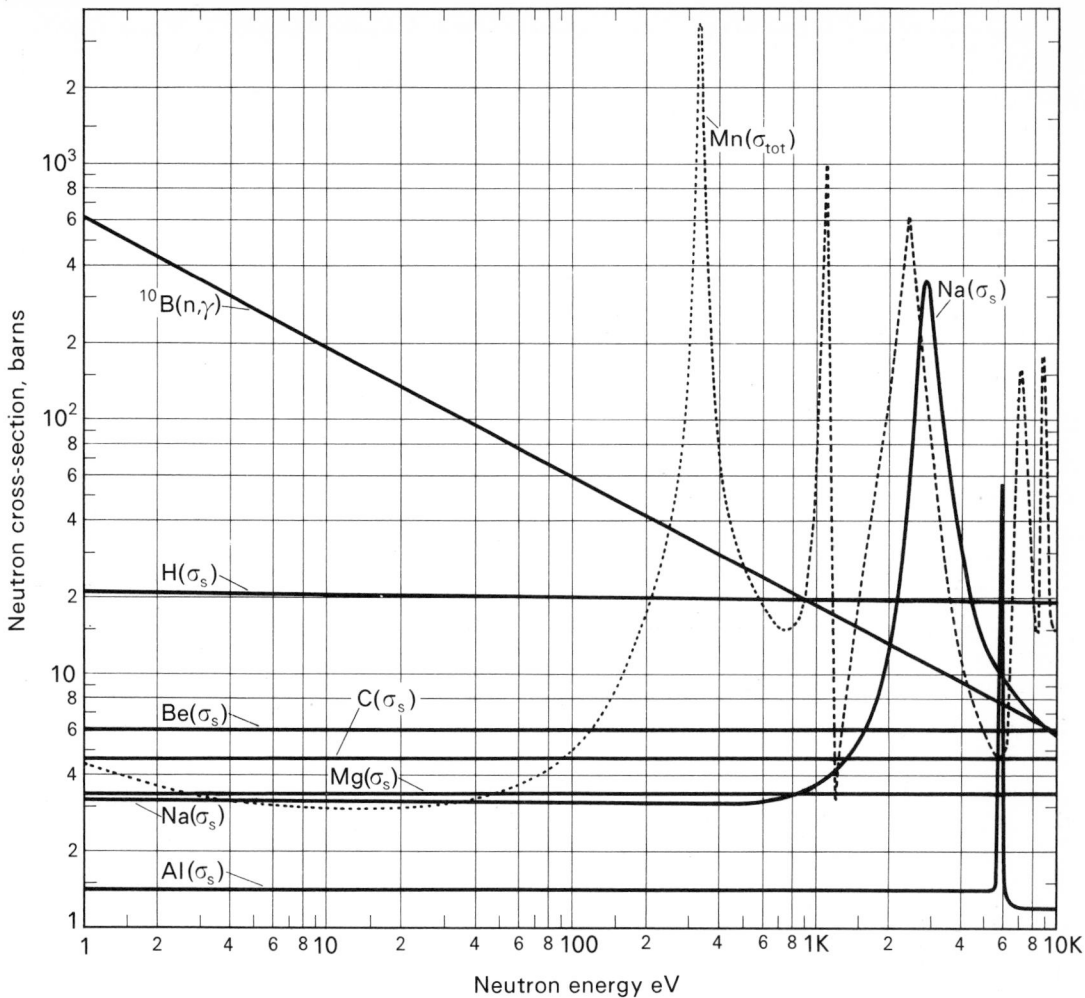

C Cross-section of light nuclides in the slowing-down region

removal from the flux the radioactivity can be measured. The activity $A(t)$ (particles s^{-1}) at a time t seconds after the removal of a thin foil from a neutron flux having an energy spectrum $\varphi(E)$ in which the foil has been irradiated for T seconds is given by

$$A(t) = \frac{V_\rho N_A}{A} (1 - \exp(-\lambda T)) \exp(-\lambda t) \int \sigma_{act}(E) \varphi(E) \, dE$$

where V, ρ and A are the volume, density and atomic weight of the foil material respectively, N_A is Avogadro's constant, λ (s^{-1}) is the decay constant of the radioactive species produced by the irradiation and $\sigma_{act}(E)$ is the activation cross-section of the material. Values of σ_{act} averaged over the thermal and fission energy regions are given in the table together with the activation resonance integral. The materials listed either have a predominant resonance or a threshold and may be used to detect neutrons over a narrow energy region. Additional decay properties of the isotopes listed may be found in the Table of Nuclides (p. 286).

The values in both tables are mostly taken from the UK Nuclear Data Library and the average values have been computed by A. L. Pope.

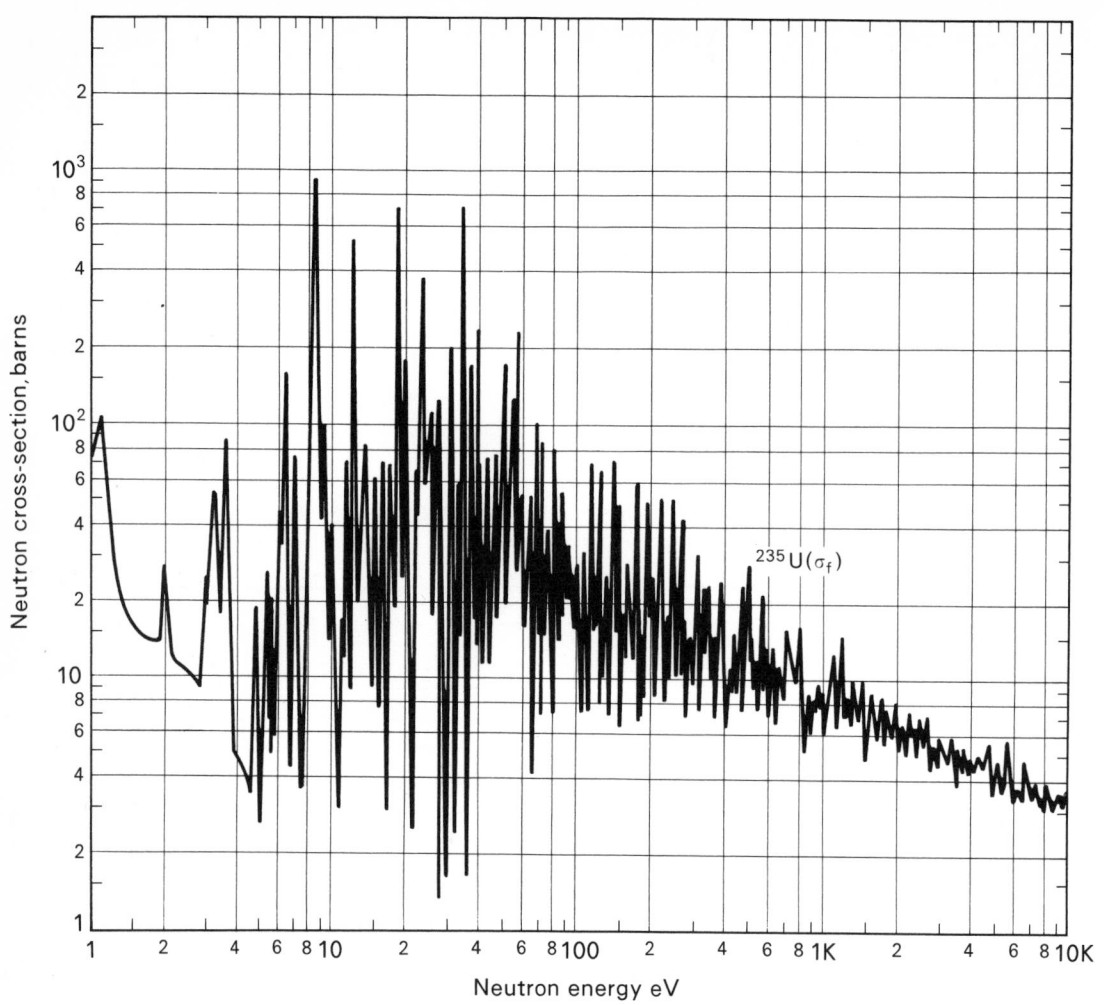

D   Cross-section of ^{235}U in slowing-down region

(Above 20 eV the resonances are too numerous to be shown correctly on this scale.)

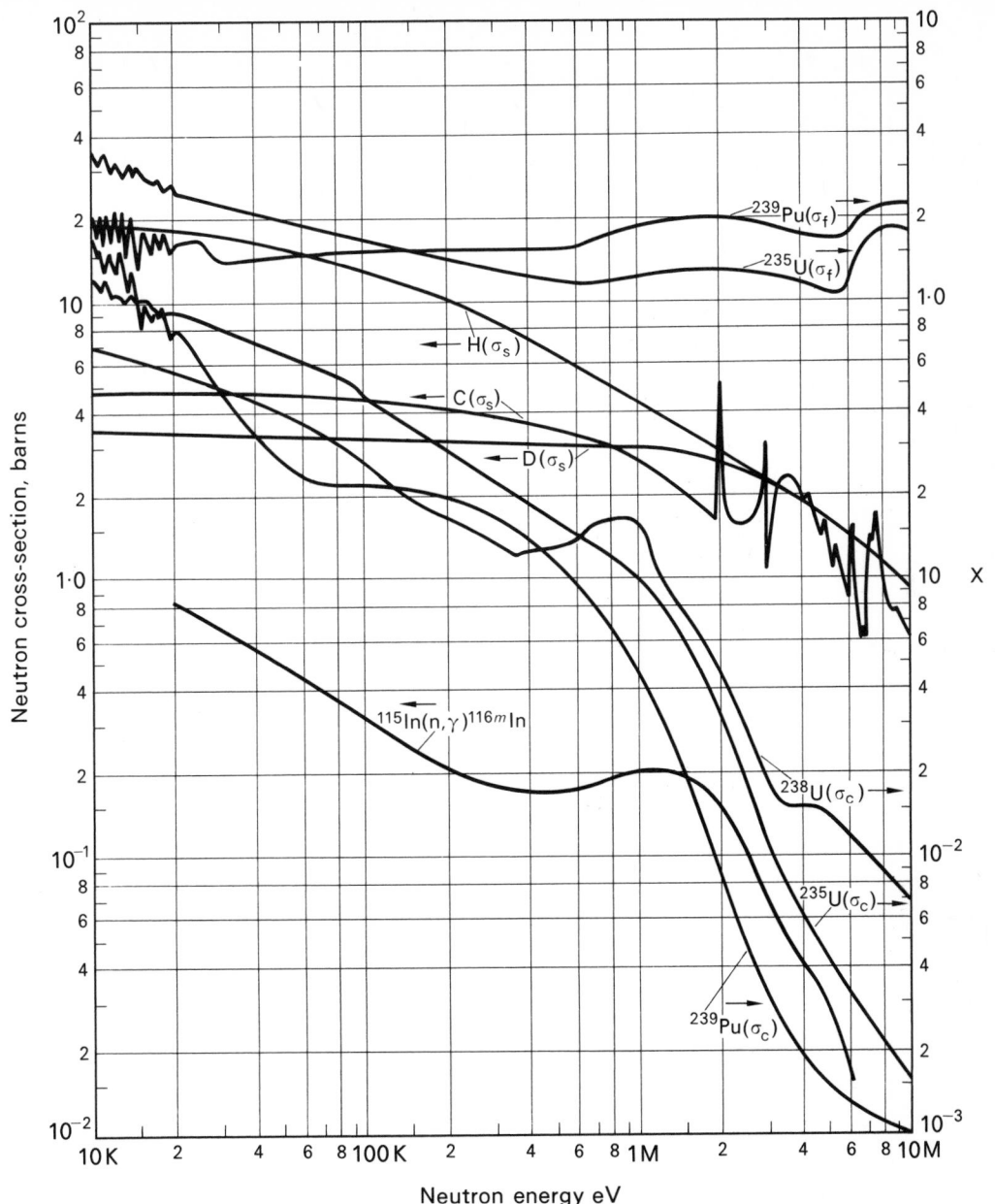

E   Non-threshold cross-sections in fast region
(Note different scales on left and right.)

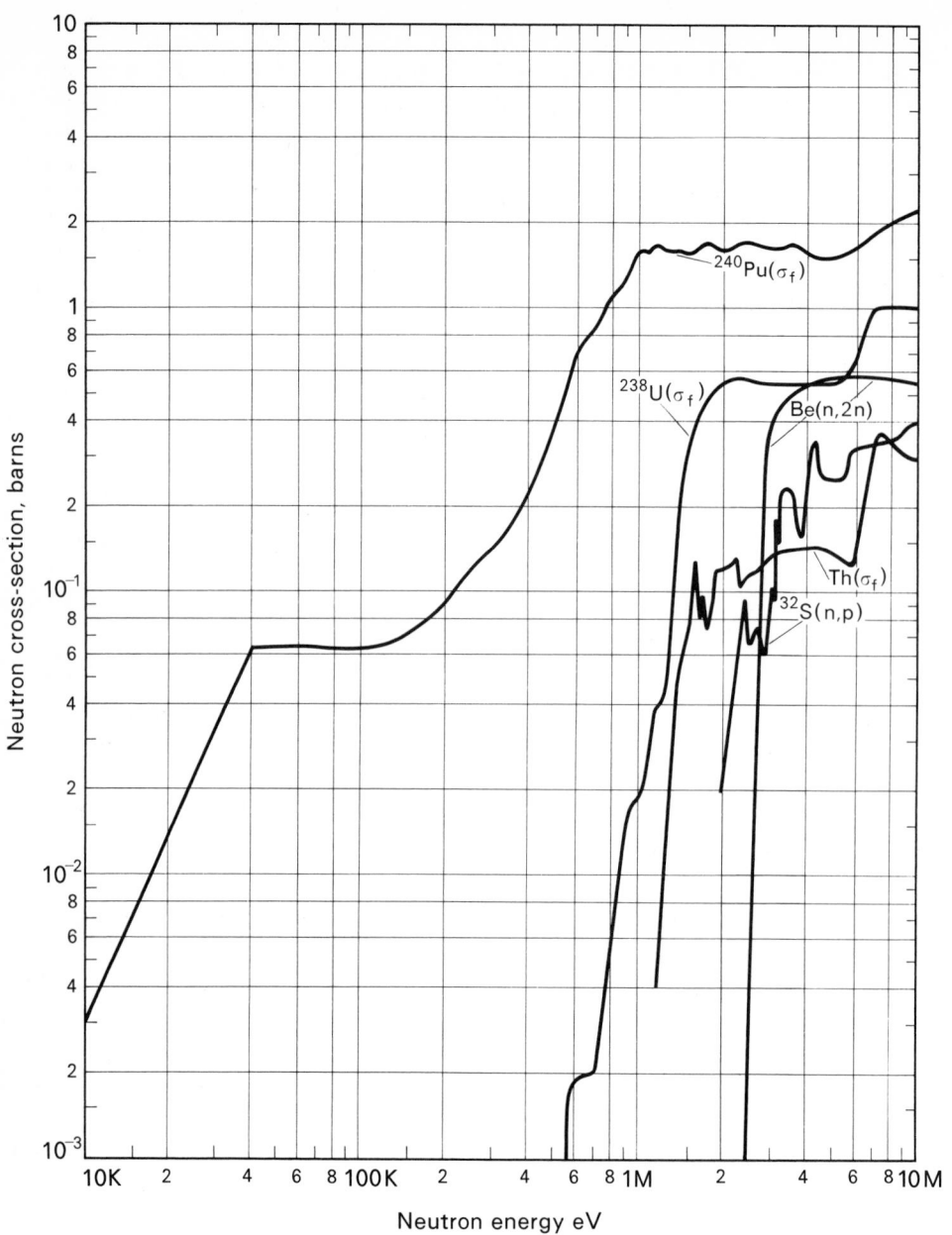

F   Threshold reactions in fast region

D.J.

3.7.3 Attenuation of fast neutrons: neutron moderation and diffusion

In materials containing atoms of low atomic mass, neutrons of all energies can lose a significant fraction of their energy in a single elastic collision and such materials are referred to as moderators. In heavy nuclei appreciable energy loss in a collision is only possible at high energies where inelastic scattering can occur. The neutron dose rate from a point source of fast neutrons falls off with distance r

approximately as $\exp(-\Sigma_{rem} r)/4\pi r^2$, where Σ_{rem} depends on the medium. This macroscopic cross-section is called the removal cross-section and since all interactions tend to remove energy from the beam its value is not too different from the total macroscopic section $(N\sigma_{tot})$ of the material, but is slightly lower. This exponential fall off is only approximate and holds less well for media in which hydrogen is the principal fast neutron attenuator. In the following table the removal cross-section refers to a fission neutron source.

In the slowing-down region the average number of collisions, $\bar{n}$, to slow a neutron from energy E_1 to energy E_2 is equal to $\ln(E_1/E_2)/\xi$, where ξ is the average change per collision in the logarithm of the energy. At energies below that at which scattering becomes entirely elastic, ξ is independent of energy and is approximately equal to $2/(A+\frac{2}{3})$. The spatial distribution of neutrons of energy E_2 which have slowed down from a point source of energy E_1 is of the form $\exp(-r^2/4\tau)^2$ where τ is referred to as the Fermi Age and is the mean square distance a neutron migrates in slowing down from E_1 to E_2. It is given by:

$$\tau = \int_{E_2}^{E_1} \frac{D \, dE}{E \xi \Sigma_s}$$

where D is the diffusion coefficient and equal to $(3\Sigma_{tot} - 3b\Sigma_s)^{-1}$ and b is the average value of $\cos\psi$ where ψ is the angle of scatter of a neutron in a collision. The table refers to the age of neutrons from a fission source slowing down to an energy of 1.46 eV. This value, which is just above the thermal region, is appropriate to the age determined from the measured spatial distribution of the resonance neutrons detected by indium foils.

The root mean square distance a neutron travels from the position where it is thermalized to the point where it is absorbed is the thermal diffusion length, L, and is equal to

$$(4/\pi)^{1/4} (D_{th}/\Sigma_a(kT_n))^{1/2}$$

where D_{th} is the value of the diffusion coefficient averaged over the thermal neutron spectrum and Σ_a is assumed to have a $1/v$ dependence and is evaluated at an energy kT_n where T_n is the temperature of the medium.

Properties of moderators and shielding materials†

(At 20 °C unless stated otherwise)

Material	$\dfrac{Density}{10^3 \text{ kg m}^{-3}}$	$\Sigma_{rem}/\text{m}^{-1}$	ξ	$\tau/(10^3 \text{ mm}^2)$	L/mm	D_{th}/mm	$t_s/\mu s$	$t_{th}/\mu s$	$\bar{n}$
H_2O	1.00	9.0[a]	0.948	2.67[c]	27[a]	1.4[a]	6	205	20
D_2O (pure).	1.10	9.1[a]	0.570	11.7[c]	940	8.4	53	~10⁵	33
Diphenyl ($C_{12}H_{10}$) 85 °C .	0.99	7.1[b]	0.812	4.6[d]	48	2.6	13	354	23
Paraffin Wax ($C_{30}H_{62}$) .	0.89	10.9[b]	0.913	1.8	21	1.1	7	160	21
Be	1.85	13.0[a]	0.209	7.32[e]	208[a]	5.0[a]	50	3460	90
BeO	3.00	14.3[b]	0.173	9.38[e]	290[a]	5.0[a]	102	7 000	109
Graphite	1.67	8.1[a]	0.158	29.8[c]	520[a]	8.5[a]	140	13 000	119
Concrete‡	2.3	8.8[b]	0.55	10.0	770	6.0	30	400	30
Al	2.70	7.9[a]	0.072	430	200	55	900	8 800	262
Fe	7.86	16.8[a]	0.035	33.0	12.7	3.4	360	19	540
Pb	11.35	11.6[a]	0.0096	600	121.8	9.2	2720	640	1960
Bi	9.75	9.8[a]	0.0095	800	320	11.2	3000	3 660	1990
U	18.9	1.7[a]	0.0084	§	13.7	7.0	2040	11	2250

† Data obtained from various sources: a, experimental value; b, derived from components; c, UK Nuclear Data File; d, taken from ANL 5800; e, ENDF/B data (this is a US data set—see BNL 50066). Values not marked are obtained from approximate formulae.

‡ Composition in 10^3 kg m^{-3}: H, 0.023; O, 1.22; C, 0.0023; Mg, 0.005; Al, 0.078; Si, 0.775; K, 0.03; Ca, 0.1; Fe, 0.03; Na, 0.037.

§ Because of its large absorption resonance integral and its small value of ξ, almost no neutrons slowing down in uranium reach thermal energies.

The slowing down time, t_s, is the time taken for neutrons to slow down from an energy E_s to the thermal energy E_0 and is independent of E_s when $E_s \gg E_0$. The mean thermal neutron lifetime, t_{th}, refers to thermal diffusion before the neutron is absorbed in the medium. In the table values of t_s and t_{th} have been obtained from the simple formulae $t_s = 2\overline{(1/\xi\Sigma_s)}/v_0$ and $t_{th} = 1/\overline{(\Sigma_a v)}$ where the averages are taken over the slowing-down and thermal regions respectively.

D.J.

3.8 Nuclei and particles

3.8.1 Size of atomic nuclei

The majority of atomic nuclei are approximately spherical in shape, and it is sufficient for many purposes to use spherically symmetrical distributions to describe them. The mean square radius of the matter distribution is given approximately by $r_0 A^{1/3}$, where $r_0 \simeq 1.2$ fm and A is the mass number. For medium and heavy nuclei the density distribution ρ as a function of distance r from the nuclear centre is customarily approximated by the form

$$\rho = \rho_0 \left(1 + \exp\left(\frac{r-c}{a}\right)\right)^{-1}$$

where c, the radius at which the density drops to $\frac{1}{2}\rho_0$, is $\simeq 1.1 A^{1/3}$ fm, the nuclear surface thickness parameter $a \simeq 0.6$ fm is little dependent on nuclear size, and $\rho_0 \simeq 0.17$ nucleon fm^{-3}.

See A. Bohr and B. Mottelson, *Nuclear Structure*, 1969, Vol. 1 (Benjamin, New York).

B.R.

3.8.2 Stable particles

These data are restricted to particles immune to decay via the strong interaction and are derived from the Particle Data Group in *Rev. Mod. Phys.*, 1971, **43**, No. 2, part II, which contains a considerable expansion of the data of the table together with data on unstable mesons and baryons. At present these data are updated annually by the Particle Data Group.

Stable particles

Only decay modes having branching ratios in excess of 10% are shown.

The observed mean life of a particle in flight is longer than its mean life at rest by a factor

$$(1 - \beta^2)^{-1/2} = 1 + \frac{\text{Kinetic energy}}{\text{Rest energy}}$$

Name and symbol		Charge	Spin	Strangeness	Mass/MeV	Mean life/s	Principal modes of decay
Photon	γ	0	1	0	0	Stable	Stable
Neutrino	ν	0	$\frac{1}{2}$	0	$< 6 \times 10^{-5}$	Stable	Stable
Electron	e	$\pm e$	$\frac{1}{2}$	0	0.511 0041 $\pm$0.000 0016	Stable $\tau_{e^-} > 6 \times 10^{28}$	Stable
Mu-meson	μ	$\pm e$	$\frac{1}{2}$	0	105.6659 $\pm$0.0014	2.1983×10^{-6} $\pm$0.0008	$e^{\pm}\nu\bar{\nu}$
Pi-meson	$\pi^{\pm}$	$\pm e$	0	0	139.576 $\pm$0.011	2.6024×10^{-8} $\pm$0.0024	$\mu^{\pm}\nu$
	π^0	0	0	0	134.972 $\pm$0.012	0.84×10^{-16} $\pm$0.10	$\gamma\gamma$

Stable particles (*contd*)

Name and symbol		Charge	Spin	Strangeness	Mass/MeV	Mean life/s	Principal modes of decay
K-mesons	K	$\pm e$	$\frac{1}{2}$	± 1	493.84 ± 0.11	1.2371×10^{-8} ± 0.0026	$\mu^{\pm}\nu$, $\pi^{\pm}\pi^0$
	K_S^0	0	$\frac{1}{2}$			0.862×10^{-10} ± 0.006	$\pi^+\pi^-$, $\pi^0\pi^0$
	K_L^0	0	$\frac{1}{2}$	$\begin{cases} \sim 50\% +1 \\ \sim 50\% -1 \end{cases}$	$\left\} 497.79 \atop \pm 0.15 \right.$	5.172×10^{-8} ± 0.043	$\pi^0\pi^0\pi^0$, $\pi^{\pm}\mu^{\mp}\nu$ $\pi^+\pi^-\pi^0$, $\pi^{\pm}e^{\mp}\nu$
Eta	η	0	0	0	548.8 ± 0.6	2.50×10^{-19} ± 0.56	$\gamma\gamma$, $\pi^0\pi^0\pi^0$ $\pi^+\pi^-\pi^0$
Proton	p	$+e$	$\frac{1}{2}$	0	938.2592 ± 0.0052	Stable $\tau_p > 6 \times 10^{35}$	Stable
Neutron	n	0	$\frac{1}{2}$	0	939.5527 ± 0.0052	0.932×10^3 ± 0.014	$pe^-\bar{\nu}$
Hyperons	Λ	0	$\frac{1}{2}$	-1	1115.59 ± 0.06	2.517×10^{-10} ± 0.024	$p\pi^-$ $n\pi^0$
	Σ^+	$+e$	$\frac{1}{2}$	-1	1189.42 ± 0.11	0.800×10^{-10} ± 0.006	$p\pi^0$ $n\pi^+$
	Σ^-	$-e$	$\frac{1}{2}$	-1	1197.37 ± 0.07	1.489×10^{-10} ± 0.022	$n\pi^-$
	Σ^0	0	$\frac{1}{2}$	-1	1192.51 ± 0.10	$< 1.0 \times 10^{-14}$	$\Lambda\gamma$
	Ξ^-	$-e$	$\frac{1}{2}$	-2	1321.31 ± 0.17	1.660×10^{-10} ± 0.037	$\Lambda\pi^-$
	Ξ^0	0	$\frac{1}{2}$	-2	1314.7 ± 0.7	3.03×10^{-10} ± 0.18	$\Lambda\pi^0$
	Ω^-	$-e$	$\frac{3}{2}$	-3	1672.5 ± 0.5	1.3×10^{-10} $+0.4$ -0.3	$\Xi^0\pi^-$, $\Xi^-\pi^0$ ΛK^-

C.W.

Mathematical Tables

4.1 Functions

The main purpose of this book is to tabulate values of physical and chemical constants. For convenience we also include short tables of some elementary mathematical functions useful in calculations. The following pages contain tables of logarithms and antilogarithms, reciprocals and squares, trigonometrical functions, degrees to radians, and exponentials. The majority of these are four-figure tables. References to more extensive tables are given below.

In addition to their uses in numerical calculations, exponential and trigonometric functions appear as the solutions of linear differential equations with constant coefficients. Many of the other functions of mathematical physics also appear as the solutions of differential equations, often in the determination of fields (electromagnetic, hydrodynamic or gravitational, for example) with given boundary conditions. Although direct numerical solution with the aid of a computer is increasingly used, there are still many occasions when it is convenient to express the solution in terms of familiar tabulated functions. *Bessel functions* occur in the solution of potential problems in two dimensions, or in three dimensions with cylindrical symmetry, and in a great variety of other problems. In three-dimensional potential problems involving spheres—the Earth's gravitational and magnetic fields, for example—it is convenient to work in terms of *surface harmonics* and *Legendre functions*, as on pp. 134 and 136. *Elliptic functions* and related integrals occur in the solution of certain non-linear differential equations. Space does not permit tabulation of these functions here, but tables where they can be found are included in the references below.

Certain constants and their logarithms

Constant	Value	Logarithm	Constant	Value	Logarithm
π	3.141 59	0.497 15	$\sqrt{3}$	1.732 05	0.238 56
π^2	9.869 60	0.994 30	$\sqrt{10}$	3.162 28	0.500 00
$1/\pi$	0.318 31	$\bar{1}$.502 85	e	2.718 28	0.434 29
$1/\pi^2$	0.101 32	$\bar{1}$.005 70	$\log_e 10$	2.302 59	0.362 22
$\sqrt{\pi}$	1.772 45	0.248 57	1 rad	57.295 78°	1.758 12
$\sqrt{2}$	1.414 21	0.150 51	1°	0.017 45 rad	$\bar{2}$.241 88

References

(1) J. Pryde (ed.), *Chambers's Seven-Figure Mathematical Tables*, 1930 (Chambers). Includes logarithms to base 10 and natural and logarithmic trigonometrical functions.

(2) Jahnke and Emde (revised by Lösch), *Tables of Higher Functions*, 1960 (McGraw-Hill). Includes, among others, gamma function, exponential, sine and cosine integrals, elliptic functions, orthogonal polynomials, Legendre functions, Bessel functions, Mathieu functions and confluent hypergeometric functions.

(3) G. F. Becker and C. E. Van Orstrand, *Hyperbolic Functions: Smithsonian Mathematical Tables*, 1942. To five places.

(4) National Bureau of Standards (Federal Works Agency, Works Projects Administration of the City of New York):
    Tables of the Exponential Function, 1939 (12–18 decimal places);
    Tables of Sines and Cosines, 1940 (radian arguments, 8 places, increasing to 15 for small arguments);
    Tables of Sine, Cosine and Exponential Integrals, 1940 (9 places).

(5) M. Abramowitz and I. A. Stegun (eds), *Handbook of Mathematical Functions with Formulas, Graphs and Mathematical Tables*, NBS Applied Mathematics Series No. 55, 1965. Includes extensive tables of functions to many decimal places—e.g. exponentials to 15 places, sines and cosines to 23 places.

A.E.B.

	0	1	2	3	4	5	6	7	8	9	1	2	3	4	5	6	7	8	9
10	·0000	0043	0086	0128	0170	0212	0253	0294	0334	0374									
11	·0414	0453	0492	0531	0569	0607	0645	0682	0719	0755									
12	·0792	0828	0864	0899	0934	0969	1004	1038	1072	1106									
13	·1139	1173	1206	1239	1271	1303	1335	1367	1399	1430									
14	·1461	1492	1523	1553	1584	1614	1644	1673	1703	1732									
15	·1761	1790	1818	1847	1875	1903	1931	1959	1987	2014									
16	·2041	2068	2095	2122	2148	2175	2201	2227	2253	2279									
17	·2304	2330	2355	2380	2405	2430	2455	2480	2504	2529									
18	·2553	2577	2601	2625	2648	2672	2695	2718	2742	2765									
19	·2788	2810	2833	2856	2878	2900	2923	2945	2967	2989									
20	·3010	3032	3054	3075	3096	3118	3139	3160	3181	3201	2	4	6	8	11	13	15	17	19
21	·3222	3243	3263	3284	3304	3324	3345	3365	3385	3404	2	4	6	8	10	12	14	16	18
22	·3424	3444	3464	3483	3502	3522	3541	3560	3579	3598	2	4	6	8	10	12	14	15	17
23	·3617	3636	3655	3674	3692	3711	3729	3747	3766	3784	2	4	6	7	9	11	13	15	17
24	·3802	3820	3838	3856	3874	3892	3909	3927	3945	3962	2	4	5	7	9	11	12	14	16
25	·3979	3997	4014	4031	4048	4065	4082	4099	4116	4133	2	3	5	7	9	10	12	14	15
26	·4150	4166	4183	4200	4216	4232	4249	4265	4281	4298	2	3	5	7	8	10	11	13	15
27	·4314	4330	4346	4362	4378	4393	4409	4425	4440	4456	2	3	5	6	8	9	11	13	14
28	·4472	4487	4502	4518	4533	4548	4564	4579	4594	4609	2	3	5	6	8	9	11	12	14
29	·4624	4639	4654	4669	4683	4698	4713	4728	4742	4757	1	3	4	6	7	9	10	12	13
30	·4771	4786	4800	4814	4829	4843	4857	4871	4886	4900	1	3	4	6	7	9	10	11	13
31	·4914	4928	4942	4955	4969	4983	4997	5011	5024	5038	1	3	4	6	7	8	10	11	12
32	·5051	5065	5079	5092	5105	5119	5132	5145	5159	5172	1	3	4	5	7	8	9	11	12
33	·5185	5198	5211	5224	5237	5250	5263	5276	5289	5302	1	3	4	5	7	8	9	10	12
34	·5315	5328	5340	5353	5366	5378	5391	5403	5416	5428	1	3	4	5	6	8	9	10	11
35	·5441	5453	5465	5478	5490	5502	5514	5527	5539	5551	1	2	4	5	6	7	9	10	11
36	·5563	5575	5587	5599	5611	5623	5635	5647	5658	5670	1	2	4	5	6	7	8	10	11
37	·5682	5694	5705	5717	5729	5740	5752	5763	5775	5786	1	2	3	5	6	7	8	9	10
38	·5798	5809	5821	5832	5843	5855	5866	5877	5888	5899	1	2	3	5	6	7	8	9	10
39	·5911	5922	5933	5944	5955	5966	5977	5988	5999	6010	1	2	3	4	6	7	8	9	10
40	·6021	6031	6042	6053	6064	6075	6085	6096	6107	6117	1	2	3	4	5	6	7	9	10
41	·6128	6138	6149	6160	6170	6180	6191	6201	6212	6222	1	2	3	4	5	6	7	8	9
42	·6232	6243	6253	6263	6274	6284	6294	6304	6314	6325	1	2	3	4	5	6	7	8	9
43	·6335	6345	6355	6365	6375	6385	6395	6405	6415	6425	1	2	3	4	5	6	7	8	9
44	·6435	6444	6454	6464	6474	6484	6493	6503	6513	6522	1	2	3	4	5	6	7	8	9
45	·6532	6542	6551	6561	6571	6580	6590	6599	6609	6618	1	2	3	4	5	6	7	8	9
46	·6628	6637	6646	6656	6665	6675	6684	6693	6702	6712	1	2	3	4	5	6	7	7	8
47	·6721	6730	6739	6749	6758	6767	6776	6785	6794	6803	1	2	3	4	5	5	6	7	8
48	·6812	6821	6830	6839	6848	6857	6866	6875	6884	6893	1	2	3	4	4	5	6	7	8
49	·6902	6911	6920	6928	6937	6946	6955	6964	6972	6981	1	2	3	4	4	5	6	7	8
50	·6990	6998	7007	7016	7024	7033	7042	7050	7059	7067	1	2	3	3	4	5	6	7	8
51	·7076	7084	7093	7101	7110	7118	7126	7135	7143	7152	1	2	3	3	4	5	6	7	8
52	·7160	7168	7177	7185	7193	7202	7210	7218	7226	7235	1	2	2	3	4	5	6	7	7
53	·7243	7251	7259	7267	7275	7284	7292	7300	7308	7316	1	2	2	3	4	5	6	6	7
54	·7324	7332	7340	7348	7356	7364	7372	7380	7388	7396	1	2	2	3	4	5	6	6	7
	0	1	2	3	4	5	6	7	8	9	1	2	3	4	5	6	7	8	9

Mean Differences not
sufficiently accurate
(see pp. 352, 353)

	0	1	2	3	4	5	6	7	8	9	1	2	3	4	5	6	7	8	9
55	·7404	7412	7419	7427	7435	7443	7451	7459	7466	7474	1	2	2	3	4	5	5	6	7
56	·7482	7490	7497	7505	7513	7520	7528	7536	7543	7551	1	2	2	3	4	5	5	6	7
57	·7559	7566	7574	7582	7589	7597	7604	7612	7619	7627	1	2	2	3	4	5	5	6	7
58	·7634	7642	7649	7657	7664	7672	7679	7686	7694	7701	1	2	2	3	4	5	5	6	7
59	·7709	7716	7723	7731	7738	7745	7752	7760	7767	7774	1	1	2	3	4	4	5	6	7
60	·7782	7789	7796	7803	7810	7818	7825	7832	7839	7846	1	1	2	3	4	4	5	6	6
61	·7853	7860	7868	7875	7882	7889	7896	7903	7910	7917	1	1	2	3	4	4	5	6	6
62	·7924	7931	7938	7945	7952	7959	7966	7973	7980	7987	1	1	2	3	3	4	5	6	6
63	·7993	8000	8007	8014	8021	8028	8035	8041	8048	8055	1	1	2	3	3	4	5	6	6
64	·8062	8069	8075	8082	8089	8096	8102	8109	8116	8122	1	1	2	3	3	4	5	5	6
65	·8129	8136	8142	8149	8156	8162	8169	8176	8182	8189	1	1	2	3	3	4	5	5	6
66	·8195	8202	8209	8215	8222	8228	8235	8241	8248	8254	1	1	2	3	3	4	5	5	6
67	·8261	8267	8274	8280	8287	8293	8299	8306	8312	8319	1	1	2	3	3	4	4	5	6
68	·8325	8331	8338	8344	8351	8357	8363	8370	8376	8382	1	1	2	3	3	4	4	5	6
69	·8388	8395	8401	8407	8414	8420	8426	8432	8439	8445	1	1	2	3	3	4	4	5	6
70	·8451	8457	8463	8470	8476	8482	8488	8494	8500	8506	1	1	2	2	3	4	4	5	6
71	·8513	8519	8525	8531	8537	8543	8549	8555	8561	8567	1	1	2	2	3	4	4	5	5
72	·8573	8579	8585	8591	8597	8603	8609	8615	8621	8627	1	1	2	2	3	4	4	5	5
73	·8633	8639	8645	8651	8657	8663	8669	8675	8681	8686	1	1	2	2	3	4	4	5	5
74	·8692	8698	8704	8710	8716	8722	8727	8733	8739	8745	1	1	2	2	3	4	4	5	5
75	·8751	8756	8762	8768	8774	8779	8785	8791	8797	8802	1	1	2	2	3	3	4	5	5
76	·8808	8814	8820	8825	8831	8837	8842	8848	8854	8859	1	1	2	2	3	3	4	5	5
77	·8865	8871	8876	8882	8887	8893	8899	8904	8910	8915	1	1	2	2	3	3	4	4	5
78	·8921	8927	8932	8938	8943	8949	8954	8960	8965	8971	1	1	2	2	3	3	4	4	5
79	·8976	8982	8987	8993	8998	9004	9009	9015	9020	9025	1	1	2	2	3	3	4	4	5
80	·9031	9036	9042	9047	9053	9058	9063	9069	9074	9079	1	1	2	2	3	3	4	4	5
81	·9085	9090	9096	9101	9106	9112	9117	9122	9128	9133	1	1	2	2	3	3	4	4	5
82	·9138	9143	9149	9154	9159	9165	9170	9175	9180	9186	1	1	2	2	3	3	4	4	5
83	·9191	9196	9201	9206	9212	9217	9222	9227	9232	9238	1	1	2	2	3	3	4	4	5
84	·9243	9248	9253	9258	9263	9269	9274	9279	9284	9289	1	1	2	2	3	3	4	4	5
85	·9294	9299	9304	9309	9315	9320	9325	9330	9335	9340	1	1	2	2	3	3	4	4	5
86	·9345	9350	9355	9360	9365	9370	9375	9380	9385	9390	1	1	2	2	3	3	4	4	5
87	·9395	9400	9405	9410	9415	9420	9425	9430	9435	9440	1	1	2	2	3	3	4	4	5
88	·9445	9450	9455	9460	9465	9469	9474	9479	9484	9489	0	1	1	2	2	3	3	4	4
89	·9494	9499	9504	9509	9513	9518	9523	9528	9533	9538	0	1	1	2	2	3	3	4	4
90	·9542	9547	9552	9557	9562	9566	9571	9576	9581	9586	0	1	1	2	2	3	3	4	4
91	·9590	9595	9600	9605	9609	9614	9619	9624	9628	9633	0	1	1	2	2	3	3	4	4
92	·9638	9643	9647	9652	9657	9661	9666	9671	9675	9680	0	1	1	2	2	3	3	4	4
93	·9685	9689	9694	9699	9703	9708	9713	9717	9722	9727	0	1	1	2	2	3	3	4	4
94	·9731	9736	9741	9745	9750	9754	9759	9763	9768	9773	0	1	1	2	2	3	3	4	4
95	·9777	9782	9786	9791	9795	9800	9805	9809	9814	9818	0	1	1	2	2	3	3	4	4
96	·9823	9827	9832	9836	9841	9845	9850	9854	9859	9863	0	1	1	2	2	3	3	4	4
97	·9868	9872	9877	9881	9886	9890	9894	9899	9903	9908	0	1	1	2	2	3	3	4	4
98	·9912	9917	9921	9926	9930	9934	9939	9943	9948	9952	0	1	1	2	2	3	3	4	4
99	·9956	9961	9965	9969	9974	9978	9983	9987	9991	9996	0	1	1	2	2	3	3	4	4
	0	1	2	3	4	5	6	7	8	9	1	2	3	4	5	6	7	8	9

	0	1	2	3	4	5	6	7	8	9
100	·0000	0004	0009	0013	0017	0022	0026	0030	0035	0039
101	·0043	0048	0052	0056	0060	0065	0069	0073	0077	0082
102	·0086	0090	0095	0099	0103	0107	0111	0116	0120	0124
103	·0128	0133	0137	0141	0145	0149	0154	0158	0162	0166
104	·0170	0175	0179	0183	0187	0191	0195	0199	0204	0208
105	·0212	0216	0220	0224	0228	0233	0237	0241	0245	0249
106	·0253	0257	0261	0265	0269	0273	0278	0282	0286	0290
107	·0294	0298	0302	0306	0310	0314	0318	0322	0326	0330
108	·0334	0338	0342	0346	0350	0354	0358	0362	0366	0370
109	·0374	0378	0382	0386	0390	0394	0398	0402	0406	0410
110	·0414	0418	0422	0426	0430	0434	0438	0441	0445	0449
111	·0453	0457	0461	0465	0469	0473	0477	0481	0484	0488
112	·0492	0496	0500	0504	0508	0512	0515	0519	0523	0527
113	·0531	0535	0538	0542	0546	0550	0554	0558	0561	0565
114	·0569	0573	0577	0580	0584	0588	0592	0596	0599	0603
115	·0607	0611	0615	0618	0622	0626	0630	0633	0637	0641
116	·0645	0648	0652	0656	0660	0663	0667	0671	0674	0678
117	·0682	0686	0689	0693	0697	0700	0704	0708	0711	0715
118	·0719	0722	0726	0730	0734	0737	0741	0745	0748	0752
119	·0755	0759	0763	0766	0770	0774	0777	0781	0785	0788
120	·0792	0795	0799	0803	0806	0810	0813	0817	0821	0824
121	·0828	0831	0835	0839	0842	0846	0849	0853	0856	0860
122	·0864	0867	0871	0874	0878	0881	0885	0888	0892	0896
123	·0899	0903	0906	0910	0913	0917	0920	0924	0927	0931
124	·0934	0938	0941	0945	0948	0952	0955	0959	0962	0966
125	·0969	0973	0976	0980	0983	0986	0990	0993	0997	1000
126	·1004	1007	1011	1014	1017	1021	1024	1028	1031	1035
127	·1038	1041	1045	1048	1052	1055	1059	1062	1065	1069
128	·1072	1075	1079	1082	1086	1089	1092	1096	1099	1103
129	·1106	1109	1113	1116	1119	1123	1126	1129	1133	1136
130	·1139	1143	1146	1149	1153	1156	1159	1163	1166	1169
131	·1173	1176	1179	1183	1186	1189	1193	1196	1199	1202
132	·1206	1209	1212	1216	1219	1222	1225	1229	1232	1235
133	·1239	1242	1245	1248	1252	1255	1258	1261	1265	1268
134	·1271	1274	1278	1281	1284	1287	1290	1294	1297	1300
135	·1303	1307	1310	1313	1316	1319	1323	1326	1329	1332
136	·1335	1339	1342	1345	1348	1351	1355	1358	1361	1364
137	·1367	1370	1374	1377	1380	1383	1386	1389	1392	1396
138	·1399	1402	1405	1408	1411	1414	1418	1421	1424	1427
139	·1430	1433	1436	1440	1443	1446	1449	1452	1455	1458
140	·1461	1464	1467	1471	1474	1477	1480	1483	1486	1489
141	·1492	1495	1498	1501	1504	1508	1511	1514	1517	1520
142	·1523	1526	1529	1532	1535	1538	1541	1544	1547	1550
143	·1553	1556	1559	1562	1565	1569	1572	1575	1578	1581
144	·1584	1587	1590	1593	1596	1599	1602	1605	1608	1611
145	·1614	1617	1620	1623	1626	1629	1632	1635	1638	1641
146	·1644	1647	1649	1652	1655	1658	1661	1664	1667	1670
147	·1673	1676	1679	1682	1685	1688	1691	1694	1697	1700
148	·1703	1706	1708	1711	1714	1717	1720	1723	1726	1729
149	·1732	1735	1738	1741	1744	1746	1749	1752	1755	1758
	0	1	2	3	4	5	6	7	8	9

	0	1	2	3	4	5	6	7	8	9
150	·1761	1764	1767	1770	1772	1775	1778	1781	1784	1787
151	·1790	1793	1796	1798	1801	1804	1807	1810	1813	1816
152	·1818	1821	1824	1827	1830	1833	1836	1838	1841	1844
153	·1847	1850	1853	1855	1858	1861	1864	1867	1870	1872
154	·1875	1878	1881	1884	1886	1889	1892	1895	1898	1901
155	·1903	1906	1909	1912	1915	1917	1920	1923	1926	1928
156	·1931	1934	1937	1940	1942	1945	1948	1951	1953	1956
157	·1959	1962	1965	1967	1970	1973	1976	1978	1981	1984
158	·1987	1989	1992	1995	1998	2000	2003	2006	2009	2011
159	·2014	2017	2019	2022	2025	2028	2030	2033	2036	2038
160	·2041	2044	2047	2049	2052	2055	2057	2060	2063	2066
161	·2068	2071	2074	2076	2079	2082	2084	2087	2090	2092
162	·2095	2098	2101	2103	2106	2109	2111	2114	2117	2119
163	·2122	2125	2127	2130	2133	2135	2138	2140	2143	2146
164	·2148	2151	2154	2156	2159	2162	2164	2167	2170	2172
165	·2175	2177	2180	2183	2185	2188	2191	2193	2196	2198
166	·2201	2204	2206	2209	2212	2214	2217	2219	2222	2225
167	·2227	2230	2232	2235	2238	2240	2243	2245	2248	2251
168	·2253	2256	2258	2261	2263	2266	2269	2271	2274	2276
169	·2279	2281	2284	2287	2289	2292	2294	2297	2299	2302
170	·2304	2307	2310	2312	2315	2317	2320	2322	2325	2327
171	·2330	2333	2335	2338	2340	2343	2345	2348	2350	2353
172	·2355	2358	2360	2363	2365	2368	2370	2373	2375	2378
173	·2380	2383	2385	2388	2390	2393	2395	2398	2400	2403
174	·2405	2408	2410	2413	2415	2418	2420	2423	2425	2428
175	·2430	2433	2435	2438	2440	2443	2445	2448	2450	2453
176	·2455	2458	2460	2463	2465	2467	2470	2472	2475	2477
177	·2480	2482	2485	2487	2490	2492	2494	2497	2499	2502
178	·2504	2507	2509	2512	2514	2516	2519	2521	2524	2526
179	·2529	2531	2533	2536	2538	2541	2543	2545	2548	2550
180	·2553	2555	2558	2560	2562	2565	2567	2570	2572	2574
181	·2577	2579	2582	2584	2586	2589	2591	2594	2596	2598
182	·2601	2603	2605	2608	2610	2613	2615	2617	2620	2622
183	·2625	2627	2629	2632	2634	2636	2639	2641	2643	2646
184	·2648	2651	2653	2655	2658	2660	2662	2665	2667	2669
185	·2672	2674	2676	2679	2681	2683	2686	2688	2690	2693
186	·2695	2697	2700	2702	2704	2707	2709	2711	2714	2716
187	·2718	2721	2723	2725	2728	2730	2732	2735	2737	2739
188	·2742	2744	2746	2749	2751	2753	2755	2758	2760	2762
189	·2765	2767	2769	2772	2774	2776	2778	2781	2783	2785
190	·2788	2790	2792	2794	2797	2799	2801	2804	2806	2808
191	·2810	2813	2815	2817	2819	2822	2824	2826	2828	2831
192	·2833	2835	2838	2840	2842	2844	2847	2849	2851	2853
193	·2856	2858	2860	2862	2865	2867	2869	2871	2874	2876
194	·2878	2880	2882	2885	2887	2889	2891	2894	2896	2898
195	·2900	2903	2905	2907	2909	2911	2914	2916	2918	2920
196	·2923	2925	2927	2929	2931	2934	2936	2938	2940	2942
197	·2945	2947	2949	2951	2953	2956	2958	2960	2962	2964
198	·2967	2969	2971	2973	2975	2978	2980	2982	2984	2986
199	·2989	2991	2993	2995	2997	2999	3002	3004	3006	3008
	0	1	2	3	4	5	6	7	8	9

	0	1	2	3	4	5	6	7	8	9	1	2	3	4	5	6	7	8	9
·00	1000	1002	1005	1007	1009	1012	1014	1016	1019	1021	0	0	1	1	1	1	2	2	2
·01	1023	1026	1028	1030	1033	1035	1038	1040	1042	1045	0	0	1	1	1	1	2	2	2
·02	1047	1050	1052	1054	1057	1059	1062	1064	1067	1069	0	0	1	1	1	1	2	2	2
·03	1072	1074	1076	1079	1081	1084	1086	1089	1091	1094	0	0	1	1	1	1	2	2	2
·04	1096	1099	1102	1104	1107	1109	1112	1114	1117	1119	0	1	1	1	1	2	2	2	2
·05	1122	1125	1127	1130	1132	1135	1138	1140	1143	1146	0	1	1	1	1	2	2	2	2
·06	1148	1151	1153	1156	1159	1161	1164	1167	1169	1172	0	1	1	1	1	2	2	2	2
·07	1175	1178	1180	1183	1186	1189	1191	1194	1197	1199	0	1	1	1	1	2	2	2	2
·08	1202	1205	1208	1211	1213	1216	1219	1222	1225	1227	0	1	1	1	1	2	2	2	3
·09	1230	1233	1236	1239	1242	1245	1247	1250	1253	1256	0	1	1	1	1	2	2	2	3
·10	1259	1262	1265	1268	1271	1274	1276	1279	1282	1285	0	1	1	1	1	2	2	2	3
·11	1288	1291	1294	1297	1300	1303	1306	1309	1312	1315	0	1	1	1	2	2	2	2	3
·12	1318	1321	1324	1327	1330	1334	1337	1340	1343	1346	0	1	1	1	2	2	2	3	3
·13	1349	1352	1355	1358	1361	1365	1368	1371	1374	1377	0	1	1	1	2	2	2	3	3
·14	1380	1384	1387	1390	1393	1396	1400	1403	1406	1409	0	1	1	1	2	2	2	3	3
·15	1413	1416	1419	1422	1426	1429	1432	1435	1439	1442	0	1	1	1	2	2	2	3	3
·16	1445	1449	1452	1455	1459	1462	1466	1469	1472	1476	0	1	1	1	2	2	2	3	3
·17	1479	1483	1486	1489	1493	1496	1500	1503	1507	1510	0	1	1	1	2	2	2	3	3
·18	1514	1517	1521	1524	1528	1531	1535	1538	1542	1545	0	1	1	1	2	2	2	3	3
·19	1549	1552	1556	1560	1563	1567	1570	1574	1578	1581	0	1	1	1	2	2	3	3	3
·20	1585	1589	1592	1596	1600	1603	1607	1611	1614	1618	0	1	1	1	2	2	3	3	3
·21	1622	1626	1629	1633	1637	1641	1644	1648	1652	1656	0	1	1	2	2	2	3	3	3
·22	1660	1663	1667	1671	1675	1679	1683	1687	1690	1694	0	1	1	2	2	2	3	3	3
·23	1698	1702	1706	1710	1714	1718	1722	1726	1730	1734	0	1	1	2	2	2	3	3	4
·24	1738	1742	1746	1750	1754	1758	1762	1766	1770	1774	0	1	1	2	2	2	3	3	4
·25	1778	1782	1786	1791	1795	1799	1803	1807	1811	1816	0	1	1	2	2	3	3	3	4
·26	1820	1824	1828	1832	1837	1841	1845	1849	1854	1858	0	1	1	2	2	3	3	3	4
·27	1862	1866	1871	1875	1879	1884	1888	1892	1897	1901	0	1	1	2	2	3	3	3	4
·28	1905	1910	1914	1919	1923	1928	1932	1936	1941	1945	0	1	1	2	2	3	3	4	4
·29	1950	1954	1959	1963	1968	1972	1977	1982	1986	1991	0	1	1	2	2	3	3	4	4
·30	1995	2000	2004	2009	2014	2018	2023	2028	2032	2037	0	1	1	2	2	3	3	4	4
·31	2042	2046	2051	2056	2061	2065	2070	2075	2080	2084	0	1	1	2	2	3	3	4	4
·32	2089	2094	2099	2104	2109	2113	2118	2123	2128	2133	0	1	1	2	2	3	3	4	4
·33	2138	2143	2148	2153	2158	2163	2168	2173	2178	2183	1	1	2	2	3	3	4	4	5
·34	2188	2193	2198	2203	2208	2213	2218	2223	2228	2234	1	1	2	2	3	3	4	4	5
·35	2239	2244	2249	2254	2259	2265	2270	2275	2280	2286	1	1	2	2	3	3	4	4	5
·36	2291	2296	2301	2307	2312	2317	2323	2328	2333	2339	1	1	2	2	3	3	4	4	5
·37	2344	2350	2355	2360	2366	2371	2377	2382	2388	2393	1	1	2	2	3	3	4	4	5
·38	2399	2404	2410	2415	2421	2427	2432	2438	2443	2449	1	1	2	2	3	3	4	4	5
·39	2455	2460	2466	2472	2477	2483	2489	2495	2500	2506	1	1	2	2	3	3	4	5	5
·40	2512	2518	2523	2529	2535	2541	2547	2553	2559	2564	1	1	2	2	3	3	4	5	5
·41	2570	2576	2582	2588	2594	2600	2606	2612	2618	2624	1	1	2	2	3	4	4	5	5
·42	2630	2636	2642	2649	2655	2661	2667	2673	2679	2685	1	1	2	2	3	4	4	5	6
·43	2692	2698	2704	2710	2716	2723	2729	2735	2742	2748	1	1	2	2	3	4	4	5	6
·44	2754	2761	2767	2773	2780	2786	2793	2799	2805	2812	1	1	2	3	3	4	4	5	6
·45	2818	2825	2831	2838	2844	2851	2858	2864	2871	2877	1	1	2	3	3	4	5	5	6
·46	2884	2891	2897	2904	2911	2917	2924	2931	2938	2944	1	1	2	3	3	4	5	5	6
·47	2951	2958	2965	2972	2979	2985	2992	2999	3006	3013	1	1	2	3	3	4	5	6	6
·48	3020	3027	3034	3041	3048	3055	3062	3069	3076	3083	1	1	2	3	4	4	5	6	6
·49	3090	3097	3105	3112	3119	3126	3133	3141	3148	3155	1	1	2	3	4	4	5	6	6
	0	1	2	3	4	5	6	7	8	9	1	2	3	4	5	6	7	8	9

	0	1	2	3	4	5	6	7	8	9	1	2	3	4	5	6	7	8	9
·50	3162	3170	3177	3184	3192	3199	3206	3214	3221	3228	1	1	2	3	4	4	5	6	7
·51	3236	3243	3251	3258	3266	3273	3281	3289	3296	3304	1	2	2	3	4	5	5	6	7
·52	3311	3319	3327	3334	3342	3350	3357	3365	3373	3381	1	2	2	3	4	5	5	6	7
·53	3388	3396	3404	3412	3420	3428	3436	3443	3451	3459	1	2	2	3	4	5	6	6	7
·54	3467	3475	3483	3491	3499	3508	3516	3524	3532	3540	1	2	2	3	4	5	6	6	7
·55	3548	3556	3565	3573	3581	3589	3597	3606	3614	3622	1	2	2	3	4	5	6	7	7
·56	3631	3639	3648	3656	3664	3673	3681	3690	3698	3707	1	2	3	3	4	5	6	7	8
·57	3715	3724	3733	3741	3750	3758	3767	3776	3784	3793	1	2	3	3	4	5	6	7	8
·58	3802	3811	3819	3828	3837	3846	3855	3864	3873	3882	1	2	3	4	4	5	6	7	8
·59	3890	3899	3908	3917	3926	3936	3945	3954	3963	3972	1	2	3	4	5	5	6	7	8
·60	3981	3990	3999	4009	4018	4027	4036	4046	4055	4064	1	2	3	4	5	6	7	7	8
·61	4074	4083	4093	4102	4111	4121	4130	4140	4150	4159	1	2	3	4	5	6	7	8	9
·62	4169	4178	4188	4198	4207	4217	4227	4236	4246	4256	1	2	3	4	5	6	7	8	9
·63	4266	4276	4285	4295	4305	4315	4325	4335	4345	4355	1	2	3	4	5	6	7	8	9
·64	4365	4375	4385	4395	4406	4416	4426	4436	4446	4457	1	2	3	4	5	6	7	8	9
·65	4467	4477	4487	4498	4508	4519	4529	4539	4550	4560	1	2	3	4	5	6	7	8	9
·66	4571	4581	4592	4603	4613	4624	4634	4645	4656	4667	1	2	3	4	5	6	7	8	10
·67	4677	4688	4699	4710	4721	4732	4742	4753	4764	4775	1	2	3	4	5	7	8	9	10
·68	4786	4797	4808	4819	4831	4842	4853	4864	4875	4887	1	2	3	4	6	7	8	9	10
·69	4898	4909	4920	4932	4943	4955	4966	4977	4989	5000	1	2	3	5	6	7	8	9	10
·70	5012	5023	5035	5047	5058	5070	5082	5093	5105	5117	1	2	4	5	6	7	8	9	11
·71	5129	5140	5152	5164	5176	5188	5200	5212	5224	5236	1	2	4	5	6	7	8	10	11
·72	5248	5260	5272	5284	5297	5309	5321	5333	5346	5358	1	2	4	5	6	7	9	10	11
·73	5370	5383	5395	5408	5420	5433	5445	5458	5470	5483	1	3	4	5	6	8	9	10	11
·74	5495	5508	5521	5534	5546	5559	5572	5585	5598	5610	1	3	4	5	6	8	9	10	12
·75	5623	5636	5649	5662	5675	5689	5702	5715	5728	5741	1	3	4	5	7	8	9	10	12
·76	5754	5768	5781	5794	5808	5821	5834	5848	5861	5875	1	3	4	5	7	8	9	11	12
·77	5888	5902	5916	5929	5943	5957	5970	5984	5998	6012	1	3	4	6	7	8	10	11	12
·78	6026	6039	6053	6067	6081	6095	6109	6124	6138	6152	1	3	4	6	7	8	10	11	13
·79	6166	6180	6194	6209	6223	6237	6252	6266	6281	6295	1	3	4	6	7	9	10	12	13
·80	6310	6324	6339	6353	6368	6383	6397	6412	6427	6442	1	3	4	6	7	9	10	12	13
·81	6457	6471	6486	6501	6516	6531	6546	6561	6577	6592	2	3	5	6	8	9	11	12	14
·82	6607	6622	6637	6653	6668	6683	6699	6714	6730	6745	2	3	5	6	8	9	11	12	14
·83	6761	6776	6792	6808	6823	6839	6855	6871	6887	6902	2	3	5	6	8	9	11	13	14
·84	6918	6934	6950	6966	6982	6998	7015	7031	7047	7063	2	3	5	6	8	10	11	13	14
·85	7079	7096	7112	7129	7145	7161	7178	7194	7211	7228	2	3	5	7	8	10	12	13	15
·86	7244	7261	7278	7295	7311	7328	7345	7362	7379	7396	2	3	5	7	8	10	12	14	15
·87	7413	7430	7447	7464	7482	7499	7516	7534	7551	7568	2	3	5	7	9	10	12	14	16
·88	7586	7603	7621	7638	7656	7674	7691	7709	7727	7745	2	4	5	7	9	11	12	14	16
·89	7762	7780	7798	7816	7834	7852	7870	7889	7907	7925	2	4	5	7	9	11	13	14	16
·90	7943	7962	7980	7998	8017	8035	8054	8072	8091	8110	2	4	6	7	9	11	13	15	17
·91	8128	8147	8166	8185	8204	8222	8241	8260	8279	8299	2	4	6	8	10	11	13	15	17
·92	8318	8337	8356	8375	8395	8414	8433	8453	8472	8492	2	4	6	8	10	12	14	15	17
·93	8511	8531	8551	8570	8590	8610	8630	8650	8670	8690	2	4	6	8	10	12	14	16	18
·94	8710	8730	8750	8770	8790	8810	8831	8851	8872	8892	2	4	6	8	10	12	14	16	18
·95	8913	8933	8954	8974	8995	9016	9036	9057	9078	9099	2	4	6	8	10	12	14	17	19
·96	9120	9141	9162	9183	9204	9226	9247	9268	9290	9311	2	4	6	9	11	13	15	17	19
·97	9333	9354	9376	9397	9419	9441	9462	9484	9506	9528	2	4	7	9	11	13	15	17	20
·98	9550	9572	9594	9616	9638	9661	9683	9705	9727	9750	2	4	7	9	11	13	16	18	20
·99	9772	9795	9817	9840	9863	9886	9908	9931	9954	9977	2	5	7	9	11	14	16	18	21
	0	1	2	3	4	5	6	7	8	9	1	2	3	4	5	6	7	8	9

	0	1	2	3	4	5	6	7	8	9	1	2	3	4	5	6	7	8	9
10	10000	9901	9804	9709	9615	9524	9434	9346	9259	9174									
11	9091	9009	8929	8850	8772	8696	8621	8547	8475	8403									
12	8333	8264	8197	8130	8065	8000	7937	7874	7812	7752									
13	7692	7634	7576	7519	7463	7407	7353	7299	7246	7194									
14	7143	7092	7042	6993	6944	6897	6849	6803	6757	6711									
15	6667	6623	6579	6536	6494	6452	6410	6369	6329	6289									
16	6250	6211	6173	6135	6098	6061	6024	5988	5952	5917									
17	5882	5848	5814	5780	5747	5714	5682	5650	5618	5587									
18	5556	5525	5495	5464	5435	5405	5376	5348	5319	5291									
19	5263	5236	5208	5181	5155	5128	5102	5076	5051	5025									
20	5000	4975	4950	4926	4902	4878	4854	4831	4808	4785	2	5	7	10	12	14	17	19	21
21	4762	4739	4717	4695	4673	4651	4630	4608	4587	4566	2	4	7	9	11	13	15	17	20
22	4545	4525	4505	4484	4464	4444	4425	4405	4386	4367	2	4	6	8	10	12	14	16	18
23	4348	4329	4310	4292	4274	4255	4237	4219	4202	4184	2	4	5	7	9	11	13	14	16
24	4167	4149	4132	4115	4098	4082	4065	4049	4032	4016	2	3	5	7	8	10	12	13	15
25	4000	3984	3968	3953	3937	3922	3906	3891	3876	3861	2	3	5	6	8	9	11	12	14
26	3846	3831	3817	3802	3788	3774	3759	3745	3731	3717	1	3	4	6	7	9	10	11	13
27	3704	3690	3676	3663	3650	3636	3623	3610	3597	3584	1	3	4	5	7	8	9	11	12
28	3571	3559	3546	3534	3521	3509	3497	3484	3472	3460	1	2	4	5	6	7	9	10	11
29	3448	3436	3425	3413	3401	3390	3378	3367	3356	3344	1	2	3	5	6	7	8	9	10
30	3333	3322	3311	3300	3289	3279	3268	3257	3247	3236	1	2	3	4	5	6	7	9	10
31	3226	3215	3205	3195	3185	3175	3165	3155	3145	3135	1	2	3	4	5	6	7	8	9
32	3125	3115	3106	3096	3086	3077	3067	3058	3049	3040	1	2	3	4	5	6	7	8	9
33	3030	3021	3012	3003	2994	2985	2976	2967	2959	2950	1	2	3	4	4	5	6	7	8
34	2941	2933	2924	2915	2907	2899	2890	2882	2874	2865	1	2	3	3	4	5	6	7	8
35	2857	2849	2841	2833	2825	2817	2809	2801	2793	2786	1	2	2	3	4	5	6	6	7
36	2778	2770	2762	2755	2747	2740	2732	2725	2717	2710	1	2	2	3	4	5	5	6	7
37	2703	2695	2688	2681	2674	2667	2660	2653	2646	2639	1	1	2	3	4	4	5	6	6
38	2632	2625	2618	2611	2604	2597	2591	2584	2577	2571	1	1	2	3	3	4	5	5	6
39	2564	2558	2551	2545	2538	2532	2525	2519	2513	2506	1	1	2	3	3	4	4	5	6
40	2500	2494	2488	2481	2475	2469	2463	2457	2451	2445	1	1	2	2	3	4	4	5	5
41	2439	2433	2427	2421	2415	2410	2404	2398	2392	2387	1	1	2	2	3	3	4	5	5
42	2381	2375	2370	2364	2358	2353	2347	2342	2336	2331	1	1	2	2	3	3	4	4	5
43	2326	2320	2315	2309	2304	2299	2294	2288	2283	2278	1	1	2	2	3	3	4	4	5
44	2273	2268	2262	2257	2252	2247	2242	2237	2232	2227	1	1	2	2	3	3	4	4	5
45	2222	2217	2212	2208	2203	2198	2193	2188	2183	2179	0	1	1	2	2	3	3	4	4
46	2174	2169	2165	2160	2155	2151	2146	2141	2137	2132	0	1	1	2	2	3	3	4	4
47	2128	2123	2119	2114	2110	2105	2101	2096	2092	2088	0	1	1	2	2	3	3	4	4
48	2083	2079	2075	2070	2066	2062	2058	2053	2049	2045	0	1	1	2	2	3	3	3	4
49	2041	2037	2033	2028	2024	2020	2016	2012	2008	2004	0	1	1	2	2	2	3	3	4
50	2000	1996	1992	1988	1984	1980	1976	1972	1969	1965	0	1	1	2	2	2	3	3	4
51	1961	1957	1953	1949	1946	1942	1938	1934	1931	1927	0	1	1	2	2	2	3	3	3
52	1923	1919	1916	1912	1908	1905	1901	1898	1894	1890	0	1	1	1	2	2	3	3	3
53	1887	1883	1880	1876	1873	1869	1866	1862	1859	1855	0	1	1	1	2	2	2	3	3
54	1852	1848	1845	1842	1838	1835	1832	1828	1825	1821	0	1	1	1	2	2	2	3	3

Mean Differences not sufficiently accurate

	0	1	2	3	4	5	6	7	8	9	1	2	3	4	5	6	7	8	9

	0	1	2	3	4	5	6	7	8	9	1	2	3	4	5	6	7	8	9
55	1818	1815	1812	1808	1805	1802	1799	1795	1792	1789	0	1	1	1	2	2	2	3	3
56	1786	1783	1779	1776	1773	1770	1767	1764	1761	1757	0	1	1	1	2	2	2	3	3
57	1754	1751	1748	1745	1742	1739	1736	1733	1730	1727	0	1	1	1	2	2	2	2	3
58	1724	1721	1718	1715	1712	1709	1706	1704	1701	1698	0	1	1	1	1	2	2	2	3
59	1695	1692	1689	1686	1684	1681	1678	1675	1672	1669	0	1	1	1	1	2	2	2	3
60	1667	1664	1661	1658	1656	1653	1650	1647	1645	1642	0	1	1	1	1	2	2	2	3
61	1639	1637	1634	1631	1629	1626	1623	1621	1618	1616	0	1	1	1	1	2	2	2	2
62	1613	1610	1608	1605	1603	1600	1597	1595	1592	1590	0	1	1	1	1	2	2	2	2
63	1587	1585	1582	1580	1577	1575	1572	1570	1567	1565	0	0	1	1	1	1	2	2	2
64	1562	1560	1558	1555	1553	1550	1548	1546	1543	1541	0	0	1	1	1	1	2	2	2
65	1538	1536	1534	1531	1529	1527	1524	1522	1520	1517	0	0	1	1	1	1	2	2	2
66	1515	1513	1511	1508	1506	1504	1502	1499	1497	1495	0	0	1	1	1	1	2	2	2
67	1493	1490	1488	1486	1484	1481	1479	1477	1475	1473	0	0	1	1	1	1	2	2	2
68	1471	1468	1466	1464	1462	1460	1458	1456	1453	1451	0	0	1	1	1	1	2	2	2
69	1449	1447	1445	1443	1441	1439	1437	1435	1433	1431	0	0	1	1	1	1	1	2	2
70	1429	1427	1425	1422	1420	1418	1416	1414	1412	1410	0	0	1	1	1	1	1	2	2
71	1408	1406	1404	1403	1401	1399	1397	1395	1393	1391	0	0	1	1	1	1	1	2	2
72	1389	1387	1385	1383	1381	1379	1377	1376	1374	1372	0	0	1	1	1	1	1	2	2
73	1370	1368	1366	1364	1362	1361	1359	1357	1355	1353	0	0	1	1	1	1	1	2	2
74	1351	1350	1348	1346	1344	1342	1340	1339	1337	1335	0	0	1	1	1	1	1	1	2
75	1333	1332	1330	1328	1326	1325	1323	1321	1319	1318	0	0	1	1	1	1	1	1	2
76	1316	1314	1312	1311	1309	1307	1305	1304	1302	1300	0	0	1	1	1	1	1	1	2
77	1299	1297	1295	1294	1292	1290	1289	1287	1285	1284	0	0	1	1	1	1	1	1	2
78	1282	1280	1279	1277	1276	1274	1272	1271	1269	1267	0	0	0	1	1	1	1	1	1
79	1266	1264	1263	1261	1259	1258	1256	1255	1253	1252	0	0	0	1	1	1	1	1	1
80	1250	1248	1247	1245	1244	1242	1241	1239	1238	1236	0	0	0	1	1	1	1	1	1
81	1235	1233	1232	1230	1229	1227	1225	1224	1222	1221	0	0	0	1	1	1	1	1	1
82	1220	1218	1217	1215	1214	1212	1211	1209	1208	1206	0	0	0	1	1	1	1	1	1
83	1205	1203	1202	1200	1199	1198	1196	1195	1193	1192	0	0	0	1	1	1	1	1	1
84	1190	1189	1188	1186	1185	1183	1182	1181	1179	1178	0	0	0	1	1	1	1	1	1
85	1176	1175	1174	1172	1171	1170	1168	1167	1166	1164	0	0	0	1	1	1	1	1	1
86	1163	1161	1160	1159	1157	1156	1155	1153	1152	1151	0	0	0	1	1	1	1	1	1
87	1149	1148	1147	1145	1144	1143	1142	1140	1139	1138	0	0	0	1	1	1	1	1	1
88	1136	1135	1134	1133	1131	1130	1129	1127	1126	1125	0	0	0	1	1	1	1	1	1
89	1124	1122	1121	1120	1119	1117	1116	1115	1114	1112	0	0	0	1	1	1	1	1	1
90	1111	1110	1109	1107	1106	1105	1104	1103	1101	1100	0	0	0	0	1	1	1	1	1
91	1099	1098	1096	1095	1094	1093	1092	1091	1089	1088	0	0	0	0	1	1	1	1	1
92	1087	1086	1085	1083	1082	1081	1080	1079	1078	1076	0	0	0	0	1	1	1	1	1
93	1075	1074	1073	1072	1071	1070	1068	1067	1066	1065	0	0	0	0	1	1	1	1	1
94	1064	1063	1062	1060	1059	1058	1057	1056	1055	1054	0	0	0	0	1	1	1	1	1
95	1053	1052	1050	1049	1048	1047	1046	1045	1044	1043	0	0	0	0	1	1	1	1	1
96	1042	1041	1040	1038	1037	1036	1035	1034	1033	1032	0	0	0	0	1	1	1	1	1
97	1031	1030	1029	1028	1027	1026	1025	1024	1022	1021	0	0	0	0	1	1	1	1	1
98	1020	1019	1018	1017	1016	1015	1014	1013	1012	1011	0	0	0	0	1	1	1	1	1
99	1010	1009	1008	1007	1006	1005	1004	1003	1002	1001	0	0	0	0	1	1	1	1	1
	0	1	2	3	4	5	6	7	8	9	1	2	3	4	5	6	7	8	9

	0	1	2	3	4	5	6	7	8	9	1	2	3	4	5	6	7	8	9
1.0	1.000	1.020	1.040	1.061	1.082	1.102	1.124	1.145	1.166	1.188	2	4	6	8	11	13	15	17	19
1.1	1.210	1.232	1.254	1.277	1.300	1.322	1.346	1.369	1.392	1.416	2	5	7	9	12	14	16	18	21
1.2	1.440	1.464	1.488	1.513	1.538	1.562	1.588	1.613	1.638	1.664	3	5	8	10	13	15	18	20	23
1.3	1.690	1.716	1.742	1.769	1.796	1.822	1.850	1.877	1.904	1.932	3	5	8	11	14	16	19	22	24
1.4	1.960	1.988	2.016	2.045	2.074	2.102	2.132	2.161	2.190	2.220	3	6	9	12	15	17	20	23	26
1.5	2.250	2.280	2.310	2.341	2.372	2.402	2.434	2.465	2.496	2.528	3	6	9	12	16	19	22	25	28
1.6	2.560	2.592	2.624	2.657	2.690	2.722	2.756	2.789	2.822	2.856	3	7	10	13	17	20	23	26	30
1.7	2.890	2.924	2.958	2.993	3.028	3.062	3.098	3.133	3.168	3.204	4	7	11	14	18	21	25	28	32
1.8	3.240	3.276	3.312	3.349	3.386	3.422	3.460	3.497	3.534	3.572	4	7	11	15	19	22	26	30	33
1.9	3.610	3.648	3.686	3.725	3.764	3.802	3.842	3.881	3.920	3.960	4	8	12	16	20	23	27	31	35
2.0	4.000	4.040	4.080	4.121	4.162	4.202	4.244	4.285	4.326	4.368	4	8	12	16	21	25	29	33	37
2.1	4.410	4.452	4.494	4.537	4.580	4.622	4.666	4.709	4.752	4.796	4	9	13	17	22	26	30	34	39
2.2	4.840	4.884	4.928	4.973	5.018	5.062	5.108	5.153	5.198	5.244	5	9	14	18	23	27	32	36	41
2.3	5.290	5.336	5.382	5.429	5.476	5.522	5.570	5.617	5.664	5.712	5	9	14	19	24	28	33	38	42
2.4	5.760	5.808	5.856	5.905	5.954	6.002	6.052	6.101	6.150	6.200	5	10	15	20	25	29	34	39	44
2.5	6.250	6.300	6.350	6.401	6.452	6.502	6.554	6.605	6.656	6.708	5	10	15	20	26	31	36	41	46
2.6	6.760	6.812	6.864	6.917	6.970	7.022	7.076	7.129	7.182	7.236	5	11	16	21	27	32	37	42	48
2.7	7.290	7.344	7.398	7.453	7.508	7.562	7.618	7.673	7.728	7.784	6	11	17	22	28	33	39	44	50
2.8	7.840	7.896	7.952	8.009	8.066	8.122	8.180	8.237	8.294	8.352	6	11	17	23	29	34	40	46	51
2.9	8.410	8.468	8.526	8.585	8.644	8.702	8.762	8.821	8.880	8.940	6	12	18	24	30	35	41	47	53
3.0	9.00	9.06	9.12	9.18	9.24	9.30	9.36	9.42	9.49	9.55	1	1	2	2	3	4	4	5	5
3.1	9.61	9.67	9.73	9.80	9.86	9.92	9.99	10.05	10.11	10.18	1	1	2	3	3	4	4	5	6
3.2	10.24	10.30	10.37	10.43	10.50	10.56	10.63	10.69	10.76	10.82	1	1	2	3	3	4	5	5	6
3.3	10.89	10.96	11.02	11.09	11.16	11.22	11.29	11.36	11.42	11.49	1	1	2	3	3	4	5	5	6
3.4	11.56	11.63	11.70	11.76	11.83	11.90	11.97	12.04	12.11	12.18	1	1	2	3	3	4	5	6	6
3.5	12.25	12.32	12.39	12.46	12.53	12.60	12.67	12.74	12.82	12.89	1	1	2	3	4	4	5	6	6
3.6	12.96	13.03	13.10	13.18	13.25	13.32	13.40	13.47	13.54	13.62	1	1	2	3	4	4	5	6	7
3.7	13.69	13.76	13.84	13.91	13.99	14.06	14.14	14.21	14.29	14.36	1	2	2	3	4	5	5	6	7
3.8	14.44	14.52	14.59	14.67	14.75	14.82	14.90	14.98	15.05	15.13	1	2	2	3	4	5	5	6	7
3.9	15.21	15.29	15.37	15.44	15.52	15.60	15.68	15.76	15.84	15.92	1	2	2	3	4	5	6	6	7
4.0	16.00	16.08	16.16	16.24	16.32	16.40	16.48	16.56	16.65	16.73	1	2	2	3	4	5	6	6	7
4.1	16.81	16.89	16.97	17.06	17.14	17.22	17.31	17.39	17.47	17.56	1	2	2	3	4	5	6	7	7
4.2	17.64	17.72	17.81	17.89	17.98	18.06	18.15	18.23	18.32	18.40	1	2	3	3	4	5	6	7	8
4.3	18.49	18.58	18.66	18.75	18.84	18.92	19.01	19.10	19.18	19.27	1	2	3	3	4	5	6	7	8
4.4	19.36	19.45	19.54	19.62	19.71	19.80	19.89	19.98	20.07	20.16	1	2	3	4	4	5	6	7	8
4.5	20.25	20.34	20.43	20.52	20.61	20.70	20.79	20.88	20.98	21.07	1	2	3	4	5	5	6	7	8
4.6	21.16	21.25	21.34	21.44	21.53	21.62	21.72	21.81	21.90	22.00	1	2	3	4	5	6	7	7	8
4.7	22.09	22.18	22.28	22.37	22.47	22.56	22.66	22.75	22.85	22.94	1	2	3	4	5	6	7	8	9
4.8	23.04	23.14	23.23	23.33	23.43	23.52	23.62	23.72	23.81	23.91	1	2	3	4	5	6	7	8	9
4.9	24.01	24.11	24.21	24.30	24.40	24.50	24.60	24.70	24.80	24.90	1	2	3	4	5	6	7	8	9
5.0	25.00	25.10	25.20	25.30	25.40	25.50	25.60	25.70	25.81	25.91	1	2	3	4	5	6	7	8	9
5.1	26.01	26.11	26.21	26.32	26.42	26.52	26.63	26.73	26.83	26.94	1	2	3	4	5	6	7	8	9
5.2	27.04	27.14	27.25	27.35	27.46	27.56	27.67	27.77	27.88	27.98	1	2	3	4	5	6	7	8	9
5.3	28.09	28.20	28.30	28.41	28.52	28.62	28.73	28.84	28.94	29.05	1	2	3	4	5	6	7	9	10
5.4	29.16	29.27	29.38	29.48	29.59	29.70	29.81	29.92	30.03	30.14	1	2	3	4	5	7	8	9	10
	0	1	2	3	4	5	6	7	8	9	1	2	3	4	5	6	7	8	9

	0	1	2	3	4	5	6	7	8	9	1	2	3	4	5	6	7	8	9
5·5	30·25	30·36	30·47	30·58	30·69	30·80	30·91	31·02	31·14	31·25	1	2	3	4	6	7	8	9	10
5·6	31·36	31·47	31·58	31·70	31·81	31·92	32·04	32·15	32·26	32·38	1	2	3	5	6	7	8	9	10
5·7	32·49	32·60	32·72	32·83	32·95	33·06	33·18	33·29	33·41	33·52	1	2	3	5	6	7	8	9	10
5·8	33·64	33·76	33·87	33·99	34·11	34·22	34·34	34·46	34·57	34·69	1	2	4	5	6	7	8	9	11
5·9	34·81	34·93	35·05	35·16	35·28	35·40	35·52	35·64	35·76	35·88	1	2	4	5	6	7	8	10	11
6·0	36·00	36·12	36·24	36·36	36·48	36·60	36·72	36·84	36·97	37·09	1	2	4	5	6	7	8	10	11
6·1	37·21	37·33	37·45	37·58	37·70	37·82	37·95	38·07	38·19	38·32	1	2	4	5	6	7	9	10	11
6·2	38·44	38·56	38·69	38·81	38·94	39·06	39·19	39·31	39·44	39·56	1	3	4	5	6	8	9	10	11
6·3	39·69	39·82	39·94	40·07	40·20	40·32	40·45	40·58	40·70	40·83	1	3	4	5	6	8	9	10	11
6·4	40·96	41·09	41·22	41·34	41·47	41·60	41·73	41·86	41·99	42·12	1	3	4	5	6	8	9	10	12
6·5	42·25	42·38	42·51	42·64	42·77	42·90	43·03	43·16	43·30	43·43	1	3	4	5	7	8	9	10	12
6·6	43·56	43·69	43·82	43·96	44·09	44·22	44·36	44·49	44·62	44·76	1	3	4	5	7	8	9	11	12
6·7	44·89	45·02	45·16	45·29	45·43	45·56	45·70	45·83	45·97	46·10	1	3	4	5	7	8	9	11	12
6·8	46·24	46·38	46·51	46·65	46·79	46·92	47·06	47·20	47·33	47·47	1	3	4	5	7	8	10	11	12
6·9	47·61	47·75	47·89	48·02	48·16	48·30	48·44	48·58	48·72	48·86	1	3	4	6	7	8	10	11	13
7·0	49·00	49·14	49·28	49·42	49·56	49·70	49·84	49·98	50·13	50·27	1	3	4	6	7	8	10	11	13
7·1	50·41	50·55	50·69	50·84	50·98	51·12	51·27	51·41	51·55	51·70	1	3	4	6	7	9	10	11	13
7·2	51·84	51·98	52·13	52·27	52·42	52·56	52·71	52·85	53·00	53·14	1	3	4	6	7	9	10	12	13
7·3	53·29	53·44	53·58	53·73	53·88	54·02	54·17	54·32	54·46	54·61	1	3	4	6	7	9	10	12	13
7·4	54·76	54·91	55·06	55·20	55·35	55·50	55·65	55·80	55·95	56·10	1	3	4	6	7	9	10	12	13
7·5	56·25	56·40	56·55	56·70	56·85	57·00	57·15	57·30	57·46	57·61	2	3	5	6	8	9	11	12	14
7·6	57·76	57·91	58·06	58·22	58·37	58·52	58·68	58·83	58·98	59·14	2	3	5	6	8	9	11	12	14
7·7	59·29	59·44	59·60	59·75	59·91	60·06	60·22	60·37	60·53	60·68	2	3	5	6	8	9	11	12	14
7·8	60·84	61·00	61·15	61·31	61·47	61·62	61·78	61·94	62·09	62·25	2	3	5	6	8	9	11	13	14
7·9	62·41	62·57	62·73	62·88	63·04	63·20	63·36	63·52	63·68	63·84	2	3	5	6	8	10	11	13	14
8·0	64·00	64·16	64·32	64·48	64·64	64·80	64·96	65·12	65·29	65·45	2	3	5	6	8	10	11	13	14
8·1	65·61	65·77	65·93	66·10	66·26	66·42	66·59	66·75	66·91	67·08	2	3	5	7	8	10	11	13	15
8·2	67·24	67·40	67·57	67·73	67·90	68·06	68·23	68·39	68·56	68·72	2	3	5	7	8	10	12	13	15
8·3	68·89	69·06	69·22	69·39	69·56	69·72	69·89	70·06	70·22	70·39	2	3	5	7	8	10	12	13	15
8·4	70·56	70·73	70·90	71·06	71·23	71·40	71·57	71·74	71·91	72·08	2	3	5	7	8	10	12	14	15
8·5	72·25	72·42	72·59	72·76	72·93	73·10	73·27	73·44	73·62	73·79	2	3	5	7	9	10	12	14	15
8·6	73·96	74·13	74·30	74·48	74·65	74·82	75·00	75·17	75·34	75·52	2	3	5	7	9	10	12	14	16
8·7	75·69	75·86	76·04	76·21	76·39	76·56	76·74	76·91	77·09	77·26	2	4	5	7	9	11	12	14	16
8·8	77·44	77·62	77·79	77·97	78·15	78·32	78·50	78·68	78·85	79·03	2	4	5	7	9	11	12	14	16
8·9	79·21	79·39	79·57	79·74	79·92	80·10	80·28	80·46	80·64	80·82	2	4	5	7	9	11	13	14	16
9·0	81·00	81·18	81·36	81·54	81·72	81·90	82·08	82·26	82·45	82·63	2	4	5	7	9	11	13	14	16
9·1	82·81	82·99	83·17	83·36	83·54	83·72	83·91	84·09	84·27	84·46	2	4	5	7	9	11	13	15	16
9·2	84·64	84·82	85·01	85·19	85·38	85·56	85·75	85·93	86·12	86·30	2	4	6	7	9	11	13	15	17
9·3	86·49	86·68	86·86	87·05	87·24	87·42	87·61	87·80	87·98	88·17	2	4	6	7	9	11	13	15	17
9·4	88·36	88·55	88·74	88·92	89·11	89·30	89·49	89·68	89·87	90·06	2	4	6	8	9	11	13	15	17
9·5	90·25	90·44	90·63	90·82	91·01	91·20	91·39	91·58	91·78	91·97	2	4	6	8	10	11	13	15	17
9·6	92·16	92·35	92·54	92·74	92·93	93·12	93·32	93·51	93·70	93·90	2	4	6	8	10	12	14	15	17
9·7	94·09	94·28	94·48	94·67	94·87	95·06	95·26	95·45	95·65	95·84	2	4	6	8	10	12	14	16	18
9·8	96·04	96·24	96·43	96·63	96·83	97·02	97·22	97·42	97·61	97·81	2	4	6	8	10	12	14	16	18
9·9	98·01	98·21	98·41	98·60	98·80	99·00	99·20	99·40	99·60	99·80	2	4	6	8	10	12	14	16	18
	0	1	2	3	4	5	6	7	8	9	1	2	3	4	5	6	7	8	9

	0′	6′	12′	18′	24′	30′	36′	42′	48′	54′	1′	2′	3′	4′	5′
0°	0·0000	0017	0035	0052	0070	0087	0105	0122	0140	0157	3	6	9	12	15
1	0·0175	0192	0209	0227	0244	0262	0279	0297	0314	0332	3	6	9	12	15
2	0·0349	0366	0384	0401	0419	0436	0454	0471	0488	0506	3	6	9	12	15
3	0·0523	0541	0558	0576	0593	0610	0628	0645	0663	0680	3	6	9	12	15
4	0·0698	0715	0732	0750	0767	0785	0802	0819	0837	0854	3	6	9	12	14
5	0·0872	0889	0906	0924	0941	0958	0976	0993	1011	1028	3	6	9	12	14
6	0·1045	1063	1080	1097	1115	1132	1149	1167	1184	1201	3	6	9	12	14
7	0·1219	1236	1253	1271	1288	1305	1323	1340	1357	1374	3	6	9	12	14
8	0·1392	1409	1426	1444	1461	1478	1495	1513	1530	1547	3	6	9	11	14
9	0·1564	1582	1599	1616	1633	1650	1668	1685	1702	1719	3	6	9	11	14
10	0·1736	1754	1771	1788	1805	1822	1840	1857	1874	1891	3	6	9	11	14
11	0·1908	1925	1942	1959	1977	1994	2011	2028	2045	2062	3	6	9	11	14
12	0·2079	2096	2113	2130	2147	2164	2181	2198	2215	2233	3	6	9	11	14
13	0·2250	2267	2284	2300	2317	2334	2351	2368	2385	2402	3	6	8	11	14
14	0·2419	2436	2453	2470	2487	2504	2521	2538	2554	2571	3	6	8	11	14
15	0·2588	2605	2622	2639	2656	2672	2689	2706	2723	2740	3	6	8	11	14
16	0·2756	2773	2790	2807	2823	2840	2857	2874	2890	2907	3	6	8	11	14
17	0·2924	2940	2957	2974	2990	3007	3024	3040	3057	3074	3	6	8	11	14
18	0·3090	3107	3123	3140	3156	3173	3190	3206	3223	3239	3	6	8	11	14
19	0·3256	3272	3289	3305	3322	3338	3355	3371	3387	3404	3	5	8	11	14
20	0·3420	3437	3453	3469	3486	3502	3518	3535	3551	3567	3	5	8	11	14
21	0·3584	3600	3616	3633	3649	3665	3681	3697	3714	3730	3	5	8	11	13
22	0·3746	3762	3778	3795	3811	3827	3843	3859	3875	3891	3	5	8	11	13
23	0·3907	3923	3939	3955	3971	3987	4003	4019	4035	4051	3	5	8	11	13
24	0·4067	4083	4099	4115	4131	4147	4163	4179	4195	4210	3	5	8	11	13
25	0·4226	4242	4258	4274	4289	4305	4321	4337	4352	4368	3	5	8	11	13
26	0·4384	4399	4415	4431	4446	4462	4478	4493	4509	4524	3	5	8	10	13
27	0·4540	4555	4571	4586	4602	4617	4633	4648	4664	4679	3	5	8	10	13
28	0·4695	4710	4726	4741	4756	4772	4787	4802	4818	4833	3	5	8	10	13
29	0·4848	4863	4879	4894	4909	4924	4939	4955	4970	4985	3	5	8	10	13
30	0·5000	5015	5030	5045	5060	5075	5090	5105	5120	5135	3	5	8	10	12
31	0·5150	5165	5180	5195	5210	5225	5240	5255	5270	5284	2	5	7	10	12
32	0·5299	5314	5329	5344	5358	5373	5388	5402	5417	5432	2	5	7	10	12
33	0·5446	5461	5476	5490	5505	5519	5534	5548	5563	5577	2	5	7	10	12
34	0·5592	5606	5621	5635	5650	5664	5678	5693	5707	5721	2	5	7	10	12
35	0·5736	5750	5764	5779	5793	5807	5821	5835	5850	5864	2	5	7	9	12
36	0·5878	5892	5906	5920	5934	5948	5962	5976	5990	6004	2	5	7	9	12
37	0·6018	6032	6046	6060	6074	6088	6101	6115	6129	6143	2	5	7	9	12
38	0·6157	6170	6184	6198	6211	6225	6239	6252	6266	6280	2	5	7	9	11
39	0·6293	6307	6320	6334	6347	6361	6374	6388	6401	6414	2	4	7	9	11
40	0·6428	6441	6455	6468	6481	6494	6508	6521	6534	6547	2	4	7	9	11
41	0·6561	6574	6587	6600	6613	6626	6639	6652	6665	6678	2	4	7	9	11
42	0·6691	6704	6717	6730	6743	6756	6769	6782	6794	6807	2	4	6	9	11
43	0·6820	6833	6845	6858	6871	6884	6896	6909	6921	6934	2	4	6	8	11
44	0·6947	6959	6972	6984	6997	7009	7022	7034	7046	7059	2	4	6	8	10
	0′	6′	12′	18′	24′	30′	36′	42′	48′	54′	1′	2′	3′	4′	5′

	0′	6′	12′	18′	24′	30′	36′	42′	48′	54′	1′	2′	3′	4′	5′
45°	0·7071	7083	7096	7108	7120	7133	7145	7157	7169	7181	2	4	6	8	10
46	0·7193	7206	7218	7230	7242	7254	7266	7278	7290	7302	2	4	6	8	10
47	0·7314	7325	7337	7349	7361	7373	7385	7396	7408	7420	2	4	6	8	10
48	0·7431	7443	7455	7466	7478	7490	7501	7513	7524	7536	2	4	6	8	10
49	0·7547	7559	7570	7581	7593	7604	7615	7627	7638	7649	2	4	6	8	9
50	0·7660	7672	7683	7694	7705	7716	7727	7738	7749	7760	2	4	6	7	9
51	0·7771	7782	7793	7804	7815	7826	7837	7848	7859	7869	2	4	5	7	9
52	0·7880	7891	7902	7912	7923	7934	7944	7955	7965	7976	2	4	5	7	9
53	0·7986	7997	8007	8018	8028	8039	8049	8059	8070	8080	2	3	5	7	9
54	0·8090	8100	8111	8121	8131	8141	8151	8161	8171	8181	2	3	5	7	8
55	0·8192	8202	8211	8221	8231	8241	8251	8261	8271	8281	2	3	5	7	8
56	0·8290	8300	8310	8320	8329	8339	8348	8358	8368	8377	2	3	5	6	8
57	0·8387	8396	8406	8415	8425	8434	8443	8453	8462	8471	2	3	5	6	8
58	0·8480	8490	8499	8508	8517	8526	8536	8545	8554	8563	2	3	5	6	8
59	0·8572	8581	8590	8599	8607	8616	8625	8634	8643	8652	1	3	4	6	7
60	0·8660	8669	8678	8686	8695	8704	8712	8721	8729	8738	1	3	4	6	7
61	0·8746	8755	8763	8771	8780	8788	8796	8805	8813	8821	1	3	4	6	7
62	0·8829	8838	8846	8854	8862	8870	8878	8886	8894	8902	1	3	4	5	7
63	0·8910	8918	8926	8934	8942	8949	8957	8965	8973	8980	1	3	4	5	6
64	0·8988	8996	9003	9011	9018	9026	9033	9041	9048	9056	1	2	4	5	6
65	0·9063	9070	9078	9085	9092	9100	9107	9114	9121	9128	1	2	4	5	6
66	0·9135	9143	9150	9157	9164	9171	9178	9184	9191	9198	1	2	4	5	6
67	0·9205	9212	9219	9225	9232	9239	9245	9252	9259	9265	1	2	3	4	6
68	0·9272	9278	9285	9291	9298	9304	9311	9317	9323	9330	1	2	3	4	5
69	0·9336	9342	9348	9354	9361	9367	9373	9379	9385	9391	1	2	3	4	5
70	0·9397	9403	9409	9415	9421	9426	9432	9438	9444	9449	1	2	3	4	5
71	0·9455	9461	9466	9472	9478	9483	9489	9494	9500	9505	1	2	3	4	5
72	0·9511	9516	9521	9527	9532	9537	9542	9548	9553	9558	1	2	3	3	4
73	0·9563	9568	9573	9578	9583	9588	9593	9598	9603	9608	1	2	3	3	4
74	0·9613	9617	9622	9627	9632	9636	9641	9646	9650	9655	1	2	2	3	4
75	0·9659	9664	9668	9673	9677	9681	9686	9690	9694	9699	1	1	2	3	4
76	0·9703	9707	9711	9715	9720	9724	9728	9732	9736	9740	1	1	2	3	3
77	0·9744	9748	9751	9755	9759	9763	9767	9770	9774	9778	1	1	2	2	3
78	0·9781	9785	9789	9792	9796	9799	9803	9806	9810	9813	1	1	2	2	3
79	0·9816	9820	9823	9826	9829	9833	9836	9839	9842	9845	1	1	2	2	3
80	0·9848	9851	9854	9857	9860	9863	9866	9869	9871	9874	0	1	1	2	2
81	0·9877	9880	9882	9885	9888	9890	9893	9895	9898	9900	0	1	1	2	2
82	0·9903	9905	9907	9910	9912	9914	9917	9919	9921	9923	0	1	1	1	2
83	0·9925	9928	9930	9932	9934	9936	9938	9940	9942	9943	0	1	1	1	2
84	0·9945	9947	9949	9951	9952	9954	9956	9957	9959	9960	0	1	1	1	1
85	0·9962	9963	9965	9966	9968	9969	9971	9972	9973	9974	0	0	1	1	1
86	0·9976	9977	9978	9979	9980	9981	9982	9983	9984	9985	0	0	1	1	1
87	0·9986	9987	9988	9989	9990	9990	9991	9992	9993	9993	0	0	0	1	1
88	0·9994	9995	9995	9996	9996	9997	9997	9997	9998	9998	0	0	0	0	0
89	0·9998	9999	9999	9999	9999	0000	0000	0000	0000	0000	0	0	0	0	0
	0′	6′	12′	18′	24′	30′	36′	42′	48′	54′	1′	2′	3′	4′	5′

	0′	6′	12′	18′	24′	30′	36′	42′	48′	54′	1′	2′	3′	4′	5′
0°	1·0000	0000	0000	0000	0000	0000	9999	9999	9999	9999	0	0	0	0	0
1	0·9998	9998	9998	9997	9997	9997	9996	9996	9995	9995	0	0	0	0	0
2	0·9994	9993	9993	9992	9991	9990	9990	9989	9988	9987	0	0	0	1	1
3	0·9986	9985	9984	9983	9982	9981	9980	9979	9978	9977	0	0	1	1	1
4	0·9976	9974	9973	9972	9971	9969	9968	9966	9965	9963	0	0	1	1	1
5	0·9962	9960	9959	9957	9956	9954	9952	9951	9949	9947	0	1	1	1	1
6	0·9945	9943	9942	9940	9938	9936	9934	9932	9930	9928	0	1	1	1	2
7	0·9925	9923	9921	9919	9917	9914	9912	9910	9907	9905	0	1	1	1	2
8	0·9903	9900	9898	9895	9893	9890	9888	9885	9882	9880	0	1	1	2	2
9	0·9877	9874	9871	9869	9866	9863	9860	9857	9854	9851	0	1	1	2	2
10	0·9848	9845	9842	9839	9836	9833	9829	9826	9823	9820	1	1	2	2	3
11	0·9816	9813	9810	9806	9803	9799	9796	9792	9789	9785	1	1	2	2	3
12	0·9781	9778	9774	9770	9767	9763	9759	9755	9751	9748	1	1	2	2	3
13	0·9744	9740	9736	9732	9728	9724	9720	9715	9711	9707	1	1	2	3	3
14	0·9703	9699	9694	9690	9686	9681	9677	9673	9668	9664	1	1	2	3	4
15	0·9659	9655	9650	9646	9641	9636	9632	9627	9622	9617	1	2	2	3	4
16	0·9613	9608	9603	9598	9593	9588	9583	9578	9573	9568	1	2	3	3	4
17	0·9563	9558	9553	9548	9542	9537	9532	9527	9521	9516	1	2	3	3	4
18	0·9511	9505	9500	9494	9489	9483	9478	9472	9466	9461	1	2	3	4	5
19	0·9455	9449	9444	9438	9432	9426	9421	9415	9409	9403	1	2	3	4	5
20	0·9397	9391	9385	9379	9373	9367	9361	9354	9348	9342	1	2	3	4	5
21	0·9336	9330	9323	9317	9311	9304	9298	9291	9285	9278	1	2	3	4	5
22	0·9272	9265	9259	9252	9245	9239	9232	9225	9219	9212	1	2	3	4	6
23	0·9205	9198	9191	9184	9178	9171	9164	9157	9150	9143	1	2	4	5	6
24	0·9135	9128	9121	9114	9107	9100	9092	9085	9078	9070	1	2	4	5	6
25	0·9063	9056	9048	9041	9033	9026	9018	9011	9003	8996	1	2	4	5	6
26	0·8988	8980	8973	8965	8957	8949	8942	8934	8926	8918	1	3	4	5	6
27	0·8910	8902	8894	8886	8878	8870	8862	8854	8846	8838	1	3	4	5	7
28	0·8829	8821	8813	8805	8796	8788	8780	8771	8763	8755	1	3	4	6	7
29	0·8746	8738	8729	8721	8712	8704	8695	8686	8678	8669	1	3	4	6	7
30	0·8660	8652	8643	8634	8625	8616	8607	8599	8590	8581	1	3	4	6	7
31	0·8572	8563	8554	8545	8536	8526	8517	8508	8499	8490	2	3	5	6	8
32	0·8480	8471	8462	8453	8443	8434	8425	8415	8406	8396	2	3	5	6	8
33	0·8387	8377	8368	8358	8348	8339	8329	8320	8310	8300	2	3	5	6	8
34	0·8290	8281	8271	8261	8251	8241	8231	8221	8211	8202	2	3	5	7	8
35	0·8192	8181	8171	8161	8151	8141	8131	8121	8111	8100	2	3	5	7	8
36	0·8090	8080	8070	8059	8049	8039	8028	8018	8007	7997	2	3	5	7	9
37	0·7986	7976	7965	7955	7944	7934	7923	7912	7902	7891	2	4	5	7	9
38	0·7880	7869	7859	7848	7837	7826	7815	7804	7793	7782	2	4	5	7	9
39	0·7771	7760	7749	7738	7727	7716	7705	7694	7683	7672	2	4	6	7	9
40	0·7660	7649	7638	7627	7615	7604	7593	7581	7570	7559	2	4	6	8	9
41	0·7547	7536	7524	7513	7501	7490	7478	7466	7455	7443	2	4	6	8	10
42	0·7431	7420	7408	7396	7385	7373	7361	7349	7337	7325	2	4	6	8	10
43	0·7314	7302	7290	7278	7266	7254	7242	7230	7218	7206	2	4	6	8	10
44	0·7193	7181	7169	7157	7145	7133	7120	7108	7096	7083	2	4	6	8	10
	0′	6′	12′	18′	24′	30′	36′	42′	48′	54′	1′	2′	3′	4′	5′

Subtract Differences

	0′	6′	12′	18′	24′	30′	36′	42′	48′	54′	1′	2′	3′	4′	5′
45°	0·7071	7059	7046	7034	7022	7009	6997	6984	6972	6959	2	4	6	8	10
46	0·6947	6934	6921	6909	6896	6884	6871	6858	6845	6833	2	4	6	8	11
47	0·6820	6807	6794	6782	6769	6756	6743	6730	6717	6704	2	4	6	9	11
48	0·6691	6678	6665	6652	6639	6626	6613	6600	6587	6574	2	4	7	9	11
49	0·6561	6547	6534	6521	6508	6494	6481	6468	6455	6441	2	4	7	9	11
50	0·6428	6414	6401	6388	6374	6361	6347	6334	6320	6307	2	4	7	9	11
51	0·6293	6280	6266	6252	6239	6225	6211	6198	6184	6170	2	5	7	9	11
52	0·6157	6143	6129	6115	6101	6088	6074	6060	6046	6032	2	5	7	9	12
53	0·6018	6004	5990	5976	5962	5948	5934	5920	5906	5892	2	5	7	9	12
54	0·5878	5864	5850	5835	5821	5807	5793	5779	5764	5750	2	5	7	9	12
55	0·5736	5721	5707	5693	5678	5664	5650	5635	5621	5606	2	5	7	10	12
56	0·5592	5577	5563	5548	5534	5519	5505	5490	5476	5461	2	5	7	10	12
57	0·5446	5432	5417	5402	5388	5373	5358	5344	5329	5314	2	5	7	10	12
58	0·5299	5284	5270	5255	5240	5225	5210	5195	5180	5165	2	5	7	10	12
59	0·5150	5135	5120	5105	5090	5075	5060	5045	5030	5015	3	5	8	10	12
60	0·5000	4985	4970	4955	4939	4924	4909	4894	4879	4863	3	5	8	10	13
61	0·4848	4833	4818	4802	4787	4772	4756	4741	4726	4710	3	5	8	10	13
62	0·4695	4679	4664	4648	4633	4617	4602	4586	4571	4555	3	5	8	10	13
63	0·4540	4524	4509	4493	4478	4462	4446	4431	4415	4399	3	5	8	10	13
64	0·4384	4368	4352	4337	4321	4305	4289	4274	4258	4242	3	5	8	11	13
65	0·4226	4210	4195	4179	4163	4147	4131	4115	4099	4083	3	5	8	11	13
66	0·4067	4051	4035	4019	4003	3987	3971	3955	3939	3923	3	5	8	11	13
67	0·3907	3891	3875	3859	3843	3827	3811	3795	3778	3762	3	5	8	11	13
68	0·3746	3730	3714	3697	3681	3665	3649	3633	3616	3600	3	5	8	11	13
69	0·3584	3567	3551	3535	3518	3502	3486	3469	3453	3437	3	5	8	11	14
70	0·3420	3404	3387	3371	3355	3338	3322	3305	3289	3272	3	5	8	11	14
71	0·3256	3239	3223	3206	3190	3173	3156	3140	3123	3107	3	6	8	11	14
72	0·3090	3074	3057	3040	3024	3007	2990	2974	2957	2940	3	6	8	11	14
73	0·2924	2907	2890	2874	2857	2840	2823	2807	2790	2773	3	6	8	11	14
74	0·2756	2740	2723	2706	2689	2672	2656	2639	2622	2605	3	6	8	11	14
75	0·2588	2571	2554	2538	2521	2504	2487	2470	2453	2436	3	6	8	11	14
76	0·2419	2402	2385	2368	2351	2334	2317	2300	2284	2267	3	6	8	11	14
77	0·2250	2233	2215	2198	2181	2164	2147	2130	2113	2096	3	6	9	11	14
78	0·2079	2062	2045	2028	2011	1994	1977	1959	1942	1925	3	6	9	11	14
79	0·1908	1891	1874	1857	1840	1822	1805	1788	1771	1754	3	6	9	11	14
80	0·1736	1719	1702	1685	1668	1650	1633	1616	1599	1582	3	6	9	11	14
81	0·1564	1547	1530	1513	1495	1478	1461	1444	1426	1409	3	6	9	11	14
82	0·1392	1374	1357	1340	1323	1305	1288	1271	1253	1236	3	6	9	12	14
83	0·1219	1201	1184	1167	1149	1132	1115	1097	1080	1063	3	6	9	12	14
84	0·1045	1028	1011	0993	0976	0958	0941	0924	0906	0889	3	6	9	12	14
85	0·0872	0854	0837	0819	0802	0785	0767	0750	0732	0715	3	6	9	12	15
86	0·0698	0680	0663	0645	0628	0610	0593	0576	0558	0541	3	6	9	12	15
87	0·0523	0506	0488	0471	0454	0436	0419	0401	0384	0366	3	6	9	12	15
88	0·0349	0332	0314	0297	0279	0262	0244	0227	0209	0192	3	6	9	12	15
89	0·0175	0157	0140	0122	0105	0087	0070	0052	0035	0017	3	6	9	12	15
	0′	6′	12′	18′	24′	30′	36′	42′	48′	54′	1′	2′	3′	4′	5′

	0′	6′	12′	18′	24′	30′	36′	42′	48′	54′	1′	2′	3′	4′	5′
0°	0·0000	0017	0035	0052	0070	0087	0105	0122	0140	0157	3	6	9	12	15
1	0·0175	0192	0209	0227	0244	0262	0279	0297	0314	0332	3	6	9	12	15
2	0·0349	0367	0384	0402	0419	0437	0454	0472	0489	0507	3	6	9	12	15
3	0·0524	0542	0559	0577	0594	0612	0629	0647	0664	0682	3	6	9	12	15
4	0·0699	0717	0734	0752	0769	0787	0805	0822	0840	0857	3	6	9	12	15
5	0·0875	0892	0910	0928	0945	0963	0981	0998	1016	1033	3	6	9	12	15
6	0·1051	1069	1086	1104	1122	1139	1157	1175	1192	1210	3	6	9	12	15
7	0·1228	1246	1263	1281	1299	1317	1334	1352	1370	1388	3	6	9	12	15
8	0·1405	1423	1441	1459	1477	1495	1512	1530	1548	1566	3	6	9	12	15
9	0·1584	1602	1620	1638	1655	1673	1691	1709	1727	1745	3	6	9	12	15
10	0·1763	1781	1799	1817	1835	1853	1871	1890	1908	1926	3	6	9	12	15
11	0·1944	1962	1980	1998	2016	2035	2053	2071	2089	2107	3	6	9	12	15
12	0·2126	2144	2162	2180	2199	2217	2235	2254	2272	2290	3	6	9	12	15
13	0·2309	2327	2345	2364	2382	2401	2419	2438	2456	2475	3	6	9	12	15
14	0·2493	2512	2530	2549	2568	2586	2605	2623	2642	2661	3	6	9	12	15
15	0·2679	2698	2717	2736	2754	2773	2792	2811	2830	2849	3	6	9	13	16
16	0·2867	2886	2905	2924	2943	2962	2981	3000	3019	3038	3	6	10	13	16
17	0·3057	3076	3096	3115	3134	3153	3172	3191	3211	3230	3	6	10	13	16
18	0·3249	3269	3288	3307	3327	3346	3365	3385	3404	3424	3	6	10	13	16
19	0·3443	3463	3482	3502	3522	3541	3561	3581	3600	3620	3	7	10	13	16
20	0·3640	3659	3679	3699	3719	3739	3759	3779	3799	3819	3	7	10	13	17
21	0·3839	3859	3879	3899	3919	3939	3959	3979	4000	4020	3	7	10	13	17
22	0·4040	4061	4081	4101	4122	4142	4163	4183	4204	4224	3	7	10	14	17
23	0·4245	4265	4286	4307	4327	4348	4369	4390	4411	4431	3	7	10	14	17
24	0·4452	4473	4494	4515	4536	4557	4578	4599	4621	4642	4	7	11	14	18
25	0·4663	4684	4706	4727	4748	4770	4791	4813	4834	4856	4	7	11	14	18
26	0·4877	4899	4921	4942	4964	4986	5008	5029	5051	5073	4	7	11	15	18
27	0·5095	5117	5139	5161	5184	5206	5228	5250	5272	5295	4	7	11	15	18
28	0·5317	5340	5362	5384	5407	5430	5452	5475	5498	5520	4	8	11	15	19
29	0·5543	5566	5589	5612	5635	5658	5681	5704	5727	5750	4	8	12	15	19
30	0·5774	5797	5820	5844	5867	5890	5914	5938	5961	5985	4	8	12	16	20
31	0·6009	6032	6056	6080	6104	6128	6152	6176	6200	6224	4	8	12	16	20
32	0·6249	6273	6297	6322	6346	6371	6395	6420	6445	6469	4	8	12	16	20
33	0·6494	6519	6544	6569	6594	6619	6644	6669	6694	6720	4	8	13	17	21
34	0·6745	6771	6796	6822	6847	6873	6899	6924	6950	6976	4	9	13	17	21
35	0·7002	7028	7054	7080	7107	7133	7159	7186	7212	7239	4	9	13	18	22
36	0·7265	7292	7319	7346	7373	7400	7427	7454	7481	7508	5	9	14	18	23
37	0·7536	7563	7590	7618	7646	7673	7701	7729	7757	7785	5	9	14	18	23
38	0·7813	7841	7869	7898	7926	7954	7983	8012	8040	8069	5	9	14	19	24
39	0·8098	8127	8156	8185	8214	8243	8273	8302	8332	8361	5	10	15	20	24
40	0·8391	8421	8451	8481	8511	8541	8571	8601	8632	8662	5	10	15	20	25
41	0·8693	8724	8754	8785	8816	8847	8878	8910	8941	8972	5	10	16	21	26
42	0·9004	9036	9067	9099	9131	9163	9195	9228	9260	9293	5	11	16	21	27
43	0·9325	9358	9391	9424	9457	9490	9523	9556	9590	9623	6	11	17	22	28
44	0·9657	9691	9725	9759	9793	9827	9861	9896	9930	9965	6	11	17	23	29
	0′	6′	12′	18′	24′	30′	36′	42′	48′	54′	1′	2′	3′	4′	5′

	0′	6′	12′	18′	24′	30′	36′	42′	48′	54′	1′	2′	3′	4′	5′
45°	1·0000	0035	0070	0105	0141	0176	0212	0247	0283	0319	6	12	18	24	30
46	1·0355	0392	0428	0464	0501	0538	0575	0612	0649	0686	6	12	18	25	31
47	1·0724	0761	0799	0837	0875	0913	0951	0990	1028	1067	6	13	19	25	32
48	1·1106	1145	1184	1224	1263	1303	1343	1383	1423	1463	7	13	20	27	33
49	1·1504	1544	1585	1626	1667	1708	1750	1792	1833	1875	7	14	21	28	34
50	1·1918	1960	2002	2045	2088	2131	2174	2218	2261	2305	7	14	22	29	36
51	1·2349	2393	2437	2482	2527	2572	2617	2662	2708	2753	8	15	23	30	37
52	1·2799	2846	2892	2938	2985	3032	3079	3127	3175	3222	8	16	24	31	39
53	1·3270	3319	3367	3416	3465	3514	3564	3613	3663	3713	8	16	25	33	41
54	1·3764	3814	3865	3916	3968	4019	4071	4124	4176	4229	9	17	26	34	43
55	1·4281	4335	4388	4442	4496	4550	4605	4659	4715	4770	9	18	27	36	45
56	1·4826	4882	4938	4994	5051	5108	5166	5224	5282	5340	10	19	29	38	48
57	1·5399	5458	5517	5577	5637	5697	5757	5818	5880	5941	10	20	30	40	50
58	1·6003	6066	6128	6191	6255	6319	6383	6447	6512	6577	11	21	32	43	53
59	1·6643	6709	6775	6842	6909	6977	7045	7113	7182	7251	11	23	34	45	57
60	1·732	1·739	1·746	1·753	1·760	1·767	1·775	1·782	1·789	1·797	1	2	4	5	6
61	1·804	1·811	1·819	1·827	1·834	1·842	1·849	1·857	1·865	1·873	1	3	4	5	6
62	1·881	1·889	1·897	1·905	1·913	1·921	1·929	1·937	1·946	1·954	1	3	4	5	7
63	1·963	1·971	1·980	1·988	1·997	2·006	2·014	2·023	2·032	2·041	1	3	4	6	7
64	2·050	2·059	2·069	2·078	2·087	2·097	2·106	2·116	2·125	2·135	2	3	5	6	8
65	2·145	2·154	2·164	2·174	2·184	2·194	2·204	2·215	2·225	2·236	2	3	5	7	8
66	2·246	2·257	2·267	2·278	2·289	2·300	2·311	2·322	2·333	2·344	2	4	6	7	9
67	2·356	2·367	2·379	2·391	2·402	2·414	2·426	2·438	2·450	2·463	2	4	6	8	10
68	2·475	2·488	2·500	2·513	2·526	2·539	2·552	2·565	2·578	2·592	2	4	7	9	11
69	2·605	2·619	2·633	2·646	2·660	2·675	2·689	2·703	2·718	2·733	2	5	7	9	12
70	2·747	2·762	2·778	2·793	2·808	2·824	2·840	2·856	2·872	2·888	3	5	8	10	13
71	2·904	2·921	2·937	2·954	2·971	2·989	3·006	3·024	3·042	3·060	3	6	9	12	14
72	3·078	3·096	3·115	3·133	3·152	3·172	3·191	3·211	3·230	3·251	3	6	10	13	16
73	3·271	3·291	3·312	3·333	3·354	3·376	3·398	3·420	3·442	3·465	4	7	11	14	18
74	3·487	3·511	3·534	3·558	3·582	3·606	3·630	3·655	3·681	3·706	4	8	12	16	20
75	3·732	3·758	3·785	3·812	3·839	3·867	3·895	3·923	3·952	3·981					
76	4·011	4·041	4·071	4·102	4·134	4·165	4·198	4·230	4·264	4·297					
77	4·331	4·366	4·402	4·437	4·474	4·511	4·548	4·586	4·625	4·665					
78	4·705	4·745	4·787	4·829	4·872	4·915	4·959	5·005	5·050	5·097					
79	5·145	5·193	5·242	5·292	5·343	5·396	5·449	5·503	5·558	5·614					
80	5·671	5·730	5·789	5·850	5·912	5·976	6·041	6·107	6·174	6·243					
81	6·314	6·386	6·460	6·535	6·612	6·691	6·772	6·855	6·940	7·026		Mean Differences			
82	7·115	7·207	7·300	7·396	7·495	7·596	7·700	7·806	7·916	8·028		no longer sufficiently			
83	8·144	8·264	8·386	8·513	8·643	8·777	8·915	9·058	9·205	9·357		accurate			
84	9·514	9·677	9·845	10·02	10·20	10·39	10·58	10·78	10·99	11·20					
85	11·43	11·66	11·91	12·16	12·43	12·71	13·00	13·30	13·62	13·95					
86	14·30	14·67	15·06	15·46	15·89	16·35	16·83	17·34	17·89	18·46					
87	19·08	19·74	20·45	21·20	22·02	22·90	23·86	24·90	26·03	27·27					
88	28·64	30·14	31·82	33·69	35·80	38·19	40·92	44·07	47·74	52·08					
89	57·29	63·66	71·62	81·85	95·49	114·6	143·2	191·0	286·5	573·0					
	0′	6′	12′	18′	24′	30′	36′	42′	48′	54′	1′	2′	3′	4′	5′

RADIANS

	0′	6′	12′	18′	24′	30′	36′	42′	48′	54′	1′	2′	3′	4′	5′
0°	0·0000	0017	0035	0052	0070	0087	0105	0122	0140	0157	3	6	9	12	15
1	0·0175	0192	0209	0227	0244	0262	0279	0297	0314	0332	3	6	9	12	15
2	0·0349	0367	0384	0401	0419	0436	0454	0471	0489	0506	3	6	9	12	15
3	0·0524	0541	0559	0576	0593	0611	0628	0646	0663	0681	3	6	9	12	15
4	0·0698	0716	0733	0750	0768	0785	0803	0820	0838	0855	3	6	9	12	15
5	0·0873	0890	0908	0925	0942	0960	0977	0995	1012	1030	3	6	9	12	15
6	0·1047	1065	1082	1100	1117	1134	1152	1169	1187	1204	3	6	9	12	15
7	0·1222	1239	1257	1274	1292	1309	1326	1344	1361	1379	3	6	9	12	15
8	0·1396	1414	1431	1449	1466	1484	1501	1518	1536	1553	3	6	9	12	15
9	0·1571	1588	1606	1623	1641	1658	1676	1693	1710	1728	3	6	9	12	15
10	0·1745	1763	1780	1798	1815	1833	1850	1868	1885	1902	3	6	9	12	15
11	0·1920	1937	1955	1972	1990	2007	2025	2042	2059	2077	3	6	9	12	ì5
12	0·2094	2112	2129	2147	2164	2182	2199	2217	2234	2251	3	6	9	12	15
13	0·2269	2286	2304	2321	2339	2356	2374	2391	2409	2426	3	6	9	12	15
14	0·2443	2461	2478	2496	2513	2531	2548	2566	2583	2601	3	6	9	12	15
15	0·2618	2635	2653	2670	2688	2705	2723	2740	2758	2775	3	6	9	12	15
16	0·2793	2810	2827	2845	2862	2880	2897	2915	2932	2950	3	6	9	12	15
17	0·2967	2985	3002	3019	3037	3054	3072	3089	3107	3124	3	6	9	12	15
18	0·3142	3159	3176	3194	3211	3229	3246	3264	3281	3299	3	6	9	12	15
19	0·3316	3334	3351	3368	3386	3403	3421	3438	3456	3473	3	6	9	12	15
20	0·3491	3508	3526	3543	3560	3578	3595	3613	3630	3648	3	6	9	12	15
21	0·3665	3683	3700	3718	3735	3752	3770	3787	3805	3822	3	6	9	12	15
22	0·3840	3857	3875	3892	3910	3927	3944	3962	3979	3997	3	6	9	12	15
23	0·4014	4032	4049	4067	4084	4102	4119	4136	4154	4171	3	6	9	12	15
24	0·4189	4206	4224	4241	4259	4276	4294	4311	4328	4346	3	6	9	12	15
25	0·4363	4381	4398	4416	4433	4451	4468	4485	4503	4520	3	6	9	12	15
26	0·4538	4555	4573	4590	4608	4625	4643	4660	4677	4695	3	6	9	12	15
27	0·4712	4730	4747	4765	4782	4800	4817	4835	4852	4869	3	6	9	12	15
28	0·4887	4904	4922	4939	4957	4974	4992	5009	5027	5044	3	6	9	12	15
29	0·5061	5079	5096	5114	5131	5149	5166	5184	5201	5219	3	6	9	12	15
30	0·5236	5253	5271	5288	5306	5323	5341	5358	5376	5393	3	6	9	12	15
31	0·5411	5428	5445	5463	5480	5498	5515	5533	5550	5568	3	6	9	12	15
32	0·5585	5603	5620	5637	5655	5672	5690	5707	5725	5742	3	6	9	12	15
33	0·5760	5777	5794	5812	5829	5847	5864	5882	5899	5917	3	6	9	12	15
34	0·5934	5952	5969	5986	6004	6021	6039	6056	6074	6091	3	6	9	12	15
35	0·6109	6126	6144	6161	6178	6196	6213	6231	6248	6266	3	6	9	12	15
36	0·6283	6301	6318	6336	6353	6370	6388	6405	6423	6440	3	6	9	12	15
37	0·6458	6475	6493	6510	6528	6545	6562	6580	6597	6615	3	6	9	12	15
38	0·6632	6650	6667	6685	6702	6720	6737	6754	6772	6789	3	6	9	12	15
39	0·6807	6824	6842	6859	6877	6894	6912	6929	6946	6964	3	6	9	12	15
40	0·6981	6999	7016	7034	7051	7069	7086	7103	7121	7138	3	6	9	12	15
41	0·7156	7173	7191	7208	7226	7243	7261	7278	7295	7313	3	6	9	12	15
42	0·7330	7348	7365	7383	7400	7418	7435	7453	7470	7487	3	6	9	12	15
43	0·7505	7522	7540	7557	7575	7592	7610	7627	7645	7662	3	6	9	12	15
44	0·7679	7697	7714	7732	7749	7767	7784	7802	7819	7837	3	6	9	12	15
	0′	6′	12′	18′	24′	30′	36′	42′	48′	54′	1′	2′	3′	4′	5′

	0′	6′	12′	18′	24′	30′	36′	42′	48′	54′	1′	2′	3′	4′	5′
45°	0·7854	7871	7889	7906	7924	7941	7959	7976	7994	8011	3	6	9	12	15
46	0·8029	8046	8063	8081	8098	8116	8133	8151	8168	8186	3	6	9	12	15
47	0·8203	8221	8238	8255	8273	8290	8308	8325	8343	8360	3	6	9	12	15
48	0·8378	8395	8412	8430	8447	8465	8482	8500	8517	8535	3	6	9	12	15
49	0·8552	8570	8587	8604	8622	8639	8657	8674	8692	8709	3	6	9	12	15
50	0·8727	8744	8762	8779	8796	8814	8831	8849	8866	8884	3	6	9	12	15
51	0·8901	8919	8936	8954	8971	8988	9006	9023	9041	9058	3	6	9	12	15
52	0·9076	9093	9111	9128	9146	9163	9180	9198	9215	9233	3	6	9	12	15
53	0·9250	9268	9285	9303	9320	9338	9355	9372	9390	9407	3	6	9	12	15
54	0·9425	9442	9460	9477	9495	9512	9529	9547	9564	9582	3	6	3	12	15
55	0·9599	9617	9634	9652	9669	9687	9704	9721	9739	9756	3	6	9	12	15
56	0·9774	9791	9809	9826	9844	9861	9879	9896	9913	9931	3	6	9	12	15
57	0·9948	9966	9983	0001	0018	0036	0053	0071	0088	0105	3	6	9	12	15
58	1·0123	0140	0158	0175	0193	0210	0228	0245	0263	0280	3	6	9	12	15
59	1·0297	0315	0332	0350	0367	0385	0402	0420	0437	0455	3	6	9	12	15
60	1·0472	0489	0507	0524	0542	0559	0577	0594	0612	0629	3	6	9	12	15
61	1·0647	0664	0681	0699	0716	0734	0751	0769	0786	0804	3	6	9	12	15
62	1·0821	0838	0856	0873	0891	0908	0926	0943	0961	0978	3	6	9	12	15
63	1·0996	1013	1030	1048	1065	1083	1100	1118	1135	1153	3	6	9	12	15
64	1·1170	1188	1205	1222	1240	1257	1275	1292	1310	1327	3	6	9	12	15
65	1·1345	1362	1380	1397	1414	1432	1449	1467	1484	1502	3	6	9	12	15
66	1·1519	1537	1554	1572	1589	1606	1624	1641	1659	1676	3	6	9	12	15
67	1·1694	1711	1729	1746	1764	1781	1798	1816	1833	1851	3	6	9	12	15
68	1·1868	1886	1903	1921	1938	1956	1973	1990	2008	2025	3	6	9	12	15
69	1·2043	2060	2078	2095	2113	2130	2147	2165	2182	2200	3	6	9	12	15
70	1·2217	2235	2252	2270	2287	2305	2322	2339	2357	2374	3	6	9	12	15
71	1·2392	2409	2427	2444	2462	2479	2497	2514	2531	2549	3	6	9	12	15
72	1·2566	2584	2601	2619	2636	2654	2671	2689	2706	2723	3	6	9	12	15
73	1·2741	2758	2776	2793	2811	2828	2846	2863	2881	2898	3	6	9	12	15
74	1·2915	2933	2950	2968	2985	3003	3020	3038	3055	3073	3	6	9	12	15
75	1·3090	3107	3125	3142	3160	3177	3195	3212	3230	3247	3	6	9	12	15
76	1·3265	3282	3299	3317	3334	3352	3369	3387	3404	3422	3	6	9	12	15
77	1·3439	3456	3474	3491	3509	3526	3544	3561	3579	3596	3	6	9	12	15
78	1·3614	3631	3648	3666	3683	3701	3718	3736	3753	3771	3	6	9	12	15
79	1·3788	3806	3823	3840	3858	3875	3893	3910	3928	3945	3	6	9	12	15
80	1·3963	3980	3998	4015	4032	4050	4067	4085	4102	4120	3	6	9	12	15
81	1·4137	4155	4172	4190	4207	4224	4242	4259	4277	4294	3	6	9	12	15
82	1·4312	4329	4347	4364	4382	4399	4416	4434	4451	4469	3	6	9	12	15
83	1·4486	4504	4521	4539	4556	4573	4591	4608	4626	4643	3	6	9	12	15
84	1·4661	4678	4696	4713	4731	4748	4765	4783	4800	4818	3	6	9	12	15
85	1·4835	4853	4870	4888	4905	4923	4940	4957	4975	4992	3	6	9	12	15
86	1·5010	5027	5045	5062	5080	5097	5115	5132	5149	5167	3	6	9	12	15
87	1·5184	5202	5219	5237	5254	5272	5289	5307	5324	5341	3	6	9	12	15
88	1·5359	5376	5394	5411	5429	5446	5464	5481	5499	5516	3	6	9	12	15
89	1·5533	5551	5568	5586	5603	5621	5638	5656	5673	5691	3	6	9	12	15
	0′	6′	12′	18′	24′	30′	36′	42′	48′	54′	1′	2′	3′	4′	5′

THE EXPONENTIAL e^{-x}

	·00	·01	·02	·03	·04	·05	·06	·07	·08	·09
0·0	1·00000	99005	98020	97045	96079	95123	94176	93239	92312	91393
0·1	0·90484	89583	88692	87810	86936	86071	85214	84366	83527	82696
0·2	0·81873	81058	80252	79453	78663	77880	77105	76338	75578	74826
0·3	0·74082	73345	72615	71892	71177	70469	69768	69073	68386	67706
0·4	0·67032	66365	65705	65051	64404	63763	63128	62500	61878	61263
0·5	0·60653	60050	59452	58860	58275	57695	57121	56553	55990	55433
0·6	0·54881	54335	53794	53259	52729	52205	51685	51171	50662	50158
0·7	0·49659	49164	48675	48191	47711	47237	46767	46301	45841	45384
0·8	0·44933	44486	44043	43605	43171	42741	42316	41895	41478	41066
0·9	0·40657	40252	39852	39455	39063	38674	38289	37908	37531	37158
1·0	0·36788	36422	36059	35701	35345	34994	34646	34301	33960	33622
1·1	0·33287	32956	32628	32303	31982	31664	31349	31037	30728	30422
1·2	0·30119	29820	29523	29229	28938	28650	28365	28083	27804	27527
1·3	0·27253	26982	26714	26448	26185	25924	25666	25411	25158	24908
1·4	0·24660	24414	24171	23931	23693	23457	23224	22993	22764	22537
1·5	0·22313	22091	21871	21654	21438	21225	21014	20805	20598	20393
1·6	0·20190	19989	19790	19593	19398	19205	19014	18825	18637	18452
1·7	0·18268	18087	17907	17728	17552	17377	17204	17033	16864	16696
1·8	0·16530	16365	16203	16041	15882	15724	15567	15412	15259	15107
1·9	0·14957	14808	14661	14515	14370	14227	14086	13946	13807	13670
2·0	0·13534	13399	13266	13134	13003	12873	12745	12619	12493	12369
2·1	0·12246	12124	12003	11884	11765	11648	11533	11418	11304	11192
2·2	0·11080	10970	10861	10753	10646	10540	10435	10331	10228	10127
2·3	0·10026	09926	09827	09730	09633	09537	09442	09348	09255	09163
2·4	0·09072	08982	08892	08804	08716	08629	08543	08458	08374	08291
2·5	0·08208	08127	08046	07966	07887	07808	07730	07654	07577	07502
2·6	0·07427	07353	07280	07208	07136	07065	06995	06925	06856	06788
2·7	0·06721	06654	06587	06522	06457	06393	06329	06266	06204	06142
2·8	0·06081	06020	05961	05901	05843	05784	05727	05670	05613	05558
2·9	0·05502	05448	05393	05340	05287	05234	05182	05130	05079	05029
3·0	0·04979	04929	04880	04832	04783	04736	04689	04642	04596	04550
3·1	0·04505	04460	04416	04372	04328	04285	04243	04200	04159	04117
3·2	0·04076	04036	03996	03956	03916	03877	03839	03801	03763	03725
3·3	0·03688	03652	03615	03579	03544	03508	03474	03439	03405	03371
3·4	0·03337	03304	03271	03239	03206	03175	03143	03112	03081	03050
3·5	0·03020	02990	02960	02930	02901	02872	02844	02816	02788	02760
3·6	0·02732	02705	02678	02652	02625	02599	02573	02548	02522	02497
3·7	0·02472	02448	02423	02399	02375	02352	02328	02305	02282	02260
3·8	0·02237	02215	02193	02171	02149	02128	02107	02086	02065	02045
3·9	0·02024	02004	01984	01964	01945	01925	01906	01887	01869	01850
4·0	0·01832	01813	01795	01777	01760	01742	01725	01708	01691	01674

	·0000	·0001	·0002	·0003	·0004	·0005	·0006	·0007	·0008	·0009
0·000	1·00000	99990	99980	99970	99960	99950	99940	99930	99920	99910
0·001	0·99900	99890	99880	99870	99860	99850	99840	99830	99820	99810
0·002	0·99800	99790	99780	99770	99760	99750	99740	99730	99720	99710
0·003	0·99700	99690	99681	99671	99661	99651	99641	99631	99621	99611
0·004	0·99601	99591	99581	99571	99561	99551	99541	99531	99521	99511
0·005	0·99501	99491	99481	99471	99461	99452	99442	99432	99422	99412
0·006	0·99402	99392	99382	99372	99362	99352	99342	99332	99322	99312
0·007	0·99302	99293	99283	99273	99263	99253	99243	99233	99223	99213
0·008	0·99203	99193	99183	99173	99164	99154	99144	99134	99124	99114
0·009	0·99104	99094	99084	99074	99064	99054	99045	99035	99025	99015

Obtain values not tabulated by multiplication, for example:

$$e^{-2·7567} = e^{-2·75} \cdot e^{-0·0067}$$
$$e^{-4·75} = e^{-2·75} \cdot e^{-2·00}$$
$$e^{-4·7567} = e^{-2·75} \cdot e^{-2·00} \cdot e^{-0·0067}$$

E. T. G.

4.2 Statistical methods for the treatment of experimental data

The result of an experiment to measure a physical quantity is never exact but is subject to errors. It is convenient, though sometimes an oversimplification, to divide these into two classes: (a) *random errors*, which vary unpredictably from one repetition of the measurement to another, and (b) *systematic errors* due to consistent but unknown inaccuracies in the equipment used or in the calculations. The effect of random errors can be reduced by repeating the experiment. Statistical methods for treating the data are well known, although their logical bases are still to some extent a subject of controversy. Some elementary formulae are given below. More detailed treatments can be found in the references. Systematic errors are more difficult to handle: each case has to be treated on its merits. Sometimes it is straightforward, as when a constant (such as the velocity of light) used in the calculations has a known uncertainty. Often a detailed analysis of the possible sources of error in the apparatus is needed and may involve subsidiary experiments.

Suppose that the object of the experiment is to measure the value of some quantity, x. If the experiment is carried out n times, there will be n results, $x_1 \ldots x_i \ldots x_n$, varying in value because of random errors. Merely to list these results is not very informative: the purpose of statistical treatment is to reduce them to a form whose significance can be readily appreciated. A first step which is often useful when n is large is to plot a *histogram* of the results, dividing the range of values taken by the x_i into a number of equal intervals and plotting the number of readings falling in each interval. In a typical case the histogram will have a single peak, which we may assume to be somewhere near the true value of x, and will have a spread about this peak which is an indication of the precision of the measurements.

As best estimate of the true value of x it is usual to take the *mean* of the x_i, given by

$$\bar{x} = n^{-1} \sum_{i=1}^{n} x_i$$

The *standard deviation*, s, is taken as a quantitative measure of the spread of the readings. Its square is known as the *variance* and is given by

$$v = s^2 = (n-1)^{-1} \sum_{i=1}^{n} (x_i - \bar{x})^2$$

Another important quantity is the *standard deviation of the mean*, sometimes called the *standard error* and equal to $n^{-1/2}s$. The mean, $\bar{x}$, and either s or $n^{-1/2}s$ are quantities which, with an estimate of the systematic errors, are often all that is needed to summarize the results of a physical measurement. To go further, and fully to appreciate their significance, needs some underlying theory. The unqualified statement that the result of a series of experiments is $\bar{x} \pm a$ is meaningless because the reader does not know if a is to be taken as s, $n^{-1/2}s$ or some other quantity. It is essential that what is stated should be made clear.

In biology and the social sciences it is often the object of an investigation to make an estimate of some property of a large but finite *population* of individuals, too numerous for each to be measured, by means of measurements on a limited *sample*. There is a unique true value to the property being estimated, which could be ascertained given sufficient time and effort. A range of statistical techniques is available to estimate the properties of the parent population from sample measurements with calculable uncertainties and confidence limits.

In the physical sciences a finite parent population is relatively uncommon. Instead it is convenient to postulate that our series of measurements forms a finite sample from an infinite population. We can speak of this population having a *probability distribution function* $F(x)$ with the following properties:

(i)
$$\int_{-\infty}^{\infty} F(x)\, dx = 1$$

(ii) The true value of the quantity being measured is the mean of the population,

$$\mu = \int_{-\infty}^{\infty} x\, F(x)\, dx$$

(iii) A measure of the 'true' uncertainty of the measurement (due to random errors) is given by the standard deviation, σ, of the population; its square, the variance, is given by

$$\sigma^2 = \int_{-\infty}^{\infty} (x-\mu)^2\, F(x)\, dx$$

The histogram, mean and standard deviation derived from our set of measurements can be regarded as sample approximations to the probability distribution, mean and standard deviation of the parent population.

It is sometimes necessary to combine the results of a number of independent determinations of a quantity into a single estimate. The best precision (minimum variance) is obtained if the individual results are 'weighted' inversely to their variances. Suppose that there are m separate results X_j, each the mean of n_j readings with variance s_j^2. Then the best estimate of the quantity is

$$\bar{X} = \sum_{j=1}^{m} X_j n_j s_j^{-2} \Big/ \sum_{j=1}^{m} n_j s_j^{-2}$$

and the expected variance of this will be

$$\sum_{j=1}^{m} n_j s_j^{-2}$$

In most physical measurements, where the random errors can be thought of as made up of a number of small independent contributions, the probability distribution has the form of the *normal error function* or *Gaussian*,

$$F(x)\, dx = (2\pi)^{-1/2} \sigma^{-1} \exp\left\{ -(x-\mu)^2/2\sigma^2 \right\} dx$$

Much of the further statistical treatment of the data is based on theory which assumes that the distribution function is Gaussian. If the histogram has a form which differs widely from the Gaussian it is a warning to proceed with caution. However, the *Central Limit Theorem* states that the sample means from a non-Gaussian population have a distribution which approximates to the Gaussian, and the larger the number of observations the better the approximation. Consequently, many tests are valid even with non-Gaussian populations.

The Gaussian and its integral, commonly known as the error function, are tabulated in a number of the references below.

In an important class of experiments the object is to test the *significance* of an observation or some combination of observations. As a very simple example we may take the following: given a sample observation x, test the hypothesis that it comes from a parent population with Gaussian distribution having mean μ and standard deviation σ. We form the test function $c = |x-\mu|/\sigma$. By integrating the Gaussian function we can calculate the probability $p(C)$ that a sample observation taken at random will have a value of c less than C, and construct the following table:

C	1.65	1.96	2.58	3.29
$p(C)$	90%	95%	99%	99.9%

If in our experiment $c > 2.58$ we can say that the difference between the observation and the assumed distribution is significant at the 1% level of probability, meaning that there is less than a 1% chance that the result is due to random causes.

In practice we do not usually know the standard deviation σ of the parent population but have to work with the standard deviation s of the observations. This causes an important change in the method as the following illustration will show: given a set of n observations with mean $\bar{x}$ and standard deviation s, test the hypothesis that the true value is μ (i.e. that they are a sample from a parent population with Gaussian distribution having mean μ). We now form the test function

$$t = |\bar{x} - \mu|/(s/n^{1/2})$$

This has a distribution which can be calculated for a Gaussian population and is known as the '*Student*' *t distribution*. It involves a parameter known as the number of *degrees of freedom* which is, loosely, the number of independent observations, $n-1$ in our case (not n because for a given $\bar{x}$, once $n-1$ values are known, the nth is determinate). Values of t^2 for given significance levels P and numbers of degrees of freedom are tabulated in the literature and in the following short table where values of t^2 are given against parameter values $N_1 = 1$, $N_2 = n-1$ for three significance levels $P = 0.05$, 0.01 and 0.001 (corresponding to values of our earlier p of 95%, 99% and 99.9%).

As an illustration of the use of this we may return to our original experiment, to measure x, and improve on our earlier statement that $\bar{x}$ is an estimate of the true value of x by adding *confidence limits*. We can say that the value of x is

$$\bar{x} \pm tsn^{-1/2}$$

at the confidence level $1 - P$, where t^2 is the entry in the following table at the appropriate values of P and the other parameters. We are then asserting that the probability is $1 - P$ that the true value of x lies between the limits specified. A statement of the final result of our measurement would then include the following items:

   (i) The estimated value $\bar{x}$.

  (ii) The confidence limits due to random errors (calculated as above from the data and desirably including a statement that the Gaussian nature of the distribution has been checked, perhaps by plotting a histogram).

 (iii) The confidence limits due to systematic errors (the way in which these are calculated will depend on the circumstances of the experiment).

In some cases it may be necessary or desirable to state results in a less complete form. However they are presented, exactly what is being stated should be made clear.

In a similar way, the estimate of the standard deviation σ of the parent population can be improved by using the test function

$$\chi^2 = (n-1)s^2/\sigma^2$$

The distribution function of χ^2, the *chi-squared distribution*, is tabulated in the literature and can be used to set confidence limits for σ. It can also be used to check whether a series of observations agrees with an assumed distribution function. In the following table χ^2 for $k-1$ degrees of freedom is given by the entries with $N_1 = k-1$, $N_2 = \infty$.

Another test function, used in checking whether two independent estimates s_1 and s_2 (with degrees of freedom $n_1 - 1$ and $n_2 - 1$ respectively) of the standard deviation of a measured quantity are significantly different, is

$$F = s_1^2/s_2^2$$

The *F distribution* is also tabulated. Values of F can be obtained from the following table with $N_1 = n_1 - 1$, $N_2 = n_2 - 1$.

References

(1) J. T. Richardson, *The Reduction and Presentation of Experimental Results*, BS 2846: 1957.

(2) National Bureau of Standards (Federal Works Agency, Works Projects Administration of the City of New York), *Tables of Probability Functions*, 1942. Tables of the Gaussian function and its integral, the error function, up to 15 decimal places.

(3) R. A. Fisher, *Statistical Methods for Research Workers*, 1958 (Oliver and Boyd).

(4) C. Mack, *Essentials of Statistics for Scientists and Technologists*, 1967 (Plenum Press).

(5) O. L. Davies (ed.), *Statistical Methods in Research and Production*, 1957 (Oliver and Boyd).

(6) A. C. Aitken, *Statistical Mathematics*, 1952 (Oliver and Boyd).

(7) H. H. Ku (ed.), *Precision Measurement and Calibration—Statistical Concepts and Procedures*, National Bureau of Standards Special Publication 300, Vol. 1, 1969.

(8) O. L. Davies (ed.), *The Design and Analysis of Industrial Experiments*, 1960 (Oliver and Boyd).

A.E.B.

Table for significance tests

Nine commonly needed tests are listed in column 1. Use of the transformations of the observations given in column 2 enables a single table to be used. To make a test calculate the function of the

1 *To test whether:*	2 *Function of observations to be calculated*	3 *Cell of table*	
		N_1	N_2
1. A single, randomly chosen,† observation differs significantly from a given mean (μ) (standard deviation (σ) known).	$\left(\dfrac{x-\mu}{\sigma}\right)^2$	1	∞
2. The mean of n observations differs significantly from a given value (μ) (standard deviation (σ) known).	$n\left(\dfrac{\bar{x}-\mu}{\sigma}\right)^2$	1	∞
3. The mean of n observations differs significantly from a given value (μ) (standard deviation estimated (s) from sample).	$n\left(\dfrac{\bar{x}-\mu}{s}\right)^2$ $s^2 = \{\text{Sum } (x-\bar{x})^2\}/(n-1)$	1	$n-1$
4. Two means, of n_1 and n_2 observations respectively, differ significantly (standard deviation estimated (s) from samples).	$\dfrac{n_1 n_2}{n_1+n_2}\left(\dfrac{\bar{x}_1-\bar{x}_2}{s}\right)^2$ $s^2 = \dfrac{1}{n_1+n_2-2}\left\{\underset{n_1}{\underset{\text{over}}{\text{Sum}}} (x-\bar{x}_1)^2 + \underset{n_2}{\underset{\text{over}}{\text{Sum}}} (x-\bar{x}_2)^2\right\}$	1	n_1+n_2-2
5. The numbers (o) of observations falling into k classes differ significantly from expected numbers (e) (total number expected made to agree with observation).	$\underset{k}{\underset{\text{over}}{\text{Sum}}} \{(o-e)^2/e\}/(k-1)$ No e should be less than 5	$k-1$	∞
6. The numbers (o) of observations falling into k classes differ significantly from expected numbers (e) (expected numbers calculated from observations via a function of l parameters).	$\underset{k}{\underset{\text{over}}{\text{Sum}}} \{(o-e)^2/e\}/(k-l-1)$ No e should be less than 5	$k-l-1$	∞
7. One variance (v_1) estimated from n_1 observations is significantly larger than another (v_2) estimated from n_2 observations.	v_1/v_2	n_1-1	n_2-1
8. One variance estimated from n_1 observations is significantly different from another estimated from n_2 observations.	v_1/v_2 where v_1 is the larger of the two estimates, v_2 the smaller.‡	(n_1-1)	(n_2-1)
9. An observed proportion, r out of n, differs significantly from a given proportion p. (a) $p<r/n$ (b) $p>r/n$	 $(1-p)r/p(n-r+1)$ $p(n-r)/(1-p)(r+1)$	 $2(n-r+1)$ $2(r+1)$	 $2r$ $2(n-r)$

† E.g. this test would not be legitimate for testing the largest of a set.

‡ For this test the values of P are to be doubled, the three lines of the table are then for levels of significance $P=0.10, 0.02, 0.002$.

observations given in column 2 and compare its value with those given in that cell of the table identified in column 3. Greater values should be adjudged significant; smaller values are not conclusive, a larger experiment might show significance. P of the table is the *level of significance* to be quoted.

The three entries in each cell of the table are for three levels of significance ($P = 0.05, 0.01, 0.001$), P being the risk of a wrong decision *when no difference exists*; the risk of a wrong decision in other cases obviously cannot be stated generally because the differences may be of any magnitude. The smaller the value of P used the larger will a real difference have to be before it makes itself apparent by these tests. The choice of P must therefore be made by balancing this risk against the magnitude of the difference which will just escape detection, i.e. the table value.

Except for test 9, the table is calculated on the assumption that the error or random sampling variation referred to above results in observations being normally distributed. Small departures from normality will not usually affect the decisions because their effect on P is small.

This table abridged, by kind permission of the authors and publishers, from Table V of *Statistical Tables for Biological, Agricultural and Medical Research* by R. A. Fisher and F. Yates (Oliver and Boyd).

The three rows for each value of N_2 correspond to values of $P = 0.05, 0.01$ and 0.001 respectively; the values for $P = 0.01$ are printed in bold type.

Table for significance tests

N_2	N_1									
	1	2	3	4	5	6	8	12	24	∞
1	161.4	199.5	215.7	224.6	230.2	234.0	238.9	243.9	249.0	254.3
	4052	**4999**	**5403**	**5625**	**5764**	**5859**	**5981**	**6106**	**6234**	**6366**
	Values for $P = 0.001$ too large for entry									
2	18.5	19.0	19.2	19.2	19.3	19.3	19.4	19.4	19.5	19.5
	98.5	**99.0**	**99.2**	**99.2**	**99.3**	**99.3**	**99.4**	**99.4**	**99.5**	**99.5**
	998.5	999.0	999.2	999.2	999.3	999.3	999.4	999.4	999.5	999.5
3	10.1	9.6	9.3	9.1	9.0	8.9	8.8	8.7	8.6	8.5
	34.1	**30.8**	**29.5**	**28.7**	**28.2**	**27.9**	**27.5**	**27.1**	**26.6**	**26.1**
	167.5	148.5	141.1	137.1	134.6	132.8	130.6	128.3	125.9	123.5
4	7.7	6.9	6.6	6.4	6.3	6.2	6.0	5.9	5.8	5.6
	21.2	**18.0**	**16.7**	**16.0**	**15.5**	**15.2**	**14.8**	**14.4**	**13.9**	**13.5**
	74.1	61.2	56.2	53.4	51.7	50.5	49.0	47.4	45.8	44.1
5	6.6	5.8	5.4	5.2	5.1	5.0	4.8	4.7	4.5	4.4
	16.3	**13.3**	**12.1**	**11.4**	**11.0**	**10.7**	**10.3**	**9.9**	**9.5**	**9.0**
	47.0	36.6	33.2	31.1	29.7	28.8	27.6	26.4	25.1	23.8
6	6.0	5.1	4.8	4.5	4.4	4.3	4.1	4.0	3.8	3.7
	13.7	**10.9**	**9.8**	**9.1**	**8.7**	**8.5**	**8.1**	**7.7**	**7.3**	**6.9**
	35.5	27.0	23.7	21.9	20.8	20.0	19.0	18.0	16.9	15.7
7	5.6	4.7	4.3	4.1	4.0	3.9	3.7	3.6	3.4	3.2
	12.2	**9.5**	**8.5**	**7.8**	**7.5**	**7.2**	**6.8**	**6.5**	**6.1**	**5.6**
	29.2	21.7	18.8	17.2	16.2	15.5	14.6	13.7	12.7	11.7
8	5.3	4.5	4.1	3.8	3.7	3.6	3.4	3.3	3.1	2.9
	11.3	**8.6**	**7.6**	**7.0**	**6.6**	**6.4**	**6.0**	**5.7**	**5.3**	**4.9**
	25.4	18.5	15.8	14.4	13.5	12.9	12.0	11.2	10.3	9.3

Table for significance tests (contd)

N_2	N_1									
	1	2	3	4	5	6	8	12	24	∞
9	5.1	4.3	3.9	3.6	3.5	3.4	3.2	3.1	2.9	2.7
	10.6	**8.0**	**7.0**	**6.4**	**6.1**	**5.8**	**5.5**	**5.1**	**4.7**	**4.3**
	22.9	16.4	13.9	12.6	11.7	11.1	10.4	9.6	8.7	7.8
10	5.0	4.1	3.7	3.5	3.3	3.2	3.1	2.9	2.7	2.5
	10.0	**7.6**	**6.6**	**6.0**	**5.6**	**5.4**	**5.1**	**4.7**	**4.3**	**3.9**
	21.0	14.9	12.6	11.3	10.5	9.9	9.2	8.4	7.6	6.8
11	4.7	4.0	3.6	3.4	3.2	3.1	2.9	2.8	2.6	2.4
	9.6	**7.2**	**6.2**	**5.7**	**5.3**	**5.1**	**4.7**	**4.4**	**4.0**	**3.6**
	19.7	13.8	11.6	10.3	9.6	9.0	8.4	7.6	6.8	6.0
12	4.8	3.9	3.5	3.3	3.1	3.0	2.8	2.7	2.5	2.3
	9.3	**6.9**	**6.0**	**5.4**	**5.1**	**4.8**	**4.5**	**4.2**	**3.8**	**3.4**
	18.6	13.0	10.8	9.6	8.9	8.4	7.7	7.0	6.2	5.4
13	4.7	3.8	3.4	3.2	3.0	2.9	2.8	2.6	2.4	2.2
	9.1	**6.7**	**5.7**	**5.2**	**4.9**	**4.6**	**4.3**	**4.0**	**3.6**	**3.2**
	17.8	12.3	10.2	9.1	8.4	7.9	7.2	6.5	5.8	5.0
14	4.6	3.7	3.3	3.1	3.0	2.8	2.7	2.5	2.3	2.1
	8.9	**6.5**	**5.6**	**5.0**	**4.7**	**4.5**	**4.1**	**3.8**	**3.4**	**3.0**
	17.1	11.8	9.7	8.6	7.9	7.4	6.8	6.1	5.4	4.6
15	4.5	3.7	3.3	3.1	2.9	2.8	2.6	2.5	2.3	2.1
	8.7	**6.4**	**5.4**	**4.9**	**4.6**	**4.3**	**4.0**	**3.7**	**3.3**	**2.9**
	16.6	11.3	9.3	8.3	7.6	7.1	6.5	5.8	5.1	4.3
16	4.5	3.6	3.2	3.0	2.9	2.7	2.6	2.4	2.2	2.0
	8.5	**6.2**	**5.3**	**4.8**	**4.4**	**4.2**	**3.9**	**3.6**	**3.2**	**2.8**
	16.1	11.0	9.0	7.9	7.3	6.8	6.2	5.5	4.8	4.1
17	4.5	3.6	3.2	3.0	2.8	2.7	2.5	2.4	2.2	2.0
	8.4	**6.1**	**5.2**	**4.7**	**4.3**	**4.1**	**3.8**	**3.5**	**3.1**	**2.7**
	15.7	10.7	8.7	7.7	7.0	6.6	6.0	5.3	4.6	3.8
18	4.4	3.6	3.2	2.9	2.8	2.7	2.5	2.3	2.1	1.9
	8.3	**6.0**	**5.1**	**4.6**	**4.2**	**4.0**	**3.7**	**3.4**	**3.0**	**2.6**
	15.4	10.4	8.5	7.5	6.8	6.4	5.8	5.1	4.4	3.7
19	4.4	3.5	3.1	2.9	2.7	2.6	2.5	2.3	2.1	1.9
	8.2	**5.9**	**5.0**	**4.5**	**4.2**	**3.9**	**3.6**	**3.3**	**2.9**	**2.5**
	15.1	10.2	8.3	7.3	6.6	6.2	5.6	5.0	4.3	3.5
20	4.4	3.5	3.1	2.9	2.7	2.6	2.4	2.3	2.1	1.8
	8.1	**5.8**	**4.9**	**4.4**	**4.1**	**3.9**	**3.6**	**3.2**	**2.9**	**2.4**
	14.8	10.0	8.1	7.1	6.5	6.0	5.4	4.8	4.1	3.4
22	4.3	3.4	3.0	2.8	2.7	2.5	2.4	2.2	2.0	1.8
	7.9	**5.7**	**4.8**	**4.3**	**4.0**	**3.8**	**3.5**	**3.1**	**2.7**	**2.3**
	16.4	9.6	7.8	6.8	6.2	5.8	5.2	4.6	3.9	3.2

N_2	N_1									
	1	2	3	4	5	6	8	12	24	∞
24	4.3	3.4	3.0	2.8	2.6	2.5	2.4	2.2	2.0	1.7
	7.8	**5.6**	**4.7**	**4.2**	**3.9**	**3.7**	**3.4**	**3.0**	**2.7**	**2.2**
	14.0	9.3	7.6	6.6	6.0	5.6	5.0	4.4	3.7	3.0
26	4.2	3.4	3.0	2.7	2.6	2.5	2.3	2.1	1.9	1.7
	7.7	**5.5**	**4.6**	**4.1**	**3.8**	**3.6**	**3.3**	**3.0**	**2.6**	**2.1**
	13.7	9.1	7.4	6.4	5.8	5.4	4.8	4.2	3.6	2.8
28	4.2	3.3	2.9	2.7	2.6	2.4	2.3	2.1	1.9	1.7
	7.6	**5.5**	**4.6**	**4.1**	**3.8**	**3.5**	**3.2**	**2.9**	**2.5**	**2.1**
	13.5	8.9	7.2	6.3	5.7	5.2	4.7	4.1	3.5	2.7
30	4.2	3.3	2.9	2.7	2.5	2.4	2.3	2.1	1.9	1.6
	7.6	**5.4**	**4.5**	**4.0**	**3.7**	**3.5**	**3.2**	**2.8**	**2.5**	**2.0**
	13.3	8.8	7.0	6.1	5.5	5.1	4.6	4.0	3.4	2.6
40	4.1	3.2	2.8	2.6	2.4	2.3	2.2	2.0	1.8	1.5
	7.3	**5.2**	**4.3**	**3.8**	**3.5**	**3.3**	**3.0**	**2.7**	**2.3**	**1.8**
	12.6	8.3	6.6	5.7	5.1	4.7	4.2	3.6	3.0	2.2
60	4.0	3.2	2.8	2.5	2.4	2.3	2.1	1.9	1.7	1.4
	7.1	**5.0**	**4.1**	**3.6**	**3.3**	**3.1**	**2.8**	**2.5**	**2.1**	**1.6**
	12.0	7.8	6.2	5.3	4.8	4.4	3.9	3.3	2.7	1.9
120	3.9	3.1	2.7	2.4	2.3	2.2	2.0	1.8	1.6	1.3
	6.9	**4.8**	**3.9**	**3.5**	**3.2**	**3.0**	**2.7**	**2.3**	**1.9**	**1.4**
	11.4	7.3	5.8	4.9	4.4	4.0	3.6	3.0	2.4	1.6
∞	3.8	3.0	2.6	2.4	2.2	2.1	1.9	1.8	1.5	1.0
	6.6	**4.6**	**3.8**	**3.3**	**3.0**	**2.8**	**2.5**	**2.2**	**1.8**	**1.0**
	10.8	6.9	5.4	4.6	4.1	3.7	3.3	2.7	2.1	1.0

E.D.v.R.